Uncertainty and Graphing in Discovery Work

Wolff-Michael Roth

Uncertainty and Graphing in Discovery Work

Implications for and Applications in STEM Education

Springer

Wolff-Michael Roth
University of Victoria
Victoria, British Columbia
Canada

ISBN 978-94-024-0103-5 ISBN 978-94-007-7009-6 (eBook)
DOI 10.1007/978-94-007-7009-6
Springer Dordrecht Heidelberg New York London

Softcover reprint of the hardcover 1st edition 2014

Printed on acid-free paper

Springer is part of Springer Science+Business Media (www.springer.com)

Preface

> *Rule 7*. Before attributing any special quality to the mind or to the method of people, let us examine first the many ways through which inscriptions are gathered, combined, tied together, and sent back. Only if there is something to be explained once the networks have been studied shall we start to speak of cognitive factors. (Latour 1987, p. 258)

As part of an extended research program on graphing that I maintained over the past 20 years, there were also several long-term ethnographic studies of graphs and uncertainty in scientific discovery work. This book derives from one of these: a 5-year ethnographic study of one advanced scientific research laboratory associated with a 5-year ethnographic study of one fish hatchery in particular where the scientific laboratory sourced many of its specimen. Because the laboratory worked at the forefront of its field (i.e. fish vision), there were many occasions to study not only (a) the use of familiar graphs but also (b) graphs as a means to articulate and develop the thinking of the scientists and (c) the production and interpretation of data and graphs that were the ultimate outcome of the research and that the scientists did not yet know whether these would pass muster during the peer review process. Such an approach is in the spirit of the methodological proposal recently offered to learning scientists concerning the question "of how one might recognize a discovery without knowing in advance of what is available to be discovered" (Koschmann and Zemel 2009, p. 200). As these authors, I am concerned with how something initially unknown – graphs and the scientific knowledge claims they can be used to support – becomes, through the corporeal and sensual practices of research scientists, transformed into something that is available to be known to humankind more generally. I am particularly interested in graphing in the face of uncertainty, which, as I show in the present book, is a steady way companion of scientists in the course of their work. Thus, at times scientists have something that they may denote to be "a crazy idea," which subsequently turns into a stable scientific fact and at other times may just remain what it was. The story of the research reported here is of that kind. It therefore tells us a great deal about how scientists work, how they use and come to be familiar graphs, and, ultimately, how science, mathematics, and technology come to be bound up with one another in the mangle of practice that constitutes this world in the way we know it.

The core of this book consists of a close study of graphing and uncertainty in a discovery science, from the very early days of the inception of the research, through the data collection and interpretation periods, and to the ultimate publication of an article that provided massive evidence against the reigning dogma initially articulated by a Nobel Prize winner. The common way in our culture to celebrate major discoveries is to ascribe it to the skills of one or more individual scientists. In this book, I take a different tack, following the precepts of the seventh rule of method Bruno Latour had recommended for the study of scientists and engineers, and as reproduced here in the opening quotation to this preface. I came across the text from which I culled this quotation in about 1990, the same year that I was reading a number of other ethnographies of science, including *Laboratory Life: The Social Construction of Scientific Facts* (Latour and Woolgar 1986) and *The Manufacture of Knowledge: An Essay on the Constructivist and Contextual Nature of Science* (Knorr-Cetina 1981). It was also the year that I read *Cognition in Practice: Mind, Mathematics and Culture in Everyday Life* (Lave 1988). The three books turned my life around and profoundly influenced what eventually should become my academic career.

A decade before, I had completed an MSc in Physics but had turned to teaching when the jobs for research physicists were scarce during an economic stagnation period at the end of the 1970s and at the beginning of the 1980s. With a minor in applied mathematics, it was not surprising that I would later do a PhD studying mathematical cognition and the cognitive development of adults. This I did in the *College of Science and Technology* at the University of Southern Mississippi, Hattiesburg. Here, my theories of knowing and learning were shaped by neo-Piagetian information processing approaches. It was after having returned to teaching physics in a private college preparatory school that I somehow came across and read the three above-mentioned books. It was during that summer of 1990 that I decided to begin a whole new way of researching cognition and learning and to look at the data I was to collect in a different way. My own background in science (physics) and applied mathematics – I also had done a statistics minor during my PhD work – allowed me to bring together the lines of work that I had found in those three books. It eventually led to the framing of graphing as social practice for rethinking mathematics and science education (Roth 1996; Roth and McGinn 1997, 1998), to the study of mathematics among technicians and scientists (Roth 2003b), and to investigations of mathematics in different workplace settings, among fish culturists (Roth 2005b) and electricians (Roth 2012b), and to the study of graphs and other mathematical inscriptions in scientific and engineering journals (Roth 2013c; Roth et al. 1999). In the course, I arrived at phenomenological, semiotic, and cultural historical activity-theoretic theories of graphing in everyday settings, scientific and non-scientific settings alike (Roth 2003a, b; Roth et al. 2002).

The idea of studying graphing in the everyday world generally and among scientists more specifically came to me after one of my colleagues and then graduate students (Michael Bowen) and I had found out that university science graduates,[1] some with Masters of Science degrees, employed powerful mathematical approaches with a significantly smaller frequency than pairs of eight-grade students who had gone through a curriculum in which they designed and conducted experiments and reported the results thereof to their peers (Roth et al. 1998).[2] At the time, we were interested in developing curriculum that would induct school students to the "authentic" practices of science and how students mathematized their experiences during an inquiry curriculum (Roth and Bowen 1994). We therefore decided to investigate how scientists and those in the process of becoming scientists were using graphs in and as part of their discovery work. We began a series of ethnographic studies among graduate (Roth and Bowen 1999b) and undergraduate honors students (Bowen et al. 1999; Roth and Bowen 2001a); and we conducted think-aloud studies with practicing scientists reading graphs culled from introductory textbooks of their own field and from their own publications (Roth 2012; Roth and Bowen 2001b, 2003). These studies showed that scientists were not the perfect experts in graphing that they are often held out to be when looking at unfamiliar graphs, even when these graphs were from their own field. We came to understand – using the imagery of cannibals, missionaries, and converts – that there was a discontinuity in the graph-related practices at the boundary between people who engage in work that requires them to transform data into graphical representations and people who do not have had such experiences (Roth and Bowen 1999c). This boundary was independent of prior experience such that experienced scientists, technicians, and students did well when graphing was related to their ongoing work but all of them did poorly when graphs were unfamiliar or related to unfamiliar phenomena. In the case of graphs from their work, on the other hand, all of these people were highly fluent. These studies subsequently were extended to cover the mathematics used by electricians (Roth 2014b), fish culturists (Roth 2005a, b, c), or water technicians (Roth 2007). We also studied how scientists employed graphs while lecturing undergraduate students, finding out that scientists too talk about graphs in inappropriate ways (Roth 2010; Roth and Bowen 1999a).

The present book constitutes a continuation and transformation of my past research endeavours. I describe and theorize the function of graphs and graphing in scientific discovery work from a social practice perspective. In this work, graphs are topic (scientists talk *about* them), tool (scientists use graphs in the course of pursuing their research work), and ground (scientists' talk occurs over and against graphs). In discovery work, where scientists do not initially know what to make of graphs, there is a lot of uncertainty, and scientists struggle much as everyone else in trying to produce sense-making talk related to graphs. Contrary to the belief that scientists unproblematically "interpret" graphs, the chapters in Part II of this book show that uncertainty about their research object is tied to uncertainty in and of the

[1] These were graduates who subsequently enrolled in a 1-year professional development program to obtain a license for teaching science in high schools.

[2] I return to the work from which this study emerged in Part IV of this book.

graphs. That is, scientists do not just interpret graphs and thereby find something out. Rather, scientists are caught up in a double bind: (a) to know what is in the graph, they need to know what it refers to (shows) and (b) to know what is shown, they need to know what is in the graph (Roth 2009; Chap. 4). It may, as in the present study, take several years of struggle in their workplace before scientists are sufficiently familiar with their equipment and the natural phenomena to have a sense about just what their graphs are evidence of (Chap. 7). Scientists may resist what eventually comes to be known as correct interpretations – leading, as in the present case, to a discovery that overthrows what has been a 60-year scientific canon. The present research shows that graphs turn out to stand to the entire research in a part–whole relation, where scientists not only need to be highly familiar with the contexts from which their data are extracted but also with the entire process by means of which the natural world comes to be translated into a graphical form that can be used to make the natural world present again. This, as I show in Part IV of this book, has considerable implications for science, technology, engineering, and mathematics education at the secondary and tertiary level and in vocational training.

Across Chaps. 2, 3, 4, 5, 6, 7, and 8, readers are taken (a) from the stages in the research that are characterized by radical uncertainty and a lot of fitting data to beliefs (b) to the final stages of their research when scientists change their ways of looking at the data and, thereby, come to develop new ways of understanding. Rather than constituting abstract representations and skills, graphs and graphing turn out to be integral aspects of the totality of the research process. Only after everything has been said and done do graphs, in the talk of the scientists, become something independent standing *for* something else: the scientific object. That is, graphs cannot be understood independently of everything else that makes their research, including the scientific instrumentation, the source of their specimens, their work habits, and a lot of detail about the different settings including the environmental contexts in which their specimens were caught, the environment in which these were kept until processing, and the environment of the laboratory itself.

There is quite a bit of auto/ethnography involved in this book (Roth 2005c), as I had been both a member of the scientific research team and studying it for the purposes of better understanding the sciences. The "auto-" in the term refers to the fact that the ethnographer *also* is a member of the "tribe" studied. The practices studied and described are not merely those of others but are the very practices I accomplished as participant in the scientific research. But this does not put me in a privileged position, for, as Bakhtin (1981) notes, whoever the author is, the narrative form (genre) imposes the same kind of constraints and is associated with the same kind of expectancies on the part of the reader. But this membership provides me with a particular high degree of familiarity that is central to the adequacy of the ethnographic account – a fact that also is apparent in the work of the natural sciences as I report it in Part II.

The chapters of this book are written such that they may stand on their own, summarizing and pointing to relevant places in other chapters that readers are enticed but do not have to consult. For those who access and read the book in its entirety, this constitutes a complete narrative of graphs and graphing in scientific discovery work from the initial inception of a study to its completion 7 years later with the published journal article. As part of a joint project, I had become part of this research team (being joint author on Temple et al. 2006, 2008), documenting the entire research process in which the canon (which had led to a Nobel Prize for the original scientific discovery) came to be overthrown.

The book, though written primarily for STEM educators, also should be of interest to those working in the fields of social studies or history of science, because it documents a scientific discovery that transcends the "social construction" of scientific knowledge. The book should also be of interest to those concerned with workplace learning, because it shows that graphing is not some abstract practice or social skill that scientists apply to new settings but that graphing is a local achievement that advances in the course of the unfolding research process as scientists become increasingly familiar with all aspects of their research. The work presented raises serious doubt about teaching graphs and graphing independent of the settings in which they are produced and used for the salient purposes at hand.

In this book, I draw on materials that also have figured in peer-reviewed research journals (Roth 2013a, b, d; Roth and Temple 2014). But all of these materials have been transformed to become integral part of a whole – the book – and this whole constituted a force that asked for reshaping the individual chapter texts. I am grateful to the publishers, including Springer, to copyrights that allow me to re-use materials. In most instances, in fact, the work by far exceeds what can be represented in a journal article because of length limitations to which journal articles are subject to. Thus, in these pages, I mobilize a lot of the original materials that never made it into the published articles – because of the page-limitations that print journals are subject to – but on which the articles were integrally based. In fact, it is only in the pages of this book that readers will really see scientists at work, for example, the ways in which they engage each other. And much of this engagement is as mundane as any other societal setting, where people go about doing the things that they normally do and in the ways they normally do them. In contrast to some popular conceptions, which are also spread in schools, science is not some special ability or endeavor – or is as special as any other practice (cooking, manufacture). It has emerged – both on cultural-historical and individual ontogenetic grounds – from everyday common sense pursuits (Husserl 1976a). It is therefore *founded upon* the ordinary ways in which we conduct our lives. This, therefore, is my guiding thread throughout this book: to acknowledge scientific work as an extension and particular kind of everyday human endeavor in the pursuit of getting the day's work done – even though, at times, this endeavor *appears to be* more esoteric than other pursuits.

No human endeavor is possible without all those others who, in one way or another, participate in relations that enable the production of a book such as this one. First and foremost I have to thank my scientific collaborator, Craig Hawryshyn, with whom I joined effort to investigate the places where coho salmon are raised, aspects of the life history of these fishes, and the knowledge exchanges between the persons who populate the places where related activities take place: the fish hatchery and scientific laboratory. I also thank those who worked on our team, including Elmar Plate, a post-doctoral fellow working with us in the early stages of the project. Theodore von Haimberger was part of the effort from its beginning and a key person in the processing of data and the writing of software. Shelby Temple did his PhD on the project; thanks to his hard work and his long hours, we accumulated all the data necessary for debunking a scientific canon. Shelby did a lot of his work with Sam Ramsden, who did his MSc thesis research within this research group and who was funded by one of my research grants. Stuart Lee (PhD), Yew-Jin Lee (PhD), and Leanna Boyer (MA) assisted me in the fish hatchery and the scientific laboratory, observing, recording, and interviewing people at work. The other individuals who were critical to the success of this study include Mike Wolfe and Erica Blake, who taught me everything I know about fish culture. As part of their work, they took me on their special assignments, such as capturing wild salmon, feeding young salmon in net pens, fertilizing lakes to assist the sockeye salmon that spawn there, or on trips counting and sampling salmon that had returned from their ocean migration and now died in the rivers where they had been born.

Victoria, BC, Canada Wolff-Michael Roth
December 2013

References

Bakhtin, M. (1981). *The dialogic imagination*. Austin: University of Texas.

Bowen, G. M., Roth, W.-M., & McGinn, M. K. (1999). Interpretations of graphs by university biology students and practicing scientists: Towards a social practice view of scientific re-presentation practices. *Journal of Research in Science Teaching, 36*, 1020–1043.

Husserl, E. (1976). *Husserliana Band III/1. Ideen zu einer reinen Phänomenologie und phänomenologischen Philosophie: Erstes Buch: Allgemeine Einführung in die reine Phänomenologie [Husserliana vol. III/1. Ideas to a pure phenomenology and phenomenological philosophy vol 1. General introduction to a pure phenomenology]*. The Hague: Martinus Nijhoff.

Knorr-Cetina, K. D. (1981). *The manufacture of knowledge: An essay on the constructivist and contextual nature of science*. Oxford: Pergamon Press.

Koschmann, T., & Zemel, A. (2009). Optical pulsars and black arrows: Discoveries as occasioned productions. *Journal of the Learning Sciences, 18*, 200–246.

Latour, B., & Woolgar, S. (1986). *Laboratory life: The social construction of scientific facts*. Princeton: Princeton University Press.

Lave, J. (1988). *Cognition in practice: Mind, mathematics and culture in everyday life*. Cambridge: Cambridge University Press.

Roth, W.-M. (1996). Where is the context in contextual word problems?: Mathematical practices and products in Grade 8 students' answers to story problems. *Cognition and Instruction, 14*, 487–527.

Roth, W.-M. (2003a). *Toward an anthropology of graphing*. Dordrecht: Kluwer Academic Publishers.

Roth, W.-M. (2003b). Competent workplace mathematics: How signs become transparent in use. *International Journal of Computers for Mathematical Learning, 8*, 161–189.

Roth, W.-M. (2005a). Mathematical inscriptions and the reflexive elaboration of understanding: An ethnography of graphing and numeracy in a fish hatchery. *Mathematical Thinking and Learning, 7*, 75–109.

Roth, W.-M. (2005b). Making classifications (at) work: Ordering practices in science. *Social Studies of Science, 35*, 581–621.

Roth, W.-M. (Ed.). (2005c). *Auto/biography and auto/ethnography: Praxis of research method*. Rotterdam: Sense Publishers.

Roth, W.-M. (2007). Graphing Hagan Creek: A case of relations in sociomaterial practice. In E. Teubal, J. Dockrell, & L. Tolchinsky (Eds.), *Notational knowledge: Historical and developmental perspectives* (pp. 179–207). Rotterdam: Sense Publishers.

Roth, W.-M. (2009). Radical uncertainty in scientific discovery work. *Science, Technology & Human Values, 34*, 313–336.

Roth, W.-M. (2010). Vygotsky's dynamic conception of the thinking-speaking relationship: A case study of science lectures. *Pedagogies: An International Journal, 5*, 49–60.

Roth, W.-M. (2014a). Learning in the discovery sciences: The history of a "radical" conceptual change or The scientific revolution that was not. *Journal of the Learning Sciences, 23*, 177–215.

Roth, W.-M. (2014b). Rules of bending, bending the rules: The geometry of conduit bending in college and workplace. *Educational Studies in Mathematics, 86*, 177–192.

Roth, W.-M. (2012). Limits to general expertise: A study of in- and out-of-field graph interpretation. In C. A. Wilhelm (Ed.), *Encyclopedia of cognitive psychology* (pp. 1–38). Hauppauge: Nova Science.

Roth, W.-M. (2013a). The social nature of representational engineering knowledge. In A. Johri & B. Olds (Eds.), *Cambridge handbook of engineering education research* (pp. 67–82). Cambridge: Cambridge University Press.

Roth, W.-M. (2013b). Contradictions and uncertainty in scientists' mathematical modeling and interpretation of data. *Journal of Mathematical Behavior, 32*, 593–612.

Roth, W.-M. (2013c). Data generation in the discovery sciences – learning from the practices in an advanced research laboratory. *Research in Science Education, 43*, 1617–1644.

Roth, W.-M. (2013d). Undoing decontextualization or how scientists come to understand their own data/graphs. *Science Education, 97*, 80–112.

Roth, W.-M., & Temple, S. (2014). On understanding variability in data: A study of graph interpretation in an advanced experimental biology laboratory. *Educational Studies in Mathematics, 86*(3), 359–376. doi:10.1007/s10649-014-9535-5.

Roth, W.-M., & Bowen, G. M. (1994). Mathematization of experience in a grade 8 open-inquiry environment: An introduction to the representational practices of science. *Journal of Research in Science Teaching, 31*, 293–318.

Roth, W.-M., & Bowen, G. M. (1999a). Digitizing lizards or the topology of vision in ecological fieldwork. *Social Studies of Science, 29*, 719–764.

Roth, W.-M., & Bowen, G. M. (1999b). Complexities of graphical representations during lectures: A phenomenological approach. *Learning and Instruction, 9*, 235–255.

Roth, W.-M., & Bowen, G. M. (1999c). Of cannibals, missionaries, and converts: Graphing competencies from grade 8 to professional science inside (classrooms) and outside (field/laboratory). *Science, Technology & Human Values, 24*, 179–212.

Roth, W.-M., & Bowen, G. M. (2001a). "Creative solutions" and "fibbing results": Enculturation in field ecology. *Social Studies of Science, 31*, 533–556.

Roth, W.-M., & Bowen, G. M. (2001b). Professionals read graphs: A semiotic analysis. *Journal for Research in Mathematics Education, 32*, 159–194.

Roth, W.-M., & Bowen, G. M. (2003). When are graphs ten thousand words worth? An expert/expert study. *Cognition and Instruction, 21*, 429–473.

Roth, W.-M., & McGinn, M. K. (1997). Graphing: A cognitive ability or cultural practice? *Science Education, 81*, 91–106.

Roth, W.-M., & McGinn, M. K. (1998). Inscriptions: a social practice approach to "representations.". *Review of Educational Research, 68*, 35–59.

Roth, W.-M., McGinn, M. K., & Bowen, G. M. (1998). How prepared are preservice teachers to teach scientific inquiry? Levels of performance in scientific representation practices. *Journal of Science Teacher Education, 9*, 25–48.

Roth, W.-M., Bowen, G. M., & McGinn, M. K. (1999). Differences in graph-related practices between high school biology textbooks and scientific ecology journals. *Journal of Research in Science Teaching, 36*, 977–1019.

Roth, W.-M., Bowen, G. M., & Masciotra, D. (2002). From thing to sign and "natural object": Toward a genetic phenomenology of graph interpretation. *Science, Technology & Human Values, 27*, 327–356.

Temple, S. E., Plate, E. M., Ramsden, S., Haimberger, T. J., Roth, W.-M., & Hawryshyn, C. W. (2006). Seasonal cycle in vitamin A1/A2-based visual pigment composition during the life history of coho salmon (*Oncorhynchus kisutch*). *Journal of Comparative Physiology A: Sensory, Neural, and Behavioral Physiology, 192*, 301–313.

Temple, S. E., Ramsden, S. D., Haimberger, T. J., Veldhoen, K. M., Veldhoen, N. J., Carter, N. L., Roth, W.-M., & Hawryshyn, C. W. (2008). Effects of exogenous thyroid hormones on visual pigment composition in coho salmon (*Oncorhynchus kisutch*). *Journal of Experimental Biology, 211*, 2134–2143.

Contents

Part V Epilogue

Part I
Introduction

Graphs and graphing are constitutive of the sciences. Although there are an increasing number of studies on graphing, few of these focus on graphs and graphing in the discovery sciences; and even fewer take a look at graphs and graphing in the face of the uncertainty with which scientists are confronted on a daily basis. The discovery sciences allow us to revisit existing psychological theories, which tend to theorize graphing as a mental skill and graphs as external representation that is imaged on the retina and subsequently interpreted by the brain. Anthropological studies that follow scientists, said to be experts in such "skills" as graphs and graphing, suggest that such a model for understanding graphs and graphing is inappropriate. More recently, graphs and graphing have been analyzed from cultural-historical activity theoretic (e.g., Roth 2003), perspectives (Roth and Bowen 1999; Roth et al. 2002), and social practice perspectives (Roth and McGinn 1998). In Chap. 1, I review existing cognitive and anthropological theories to lay the ground for an extensive study of graphs and graphing in a discovery science presented in Part II. I end the review with the proposal to take a dynamic perspective on graphs and graphing, which radically changes the ways in which we think about the topic because the categories of thought no longer are taken to be self-identical but are of a kind that theorizes graphs and graphing as continuously changing phenomena.

References

Roth, W.-M. (2003). *Toward an anthropology of graphing*. Dordrecht: Kluwer Academic.

Roth, W.-M., & Bowen, G. M. (1999). Complexities of graphical representations during lectures: A phenomenological approach. *Learning and Instruction, 9*, 235–255.

Roth, W.-M., Bowen, G. M., & Masciotra, D. (2002). From thing to sign and "natural object": Toward a genetic phenomenology of graph interpretation. *Science, Technology & Human Values, 27*, 327–356.

Roth, W.-M., & McGinn, M. K. (1998). Inscriptions: A social practice approach to "representations.". *Review of Educational Research, 68*, 35–59.

Chapter 1
Toward a Dynamic Theory of Graphing

> Positivist history of culture thus sees language as gradually shaping itself around the contours of the physical world. Romantic history of culture sees language as gradually bringing Spirit to self-consciousness. Nietzschean history of culture, and Davidsonian philosophy of language, see language as we now see evolution, as new forms of life constantly killing off old forms – *not to accomplish a higher purpose, but blindly*. Whereas the positivist sees Galileo as making a discovery – finally coming up with the words which were needed to fit the world properly, words Aristotle missed – the Davidsonian sees him as having hit upon a tool which happened to work better for certain purpose than any previous tool. (Rorty 1989, p. 19, emphasis added)

Graphs and graphing, which emerged during the Renaissance period, are quintessential to the nature of science: without these, the sciences as we know them today would not exist (Edgerton 1985). In the production of scientific knowledge claims, graphs play an important role because they depict, at a sufficiently abstract level, general tendencies in the relation of two or more variables (Latour 1987). In the introductory quotation, the pragmatist philosopher Richard Rorty refers to different ways in which scholars have come to view the history of culture generally, which also is a view of the history of the sciences – one of the ways in which the Zeitgeist of each cultural generation comes to exhibit itself – more specifically. In the natural sciences, the general view is that knowledge is a reflection of the natural world and that graphs and mathematical equations depict the ways in which the world really is – though some scientists certainly will take different views, such as a former associate director of the Harvard-Smithsonian Institute for Astrophysics, who described physics as a language that is used to invent reality (Gregory 1990). In contrast, Latour's (1993) ethnography of pedologists and botanists and our own ethnographic study among field ecologists (Roth and Bowen 1999) show how the natural matters of this world come to be transformed into scientific knowledge claims by series of embodied transformations that map, successively, one portion of the material continuum onto another, which is then mapped onto yet another portion of the continuum, and so on in a process of continuing substitution. Here, the original "portion of the continuum" may be some clump of soil, as in Latour's case, or a lizard, in my study. Soil may be placed in a horizontal array of boxes, and

W.-M. Roth, *Uncertainty and Graphing in Discovery Work*,
DOI 10.1007/978-94-007-7009-6_1, © Springer Science+Business Media Dordrecht 2014

lizards may be weighed, measured, toe-clipped, color-determined, and so on. These first arrangements come to be expressed so that a chain of translations is produced with the raw materials ("natural world") on one end and scientific knowledge claims, expressed in language, on the other end (Fig. 1.1). Between the two extremes, there may be any number of translations between intermediary states. Any two ways in which the original matter is presented again – *re*-presented by means of an inscription, itself a piece of the material continuum – are related only because of conventions and practices of translation. This is contrary to the way in which Rorty's positivists think, who view the relation of any two portions of the continuum to be natural, linked because of properties inherent to the world and its representations. These would think the relation between world and knowledge in terms of a direct correspondence: {fundamental structure [of the natural world] $\leftrightarrow$ mathematical structure}. The two studies cited above and other work in the social studies of science, which is based on careful ethnographic studies of what scientists actually do, show that the relationship between the natural world and mathematical structure or language always is a contingent achievement in and of situated work.

Figure 1.1 provides an image of the chain of translation, which replaces that of a direct relation between the natural world and knowledge about it as per Rorty's description of the positivist perspective. The figure shows that any portion of the continuum is characterized by its materiality and by its particular form. Any two neighboring portions may consist of very different material or of very different form. Thus, for example, the material on a microscopic slide and its image under the microscope consist of very different material, one being biological the other electronic. But two electronic portions, though implemented in the same material, may be radically different, such as the image of a photoreceptor and the graph representing the absorption of light in this same receptor. This, therefore, describes that between any two portions of the continuum, there is a gap (Fig. 1.1). This gap is bridged – i.e. linkages are made between them – by the cultural practices and the implicit conventions governing inscriptions. For example, a raw graph may be smoothened using the technique of Fast Fourier Transformation, which gets rid of a lot of the wiggles visible in the raw graphs, thereby better depicting what scientists treat to be the real signal from the scientific phenomenon of interest. In this way, connections come to be established between the natural world, on the one hand, and knowledge expressed in language, on the other hand (Fig. 1.1). One of the main points of this book is that scientists do not make sense of a graph – e.g. on the right of Fig. 1.1 – without a deep familiarity with the natural world from which the fish derive – on the very left of Fig. 1.1 – and the many translations that the pieces of matter (e.g., from the retina) undergo in the course of the scientists' investigations. They also have to be familiar with what is going on in their laboratory, the equipment, and the translations that they make the fish undergo until some part of it – individual photoreceptor cells from the retina – comes to lie under the microscope and the recording apparatus.

Given the importance of graphs in and to science, it comes as little surprise, then, to find graphs and graphing among the fundamentals to be taught in school science (e.g., NRC 1996). From pre-kindergarten to high school, curricula are to enable

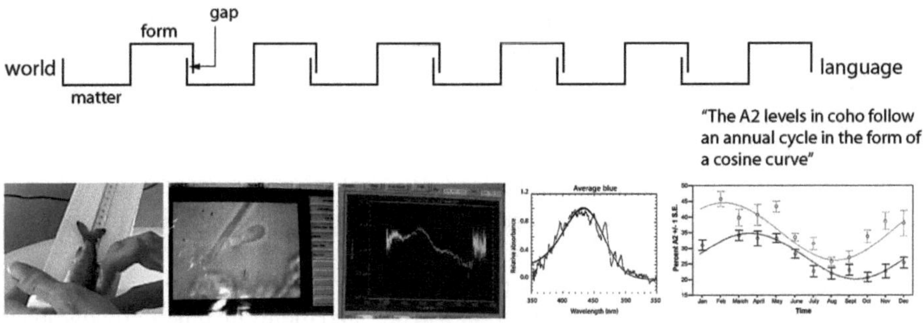

Fig. 1.1 In the social studies of science, it has come to be accepted that there is a chain of transformations that link things of the natural world, portions of the continuum, with other portions of the continuum, such as a piece of retina on a microscopic slide, the image of a photoreceptor cell under a microscope, a graph as it shows up in the laboratory, a cleaned graph in the computer laboratory, or a graph as it is ultimately published summarizing about 20,000 individual graphs that are visible in the laboratory

students to formulate questions, collect and organize data, and display them such as to provide answers to the questions posed. Students are to learn how to develop and evaluate inferences and make predictions based on their data. Curricula should enable students to "create and use representations to organize, record, and communicate mathematical ideas; select, apply, and translate among mathematical representations to solve problems; use representations to model and interpret physical, social, and mathematical phenomena" (NCTM 2000, p. 66).

Inscriptions – often referred to as "external representations" – also are central to mathematical reasoning, sense making, and knowing. It is not surprising, therefore, to find scholars suggest that "students must develop a set of skills for constructing, interpreting, transforming, and coordinating domain-specific external representations for learning and problem solving" (Stieff et al. 2011, p. 123). Mathematics educators especially are exhorted to "re-value visualization and its nature placing it as a central issue in mathematics education" (Arcavi 2003, p. 238), which includes visualization made possible by means of graphs of various kinds. Because professionals design and critique representations, it is useful to be familiar with what those in STEM-related fields do if educators want to facilitate students' learning trajectories or trajectories to develop meta-representational competencies that are required in STEM careers. It turns out, however, that accomplished scientists, too, experience trouble reading and interpreting graphs even though these may come from introductory courses in their own discipline. Where might we look for answers concerning the apparent lack of meta-representational competencies among scientists?

It has been suggested that graphs and graphing somehow depend on the familiarity of the agent with the entire context of the graph (Roth 2003b). Thus, for example, Stieff et al. (2011) asked their participants questions that – according to the researchers – required drawing on a model alone, on a model and graph, and questions that required the coordination of the previous two with a numerical equation and a general equation. The authors conclude that there was "good representational competence" among the participants in their research but also that there was "room for improvement in students' use of representations" (p. 141). The problem with such research tends to be that it does not control for the familiarity that the participants may have with the phenomena, data collection, or the pertinent representational practices. Psychologists therefore cannot know whether something constitutes relevant information unless they also investigate their research participants' horizon of awareness.

Investigating scientists' use and argumentative deployment of graphs when they do not yet know what the canonical position will be – and therefore inherently involves uncertainty – constitutes a manner of investigating the real work underlying graphs and graphing. Because during discovery work scientists often do not know the "natural object" that they are in the process of identifying – such as the astronomers discovering the first Galilean pulsar (Garfinkel et al. 1981) – and do not know whether the signals they obtain in graphical form from their data refer to anything, there is a radical uncertainty at work that does not tend to be described in research on graphs and graphing. Thus, to know whether a graph on a recording device shows something of interest, scientists need to know the natural object; and, conversely, to know that there is something like a natural object rather than an artifact, they need to know that the graphs exhibit something real rather than an artifact (Roth 2009b; Chapter 4). Although there are situations where members to a setting deliberately engage in attempts to generate uncertainty – such as in court cases, where the probabilities of incorrect DNA analyses may be amplified discursively (e.g. Lynch 1998) – the present study of graphing expertise at work[1] is focused on scientists' endogenous practices of taming uncertainties that arise in the course of and from their discovery work that produces the translational chain of inscriptions (Fig. 1.1).

In this book I am investigating graphing in the face of the uncertainties that discovery scientists inherently are confronted with because they do not yet know what they will have found out once everything has been said and done and, therefore, whether they are on some right track. Learning scientists have shown interest in uncertainty as an integral feature of mathematics and science experiences. They found that even "young children's exposure to informal inference involving uncertainty is an important learning foundation if a meaningful introduction to formal statistical tests is to take place in secondary school" (English 2012, p. 18). Uncertainty is an important aspect in the teaching of mathematics, as teachers often have to deal with unexpected situations that require deviating from lesson plans and, thus, call for improvisation in practice to cope with the uncertainties that have arisen.

The purpose of the present book then is captured by the following questions that my study had been designed to answer: What is the real work underlying graphs and

[1] "At work" should be understood in both of the ways that this expression allows to be heard, that is, at work in the scientific laboratory and graphing doing its part in the overall work of doing scientific research.

graphing in the discovery sciences when nobody knows in advance what will be the state of affairs when the research is completed? How do scientists cope with and tame the uncertainty that is an integral part of discovery work, where outcomes are not available beforehand (which was the case in the present laboratory, where the scientist thought their work would confirm what the lead scientist called the "dogma," but ended up overturning this same dogma after many years of research)? Answers to these questions are not only interesting in the context of scientific expertise, which tends to be studied when scientists are on familiar terrain, but also in the context of uncertainty as a possible characteristic in STEM education. Educators might use the findings for discussing the STEM curriculum – even though I recognize, as do others, that school science and mathematics and scientific discovery work are authentic settings in their own right.

Expertise and the Work of Graphing in Uncertainty

Educational research and practice tend to assume that scientists are experts at graphs and graphing. But even scientists are frequently at a loss even when asked to interpret introductory-level graphs in their own field (e.g., Roth and Bowen 2003). The same scientists – like everyday folk not trained in mathematics (Roth 2007) – tend to be knowledgeable when it comes to graphs from their own work. This raises questions about the nature of the expertise that tends to be ascribed to scientists when science (and mathematics) educators list graphs and graphing among the core "scientific process skills." There are studies that report on "the ease and frequency with which an expert uses multiple representations" (Tabachneck-Schijf et al. 1997, p. 312). The paradigm case taken in this particular study was an economic expert – Herbert Simon, 1978 Nobel Prize winner in economics – asked to explain to a student the economics principles of supply and demand. During this explanation, Simon was observed to be constructing and using a graph on a chalkboard in parallel with his verbal explanation. The authors suggest that the expert's performance was well accounted for by their CaMeRa (picture) model of reasoning with graphs.[2] One important problem with this and other existing studies of graphing is that these do not tease apart familiarity with graphs and graphing and processes that function independent of it.

Simon was asked to read and explain a supply/demand graph that economics students encounter in one of their first classes. In this graph (Fig. 1.2a), price is plotted against quantity of the product for both supply (price falls linearly with increasing quantity available) and demand (price rises linearly with increasing

[2] The name of the model is suggestive of processes, whereby the world comes to be imaged on the retina and in the brain as if these were operating like a camera. Such a view is inconsistent with the fact that all perception involves efferent and afferent dimensions and cannot be understood as a simple, mirror- or camera-like process.

quantity available). Without doubt, Simon is very familiar with this economics graph, having worked in the field for nearly half a century. It is not surprising, therefore, that he highlights without hesitating the intersection of the supply and demand graphs and suggests that the price would settle at this point (stable equilibrium). His response to the task is very similar to the ones that those biologists provide who teach undergraduate courses and who are asked to explain birthrate and death rate graphs (Fig. 1.1b) typical of undergraduate courses in ecology (Roth and Bowen 2003). These graphs feature two intersections one of which should be read as giving rise to a stable equilibrium the other to an unstable equilibrium. That is, the right intersection (Fig. 1.1b) is precisely of the same kind as the intersection in the economics graph (Fig. 1.1a), because the supply ("birthrate") is falling whereas demand ("death rate") is rising. My research shows that the research scientists were less than stellar dealing with the birthrate–death rate graph. However, the group was not homogeneous. Non-university research scientists were less successful at a statistically significant level than their university-based peers ($t(14) = 3.88$, $p < .002$), who were hypothesized to be much more familiar with the kind of graphs used in the study. On the seven items used, there were 5.13 correct solutions on the part of the 8 university-based scientists but only 1.75 correct solutions on the part of the 8 non-university scientists.

A related investigation in which a total of 33 scientists (16 biologists, 17 physicists) were asked to interpret graphs from introductory courses of biology shows that even with tremendous levels of training, only 9 (27 %) of the scientists gave correct answers on a graphing task that bears structural similarity to the oxygen-shrimp frequency graph in Fig. 1.2c (Roth 2009a). That other graph featured the distribution of three types of plants – distinguished by their photosynthetic mechanism (C3, C4, and CAM) – along an elevation gradient. These mechanisms differentially adapt them to the reigning climate, CAM plants being able to regulate their water losses by opening stomates at night when evaporation losses are minimized. Because of space constraints, when the density of one type of plant increases, the corresponding density of other plants decreases. When those 8 biologists who were teaching at the undergraduate levels are not considered then only 2 of 25 (8 %) scientists correctly answered the question. As in the preceding comparison, some scientists successfully talk about the graph in terms of their familiar worlds but others are less successful.

Given that scientists do not fare all too well when confronted with unfamiliar graphs or with graphs depicting relations in a domain that they are unfamiliar with, should we be surprised when students and children do not fare so well when asked to talk about unfamiliar graphs? Thus, an early study had provided middle-year students with a graph featuring the amount of oxygen and the amount of shrimp as a function of distance along a stretch of river; there was also an indication of the location with sewage effluent into the river (e.g., Fig. 1.2c). Many students had trouble with this graph, particularly failing to relate the two graphs, one featuring oxygen levels and the other one the shrimp density (Preece and Janvier 1992). Although it was concluded that the students were "poor interpreters," they did in fact provide some really good responses. For example, almost 40 % of them

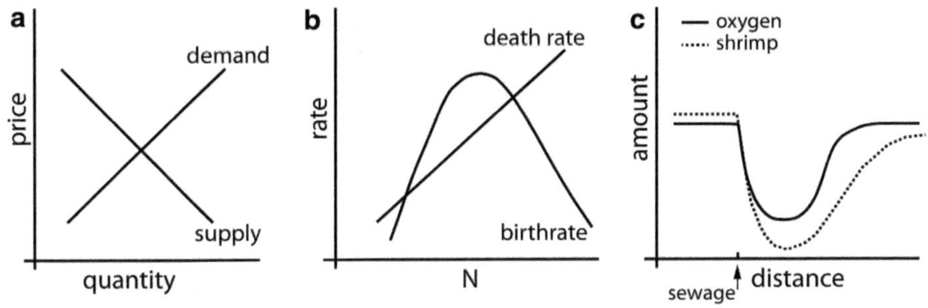

Fig. 1.2 Graphs used as part of different research projects about knowing related to graphs and graphing. (**a**) A demand–supply graph from economics. (**b**) A death rate–birthrate graph typical of undergraduate biology courses. (**c**) A graph used with children to determine their graphing competencies

suggested that the shrimp were diving or moving away to avoid the sewage. Is this not a sensible thing to say if you are not yet familiar with the concepts of respiration and the need of oxygen by certain but by far not all organisms?[3]

A problem with much of the research on graphing is that in such cases, however, we get the kinds of a posteriori explanations that also characterize university-based scientists talking about graphs that they are using in their own teaching. The experts tend to be observed when dealing with relatively simple graphs or situations from their own domain with which they are deeply familiar. The real work of seeing relationships and relating them to other things tends to become invisible. As one study shows, it *does* "*take work* to see linear relationships in messy graphs, just as it does to see the properties of light" (Lindwall and Lymer 2008, p. 218, emphasis added). What experts such as Simon do in the face of familiar graphs are not "interpretations": we should rather call them transparent readings (e.g. Roth 2003a), which differ radically from the kinds of searches that we get when the scientists are not based at the university. There is a history of research on experts externalizing themselves about graphs using think-aloud protocols. But because the experts are oriented to the think-aloud session as a specific form of activity, such work hardly accesses the forms of knowledge and the practices characteristic of mathematicians and scientists when they are working on problems in the context of their work (e.g., Livingston 1986). Not surprisingly, therefore, there has been a continuous interest in the learning sciences concerning the interactive work ("social construction") of mathematical objects generally and graphs specifically.

In the discovery sciences, graphs and graphing are tied to uncertainty (Roth 2009b). Dealing with uncertainty is *doing* work. As researchers in the field of ethnomethodology suggest, the real invisible work of everyday practice comes to the fore in case of trouble (breakdown) rather than when everything (apparently) runs smoothly (Garfinkel 1967). Such research is generally concerned with talking

[3] Anaerobic organisms do not need oxygen for growth; they may in fact die in the presence of oxygen.

about graphing as cultural-historically situated social practice (Roth and McGinn 1998). A situated social practice approach is consistent with the social-psychological approach that the Russian psychologist L. S. Vygotsky has initiated. In this approach, anything psychological is tightly related to societal relations. Thus, for example, "[a]ny higher psychological function was external; this means that it was social; before becoming a function, *it was the social relation between people*" (Vygotskij 2005, p. 1021, emphasis added) so that "*the psychological* nature of man is the totality of societal relations *shifted to the inner sphere having become function functions of the personality and forming its structure*" (p. 1023, original emphasis, underline added). Higher psychological functions in this view can be studied in societal contexts provided that the participants find themselves in situations where they have to make it appear again *as* societal relation between people. This is precisely what happens in instances of breakdown, that is, when scientists themselves recognize trouble and are struggling to get back on track. In this case, what might be assumed to be psychological and internal to their brains is actually real joint practical work that can be studied by cognitive anthropological means. The trick is to find those instances where normally smooth work comes to be troubled, at which point the by-and-large invisible work of scientific cultural practices comes to be articulated by members to the setting for members to the setting.

Graphs tend to be investigated as representational means. However, graphs, as other inscriptions used in the sciences, frequently do not have representational function, that is, as signs that stand out and stand in for something else. The mode of knowing characterized by the (mental) representation approach is derivative of forms of practical knowledgeability of how the world works in the sense that a *re*presentation requires familiarity with a more original presentation (Bourdieu 1980). Consistent with this position, there are suggestions that graphs are integral to the work process and are taken as such – as presentation rather than *re*presentation (Roth 2004). An important concept in the investigation of graph use is that of transparency. The notion is used to describe something that is not really visible and present to consciousness, such as a pair of glasses to the person wearing them. Thus, for example, when we write some notes into a research notebook, then we do not tend to be aware of the pen or pencil – unless there is some problem with the writing implement. Otherwise, it is as if the pen or pencil was not there and as if our thoughts got onto the page in image or word form. In familiar and competent use, graphs often work in the same way. Thus, when highly familiar with a form of graph – e.g., because they have produced them – child and scientific experts alike appear to be able to "see through" graphs to a reality beyond. This is what happens when people like H. Simon, highly familiar with the supply–demand graphs looks at one of these and immediately sees and talks about the equilibrium situation where the two graphs intersect and prices are stable. This also allows us to see when something appears to be odd. For example, children look at the graph in Fig. 1.3 and note that it cannot be correct because the person represented would have shrunk while she got older. Scientists may look through a graph to see a variety of retinal photoreceptors and "dark matter" on their microscopic slides. But, also shown in

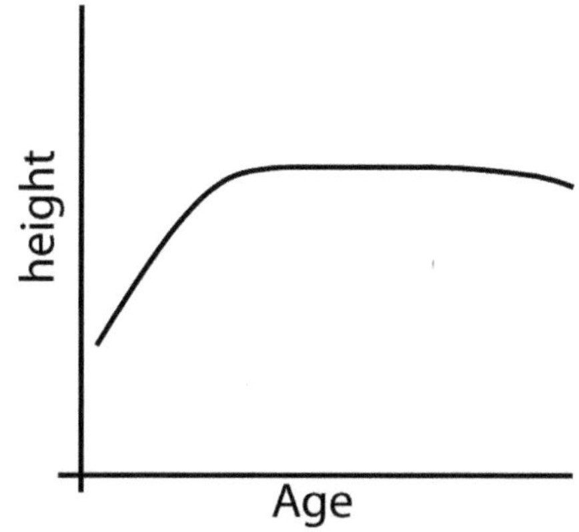

Fig. 1.3 Children suggest that this graph cannot be right, as the height of a person seems to be shrinking at a greater age

the work reported in this book, during the development of new graphs and phenomena, such familiarity does not exist and scientists struggle as much as anyone else to make anything of the data and graphs at hand.

In this vein, the researchers of one recent study report that they "spent much time trying to make sense of graphs in the absence of either scales, units to distinguish the quantities and any intimate knowledge of the process" (Noss et al. 2007, p. 373). To make anything at all, the authors suggest, "the graph reader needed to know what to look for: what were the significant quantities in the process, what might be significant in the variations in these quantities and how to recognise changes in the relationships between the quantities" (p. 373). Unsurprisingly, perhaps, another study concludes that graphs bear metonymic relations to the entirety of the work context (Roth 2004). A metonymy is the replacement of a thing by one of its properties or characteristics. It may in fact be better to characterize the relation of graphs to the entire work context as synecdochical, the use of a less comprehensive term (part) for a more comprehensive term (whole). This is beautifully illustrated in Noss et al.'s (2007) study, where the protagonist Jim uses graphs to identify trouble in the work process, but a particular "odd" graph that indicates trouble may also appear when there is no trouble at all. In the present book, I analyze one episode concerning uncertainty that arose precisely when one of the members to the setting *formulated* trouble with the determination of a specific feature – the half-maximum bandwidth of an approximately Gaussian-shaped absorption curve (Chap. 4). However, whether a feature of the graph means trouble "depended first on detecting a production problem that required some action taken" (Noss et al. 2007, p. 377). The authors note that the graphs were not traditional representations *of* something else. These were *part* of a *whole* process, and they were consulted when there was trouble in the whole process. The graphs therefore have indexical rather than representational function, they *point to* something rather than *being about* something. As long as scientists are familiar with the relations of a graph to the whole research situation, we therefore might expect to observe the kinds of use that are reported in studies of experts in their mundane environment. However, we might anticipate trouble when scientists do not yet know the significant quantities, significant variations, and the changes in the relationship between quantities.

I noted above that graphing in the sciences occurs in contexts of uncertainty, where the production of new knowledge presupposes that it cannot be the direct object towards which scientists orient. They do not know what they will find ("discover") – otherwise they would not have to do the research. Because they do not know what they will find, they cannot intentionally orient towards this finding (i.e. new knowledge). There is therefore a considerable amount of uncertainty involved in scientific discovery work. The work of dealing (coping) with uncertainty is appropriately studied from an ethnomethodological perspective, which is fundamentally concerned with the way in which social actors produce their lifeworlds in an orderly fashion. Ethnomethodological articulations of the formal structures of human praxis recognize the paired nature of (a) objects, practices, and accounts thereof and (b) the work required to lend the objects, practices, and accounts their reportable character (e.g. Garfinkel and Sacks 1986). This approach is based on the recognition that the objectivity of mathematical objects or proofs does not reside in their notational characteristics themselves but in the fact that the associated, inherently embodied and contingent practices can and do make objective structures visible over and over again (Husserl 1939). The "identifying task of ethnomethodological studies of order* is to furnish to the phenomena of order* production their genetic origins in and as immortal ordinary society" (Garfinkel and Wieder 1992, p. 180). What is of interest are neither the STEM objects in themselves (e.g., as signs) nor the associated STEM practices in themselves, but the paired nature of the work and the available texts (objects) that allows this work to be and become accountable. Moreover, of interest is not some theorist's articulation of the paired nature, the two technologies of order, but the manner in which "a phenomenon of order* is available in the lived in-courseness of its local production and natural accountability" (p. 182). It is in the pairing with the sensually embodied work that the texts/objects receive their objective nature and become independent (Galilean) objects. The phenomena exist *in and as of their accountability* in sequentially ordered laboratory talk. Thus, for example, the "extreme" data points I report on in several chapters are extreme not because of some existence prior to and outside of perception – such a position would always require the inherently unavailable recourse to some actual state (e.g., Woolgar 1990). Rather, the existence of the extremes *as extremes* depends on the power of available descriptors to make the "extreme" character of the data point or points *interactionally* adequate and accountable.

In the studies reported in Part B – where scientific work is marked through and through by uncertainty related to graphs and graphing in a laboratory during discovery work – I follow the precepts outlined by an early study that realizes an ethnomethodologically informed approach. To arrive at an adequate description of scientists' *work-in-process*, researchers are encouraged to take a "first-time-through" approach (Garfinkel et al. 1981). This is equivalent to saying that in "contrast to protocol reports, analytic just-so stories, self-reports, interviews, anecdotage, gossip, detective stories, docile records, unobtrusive measures, and other residue documents" (p. 136), researchers focus on the "analyzability and recognizability as *not yet* 'naturalized' in a reportable just so story" (p. 136). Of interest in the present research is not just the "interpretation" of the graph and the presupposed or actual relationships it displays. Rather, in the same way as Garfinkel et al.'s pursue "examin[ing] the pulsar for the way it is in hand at all times" (p. 137) in their astronomers' inquiry, I examine the graphs for the ways these are in the biologists' hands: literally so when they are drawing and making gestures over and about these. In this approach, (the ordered properties of) a graph and the work that makes it what it is form a reflexive pair just as "a proof is this inextricable *pairing* of the proof and the associated practices of its proving" (Livingston 1986, p. 14).

Science Inquiry and Graphing

> The science education literature indicates that students who are involved in collecting their own data often do not understand the fundamental reasons for doing so and are often more concerned with following laboratory protocols and getting "the right" data. As a consequence, "hands on" activities are often not "minds on" activities. (Cobb and Tzou 2009, p. 169)

In this book, I am concerned with graphs and graphing in the sciences, especially in situations of uncertainty when what eventually will be known as the result of the inquiry is not yet available. Although graphs and graphing are a topic of mathematics education, these obtain very different emphasis when used in science education. Mathematics educators may be interested in analyzing a parabolic curve, allowing students to find its slope at a range of points (Fig. 1.4). Once these slopes themselves are plotted, a line graph results (Fig. 1.4). This line graph intersects with the abscissa – i.e., $y(x) = 0$ – precisely at the same x-value where the parabolic curve has a minimum. Science educators, on the other hand, are not generally interested in the properties of such graphs but rather in the physical phenomena that are made present in and by the graph – unless of course these properties of the graphs relate to physical phenomena. These different interests reflect the different fields and the objects of their inquiries, which lead to the fact that the mathematics used and mathematical knowledgeability displayed in one or the other context – engineering versus physics versus mathematics – will be radically different (e.g., Brown et al. 1989; Roth 2005). Thus, for example, a physics teacher might have his students investigate the position of an object thrown into the air and falling back down. In this case, a motion detector would depict a parabolic graph. He might ask students to think about the physical properties of the object at its highest point, to locate the highest point. Further inquiries might pertain to questions about negative speeds and about what the slope might look like when plotted against time and the physical phenomenon that it might relate to. Here, the science and scientific concepts are of interest and the graphs figure as tools used to make present again the throwing of the object that has occurred some time earlier. It appears intuitive that the competence for talking about or argumentatively using any one graph or the graphs in their relations is tied to the physical phenomena: position and speed. When students are not familiar with respect to the relation

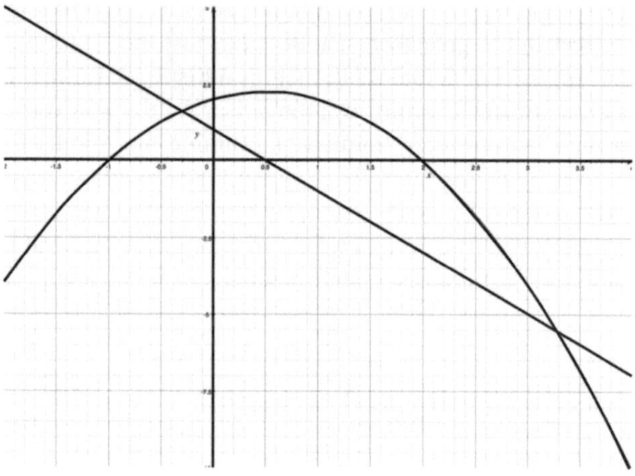

Fig. 1.4 Such a set of graphs will be of very different interests and practices when they appear in mathematics versus science courses

between the phenomenon and the graphs, we probably should not expect a lot precisely because the students are unfamiliar with the system. For many readers it may come as a surprise that scientists are in precisely the same boat – and the chapters in Part B articulate a lot of what actually happens and how it happens in a discovery science. Graphs and graphing are bound up with all aspects of scientific inquiry, including the background familiarity with the physical phenomenon.

Inquiry – both in more constrained, planned and structured investigations and in more open, unstructured real-world settings – has long been a trademark of science education generally (e.g., McElhaney and Linn 2011) and of reform-based science education particularly (e.g., Carlone et al. 2011). It is not surprising, therefore, to find (as of December 15, 2013) 958 out of the 8,186 articles in eight science education related journals included in the ISI Web of Science database (*Education and Educational Research*) using the term "inquiry" as identifier, in the title, or in the abstract. Inquiry is a trademark of science education even though in teaching practice it often is pushed to the margins in the face of high-stakes testing. But there are studies that do in fact report tremendous achievement gains when students engage in inquiry; and these gains are larger when teachers are more experienced in teaching in the inquiry mode (e.g., Fogleman et al. 2011). Even beginning teachers may find themselves surprised by the positive outcomes from their supervision of extended experimental investigations (Ritchie et al. 2013). There is further evidence that scaffolding promotes teachers' competencies to guide students through open-inquiry projects, "especially the ability to know when and how to give students a well-balanced combination of 'structure' for open-inquiry learning and sufficient 'space' for that" (van der Valk and de Jong 2009, p. 829). Others outright reject inquiry – at least the forms in which little guidance is provided. Defenders of inquiry list – among its ideal benefits – that students not only learn how science operates from designing experiments to generating data and to the ultimate reporting of results but also outperform guided-inquiry students on a number of variables (e.g. Russell and Weaver 2011). Even the youngest students learn from open inquiry. Defenders of open ("authentic") inquiry further contend that inquiry works precisely because of the high levels of control students have over the task and task definition. In a study that drew on adapted primary scientific literature, teacher education students did in fact learn by engaging in and talking about scientific inquiry, their pedagogical content knowledge, and their subsequent curriculum designs (Falk et al. 2008).

Although science educators have shown interest in inquiry approaches to science, the question of the nature of data generation is less frequently raised: only 26 of the 958 articles on inquiry also show up when the search term is "open inquiry."[4] That is, data generation appears to be of less interest even though science educators have noted the "sophisticated coordination among theories, phenomena, data, and data collection events" (Apedoe and Ford 2010, p. 165) and even though others articulated the role of anomalous data in knowledge generation (Chinn and Brewer 2010) that often are associated with open inquiry. Such coordination can been observed among more advanced ("expert") undergraduate students but not among their "novice" peers (Jordan et al. 2011). Despite Apedoe and Ford's suggestion that the complex interactions should be broken down and taught separately, ethnographic studies among research scientists suggest that these interactions between theories, phenomena, data, and data collection events are irreducible. One study did in fact report very different forms of behavior when students were provided with data only (in their mathematics class) versus when they decided about the goal of their research and generated data themselves (Roth and Barton 2004). In the controlled context of mathematics lessons with data provided, students did not engage with the task, by and large suggesting that they "don't know how to do it." On the other hand, intense engagement and highly competent practice in plotting data and talk about the plots were observed when the same students were in complete control over their investigations and the way of representing their results. Similar differences were reported in the above-mentioned study, where eighth-grade students, who had been collecting real data in investigations of their own design, outperformed pre-service teachers, who had already completed bachelors or Master's degrees in science, on data interpretation (Roth et al. 1998). This suggests that something in the data generation process allows students to know what to do with data and how to do it.

These considerations, taken up again in the later chapters of this book (e.g. Chap. 12), led me to propose a greater emphasis in schools on interdisciplinarity, especially between mathematics and the sciences (though other linkages with the arts are beneficial and fruitful). As my own work as a physics teacher showed, students can and do learn about mathematical properties and more easily use them when they have opportunities to link their mathematical investigations to scientific ones (Chap. 12). Thus, my own high school students learned a lot about data modeling using graphical and statistical software; and they intuitively came to

[4] There are 50 articles when a Boolean search is conducted in the entire Education and Educational Research database using "open inquiry" and "science" as the search terms.

use central calculus concepts – e.g. about functions and the relations between first and second derivatives and integration – when their familiar ways of getting around the physical world can be mobilized to make use of graphs and their relations in the world of inscriptions. I return to these and related issues in Parts D and E of this book.

Toward a Dynamic Theory

I entitled this chapter "Toward a Dynamic Theory of Graphing" but have not yet articulated what a dynamic theory might implicate. In the preceding sections, I describe different positions on graphs and graphing, and review some of the literature in the field. But much of the work conducted in the past takes a static perspective in the sense that it is concerned with knowledge states prior to instruction or *what* students "construct" during instruction. Even when researchers apparently investigate learning, this is still conceptualized in terms of differences between forms of knowledge at the beginning and end of some formal or informal curriculum. However, a different approach is required when we want to talk about graphical knowledgeability and graphing practices as *continuously developing* while persons, scientists or students, participate in them. To theorize continuous change and the flow associated with it, we require figures of thought – i.e., fundamental categories – that denote change itself. Using a minimum category of change is consistent with the world that we know: it is a constantly changing world, a constantly changing (evolving) life where even our languages change continuously. Once we use such categories, however, once we choose a category of change, then change and learning will become unproblematic, whereas knowledge will become problematical (Lave 1993). This is so because in a continuously changing world, stasis has to be explained rather than being taken as the figure against which change is assessed. Once we accept that the whole world is in flux, then stable (knowledge) states become problematic and, should it be argued that someone's knowledge does not change, this statement requires empirical data for its support. I begin by outlining requirements for categories of flow and then provide two brief examples of graphing and the emergent properties of graphs.

Units of Analysis and Units of Thought (Categories)

The received way of thinking about change posits some entity and some force that operates from the outside to bring about change. For example, learning and development, processes of change, are thought in terms of the difference between knowledge before and knowledge after some intervention or after a student's constructive efforts. Psychologists might then say that the curriculum or the instructional process *brought about*, *was the cause of*, the change observed.

Alternatively, constructivist educators might say that the curriculum intervention allowed the student (subject) to construct new knowledge (object). In both approaches of thinking about change, the force – the curriculum, the instruction, or the student's construction – is *external to* the knowledge. This relationship between the object and the force may be depicted analogically in terms of a change that a rectangle undergoes when subject to a process of shearing (Fig. 1.5a). The rectangle is the unit of thought, the category, and the force operates upon it to bring about a change in the object so that it becomes a parallelogram. Underlying is a theory of self-identity, because something has undergone a transformation, the *material* remaining the same while appearing in different *form*. Time, too, is external to the process; it is, consistent with Kant's analyses, taken to be an a priori that constitutes a condition for having an experience.

A different way of theorizing change uses categories of change, that is, where the minimum figure of thought is one that already *embodies* change. Cultural-historical activity theory – the approach to a concrete human (societal) psychology that L. S. Vygotsky and A. N. Leont'ev started and developed during its early phases – is one approach to learning and development that operates with such units (Roth and Lee 2007). In fact, Vygotsky wrote about replacing analysis in terms of elements, typical of the natural sciences and traditional psychology, by *unit analysis* (Vygotskij 2005). He used an analogy from chemistry. To describe the properties of water (H_2O), elemental analysis would decompose the molecule (unit) into the elements hydrogen (H_2) and oxygen (O_2). But the properties of water at room temperature (20 °C, 68 F) cannot be deduced from the properties of hydrogen and oxygen at that temperature. In fact, hydrogen burns or explodes at that temperature when ignited, whereas water is a liquid that douses a fire. The minimum unit for describing and explaining the properties of water, therefore, has to be a unit that retains all the properties of water. With respect to change, the minimum unit has to contain all the properties of the change process. We may again take the transformation of a rectangle into a parallelogram as an analogy (Fig. 1.5b). As the figure shows, the minimum unit of analysis and the minimum unit of thought (category) contains the entire change process from the beginning to the end. It is apparent from the analogy that both the force and time are now *internal* to the unit. The change is operating within a self-changing unit.

Thinking in this way has immediate consequences as to the relation between cause and effect. From *within* a unit – think of a meeting, a societal relation – we cannot ever know what the causes responsible for the final outcome will be until after the outcome is known. If we could know the operating forces, we would not have to have meetings, those of the kinds found in Part B of this book or those that we attend as part of our normal work life. This is so because knowing the forces and knowing what every individual participating in the meeting thinks prior to its beginning would allow us to enter all the givens into a (computer) algorithm that calculates the final result. But this is not how the world works. Dialogical exchanges are generative, leading to ideas that could not have been thought before (Bakhtin 1981). An example where such an approach changes the way in which we analyze is conversation. In this case, our interest in how a conversation unfolds

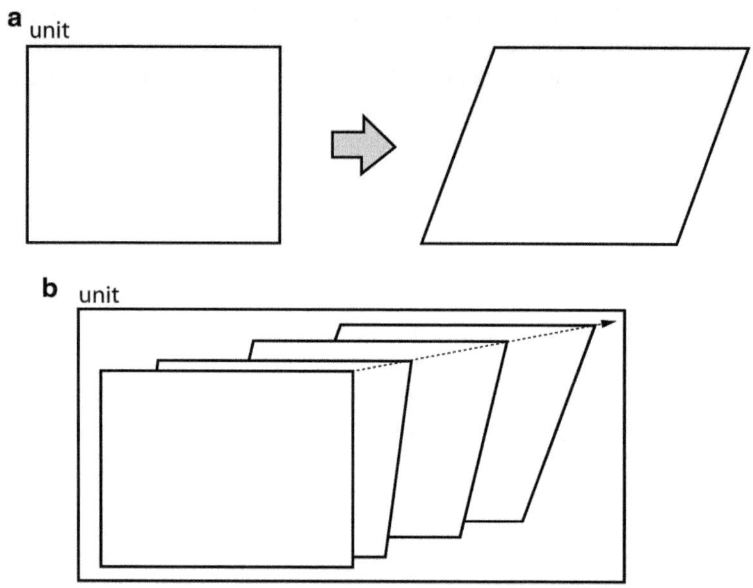

Fig. 1.5 (**a**) Shearing understood as something happening to a self-identical object, which first takes the form of a rectangle and then becomes a parallelogram. (**b**) In a dialectical view, the entire transformation is part of the same, non-self-identical unit

requires us a minimum unit that is *con*versational. Etymologically, the word comes from the Latin convertēre, to turn about, life *with*, dwell, and abide; it became, in the old French, *converser*, to pass one's life, live, or dwell *with* and, most recently, took on the sense of "to talk *with*." That is, our minimum unit for a conversation requires a unit of analysis and category of thought that reflects the *with-ness* implicit in the term *con*versation. Thus, analyzing conversations in terms of individual speakers would be doing elemental analysis; unit analysis will take two turns as a minimal unit. Take the following excerpt from a laboratory meeting of the research team that is at the heart of Part B of this book.

1 T: how about this one here;
2 (2.29)
3 T: you see right here? * ((*Points to the screen where there appears to be something in the noise*))
4 C: yea i know but the resolution is very bad
5 T: yea okay ((*goes to some window*)) i cant set the contrast to brighten it. <<p>yea we have more than one to select from it.>

Elemental analysis might begin suggesting that Theo *asked a question*: how about this one here (turn 1). Grammatically, the statement is shaped like a question because of the interrogative "how"; intonationally, however, shown by the semi-colon at the end of the transcribed statement, the pitch has fallen slightly as this tends to be the case in unfinished statements. Although Theo might have had the intention to ask a question or make an offer for which object on the monitor to consider next, conversationally thinking, this is not what is happening. That is, this is not the *effect* following an action. Instead, there is a pause, that is, there is no reply from any of the other actors. If there had been, the conversation might have continued based on the reply, taking it into account. Saying that Theo asked a question makes him an agent, a source of a cause, who brings about an effect. But there is no reply in turn 2, so, conversationally thinking, we do not have a question–reply sequence. If we conducted an analysis in this manner, we would never be able to theorize the *unfolding* nature of the conversation. So let us make the assumption that the minimum unit of analysis that has all the properties of a conversation – i.e., the give and take involving two or more people – is the turn pair.

Taking the turn pair as our unit will radically change how we analyze this excerpt. Let us take the turn 3 | turn 4 pair. Our language actually does not offer a category to take this pair *as one (sociological) unit*, forcing us to make up units. I have come to use the vertical stroke "|" as a way of combining two things into one with the agreement that this creates a whole unit that cannot be reduced further without losing its sense. So in a conversation, this makes for a pair: turn 3 | turn 4. What turn 3 does depends on turn 4, and what turn 4 does, depends on turn 3. The two turns mutually implicate each other. In our situation, there is then something like an offer to see something ("you see right here?") and its acceptance ("yea"). That is, we have an offer | acceptance pair where each part implicates the other. The offer is an offer because there is an acceptance; and the acceptance is an acceptance because there is an offer.

Let us now return to turn 2. It is made up of a pause. This pause is a conversational phenomenon, providing opportunities for anyone of the three persons in the room to speak. But nobody does for a conversationally long 2.29 s. And then, Theo speaks again (turn 3). That is, we have a sequence of two pairs: turn 1 | turn 2 and turn 2 | turn 3. It is as if Theo treated his preceding turn to have caused trouble, which he now is in the process of repairing (turn 3). Because it is Theo who speaks again in turn 3, we can therefore gloss what happens by means of the pairs offer | no reply and no reply | repair offer. That is, there is an invitation to consider something "how about this one," which, when there is a no-reply, is repaired by a finger moving to the monitor, pointing to a something, together with what we can hear as a renewed invitation, "you see right here." The adverb "here" together with the finger (see screen print in transcription) constitutes an invitation to orient to a particular location of the image displayed on the monitor. This invitation is accepted, "yea," followed by what can be heard as a descriptive statement of the object, "the resolution is too bad" (turn 4). However, to find out about its function from the perspective of the conversation, we must consider it in the context of turn 5. That turn contains two adverbs of consent (yea, okay) and then the statement "I can't set the contrast to brighten it." We now have the pair "the resolution is very bad" | "I can't set the resolution to brighten it," which can be heard as a critique | justification pair. It is as if Theo justified the state of affairs, which is a "snowy" image indicative of a "bad resolution."

In this example, therefore, by considering turn pairs as minimal units, the second part of the pair is indicative of the effect brought about by or following the first part of the pair. However, we never know what will be the second part as long as we are in the first part of a turn pair. That is, we never know the effect beforehand with 100 % certainty, so that we cannot know the causal force moving the conversation ahead until after we know the effect of a statement. It is only after the fact that we know better and best (e.g., the common practice of Monday morning quarterbacking in all walks of life, which leads people to say, "I should have done . . . instead"). In taking the turn pair as a minimal unit, as a category of thought, time and forces become internal dimensions. We therefore obtain what our initial considerations in the context of the analogy of shearing (Fig. 1.5) stated as requirements and consequences.

In the section on expertise related to graphing, I quote Vygotsky concerning the role of societal relations to higher psychological functions and the psychological nature of human beings. The psychologist considered societal *relations* as that place in which psychological functions can be found. In the mature person, we find forms of behavior even in the absence of others – such as writing a diary – that the developing person first encounters in relations with others. The writing of a diary, writing for oneself, however follows all those instances where a child speaks and writes for the purpose of acting upon others (Vygotskij 2005). In this approach, even personality is the ensemble or totality of societal relations as it plays itself out in the here-and-now of subsequent societal relations. To explain any psychological function and personality, therefore, requires us to appropriately theorize *relations*. The minimum unit of a relation has to have all the properties of relations involving at least two people. These two people do not *inter*act, that is, are not bound by actions *between* them. Instead, relations require us to theorize in terms of *trans*-actions, in other words, in terms of actions that always exceed the sum total of the parts of the relation. In relational terms, power, knowledge, teaching, or learning are not properties of individuals, who have or undergo one or the other of these phenomena. Instead, power and knowledge, for example, are not to be analyzed in terms of subjects, who know, or who have power (Foucault 1975): "the subject who knows, the objects to be known and the modalities of knowledge must be regarded *as so many effects* of these fundamental implications of power-knowledge and their historical transformations" (p. 36). The author's statement that knowledge and power mutually *implicate* one another means that they cannot be separated into entities that make sense on their own because each is involved in the definition of the other. In the same way, human relations need to be thought in terms of the mutual *implication* of the participants.

Research as Cultural-Historical Activity

Cultural-historical activity theory was created to describe and explain human psychological characteristics, development, consciousness, and personality (Leont'ev 1983). Following Vygotsky's idea that society is what distinguishes humans from other animals so that society is a specifically human characteristic, Leont'ev proposes *activity* as the minimum unit of analysis and category of thought that allows us to explain anything specifically human. It is in and through society that culture lives, that is, is reproduced and transformed even though individual lives have only a limited time span. It is in and through relations that every behavior exists when it is not somehow coded in the genes. Participation in these relations means producing society-specific cultural forms of knowing.[5] To explain anything humans do, from the tiniest perception to participation in the most complex events (such as an experiment on the Higgs boson at CERN in Geneva), requires us taking into account societal relations. Thus, even raw human perception is concrete, objective activity that subjects itself to its object. This object embodies the totality of human relations that are the source of society (Luria 2003). Any specific analysis of human behavior does not require taking into account the whole of society. Rather, it can work with the smallest unit that contains all the characteristics of society. For cultural-historical activity theorists, this smallest unit is activity. Before going on, however, I need to specify precisely what activity is because of the confusion often associated with the term.

Activity is the smallest analytic unit that bears all the characteristics of society. Thus, anything we might say about what a human being does needs to be taken in its relation to and determination by activity that produces something meeting a generalized human need. However, in each activity one can identify smaller parts that make up the whole (Fig. 1.6). For example, the production of wheat, a societally motivated activity, includes turning the soil using a plow (means), leveling the ground using a harrow (means), seeding the wheat, fertilizing the soil, harvesting and threshing the wheat using a combined harvester, and trucking the wheat to the relevant grain elevator. Each of these is constitutes a goal-directed action. Its specificity can be explained only within the context of the particular activity, because the same action, such as driving a tractor, accomplishes very different things in the context of different activities: The significance of driving a tractor as part of growing grain is different from driving a tractor as part of a demonstration of farmers against the monopoly of the wheat board, which is again different from driving a tractor at a tractor pulling competition. Thus, there is a mutually constitutive (determining) relation between the actions that realize an activity and the activity itself. It is a part–whole relation. The parts (actions) make the whole (activity), but these parts are parts only because of the whole (activity). That is,

[5] The noun culture and adjective cultural tend to be used indiscriminately to very different phenomena. Thus, there is talk about Western culture even though cultural practices even with respect to academia are very different in typically countries such as Canada, France, and Germany.

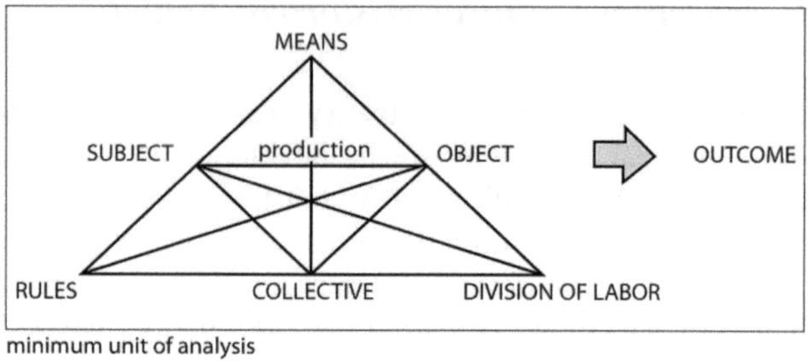

Fig. 1.6 A societal activity, which produces something (outcome) that is part of meeting a generalized human need, is the minimum unit of analysis including the beginning and endpoint of the production

the parts (actions) presuppose the whole; and the whole (activity) presupposes the parts from which it is made.

In a similar way, conscious goal-directed actions are not the smallest entities we can identify in the whole of activity. Conscious actions, such as driving a tractor, are made of operations that are not present in and to consciousness. Thus, for an experienced tractor operator, engaging a gear is such an operation: the operator does not have to think about changing and engaging gears. It happens without that the operator explicitly thinks about it. The relationship between operations and actions that they realize is mutually constitutive (determining). That is, the operations occur only because they are required to realize a specific action; and the action presupposes the operations from which it is made. Again, the relation between operations and the action that these make up is a part–whole relation. The parts, to be parts of a specific whole, presuppose this whole; and the whole only exists in and as of its concrete realization by the parts (which, therefore, are presupposed). In contrast to the operations, actions and activities have concrete, conscious goals and motives that they realize. The individual farmer drives a tractor *in order to* sow the wheat seeds; she has a specific goal for the current action. Wheat is grown to provide the raw materials for food or to produce the feed required in the production of other food sources, such as in cattle, pig, or poultry farming. That is, each activity is characterized by a generalized motive at the level of society. The object therefore is transformed to realize the motive, which is the reason why activity theorists often speak of an object/motive of activity.

In the cultural-historical activity theoretic framing, macro-societal structures and local person–person relations are connected. There is no gap between micro-social and macro-social processes or theory. Because the minimum unit of analysis is societal in nature – societal activity, in each case concretized in the things people do to contribute in the production for a generalized need – what people like scientists do inherently produces and transforms society all the while these doings are highly local, contingent, and micro-genetic. We may therefore speak of every relation or conversation within a scientific laboratory as the locus of a micro-genesis of macro-social conditions. As a result, we then get to take consciousness, personality, and higher psychological functions in *societal* terms – always at the crossroad between collective possibilities and concrete singular materializations in individual lives of these general possibilities (Roth 2014).

An activity is a living phenomenon. But to live is equivalent to saying that a phenomenon is changing. If we were to freeze it for an instant, however, we would or might be able to identify specific parts. The parts of societal activity that tend to be most emphasized include the subject, object, tools, division of labor, rules, and community (Fig. 1.6). In the activity, actions transform the object until the product is finished. The motive of activity is to get from the initial state – e.g. where there are raw materials – to the final stage, when the intended product exists materially. Taking the activity as the minimal unit, therefore, is equivalent to making time internal to the minimum unit. It also is equivalent to stating that the object never exists as a thing but is continuously in transformation, much in the way this is represented in the rectangle | parallelogram transformation (Fig. 1.5). But human actors also change when they do something: biologically, they burn energy; and psychologically, they get tired, loose attention and focus, or become more experienced. That is, the subject of activity also changes. Moreover, as the object upon which the subject acts changes, so does the consciousness of the subject in which the object specifically and relevant aspects of the system generally are reflected. The tools also change: pen tips wear, diminish the amount of ink available, and so on. In the course of activity, human relations, too, undergo change. We may be upset with a co-worker (horizontal division of labor) or with the boss (vertical division of labor). This changes our affective state; and such affective changes alter the ways in which we relate to the concerned individual specifically and to other aspects in the activity more generally. The ways in which we do our work changes: our practices change (Roth 2009c).

All of the preceding can be pictorially represented with the proviso that what appears as a structure is in fact related to activity in the way an individual frame on a movie real is related to the whole movie (Fig. 1.6). Because the activity is a whole, all its parts – the subject of activity, the object of activity, the means of production, division of labor, collective, and the operative rules – stand with it in mutually constitutive (implicative) relations. Here, the adjective mutually constitutive (implicative) expresses that we cannot consider these parts (subject, object, outcome, etc.) independently – their precise nature depends on all the other parts *when we are concerned with explaining* this *activity*. There are mutually constitutive relations between the parts themselves. The value (i.e., nature) of each moment of the activity is a function of all other moments. Producing something changes the activity itself, which means, each individual part changes; and any change in but one part entails a change in the activity as a whole and, therefore, in all other individual parts. Moreover, as evident from the preceding description, the relations between the moments are subject to change (rather than remaining stable).[6] One way in which we can think of the relation between the whole of activity and its part is by means of the documentary method (Mannheim 2004). In this approach, any

[6] In cultural-historical activity theory, as in dialectical logic generally, the term *moment* is used to denote a part of a whole that cannot be understood independently of this whole and all the other parts that can be identified. In this book, I employ the term only in this way.

aspect of the activity, whatever it may be, is viewed as a document of the whole. Each part is like a raindrop in which the whole world reflects itself, where the world is the scientific research activity in the present context. That is, every part, such as a graph, reflects the whole of the activity and, therefore, can be explained only as part of this whole. In other words, as already indicated above, each part can be viewed as the above-mentioned synecdoche, not only part of a whole but also as reflecting the character of this whole.

The pictorial representation of activity (Fig. 1.6) is often misused because researchers focus on the relation considering the parts as if these could be taken on their own and as if these were fixed. Here, I am concerned with the second fixedness. As the figure shows, activity theorists theorize the whole productive process as *one*, from the beginning to its end. Nothing smaller than this whole production can be described intelligible on its own. If we thought about the production of cars, for example, then the entire process beginning with the raw materials to the final state where the car runs off the assembly line *is the minimal unit*, just as the minimal unit of the transformation of a rectangle into a parallelogram is (Fig. 1.5b). Taking cultural-historical activity theory seriously, therefore, requires us to theorize the scientific research reported in Part B as a *one*, as a manifestation of research activity that has scientific knowledge as its research outcome. Thus, as the ethnographic research described in Part B shows, the scientific research team began with some ephemeral gestures similar to those Shelby subsequently used in one of the first research meetings (Fig. 1.7a); and it ended with a publication that featured a graph radically different in form from what the team initially conceived (Fig. 1.7b). In the process, their (rhetorical) use of graphs changed, including, among others, a major conceptual change concerning the physiology of ocean migration that those salmonid fishes undergo that the team was researching (Chap. 7). Because we consider the research activity as a whole, from beginning to the end as *one* single analytic unit or theoretical category, all the changes with respect to knowledgeable deployment of graphs and graphing practice also changes. That is, when we use this minimum unit of analysis, change in graphs and graphing are built into the theory. These are "collateral" effects of the production of scientific research results. In this way, we have arrived at a dynamic conception of graphs and graphing, because thinking about them as part of scientific research activity inherently implies that these are changing parts of a self-moving system. Because the system is moving, constantly in flux, any part of it also is in continuous flux, including the subject (Roth 2013). For this reason, I propose graphs as graphs*-in-the-making, graphing practices as graphing*-in-the-making, and the subject of activity as subject*-in-the-making.[7] This proposal is developed in greater detail in Chap. 9.

The need to think in terms of units that embody change is apparent from the following example taken from one of the meetings of our scientific research team. While the team was discussing some aspect of the data collection and processing,

[7] The asterisk is used to mark the transitional nature of the phenomenon denoted.

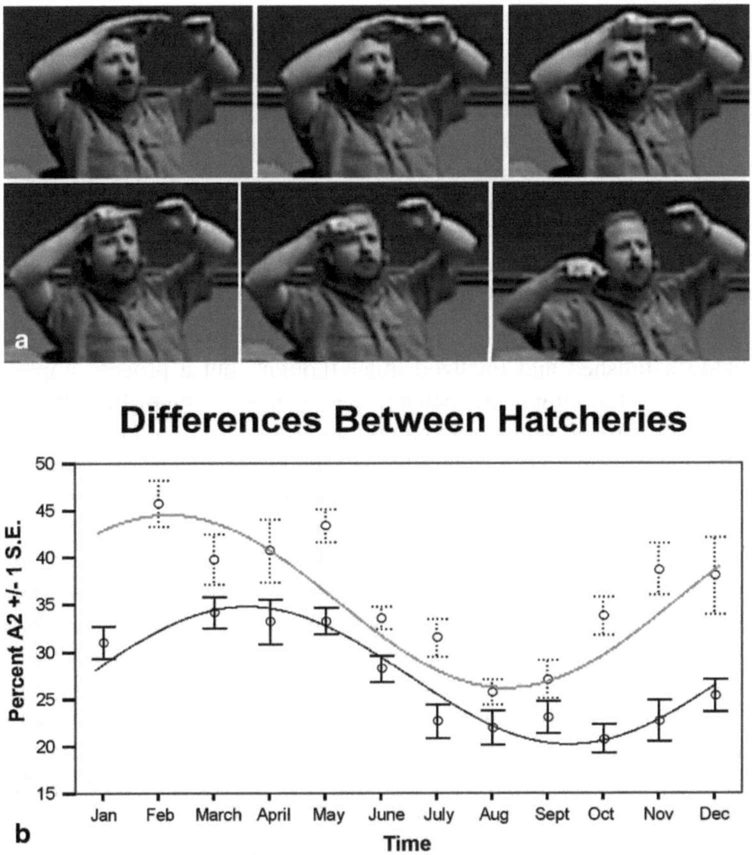

Fig. 1.7 The scientific research activity described in Part B began, in 1998, with ephemeral gestures similar to those Shelby presented in 2001 (**a**) and ended with the publication of a graph in 2006 (**b**)

Craig, the head of the team, had gotten up saying that the preconception of the graph to be expected "is something like this." Although we might anticipate this to be an announcement of the production of a graph already finished in the mind, the subsequent events show this to be an unwarranted inference. Thus, for example, standing in front of the empty chalkboard, Craig first says "If you plot a half" and then writes ½ m (Fig. 1.8); he then stares at what he has written for 5.93 s, and then erases it (Fig. 1.8). In this way, there is lots of writing and erasing going on and a lot of staring and gazing left and right at what is unfolding on the board before some graph actually materializes. Here, thinking and writing implicate one another, writing apparently bringing about changes in thinking, expressed in erasure and further writing. That is, writing itself moves thinking ahead and changes contents and process of thinking. That is, to explain graphs and graphing, we require a dynamic approach, because writing (drawing) a graph is not simply a process of

Fig. 1.8 Craig writes and then erases what will become the label of the ordinate of a graph. Prior to the appearance of the graph on the chalkboard, there are many such writing/erasing sequences

externalizing a finished and finalized inner thought, but a process implicated in another process – thinking – independently of which it cannot be explained. The two together are part of a more encompassing process – the relation of this graph specifically, the meeting more generally, and the research activity most generally.

Here at the end of this chapter, I emphasize once more how cultural-historical activity theory conceives of the relation between the micro- and macro-order of society. Thus, even an instant such as that depicted in Fig. 1.8, where Craig writes something on and erases it from the chalkboard is conceived of as societal in character (Luria 2003). Actually, Luria goes even further than I have presented the case here. He suggests that the societal organization reaches down to the physiological processes of the brain and its functional organization. Thus, "the higher forms of mental processes have a particularly complex structure," which, "as a rule are based on a series of external aides (such as language, the digital system of counting) formed in the process of societal history, are mediated by them and cannot be understood in their absence" (p. 76). As a result, the higher mental processes "are always connected with a reflection of the outer world in full activity [aktivnoj dejatel'nosti], and their conception loses its content when considered apart from this fact" (p. 76). In the present context, the graphs the scientists employ and those that they develop in and through their scientific research are reflections of societal history through and through. Moreover, because these are both means of production and product (outcome) of societally motivated activity as a whole (Fig. 1.6), any consciousness related to these is mediated by the activity. That is, graphs do not "mean" anything on their own; "meaning" is not a characteristic feature attached to them.[8] Instead, the graphs that the research team studied in Part B uses and produces are one-sided manifestations of the observed activity, which realizes (materializes, concretizes), in idiosyncratic and singular ways, general possibilities that exist at the societal (collective) level. Wittgenstein (1953/1997) proposes calling *language-game* the totality of language and the activity (*Tätigkeit*) with which it is interwoven. It is in this sense that I use the term *activity* throughout this book, as a concrete totality interwoven language-in-use.

[8] I agree with Wittgenstein (2000) that we do not need to term *meaning* because it gets into our way of appropriately describing how language actually operates and how it is deployed.

References

Apedoe, X., & Ford, M. (2010). The empirical attitude, material practice and design activities. *Science & Education, 19*, 165–186.

Arcavi, A. (2003). The role of visual representations in the learning of mathematics. *Educational Studies in Mathematics, 52*, 215–241.

Bakhtin, M. (1981). *The dialogic imagination*. Austin: University of Texas.

Bourdieu, P. (1980). *Le sens pratique* [The logic of practice]. Paris: Les Éditions de Minuit.

Brown, J. S., Collins, A., & Duguid, P. (1989). Situated cognition and the culture of learning. *Educational Researcher, 18*(1), 32–42.

Carlone, H. B., Haun-Frank, J., & Webb, A. (2011). Assessing equity beyond knowledge- and skills-based outcomes: A comparative ethnography of two fourth-grade reform-based science classrooms. *Journal of Research in Science Teaching, 48*, 459–485.

Chin, C. A., & Brewer, W. F. (2010). Models of data: A theory of how people evaluate data. *Cognition and Instruction, 19*, 323–393.

Cobb, P., & Tzou, C. (2009). Supporting students' learning about data generation. In W.-M. Roth (Ed.), *Mathematical representation at the interface of body and culture* (pp. 135–170). Charlotte: Information Age Publishing.

Edgerton, S. (1985). The renaissance development of the scientific illustration. In J. Shirley & D. Hoeniger (Eds.), *Science and the arts in the renaissance* (pp. 168–197). Washington, DC: Folger Shakespeare Library.

English, L. D. (2012). Data modeling with first-grade students. *Educational Studies in Mathematics, 81*, 15–30.

Falk, H., Brill, G., & Yarden, A. (2008). Teaching a biotechnology curriculum based on adapted primary literature. *International Journal of Science Education, 30*, 1841–1866.

Fogleman, J., McNeill, K. L., & Krajcik, J. (2011). Examining the effect of teachers' adaptations of a middle school science inquiry-oriented curriculum unit on student learning. *Journal of Research in Science Teaching, 48*, 149–169.

Foucault, M. (1975). *Surveiller et punir: Naissance de la prison* [Discipline and punish: Birth of the prison]. Paris: Gallimard.

Garfinkel, H. (1967). *Studies in ethnomethodology*. Englewood Cliffs: Prentice-Hall.

Garfinkel, H., & Sacks, H. (1986). On formal structures of practical action. In H. Garfinkel (Ed.), *Ethnomethodological studies of work* (pp. 160–193). London: Routledge & Kegan Paul.

Garfinkel, H., & Wieder, D. L. (1992). Two incommensurable, asymmetrically alternate technologies of social analysis. In G. Watson & R. M. Seiler (Eds.), *Text in context: Contributions to ethnomethodology* (pp. 175–206). Newbury Park: Sage.

Garfinkel, H., Lynch, M., & Livingston, E. (1981). The work of a discovering science construed with materials from the optically discovered pulsar. *Philosophy of the Social Sciences, 11*, 131–158.

Gregory, B. (1990). *Inventing reality: Physics as language*. New York: Wiley.

Husserl, E. (1939). *Erfahrung und Urteil: Untersuchungen zur Genealogie der Logik*. Prague: Academia Verlagsbuchhandlung.

Jordan, R. C., Ruibai-Villasenor, M., Hmelo-Silver, C. E., & Etkina, E. (2011). Laboratory materials: Affordances or constraints? *Journal of Research in Science Teaching, 48*, 1010–1025.

Latour, B. (1987). *Science in action: How to follow scientists and engineers through society*. Cambridge, MA: Harvard University Press.

Latour, B. (1993). *La clef de Berlin et d'autres leçons d'un amateur de sciences* [The key to Berlin and other lessons from a science lover]. Paris: Éditions de la Découverte.

Lave, J. (1993). The practice of learning. In S. Chaiklin & J. Lave (Eds.), *Understanding practice: Perspectives on activity and context* (pp. 3–32). Cambridge: Cambridge University Press.

Leont'ev, A. N. (1983). Dejatel'nost'. Soznanie. Ličnost'. [Activity, consciousness, personality]. In *Izbrannye psixhologičeskie proizvedenija* (Vol. 2, pp. 94–231). Moscow: Pedagogika.

Lindwal, O., & Lymer, G. (2008). The dark matter of lab work: Illuminating the negotiation of disciplined perception in mechanics. *Journal of the Learning Sciences, 17*, 180–224.
Livingston, E. (1986). *The ethnomethodological foundations of mathematics*. London: Routledge and Kegan Paul.
Luria, A. R. (2003). *Osnoby nejrolpsixologii* [Foundations of neuropsychology] Moscow: Isdatel'skij Centr «Akademija».
Lynch, M. (1998). The discursive production of uncertainty: The O. J. Simpson "dream team" and the sociology of knowledge machine. *Social Studies of Science, 28*, 829–868.
Mannheim, K. (2004). Beiträge zur Theorie der Weltanschauungs-Interpretation [Contributions to the theory of worldview interpretation]. In J. Strübing & B. Schnettler (Eds.), *Methodologie interpretativer Sozialforschung: Klassische Grundlagentexte* (pp. 103–153). Konstanz: UVK.
McElhaney, K. W., & Linn, M. C. (2011). Investigations of a complex, realistic task: Intentional, unsystematic, and exhaustive experiments. *Journal of Research in Science Teaching, 48*, 745–770.
National Council for Teaching of Mathematics (NCTM). (2000). *Principles and standards for school mathematics*. Reston: Author.
National Research Council (NRC). (1996). *National science education standards*. Washington, DC: National Academy Press.
Noss, R., Bakker, A., Hoyles, C., & Kent, P. (2007). Situating graphs as workplace knowledge. *Educational Studies in Mathematics, 65*, 367–384.
Preece, J., & Janvier, C. (1992). A study of the interpretation of trends in multiple curve graphs of ecological situations. *School Science and Mathematics, 92*, 299–306.
Ritchie, S. M., Sandhu, M., Sandhu, S., Tobin, K., Henderson, S., & Roth, W.-M. (2013). Emotional arousal of beginning physics teachers during extended experimental investigations. *Journal of Research in Science Teaching, 50*, 137–161.
Rorty, R. (1989). *Contingency, irony, and solidarity*. Cambridge: Cambridge University Press.
Roth, W.-M. (2003a). Competent workplace mathematics: How signs become transparent in use. *International Journal of Computers for Mathematical Learning, 8*, 161–189.
Roth, W.-M. (2003b). *Toward an anthropology of graphing*. Dordrecht: Kluwer Academic.
Roth, W.-M. (2004). What is the meaning of meaning? A case study from graphing. *Journal of Mathematical Behavior, 23*, 75–92.
Roth, W.-M. (2005). Mathematical inscriptions and the reflexive elaboration of understanding: An ethnography of graphing and numeracy in a fish hatchery. *Mathematical Thinking and Learning, 7*, 75–109.
Roth, W.-M. (2007). Graphing Hagan Creek: A case of relations in sociomaterial practice. In E. Teubal, J. Dockrell, & L. Tolchinsky (Eds.), *Notational knowledge: Historical and developmental perspectives* (pp. 179–207). Rotterdam: Sense Publishers.
Roth, W.-M. (2009a). Limits to general expertise: A study of in- and out-of-field graph interpretation. In S. P. Weingarten & H. O. Penat (Eds.), *Cognitive psychology research developments* (pp. 1–38). Hauppauge: Nova Science.
Roth, W.-M. (2009b). Radical uncertainty in scientific discovery work. *Science, Technology & Human Values, 34*, 313–336.
Roth, W.-M. (2009c). On the inclusion of emotions, identity, and ethico-moral dimensions of actions. In A. Sannino, H. Daniels, & K. Gutiérrez (Eds.), *Learning and expanding with activity theory* (pp. 53–71). Cambridge: Cambridge University Press.
Roth, W.-M. (2013). To event: Towards a post-constructivist approach to theorizing and researching curriculum as event*-in-the-making. *Curriculum Inquiry, 43*, 388–417.
Roth, W.-M. (2014). Reading *activity, consciousness, personality* dialectically: Cultural-historical activity theory and the centrality of society. *Mind, Culture, and Activity, 21*, 4–20. doi:10.1080/10749039.2013.771368.
Roth, W.-M., & Barton, A. C. (2004). *Rethinking scientific literacy*. New York: Routledge.
Roth, W.-M., & Bowen, G. M. (1999). Digitizing lizards or the topology of vision in ecological fieldwork. *Social Studies of Science, 29*, 719–764.

Roth, W.-M., & Bowen, G. M. (2003). When are graphs ten thousand words worth? An expert/expert study. *Cognition and Instruction, 21*, 429–473.

Roth, W.-M., & Lee, Y. J. (2007). "Vygotsky's neglected legacy": Cultural-historical activity theory. *Review of Educational Research, 77*, 186–232.

Roth, W.-M., & McGinn, M. K. (1998). Inscriptions: A social practice approach to "representations". *Review of Educational Research, 68*, 35–59.

Roth, W.-M., McGinn, M. K., & Bowen, G. M. (1998). How prepared are preservice teachers to teach scientific inquiry? Levels of performance in scientific representation practices. *Journal of Science Teacher Education, 9*, 25–48.

Russell, C. B., & Weaver, G. C. (2011). A comparative study of traditional, inquiry-based, and research-based laboratory curricula: Impacts on understanding of the nature of science. *Chemistry Education Research and Practice, 12*, 57–67.

Stieff, M., Hegarty, M., & Deslongchamps, G. (2011). Identifying representational competence with multi-representational displays. *Cognition and Instruction, 29*, 123–145.

Tabachneck-Schijf, H. J. M., Leonardo, A. M., & Simon, H. A. (1997). CaMeRa: A computational model for multiple representations. *Cognitive Science, 21*, 305–350.

van der Valk, T., & de Jong, O. (2009). Scaffolding science teachers in open-inquiry teaching. *International Journal of Science Education, 31*, 829–850.

Vygotskij, L. S. (2005). *Psychologija razvitija cheloveka* [Psychology of human development]. Moscow: Eksmo.

Wittgenstein, L. (1997). *Philosophische Untersuchungen* [Philosophical investigations] (2nd ed.). Oxford: Blackwell. (First published in 1953)

Wittgenstein, L. (2000). *Bergen text edition: Big typescript*. http://www.wittgensteinsource.org/texts/BTEn/Ts-213. Accessed 30 Nov 2013.

Woolgar, S. (1990). Time and documents in researcher interaction: Some ways of making out what is happening in experimental science. In M. Lynch & S. Woolgar (Eds.), *Representation in scientific practice* (pp. 123–152). Cambridge, MA: MIT Press.

Part II
Graphing in a Discovery Science

> *Rule 7.* Before attributing any special quality to the mind or to the method of people, let us examine first the many ways through which inscriptions are gathered, combined, tied together, and sent back. Only if there is something to be explained once the networks have been studied shall we start to speak of cognitive factors. (Latour 1987, p. 258)

Traditional lore has it that scientists are experts in a range of basic (e.g., observing and measuring) and integrated "science process skills" (e.g., interpreting data). Interpreting data means organizing data – e.g., in tables or graphs – and using them to make statements about some aspect the world that these inscriptions are employed to make present again. Communicating – one of the so-called basic science process skills – refers to the use of words and graphical symbols to describe the phenomena of interest, such as a graph that represents a growth rate. The term *skill* refers to the faculties of mind or hands, the practical knowledge that operates in concert with ability. It therefore denotes individual capacities for doing something. This received notion of ascribing what happens when scientists use graphs, as when they use other inscriptions, is called into question by the seventh rule of method that Latour states for the study of science and engineers. In the opening quotation, we are exhorted to begin by studying the public ways in which inscriptions "are gathered, combined, tied together, and sent back" before making any attribution to mental facilities. Cognitive factors enter the picture only when the ethnographic study of the witnessable actions and activities of scientists has exhausted its means. This approach is consistent with the proposal Wittgenstein (1953/1997) makes with respect to language when suggesting that we ought to look only at how words are used rather than to look for "meanings" in the heads of speakers and recipients.

In this second part of the book, I report on the entire process of the production of graphs that will appear in the publications when the findings of the research are ultimately reported. The history of this production begins with the first hazardous tracing of signification when the idea for a research project first cropped up in 1998 to the data generation process marked by radical uncertainty, scientists' coping with variance that they (initially) have no explanation for (in 2001), the reversal of a trajectory that removed all context from the samples used during measurement, the unnoticed and unresolved contradictions in the data interpretations in 2005, and a scientific revolution that was not recognized as being one. Enacting the program Latour outlines in his seventh rule of method, I treat the problem of graphs and graphing anthropologically: as a study of the things and processes involved when scientists engage in the translations of natural phenomena into series of inscriptions the endpoint of which they use in support of verbal knowledge claims. In this part of the book, I report on this entire process of a scientific research project with a special focus on graphing and uncertainty.

For an Anthropology of Graphing

The measurements scientists make and the data that they produce from them generally constitute the evidence on which they base their scientific claims. I begin this section with a chapter, "Radical Uncertainty in/of the Discovery Sciences" (Chap. 2), to exhibit what scientists do when they no longer know what they are doing. That is, I use a situation where scientists return to work one morning and find that their equipment, which still worked during the late afternoon of the day before, no longer worked. They did not know why, but – because they intended to do the data collection that they had anticipated for that day – they needed to fix whatever the problem was without knowing of what nature it might be. This uncertainty also marks the remainder of their work concerned with the data themselves. Thus, in Chap. 3, "Uncertainties in/of Data Generation," I investigate how natural scientists make decisions about the inclusion / exclusion of certain measurements in / from their data sources. I first show that there is a radical uncertainty in the discovery sciences, whereby they need to be certain that there is a natural object present that exhibits itself in a data (graph), and they need to be certain about the data (graph) to know whether there is any phenomenon. I have termed this situation *radical* uncertainty (Roth 2009), that is, an uncertainty for which there is no remedy. Moreover, I show how scientists exclude measurements from their data sources even before attempting to mathematize and explain the data. The excluded measurements therefore do not even enter the ground from and against which the scientific phenomenon emerges and therefore remain invisible to it. In these data, the phenomenon was lost; and it is this aspect, as I discuss in Part IV, that STEM students do not experience unless they design and complete open investigations.

Few studies, if any, focus on the uncertainties with which scientists wrestle before they are confident enough in the graphs they produce and in knowledgeably using these. The purpose of Chap. 4, "Coping with Graphical Variability," is to exhibit the lived work of coping with graphical variance in the laboratory's process of making a major discovery that will eventually overturn the scientific canon. I analyze an exemplifying episode from a laboratory meeting dealing with the uncertainties in the data collected and the effect these have on the graphical representations. A large number of external graphical and gesturally embodied inscriptions are used to constitute a network of interlinked inscriptions by means of which scientists cope with variance. Most importantly, when the scientists are confronted with novel graphs, their descriptions and explanations are not straightforward so that a lot of normally invisible (transactional) work is required to constitute the familiarity that gives them a sense of where they are going.[1]

The sciences have been successful in the course of recent human history because the (mathematical) inscriptions they use articulate laws and relations independent of contextual particulars and contingencies of concrete situations. This allows verification anywhere and at any time, and, therefore, the objectivity of scientific phenomena. Decontextualization, however, comes with a prize: There is evidence that scientists have difficulties explaining graphs that have been abstracted from research contexts even if these graphs constitute an integral part of undergraduate instruction in the scientists' own discipline. In Chap. 5, "Undoing Decontextualization," I investigate the role of context in the understanding of data and graphs during scientific discovery work. I exhibit the effort scientists mobilize to reconstruct the context from which their data have been abstracted. Without recontextualization, scientists struggle explaining the study results that emerge from their work. Scientists require familiarity with the settings from which the data derive and with the entire translation process that eventually produces the graphical inscriptions that scientists use in their publications for the purpose of supporting knowledge claims. Just as there is no boundary between knowing a language and knowing one's way around the world (Rorty 1989), there is no boundary between knowing a graph and knowing one's way around the world that has led to its production. Just as the speaking of a language is part of activity, a form of life (Wittgenstein 1953/1997), so is "forming and testing a hypothesis," "presenting the results of an experiment by means of tables and diagrams" (p. 12), and, relevant to the present study, the rhetorical use of graphs as evidence in support of scientific research claims.

Contradictions have been recognized as important factors in learning (conceptual change) because these are said to require individuals to engage in deep reflection that leads to accommodation and learning. The notion of contradiction is pervasive in very different theoretical approaches to learning: (a) cognitive perspectives – e.g. in theories of conceptual change, which is said to occur when scientists grapple with data that do not fit in their received schemes – and (b) many studies conducted under the auspices of cultural-historical activity theory. However, unless (logical) contradictions are thought from the perspective of the subject of activity, we do not advance in explaining of how contradictions might mediate learning. On the other hand, the dialectical perspective on (inner) contradiction would lead us to a dynamic description of graphing that explodes all thinking in terms of personal skill and procedural knowledge. In Chap. 6, "On Contradictions

[1] In regard to understanding and meanin that might be attributed to persons using graphs, I take it with pragmatic philosophy: "Das Verstehen, die Meinung, fällt aus unserer Betrachtung heraus [Understanding, meaning, disappear from our considerations]" (Wittgenstein 2000, Ts-213, 1r[1]).

in Data Interpretation," I provide an exemplifying analysis of a meeting in which the scientific research team under investigation presents its results to an informed audience. I show that with hindsight, there are logical contradictions in the mathematical (graphical, functional) models that the scientists use and the explanations that they produce, which go unnoticed from that meeting right to the ultimate publication of the results of their work.[2]

Learning scientists tend to focus on learning generally and conceptual change specifically among school-aged students. Almost inexistent are studies of knowing and learning among successful scientists. In Chap. 7, "A Scientific Revolution That Was Not," I provide a micro-genetic historical account of the conceptual change in explaining a phenomenon that occurred while the scientific research team attempted to confirm a theory but in fact ended up overthrowing it. Because the team collected its data over a 2-year period, slowing down the availability of what they would be saying if they had complete data collection, opportunities for studying the conceptual change ethnographically arose. In that chapter, I report the difficulties the team encountered because they (a) took a dogma-related perspective, (b) had to reconstruct and become familiar with the context from which they had abstracted their specimen, (c) required a biologically relevant rather than mathematically plausible explanation, and (d) exhibited aspect blindness that only disappeared as their familiarity increased.

I end this part of the book with Chap. 8 devoted to identify some of the core messages educators may want to take away from one discovery science specifically and from the discovery sciences more generally.

As part of the presentations what scientists do with and talk about graphs, I present extensive transcriptions. The extensiveness of the presentation is not for gratuitous reasons. Rather, I provide these as a test bed for other researchers interested in workplace mathematics and everyday mathematical practices generally and those related to graphing in the sciences specifically.

Ethnographic Background

The primary focus in this second part of the book is on graphs and graphing in the face of (radical) uncertainty as viewed through the lens of one discovery science. I use the term *discovery science* because the societal motive realized in the scientific laboratories of universities is the production of knowledge. In the sciences, this

[2] In dialectical logic, a clear distinction is made *between* logical contradictions and *inner* contradictions. The former refer to contradictions in the reasoning of people, for example, between two incompatible statements (e.g., "This is red" and "This is yellow"). Logical contradictions can be removed once these are discovered. *Inner* contradictions, on the other hand, cannot be removed. They are part of the essence of phenomena when these are thought with the fundamental categories (minimum units) of thought that denote change or difference in itself. I develop these ideas further in Chaps. 6 and 9.

knowledge production tends to be seen in terms of the discovery of pre-existing patterns that underlie natural phenomena.[3] In other words, it is the discovery of the second part of the couplet {natural structure ↔ mathematical structure}. In the course of the scientific research, especially as articulated in Chap. 5, it turned out that the scientific research team had difficulties talking about their data without access to detailed information about the contexts from which they sourced their specimens. This information was available to our team through the detailed ways of knowing their way around that some of its members had become familiar with while visiting and working in the field sites where the coho salmon *parr* and *smolts* were obtained for subsequent laboratory study.[4] In the following, I provide background information on the laboratory itself and on the main site from which it sourced its specimens. Any more specific information required for following the scientists in their use and talk about graphs is provided together with the respective analyses in the different chapters.

The Science Laboratory

During the 1930s and 1940s, the Nobel-prize winning biologist George Wald (e.g., 1939, 1941) had conducted research on the absorption of light in the eyes of anadromous[5] fishes with mixed visual systems. Such research reported a high prevalence of the vitamin A_2-based porphyropsin chromophore in the retinal cells while the migrating fishes reside in freshwater environments and a high prevalence of the vitamin A_1-based rhodopsin chromophore in the retinal cells during their seawater life history stages. Subsequent studies – including some conducted in a hatchery and river system of the same geographical area as the scientific team studied here – had suggested a change-over between the dominance levels of the two visual pigments and an associated metamorphosis of the visual system as the fishes were getting ready to migrate to the ocean. The suggested changes ranged from as much as 80–10 % of porphyropsin in the light absorbing rod-shaped cells of the retina (Alexander et al. 1994). When I first met the chief scientist of the team studied here (Craig), he was talking about "the dogma" suggesting changes from as much as 90–5 % porphyropsin for the fishes in freshwater and saltwater

[3] Much like Christopher Columbus' *discovery* presupposes the existence of the Americas.

[4] The life cycle of the salmon begins with the eggs, which hatch into *alevins* that still have the yolk sack attached to them. After the yolk sack is used up, the alevins begin to feed and are called *fry*. After 1 year, the salmon are known as *parr*. In this form, they stay in the freshwater for 1–4 years, depending on species and geographical location. The parr then get ready for migration, at which stage they undergo a range of physiological changes, and their scales silver. In this stage, they are referred to as *smolts*.

[5] *Anadromous* fishes, like salmon, migrate from ocean with salty water into freshwater streams to spawn. Eel is *catadromous*, because it migrates from the freshwater where it tends to live into the ocean to spawn.

environments, respectively. The idea underlying the research we were to conduct was to measure porphyropsin levels in coho salmon over the course of a year using a new instrumentation developed in his laboratory. This instrumentation permitted the collection of data in quantities that were of several orders of magnitudes greater than those in any related study conducted before. Our research was to precisely measure the changeover and, in subsequent studies, correlate the levels of porphyropsin at different release dates to return rates.

A full professor in biology, with a publication record that spanned more than 30 years, headed the lab (Craig). He had been successful throughout his career in many respects and subsequent to this study obtained an endowed chair at another university. He had received a number of awards and fellowships, had obtained continuous, often multiple concurrent funding from Canadian and U.S. agencies, and had a substantial publication record. Theodore (Theo) was a full-time research associate with a background in physics. Theo was responsible for the software, data storage, and data processing. He also participated in the collection of the data. A postdoctoral fellow (Elmar) contributed to the design of the experiments and was mostly responsible for the field settings where the specimens for the experiments were sourced. His PhD had focused on salmon. A doctoral student (Shelby) did most of the measurement together with one of the other team members, including a Masters-level student (Sam). As part of a larger project on the interaction between scientists and society, Craig and I had joined efforts to study salmon and the exchange of knowledge between a fish hatchery raising salmon and this laboratory. As a trained physicist, I was a member of the team participating in designing the studies, mathematical modeling of light absorption from source to detector, collecting data, modeling data, interpreting data, and publishing the results in the natural sciences and science education.

Our research team was interested in better modeling various aspects of the life history of salmonid fishes on the Pacific West Coast of Canada. Craig had specialized on the visual system of these fishes, its uses, and the changes it undergoes throughout the life history. An important aspect of salmonid fishes is their migration from the rivers where they hatch to the saltwater feeding grounds and back to their spawning grounds in the same river systems where they were born. Our team was interested in measuring the changes in visual pigment over time in short intervals as a possible indicator for the optimal time of releasing artificially raised juveniles into the wild. This would address the historical problem of unpredictable return rates of adult salmon, which have tremendous impacts on the local economy. High return rates are important because of the tremendous economic value of Pacific salmon to the local economy, including tourism (sports fishery), commercial fishery, the livelihood of the indigenous peoples, and any spin-off arising from these. Depending on year and hatchery, between 0.1 and 15 % of the salmon released from a hatchery return several years later without anyone knowing why these variations occur. Being able not only to model these variation but also finding practices that would keep return rates more stable than they historically have been was part of the overall agenda of our team.

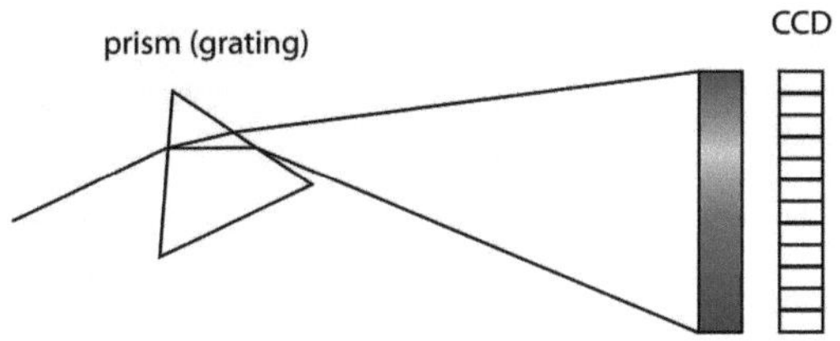

Fig. II.1 After going through a sample, the light is split into its component colors (wavelengths). In the laboratory, a grating instead of a prism was used. The charge-coupled device – basically an array of bins that are sensitive to light falling into them – records the amount of light that comes in, separately for the different colors

As noted, Craig was in the process of developing new instrumentation and new software for collecting data that exceeded the number of data points in previous studies by about two orders of magnitudes. Rather than measuring light absorption in the retina for different wavelengths one point at a time, the new apparatus allowed measuring the light absorption across the entire spectrum in "one shot" taking around 500–1,500 milliseconds per recording (Fig. II.1). This way of collecting makes use of a charge-coupled device (CCD).[6] It basically consists of an array of tubular bins that amplify when a photon (light) falls into them: the more light there is (photons there are), the more electrical charges are created and counted. In this laboratory, the light from the source was split up into its component colors after it had gone through the sample (Fig. II.1). By these means, the apparatus recorded, for all wavelength segments at the same time, how much light had been absorbed.

To maintain the excised retinal tissue in an active state, the fish specimens have to be kept in a dark container for a minimum of 2 h. Because of the light sensitivity of retinal tissue, the experiment has to be conducted at very low intensities of near-infrared light – which requires our research team members to "dark-adapt" their eyes for a period of 30–60 min – we basically sat there in the laboratory talking and waiting in the dark. The fish is anaesthetized and, immediately before removal of the eyes, sacrificed by severing the spinal cord. After removal, the eyes are "hemisected" and the retina removed. Under the microscope, the researcher cuts one piece of the retina (Fig. II.2), which he mounts on a slide whereas the remainder is stored on ice in a saline (minimal essential medium, MEM) solution. At any stage in the research process, the retinal pieces are handled only under near-infrared illumination.

The piece of retina on the microscope slide is macerated, that is, cut into small pieces using a scalpel and tweezers. Adding some saline solution, covering the preparation with a cover slip, and sealing the preparation to prevent evaporation of the solution completes the mounting process. The slide is placed under a microscope fitted with two light sources, one for the stimulus (xenon) light beam the

[6] The same (kind of) device also is at the heart of a digital camera.

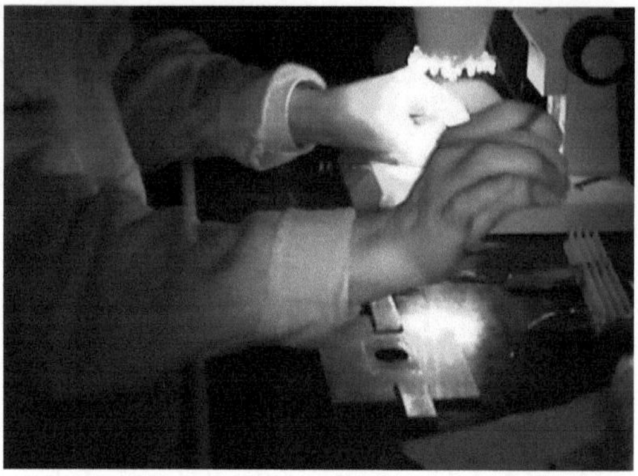

Fig. II.2 The researcher works in the dark, only with faint infrared light sources. Working under the microscope, the researcher cuts a piece of retina, to be mounted on a microscopic slide, which in turn is fixed to a steel mount – here seen below the hands of the researcher – that will be inserted into the apparatus (© Wolff-Michael Roth, used with permission)

other, an infrared lamp, for providing the background illumination to search for the objects of interest. During the time that I conducted this research, the team changed from looking at the slide contents through an optical microscope to using a CCD camera, the image of which was presented on a computer screen (Fig. II.3). In this example, a cone-shaped photoreceptor cell is clearly visible, as well as the crosshair that the team used to line up the sampling light beam with the cell (or a spot next to it for the reference recording).

Conceptually, the measurement unfolds like this: To obtain information about the photoreceptors in the retina, two measurements have to be made. In the first, a light pulse is made to traverse the slide at a spot where there are no cells (the person operating the microscope asks to take a "reference"); this yields a first graph of light intensity across the spectrum (Fig. II.4a). In the second, the pulse is made to go through the cell (the person operating the microscope asks to take a "scan"); this yields a second graph (Fig. II.4b). Because more light (normally) is absorbed in the cell than in the surrounding saline solution, the intensity difference in the two light pulses is attributed to absorption in the photoreceptor cell (Fig. II.4c). But this initially yields only the raw absorption spectrum and frequently little or nothing is to be seen. The raw absorption spectrum then is processed in a variety of ways – i.e., "massaged" – to make the signal come out as nicely as possible. For example, between the two measurements, a slight shift might occur that misaligned the two spectra. Therefore, the simple difference between the two graphs yields the wrong information. "Pixel-shifting" one graph with respect to the other realigns the two raw graphs (Fig. II.4a, b) and may show a peak where before none was visible. Generally, what scientists accept to be the true signal "sits" on background noise of various origins. Oftentimes, this noise has the form of a non-horizontal, sloped line. The team subtracts this line from the data so that their "baseline" will be as horizontal as possible.

The resulting absorption spectrum covers a range of frequencies; but depending on the type of photoreceptor cell, light in the ultraviolet, blue, green, or red part of the spectrum is maximally absorbed. In Fig. II.5, one can see the spectrum for a blue cone generated in the laboratory based on a large number of measurements. When a cell had been exposed to light before ("bleached"), no absorption spectrum is

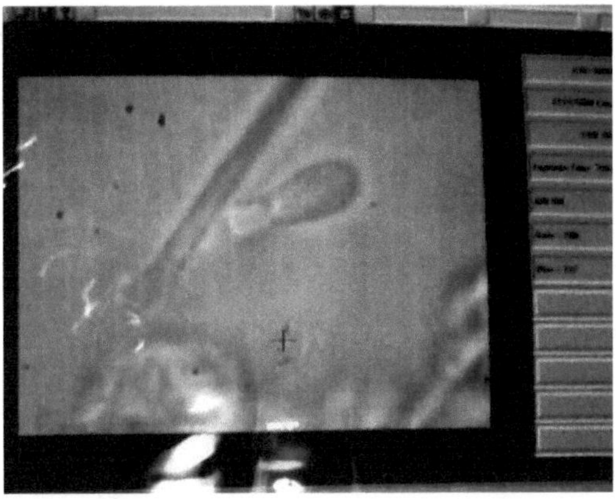

Fig. II.3 This video image shows part of the computer screen on which the contents of the microscopic slide are shown as seen through the microscope image recorded by a CCD camera. A little above the center floats a cone-shaped photoreceptor cell. The sampling light beam is lined up with the cell by means of the crosshair marked on the computer screen (© Wolff-Michael Roth, used with permission)

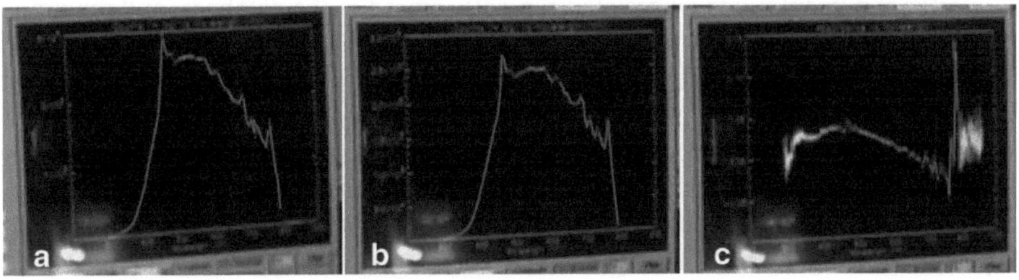

Fig. II.4 (**a**) Sending a beam of light through the microscopic slide but next to a photoreceptor cell gives a graph that represents the intensity of light across the spectrum. (**b**) Doing this again but this time through the photoreceptor cell generally decreases the light intensity, because some of it is absorbed by the receptor. (**c**) The difference between the two former spectra is attributed to the cell (© Wolff-Michael Roth, used with permission)

observed. The maximum of the absorption curve is called "lambda-max" (λ_{max}). It is used to calculate the ratio of the two vitamin-A-based chemicals that absorb light and were thought to characterize the different stages in the life history of the salmon (i.e., while living in saltwater vs. freshwater environments). That is, "the canon" held that the absorption spectrum for a blue cone shifts depending on the relative amount of vitamin-A_1- and vitamin-A_2-based photoreceptor cells in the course of the life history.

My graduate students – all with degrees in (micro-) biology and in the process of acquiring the skills of cognitive anthropology – and I videotaped (a) sessions in the wet laboratory, (b) sessions in the data analysis laboratory, and (c) meetings where our team met to discuss the data at all stages of their production. My graduate students also conducted interviews with the members of the scientific research

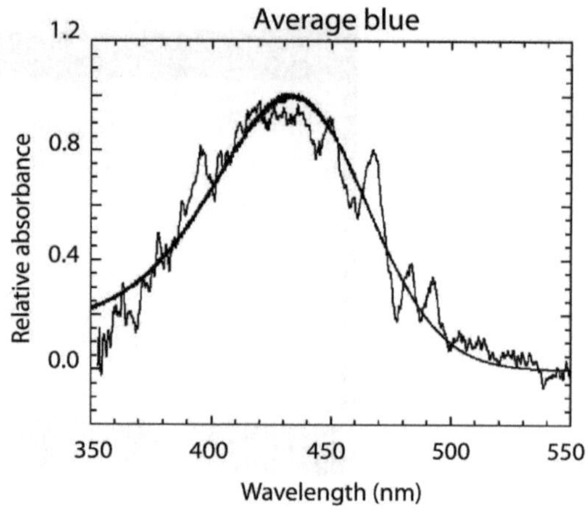

Fig. II.5 After having made selections which data to retain and which data not to include in the analysis, the scientists average all measurements for *blue cones* into one graph, which constitutes something like the signal for an "average" *blue cone*. (© Wolff-Michael Roth, used with permission)

team; these were videotaped and transcribed. Additional videotapes were recorded when a team member presented our research in the context of the national project that framed our work, at graduate student forums, or to partners (fish hatcheries). PowerPoint presentations where electronically stored in the database, and diagrams often produced during meetings were scanned and stored in digitized form. There are about 100 h of videotaped materials from each the hatcheries and the laboratory. All videotapes and audiotapes were transcribed roughly – transcriptions containing words and, in salient situations, links to video offprints – to make them available for initial reading and annotating. Where relevant, fine transcriptions featuring emphases, overlaps, pauses, accents, and so on were produced as part of the analysis and repeated viewing of the videotapes.

Fish Hatchery

To talk about how scientists used and produced graphs, I also draw on an ethnographic study of the Canadian *Salmon Enhancement Program* in general and one of its hatcheries in particular (Roth et al. 2008). The analysis of the present data made this necessary, because it turned out that the scientists could not make anything of the graphs they had produced without talking to the fish culturists, who had raised the salmon that the scientists were using in their research. The main purpose of a hatchery is to take eggs and milt from salmon after they return from their ocean journey and before they die. The eggs are fertilized with the milt and then raised in spawning channels or incubation batteries until young salmon hatch, which are raised to a size ready for their release – the optimal release size and date being determined by past experience and some scientific experimentation. The staff of the hatchery under study included two managers, five fish culturists, a maintenance supervisor, and an administrative assistant. Depending on seasonal demands, there

Fig. II.6 (**a**) The production of salmon begins with the "egg take," when eggs are taken from the sacrificed does and the fertilized with the milt from at least two bucks (males). (**b**) The fertilized eggs are stored, by the millions, in a huge hall with stacks of pans each filled with about 7,000 eggs (© Wolff-Michael Roth, used with permission). (**c**) Once the fishes have hatched and reached a certain size – about 5 g in the case of coho salmon – these are transferred into the ponds until release For coho salmon, these were earthen ponds, where the young fishes spend about 12 months prior to being released. (**d**) The fishes have to be fed daily, which, for coho, amounts to about 200 kg per day. Machines do not do a good job, so temporary workers are employed to feed the fish (© Wolff-Michael Roth, used with permission)

were also up to 30 temporary workers, assisting in the operation (e.g., feeding) and maintenance of the place.

Much of the work takes place outside the offices, though each fish culturist does have an office where records are kept, books and other work-related resources are stored, a computer with Internet connection is available, and so on. The work in the hatchery involves an important part of the lifecycle of the fish, beginning with an "egg take" and fertilization (Fig. II.6a), raising the eggs in basins stacked in the brood hall until the fish hatch and turn into alevins and then fry stage (Fig. II.6b), transferring fry to an outside pond (Fig. II.6c), feeding the young fishes until they smolt, at which time they are leased into the neighboring river (Fig. II.6d). After spending between 2 and 5 years in the ocean, the surviving adult salmon generally return to the same hatchery, where they die or are killed for the eggs and milt to be taken.

As part of this 5-year study, I collected data both (a) as a contributing participant in the major daily and seasonal tasks at the site and (b) as a mere observer of events. As part of my research agreement with the hatchery, I contributed as a helper in the various tasks that have to be done on a daily, weekly, monthly, and even yearly basis. By working in the hatchery and thereby contributing to the realization of its object/motive, I apprenticed to the different tasks and got to know the hatchery activity system from the inside. As a helper, my goals were therefore those defined by the task and motive of activity. At other moments, I engaged in observation and recording. In this mode, the data sources include observational field notes, videotapes of everyday activities, recorded and transcribed formal interviews, photographs, documents, scientific and mathematical representations, and various other notes and reports created and used as part of the everyday work in the study site. The two forms of ethnographic work – participant observer and observing participant – provide different perspectives of, and constitute complementary ways of experiencing, the productive work in the study site.

There were two individuals in particular from whom I have learned all I know about fish hatching. Erica was one of the fish culturists, responsible, among others, for the entire production of coho salmon, beginning with the taking of eggs and milt from animals that returned to the hatchery until the moment that the hatched and raised salmon are released. Erica recorded all data that she and the temporary workers generated. After high school, she had attended a business program in a local comprehensive college[7] for 2 years but then dropped out. She never enrolled in a mathematics or statistics course once she had left high school. However, more than any other fish culturist working in the facility, she liked to "play [mess] around" with the data by representing computer-based data in different ways or by correlating different pieces of information, actions that increased her familiarity with the production of coho salmon smolt. To extend her competencies using the spreadsheet, she had studied some statistics on her own and she continued finding out new ways of operating it, for example, in the context of making "neat" displays. Thus, to avoid having to put "0" in fields for which data did not yet exist so that the spreadsheet would not fill subsequent, dependent fields with the message "#VALUE!," she figured out how to use the "IF" function. I observed her use the function "IF((A6 = "-"),"-",A6*10)" to put a hyphen whenever the cell, here A6,

[7] A comprehensive college offers applied trades and business programs, university degrees in a limited number of areas, university transfer programs, and career- and academic-preparation courses.

was empty (signaled by a hyphen) but to calculate a value if there was a number in the cell.

At the time of my research, Erica had worked for over 13 years in this and one other hatchery and, by taking secondment opportunities in related sites such as a research laboratory, made her better her salmon hatching practices. She had taken several hatchery-related courses and was enrolled in an online genetics course. Erica is energetic ("I like to be progressive"), making every effort to stay up to date and questioning, "Why do we do it this way?" so that she would not "become stagnant and . . . be doing things the way we have always done." She kept personal binders with information that did not enter the official channels, or, when she had entered them in the past, they often had been removed from the official records. This information allowed her to reflect on the processes of raising fish, and in the process come to be better situated in her praxis. "I love learning," "I am sort of a geek," but also "I don't like change for the sake of change" were characteristic self-descriptions. Her co-workers regarded her not only as a "very smart" person but also as an outstanding fish culturist. All of these descriptors are consistent with the dispositions required for statistical thinking in empirical inquiry, including skepticism, curiosity and awareness, openness, a propensity to seek deeper meaning, and engagement.

The other main participant ("informant") was Mike, a fish culturist not only known widely within his city, where he often went to schools to teach children, but also within the salmon culture community across the borders of our home province (British Columbia). In fact, he had received a national award for his contributions to the field of salmon hatching, even though he had no special certification in biology, fish culture, and the like. Thus, for example, one federal fisheries biologist attributed to him the tremendous increase of chinook recruits (progeny) to the hatchery, which allowed scientific research on this species to be resumed. Trained as a plumber, with skills in carpentry and electrical work, he changed his job after 14 years. He first became a part-time park naturalist and subsequently landed a job at the fish hatchery, where he has remained ever since. Over the years he had developed tremendous knowledgeability in all areas of fish culture, including those that on the surface may be the most boring ones. For example, during the peak times of the year, more than 200 kg of fish feed have to be thrown by means of a scoop into the three ponds where about 1 million coho salmon are raised. My ethnographic study revealed that though highly repetitive, feeding the salmon is an art that machines cannot do. In fact, there had been an experiment in this hatchery with mechanical feeders. It was soon abandoned, as mechanical feeding tended to lead to uneaten food on the bottom of the ponds, creating an ideal environment for the emergence of bacterial diseases (Roth et al. 2008). The hatchery reverted to hand feeding. Like automized feeders, inexperienced feeders do not know when to stop feeding, and, when they do know, they often abandon distributing feed when the fish do not seem to take it up. Mike, however, could coax the salmon into feeding even though they initially did not "want to." Due to Mike's and others' contributions, the facility became an "indicator hatchery," a leading institution where numerous scientific studies were conducted, and which became an exemplary showcase for how a hatchery should operate. It is not that Mike and (some of) his peers *caused* the hatchery to excel; the relation between personnel and institution is *mutually constitutive*. While I accompanied him completing hatchery tasks, "feeding" a nearby lake to stimulate the growth of algae that would support juvenile sockeye salmon, or (in the fall) while collecting dead salmon and data, he would often talk about his early years in the hatchery. He described a positive working climate with a manager who supported research designed by the fish culturists for finding ways of improving hatchery practices. During those years, Mike contributed to the emergence and development of many ideas for innovation, which, in collaboration with others, were developed into formal scientifically sound experiments that compared, among other things, different ways of (a) raising fish eggs to maturity, (b) using different sterilizing agents to protect the fertilized eggs from becoming infected with disease, or (c) anesthetizing juvenile fish when they had to be handled for taking (weight, length) measurements or marking (clipping a dorsal fin, injecting a tiny pin into the nose). Near the end of my fieldwork, Mike had taken on Erica as a mentee.

Research Policy

This study used apprenticeship as ethnographic research method because I learned about the relevant biology and laboratory techniques while participating in the scientific work and in the fish hatchery. In the studies reported in this book, I do not treat my results as specifications of practice; rather, I consider these to be guides or descriptions that assist readers in locating the phenomenal field properties that make the observed practices recognizable in the way they present themselves to an observer (Garfinkel 2002). My descriptions can be read as a micro-genetic study of scientists' ethnomethods of dealing with uncertainty and using graphs, here in an advance laboratory on the neuroethology of fish vision. Here I am concerned with providing an ethnographically *adequate* description, which means that the "ethnographer must articulate the same hesitant and momentary contexts that the natives are displaying to each other and using to organize their concerted behavior" (McDermott et al. 1978, p. 246). Thus, as done by another study of graphing in a scientific laboratory, I treat my transcripts "as an occasion for construing the work done by participants in interpreting the document as it unfolds before them" (Woolgar 1990, p. 127) and before they know how it ultimately will be understood. That is, this approach gets at the ways in which reality appears to the actors themselves, and the analytic method thereby reveals a "shop floor" perspective. In the same way other research concerning the natural sciences is conceived, the present is a study *of* rather than *about* scientists' work (Garfinkel et al. 1981). The methodological precepts appropriate for such a treatment have been articulated in conversation analysis, which is in fact the analytic orientation that I describe near the end of Chap. 1.

The conversation analytic approach assumes the speaking turn pair as the minimum unit of analysis of *socio-logical processes*. That is, the minimum unit has a logic that is based on the social rather than the individual-psychological: it is a *socio-logical* approach. The effect of this approach is that it reveals the ways in which members to the conversation hear what is being said rather than the analyst's interpretation. In the following example, Craig says, "Do you want me to bleach it" (turn 019); because the intonation (pitch) is rising toward the end, a question mark is placed. Rather than considering this locution to be an emanation from Craig's mind, suggesting that Craig has asked (intended to ask) a question, the role of the statement from *within* the conversation itself is brought out by following how the subsequent speaker takes it up (turn 021). (The transcription conventions are provided in the Appendix.)

```
019   C:  okay; save that. (0.27) do you want me to blEACH
          it?
020       (0.73)
021   T:  i zINK we dont need thIS one.
022   C:  okay.
```

Theo says, "I think we do not need this one" with a strongly falling pitch (indicated by the period after the last word) as this tends to be the case in constatives. Here, then, we find a question | answer pair: Whether the preceding statement functions as a question depends on the second statement, which, technically speaking, makes available the effect. But the question actually answered may not be about bleaching but about retaining a particular data point. The next turn pair (i.e., 021–022) constitutes a proposed ("I think") constative | acceptance ("okay") pair. The team goes on not "bleaching" the receptor and therefore not capturing the bleached data point. As a consequence, this approach does not require special interpretive methods; rather, it requires the analyst to hear the participants in the manner they hear each other (Garfinkel and Sacks 1986). How someone "takes [the statement of another] is seen in the use of the word" (Wittgenstein 1953/1997, p. 14). Because I had been a member of the research team for a 5-year period, this requirement can be ascertained. When this cannot be ascertained – such as it happened in one of my studies, where a social psychologist without physics background listening to physicists did not follow their use of the minus sign and its role in describing physical phenomena such as force or rates of change – tremendous mishearing ("mis-taking") may and does occur.

To get at a shop floor perspective, where the ultimate productions are not yet known, the tapes and data were analyzed in a first-time-through approach: at no point during the analysis is it allowed to take something that happened later as a resource in explaining what happened earlier. That is, I take each instant on the tapes through the lens of the unforeseeable nature of what happened subsequently. Thus, in the episode at the center of Chap. 2, the equipment did not work. I take a look at this episode through the lens of the participants who, in that situation, did not know *why* the equipment was not functioning. This form of analysis forces the anthropologist to abandon insights that come with and from hindsight.

As part of the analysis, each episode of interest is described and analyzed, including multiple replaying of the episode. The descriptions, theories, and concepts developed during such sessions subsequently are tested for prevalence and in/consistency in the remainder of the database. To capture the sense of a first-time-through event-based perspective that is characteristic of everyday life, I move through a tape in a frame-by-frame fashion, clipping screen images of "interesting" moments and pasting them into an open document of our word processor. I then produce a description of the event, often including the rough transcript and screen prints; I also write further analytical comments. In the course of this analytic work I add even more detail to the transcriptions, such as exact pauses in speech and information about the prosody. Consistent with the analytic approach described, a pair of subsequent turns is taken as the analytical unit yielding sequentially ordered question | answer, assertion | confirmation, invitation | acceptance/rejection, etc. pairs. This description provides the participants' perspectives, because it focuses on how the recipient in the situation takes a locution – as per his/her (up-) take or reaction to the preceding locution – rather than how the analyst "interprets" it. As a consequence, the analyst has to be an insider to be able to hear participants as these hear each other. Ethnographic adequacy implies "the requirement that for the analyst to recognize, or identify, or follow the development of, or describe phenomena of order* in local production of coherent detail the analyst must be *vulgarly* competent in the local production and reflexively natural accountability of the phenomenon of order* he is 'studying'" (Garfinkel and Wieder 1992, p. 182). My membership in the scientific research team allowed me rather quickly to take part in laboratory conversations without having to ask what someone else "means."

References

Alexander, G., Sweeting, R., & McKeown, B. (1994). The shift in visual pigment dominance in the retinae of juvenile coho salmon (*Oncorhynchus kisutch*): An indicator of smolt status. *Journal of Experimental Biology, 195*, 185–197.

Garfinkel, H. (2002). *Ethnomethodology's program: Working out Durkheim's aphorism*. Lanham: Rowman & Littlefield.

Garfinkel, H., & Sacks, H. (1986). On formal structures of practical action. In H. Garfinkel (Ed.), *Ethnomethodological studies of work* (pp. 160–193). London: Routledge & Kegan Paul.

Garfinkel, H., & Wieder, D. L. (1992). Two incommensurable, asymmetrically alternate technologies of social analysis. In G. Watson & R. M. Seiler (Eds.), *Text in context: Contributions to ethnomethodology* (pp. 175–206). Newbury Park: Sage.

Garfinkel, H., Lynch, M., & Livingston, E. (1981). The work of a discovering science construed with materials from the optically discovered pulsar. *Philosophy of the Social Sciences, 11*, 131–158.

Latour, B. (1987). *Science in action: How to follow scientists and engineers through society*. Cambridge, MA: Harvard University Press.

McDermott, R. P., Gospodinoff, K., & Aron, J. (1978). Criteria for an ethnographically adequate description of concerted activities and their contexts. *Semiotica, 24*, 245–275.

Rorty, R. (1989). *Contingency, irony, and solidarity*. Cambridge: Cambridge University Press.
Roth, W.-M. (2009). Radical uncertainty in scientific discovery work. *Science, Technology & Human Values, 34*, 313–336.
Roth, W.-M., Lee, Y. J., & Boyer, L. (2008). *The eternal return: Reproduction and change in complex activity systems – The case of salmon enhancement*. Berlin: Lehmanns Media.
Wald, G. (1939). On the distribution of vitamin A1 and A2. *Journal of General Physiology, 22*, 391–415.
Wald, G. (1941). The visual systems of euryhaline fishes. *Journal of General Physiology, 25*, 235–245.
Wittgenstein, L. (1997). *Philosophische Untersuchungen/Philosophical investigations* (2nd ed.). Oxford: Blackwell. (First published in 1953)
Wittgenstein, L. (2000). *Bergen text edition: Big typescript*. http://www.wittgensteinsource.org/texts/BTEn/Ts-213. Accessed 30 Nov 2013.
Woolgar, S. (1990). Time and documents in researcher interaction: Some ways of making out what is happening in experimental science. In M. Lynch & S. Woolgar (Eds.), *Representation in scientific practice* (pp. 123–152). Cambridge, MA: MIT Press.

Chapter 2
Radical Uncertainty in/of the Discovery Sciences

Nobel Prize presentation and acceptance speeches frequently constitute narratives of heroism, rational search for the truth, and success. When released to the public, the accomplishments often sound astonishing, leading to the myth of (some) scientists as something special, a breed of people apart. They appear to deal with things that everyday folk denote as "abstract," almost otherworldly things. Thus, a recent Nobel announcement began like this:

> *Raymond Davis Jr* constructed a completely new detector, a gigantic tank filled with 600 tonnes of fluid, which was placed in a mine. Over a period of 30 years he succeeded in capturing a total of 2,000 neutrinos from the Sun and was thus able to prove that fusion provided the energy from the Sun. (Nobel e-Museum, October 8, 2002)

Such after-the-fact accounts of discovery present scientists' learning as a rational search for knowledge – in the peer-reviewed studies reporting research, the outcome of research is presented as the purpose even though the new knowledge could not have been known at the outset of the research reported.[1] Under the guise of inspiring students, science teachers tell such stories of heroism or students can find them in their science textbooks (van Eijck and Roth 2011). Being stable, structured elements of the learning environment, science teachers and textbooks thereby contribute to the production and reproduction of a citizenry with ideas about science as something different, more valued and valuable than what other people do – the popular reverence for Albert Einstein as a genius despite and because he did not do so well in school is but one example. These heroic images of nearly superhuman scientists, however, stand in contrast with the story I am telling here about the rather mundane everyday life in a scientific laboratory that becomes apparent when an ethnographer somewhat versed with the subject matter

[1] The second myth is that these scientists do it on their own, failing to recognize that without the support around them – the architects designing the laboratories, the masons building these laboratories, the cleaners and graduate students that get the work done, the bakers that feed the scientists etc. – very little of the grand and not so grand science could be done (Redfield 1996).

W.-M. Roth, *Uncertainty and Graphing in Discovery Work*,
DOI 10.1007/978-94-007-7009-6_2, © Springer Science+Business Media Dordrecht 2014

spends a few weeks, months, or even years in a place that scientists inhabit.[2] A few scientists admit that they are no more rational than other people, no more epistemic heroes than everyone else – the exceptional epistemic status is constructed after the facts, from scientists' cleaned-up accounts that hide rather than describe what they have actually done (Suzuki 1989). The stories that we anthropologists can tell are about how they do it, how scientists come to deal with the uncertainties that are always present when we grope in the dark because we do not know, whatever we do and wherever we do it. It is this knowledgeability that allows human beings to become so competent such as to make not only scientific research but also everyday life look mundane. Reporting how they make it look that way is part of my motivation to be a cognitive anthropologist of science. The following fragment takes us into the scientific laboratory described in the introduction to Part B and into the wet laboratory where Craig, Theo, and I had agreed to meet to collect data on the absorption of light in the different types of cone-shaped photoreceptors of trout.

On Being in the Dark

Together with two other individuals, Craig, the head of the laboratory and the research associate, Theo, I am in the laboratory. To protect the pieces of retina with which the scientists work, the room is kept almost completely dark. There is only a faint red light source. We had a late start: Craig had been held up and arrived an hour late; then we had to sit for 45 min in the dark before our eyes had adapted and before we could see the vaguest of outlines of instruments and the computer monitor. Craig had sacrificed a fish, excised its retina, and mounted a piece of it on a microscope slide. He placed the slide under the microscope and now audibly opens a light shutter announcing: "We should see a rod in there right now." But the window on the computer monitor – in which the image from the CCD-based camera should be displayed – remained dark (see insert next to turn 01).[3] I (M) comment that there had been something bright just as the shutter went up (turn 03), which is in stark contrast with the black screen that Theo and I currently look at and to which Craig now turns. The latter simply notes that there was something different about the field. Theo talks about the auto gain that we do not want (turn 08) and about not having sufficient light to work (turn 11). Following this exchange, Craig repeatedly refers back to the previous day, when the apparatus had worked fine for us,

[2] There is a reflexive association of inhabiting a place, which is habitat, and which shapes the habits (demeanor and clothing) and the (intellectual) habitus of those habituated (Bourdieu 1997).

[3] CCD stands for charge-coupled device (see Introduction to Part B). A CCD is the basis of the technology used in digital cameras and camcorders. It is an arrow of tubes into which light (photons) falls and generates an avalanche of electrons, which then build up a charge. This charge is then further processed and converted into pixels. The CCD therefore is in some way similar to a retina.

displaying a perfect image; he indicates that it had been better then and asks what "it [was] yesterday" (the settings). At one point, I ask the others whether they had seen a flicker that we had just observed, or whether they had used the appropriate source input for the software. At another point, Craig makes a statement about the difference in resolution between the 2 days.

Fragment 2.1

```
01   C: we should see a rod
        in there right now. *
02      (1.70)
03   M: there was eh
        something, eh, it was
        bright and then eh
        when you got back up,
        eh.
04      (1.70)
05   T: yea.
06      (0.38)
07   C: now there is (some?)
        thing different (0.35) about the field today.
08   T: no, i just think the auto gain is something we
        dont want-
09      (2.00)
10   C: yea.
11   T: because we dont have enough light to to work
        properly.
```

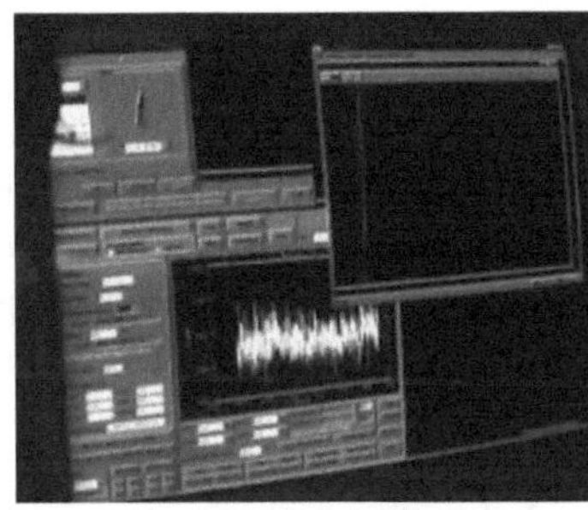

Theo, sometimes instructed by Craig, moves through different pull-down menus and we look at the different windows that pop up, sometimes making changes and sometimes merely staring at them. Sometimes we return to the same window as if we had not yet seen it before or as if we do not believe in our previous observations. Craig slowly becomes frustrated but we keep going. Over a 15-min period, we literally and metaphorically are groping in the dark, attempting to find out what is going on, making changes in parameters that appeared when selected as options in different pull-down windows. The different changes and the selections are not planned, cannot be planned actions because we do not know why the monitor is dark. None of the many actions and changes of parameters change anything. Then, all of a sudden, the searched-for image – the one resembling what we had the day before – appears.

An observer of the group over the 15-min period, although it is a world leader in its field, does not see the heroic scientists that are sometimes described in the literature. At this moment, the three of us were literally and metaphorically in the dark, moving about the software and trying to make sense. In fact, the moving about, the actions of changing parameters and seeing the feedback shared a lot with the process of trying to figure out what an unknown, absolutely darkened place is like by moving about, bumping into and feeling things and slowly getting a sense of what the place is like. The place one inhabits in such a situation does not pre-exist one's moving and groping about; it is the continuously unfolding end result of all activity up to that moment. It is in this moving and groping about that we become familiar with the place, which becomes habitat that we inhabit and that shapes our habitus (dispositions).

Although the fragment might appear negligible and dismissible, an instant better to be forgotten and unworthy to be included in tales of and about scientific research, my own research suggests that such explorations during moments of being in the dark actually provide scientists with the knowledgeability related to their phenomena, apparatus, and any scientific representation that is created in the process. They become very familiar with all those situations in which they lose their phenomenon and, simultaneously, become extremely knowledgeable about how to make it appear in the first place.[4] It is this knowledgeability that is an integral part of their competencies in reading the graphs that come out of their research. In this chapter I document and theorize scientific praxis during those everyday moments when scientists do not know what is going on. I develop the metaphor of *groping in the dark* for the knowledge-generating practices of science, and, subsequently, also for the learning of students who are asked to appropriate what they do not know how it looks like or what it means to know. We can never engage in a specific route to knowledge, for we never come to have a sense of knowing something until we actually engage in knowledgeably doing or using something. That is, *groping in the dark* takes into account the uncertainty that humans faced when they do not yet know what they will subsequently know so that they could knowledgeably orient towards that future knowledge.

Groping in the Dark

> What this being essentially brings to light, that is, making it both "open" and "bright" even for itself, was determined as care prior to all "temporal" interpretation. (Heidegger 1927/1977, p. 350)

Everyday activity is motivated by its particular concerns, which are associated with particular ways in which the world appears and the pace in which activity unfolds. It normally is rather uneventful in the sense that telling what happened in any given 15-min period of everyday activity (e.g., brushing one's teeth, drinking a cup of coffee) appears mundane, too mundane to warrant telling – it literally goes without saying. The events that I focus on here, therefore, might be characterized as "scientists' getting a software-driven monitor for a CCD-based camera to work." After the fact scientists might say something like, "We just had to reset this one value," and they might add, "but it took us a while to find the right panel." But normally, these events, though they constitute everyday life in the laboratory, are

[4] Collins (2001) shows how Western researchers could not reproduce the work of Russian researchers on the Q of sapphire and, therefore, dismissed it – when eventually it turned out that they were only losing the phenomenon. On losing a phenomenon see a description of the attempts to reproduce the Galilean experiment on the inclined plane (Garfinkel 2002).

completely left out of the reported accounts, especially the accounts that are ultimately published:

> A second CCD camera (Canadian Photonics Laboratory), mounted on the trinocular (not shown in Fig. 1), was used for viewing the microscope field and was displayed on the computer monitor. This camera was used to capture IR images of the preparation. (Hawryshyn et al. 2001, pp. 2432–2433)

This after-the-fact account from the published study portrays the use of the CCD camera as unproblematic. But the introductory fragment in this chapter already shows that making the second CCD camera work involved uncertainty and considerable work. It started with a *breakdown*, the instant when we did not know what had happened since the previous day when everything was still working. Through our practical actions, we *brought light* to the phenomenon, sometimes beginning with flicker, sometimes involving a flash of light; yet there always was a flash of hope. But there also was a lot of *groping about*, without it there could not have been a clearing; it was by groping about that scientists provided for a *first clearing*, where objects began to appear. Eventually the problem was *cleared up* and *cleared away*. In the following, I use these five themes to organize my account about how we returned to having a working monitor again.

Breakdown, Darkness

> The more urgently the missing is needed, the more it is encountered in its being unready-to-hand, the more obtrusive becomes what is present-at-hand, so that it seems to loose the character of being ready-to-hand. (Heidegger 1927/1977, p. 73)

In the opening fragment, we encounter a group of scientists attempting to begin what they considered their day's work, which was to consist of collecting more data for the planned article. They had just installed a new way of observing the preparation on the previous day, a CCD-based camera that eased the work of searching for cells. The fragment opened with the calibration of the new instrument against the existing one. We knew that the ocular worked – the microscope was focused on a cell. With this knowledge, the second, new CCD camera was calibrated. Here, then, the cell, the natural object, and a working instrument were used as mediating tools to make a new instrument work. It was its failure to work, the being unready-to-hand of the CCD image that provided the thrust, the intention to the events that followed. It was this same being unready-to-hand that turned up the being present-at-hand of the instrument and any aspect that the subsequent search for the causes of the breakdown turned up. What subsequently turned into an apparent mundane everydayness when the instrument was working again, "This camera was used to capture IR images of the preparation," the present-to-hand nature of the CCD-based IR image, will have involved considerable work on the instrument itself, work during which the instrument had the character of being present at hand.

As he turned from the ocular toward the computer monitor, Craig announced what we should have seen. He made others in the lab aware of the fact that just at this moment, he had lined up a rod in the ocular. Because we (Theo, I) had no access to the ocular, Craig's statement marked a point in time when something that we had attended to should have changed. This is the event that my own comment referred to, a bright flicker across the monitor when the shutter was opened (Fragment 2.1, turn 03). The description "there was something" was in contrast to the present state of the window, a black screen. The function of such statements was not to communicate but to align all participants to all events, or, in other words, to ensure that we are all on the same page throughout the joint activity.

Craig not only stated that there is nothing to see but also that there was something different about the field (Fragment 2.1, turn 07). The denotation "field" was potentially ambiguous, because it is normally used to describe the visible area on the slide under the microscope. At other times, this or similar expressions were used to denote a poorly mounted slide, or a slide with photoreceptors "gone bad." However, the two recipients (Theo, I) might have taken it to be a comment about the window, because it is the entity to which we all were oriented. The statement was consistent with a later one that raised the question whether the "noise" was due to the software or already originated in the lamp that provided the illumination for the slide.

In addition to the new monitor, we also had added another light-absorbing film, leading to a faint and deep-red image. The story appeared to be one where the "auto gain" darkened the image to such an extent that the computer monitor no longer emitted an image bright enough to make it through the multi-layered screen. Craig's comment, "But yesterday it was *so* bright" was descriptive of and therefore supported the search for a problem with auto gain, brightness, and absorption. This shows that the initial search was colored (and perhaps darkened) by the search to gain control over the auto gain. This intentionality was so predominant that after the fact, indeed even 2 years later, Theo suggested to me that the problem had been with the auto gain, although the videotape recorded during the event shows otherwise.

A Flash (of Light, of Wit, of Hope)

Theo was attuned the auto gain (Fragment 2.1, turn 08) and Craig to the difference between the current camera output, as seen on the monitor, and the one he remembered from the previous day (Fragment 2.1, turn 07). Theo was in control of the computer and therefore the window that should display the CCD image. In the search for dealing with the auto gain issue, he moved from pull-down menu to pull-down menu, and from window to window that could be selected within each. Craig and I periodically asked questions about entities, values, and text visible on the remainder of the monitor (e.g., Fragment 2.2, turns 03, 21). The function of these questions and comments was to make the same aspects of their world salient to everyone, that is, they established or maintained common ground in the course of searching – like children moving about in the dark holding their hands every now and to ascertain that everyone is still with the party.

Fragment 2.2

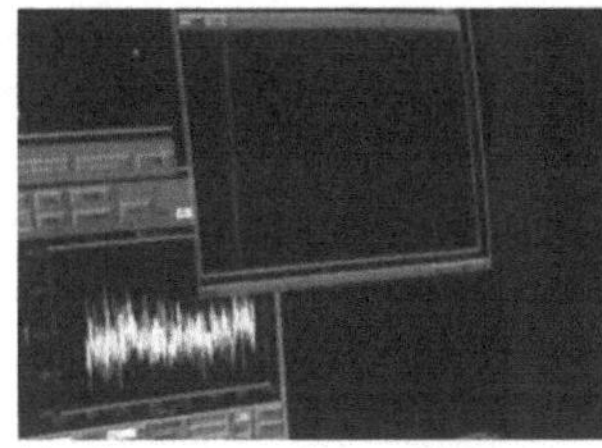

```
01          *
02          (1.36)
03    M:    dyou see that there?
04    T:    yea, every time you
            change it-
05          (0.55)
06    M:    a'hhh
07    T:    =the position
08          (0.60)
09          then it chai- then it rea- it is an auto gain.
10    M:    ah, okay
11          (0.53)
12    T:    and so it-
13          (1.49)
14          but its darkening it so
            much for us so that we
            dont see anything in
            [here].
```

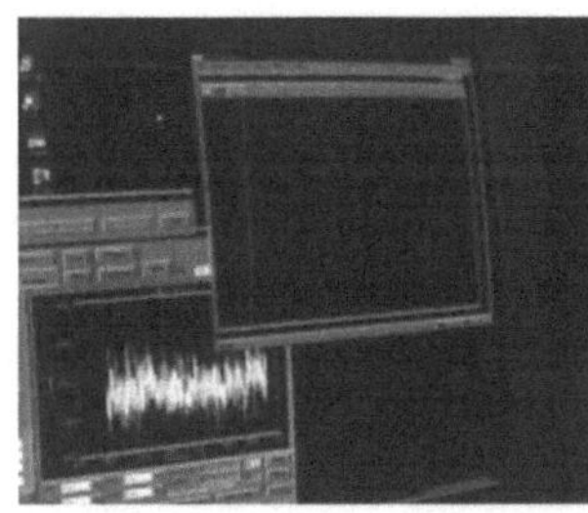

```
15    M:    [a'hh].
16          (2.93)
17    C:    yea there is something
            going on, we are not
            seeing what-
18          (3.23)
19          we saw yesterday. *
20          (9.12)
21    M:    cant you turn it off,
            this auto gain?
```

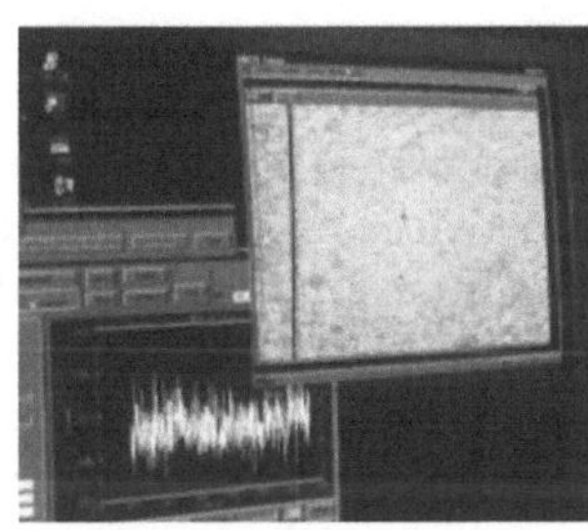

```
22          (0.90) *
23          ((The image becomes
            bright, one can hear a
            shutter of the microscope
            open or close.))
24    T:    zats what i have to find
            out how because if the
            eh-
25          (1.87)
26          if the, um-
27          (0.64)
28          the [camera might be doing it already and then you
29              [((clacking noise from shutter))
30          have to find a way to set that.
```

At the beginning of this fragment, there was a brief brightening of the monitor (turn 01). After a pause, I questioned whether the others had seen the change (turn 03). In view of the next turn that followed, the function of the statement was to ascertain whether others had been tuned to the ongoing events, particularly the changes in the same way as the speaker. It aligned the participants or assured that alignment had in fact existed. Theo moved the cursor over the pull-down windows, thereby opening them; he then moved the cursor down, highlighting different items. Craig said that there was a difference with the previous day (turn 17), but I offered up an apparent question about turning the auto gain off (turn 21). Theo then articulated what his searching movement was intended to do: "finding a way to set that [auto gain]" because it was not clear whether the software or the camera were responsible for it (turns 24–30). Theo responded without delay (turn 04), an indication that he was attuned to this change (or such changes) on the monitor. His explanatory statement focused on the auto gain – the automatic adjustment of image brightness. We can see and experience him as being attuned to the auto gain. Craig simply articulated the differences with the image of the previous day, whereas my question returned to the issue of the auto gain, whether the auto gain was something that could be turned off (turn 21). In the second situation of the brightness change (see offprint next to turn 22), nobody commented. A comment in fact may not have been necessary given the immediately preceding history of the conversation. Craig was working on the microscope, as the clacking noises indicated, and therefore was not attuned to the monitor. Theo and I were in the process of discussing auto gain, of which this new flicker may have been but another instance, or we may have both noticed the flicker as part of the opening and closing of the shutter. If we were attuned to Craig's actions in this way, then there was no need to alert others to the change in the monitor or to find out whether they had seen the changes (Roth 2004). A comment following the change and in fact announcing it to the others may have just stated the obvious, something that everyone was attuned to. Stating the obvious would have been in breach with the everyday custom to leave unsaid that which goes without saying.

Groping About

The process of groping in the dark was already evident from the previous fragment, but becomes more so in the present one, where Craig, though he did not know the software, proposed to open one of the pull-down menus ("File") and to select a particular item ("Settings") once it was pulled down (turns 01–03). Theo did just that; or more appropriately, because of the precarious nature of instructions and actions (Amerine and Bilmes 1990), what he did could afterwards be described as having followed Craig's instructions. Although there are long pauses in the text (verbal action), Theo was doing things such as moving about the screen with his cursor, thereby directing and aligning the attention of the other participants, and everyone was perceptually following the events. Theo opened the pull-down menu "File" with his cursor (turn 02), through which he then moved slowly down, thereby highlighting the different choices it offered.[5] During this time, nobody said a word for a (conversationally) long 7.22 s. Craig then stated "Settings!" with an intonation that might be heard as an order to go to the next pull-down window labeled "Settings," followed by Theo's pulling down the setting menu (off-print in turn 06).

[5] The statement "go to the file" and Theo's moving the cursor to the *File* menu are related in the same way as the statement "slab!" followed by the mason helper bringing a slab (rather than a pillar) (Wittgenstein 1953/1997). No "understanding" or "meaning" are required in our analysis, simply the sequentially organized order | order-following pair.

Fragment 2.3

```
01  C:  * okay, go to file.
02      (7.22)
03  C:  settings!
04      (4.20)
05  M:  what would
        th[e:?]-
06  C:    [CAP ]ture * settings!
        <<p>turn around yea>
07      capture * settings.
08  T:  <<p>(get?) ready for the
        auto gain>
09      (0.70)
11  C:  yea.
12      (0.88)
13      okay, now, that-
14      (0.43) * thats not it
        (0.34) go back to
        settings again.
15      (2.21)
16      how=b=t video (0.43) um:
        (0.37) format?
17      * (3.06)
18      HMm.
19      (0.83)
20      fold?
21  T:  nä:
22      (2.02)
23  C:  whats a sixforty,
        eightyeighty um::?=
24  T:  =thats the size of the
        screen here.
25      (4.70)
26      i think we just had a brighter screen yesterday,
        that's why we saw more.
```

As Theo began to move the cursor down the list of items, Craig ordered, “CAPture settings!” (turn 06) at the very moment that I had started offering up a question (turn 05). Under his breath he muttered, “turn around, yea” as if describing the action just completed or in the process. Theo, too, talked under his breath, making yet another comment about the auto gain, while moving the cursor over several of the fields and sliders (turn 08). After a while, including several false starts, Craig announced that this was not it and asked Theo to go back to the settings menu (turn 14). Theo's cursor moved across several items and, while Craig spoke, moved it to the item that Craig read out, "video format," which brought up another window (turns 16–18). After the window had popped up, Theo again moved the cursor over different buttons and places where values could be typed in. Craig read out an item, "Fold?" (turn 20) but Theo seemed to reject it (turn 21). After a pause, Craig completed a query | explanation pair concerning the "640 × 480" (turn 23), which Theo's statement explained to be the size of the screen, but which we hear to be in reference to the size of the window in which the CCD image ought to have been displayed rather than the pixel size of the total monitor. After another long pause, Theo talked about a brighter screen that they have had on the day before (turn 26).

Theo moved the cursor across the different fields and buttons and the other two participants equally were attuned to the situation. The movement makes salient the different attentional foci, which can be assumed for the other members as well. When Craig suggested that "that's not it" (turn 14), it was an expression that nothing had become salient in their visual exploration of the items in the window. His statement expressed something glossed by the statement "In this situation, there is nothing that will help us." If there had been something salient, one could have expected a comment, for there would have been a contradiction between Craig's statement and whatever someone else might have seen. But here, Theo closed the window, and Craig offered up what can be heard as an instruction to go on to another menu item. In this, Craig implicitly showed agreement with Theo's action, and in fact articulated the next action to take.

This fragment exhibits how the scientists moved from pull-down window to pull-down window and within each, from one option to another. It was as if they did not know what they were looking for but they were moving to see whether there was something salient that could be related to the missing image. That is, they were looking for something without knowing what it was. How can this be? How can we look for something without knowing what it is? In effect, we were looking not for just anything, but for something that might have to do with the lack of an image. Because we did not know what it might be, there was a possibility that we might not recognize it even if we were sitting right in front of the (eventual) solution to our problem. How do we know that what we are looking at is what we are looking for? For answering this question, we needed to know that a change in some parameter would give us the image back, or at least, bring us closer to the sought-for image. However, because we did not know why we had lost the image since the previous day, we also could not know what we needed to change to bring it back. That is, we needed to actively try changing whatever appeared to be a possible cause. We therefore were in a double-bind situation typical for learning situations when the object (knowledge) to be known has not yet revealed itself (Roth 2012). Not knowing it, all we could do was try. That is, we had to engage in actions hoping to find out something and get the instrument to work.

At this point in our work with the new instrument, we were not yet familiar with the software, its options, and how changing these might influence the display. It was through our movement, our groping about, and the responses we perceived that an instrument-related world came to be. The material (moving about the software world) and verbal actions were in fact explorations. It was through these explorations of an unfamiliar world that we came to know the world constituted by the software; more so, it was in this way that a familiar world came into being and that we therefore could relate to in an intentional way.[6] The properties of the imaging software, monitor, and CCD emerged in the course of the groping and as a consequence thereof. In fact, the groping provides access to anything like properties. It is in the process of looking at and as its outcome that which is knowable reveals itself. The actions and their perceived results allowed use to bring forth a world and come to be familiar with this world. This world cannot be said to pre-exist, though its material aspects might do so. Here, the subject–world (object) relation was a dialectical one, in which the world exists at two planes – the material world and how it appears to the acting subject on an ideal plane. With each action, the world is transformed in the sense that its appearance and therefore the opportunities that it provides for further action are reproduced and new elements are produced. That is, groping about also constitutes groping one's way, both describing the process of getting closer to the goal of a working camera and coming to know the software and its features.

Theo made reference to a brighter screen that we had on the previous day, but neither he nor Craig or I made it the focus of an inquiry (turn 26). It was noted as a fact but it was not a fact that made a difference in our search for the problem. In the time between the data collection of the previous day, Craig and Theo had decided to exchange the smaller monitor for a larger one that had been used in another one of Craig's labs. After discussions with Craig Theo had also mounted another absorbing film over the monitor so that it would maximally absorb all light other than that at the red end of the visible spectrum.[7] But these changes, though in some way made thematic by Theo, were not the focus of our solution-finding process.

In this fragment, it may be noted that Theo was not disturbed or distracted by what-Theo-and-I-knew-to-be Craig's incorrect reading of the window size, "six-forty, eighty-eighty." In a similar way, Craig had read "avis" ['eivis] from the monitor, but Theo immediately corrected the pronunciation to "a-vee-eye's" [ˈei-ˈvi-ˈais] explaining that it was a video format. That is, despite what I too heard as stumbles, Theo was sufficiently attuned to the situations to respond in an

[6] In the context of episodes from an elementary mathematics classroom, I provide a description of the process in which intentional knowledge is born (Roth 2014).

[7] Because the photoreceptors mounted on the slide are light sensitive and deactivated by the photons falling on them, the lab was faintly illuminated with red light at the near infrared part of the visible spectrum, which minimized its impact on the cells we were working with.

appropriate way, just as if Craig had said, "six-forty by four-eighty" or "A-V-Is." Craig repeatedly pointed out features that Theo and I – having worked with imaging software before – already were familiar with. Thus, neither Theo nor I picked up on the screen size (640 × 480), the AVI ("avis"), compression issues, and differences between stills and movie as possible sources of the trouble. But the fact is that in such cases, anything that participants' say has the function of bringing to the foreground anything that possibly could be the source of the trouble (breakdown).

First Clearing, First Object

This fragment, which occurred about 1 min after the preceding one, constituted a first changeover in the way we were attuned to the image before us. Until this point, brightness had been the central issue that different speakers noted and therefore made salient to each other, and for which they proposed a variety of solutions or possible causal precedents. Here, there was, for the first time, a new issue: the resolution. In normal operation, the resolution of the image withdraws to be really handy. The instrument and its constitutive parts achieve their character exactly at the moment that they do not work. Heidegger (1927/1977) called it the instant when the worldly character of the surrounding world makes itself known, comes and even forces itself into the clearing where it is accessible to our attention and to greater certainty. The resolution issue co-emerged with the first, vague image of a cell, barely visible in the noise, but enough to be pointed to and out by means of a gesture.

The fragment began when Theo attempted to bring about a change in the image by modifying brightness; but Craig suggested that the change made "is not it," that is, could not have addressed the central problem (turn 02). Theo nevertheless pointed out that there was something to be seen, but there was a conversationally long pause, and neither of the other two lab members present gave a sign that he had seen what had been referred to "see this one right here?" (turn 04). Theo evidently took the pause and lack of acknowledgment as a sign that others had not understood, and not only reiterated the indexical reference "right here" but actually used a gesture to point to and move along monitor where he saw an entity. Both words and gesture contributed to making a statement, which constituted a pointing out that communicates and defines By saying, "Yea, I know" (turn 07) Craig's statement turns out to acknowledge what Theo offered as a description and that he, too, had seen the entity but then articulated the problem in a new way, "the resolution is really bad." Theo in turn acknowledged the comment (turn 08), and began to search for something in a pull-down menu (turns 10–11). His action selected a particular window, but explained that he could not set the contrast to brighten the image – his hand moved across the position where the modifications could be made (turns 12–13).

Fragment 2.4

01 T: * okay, i brighten it right here.
02 C: thats not it.
03 T: <<p>how about this, do you see this one right here?>
04 (2.31)
05 you see right * [here.
06 [((*Moves finger back and forth along screen where there appears to be something in the noise*))
07 C: yea, i know, but the resolution is (0.69) really bad
08 T: yea.
09 (1.54)
10 okay.
11 (8.71) ((*Goes to a pull down menu, which opens a new window*))

12 i * cant set the [contrast to brighten it.
13 [((*Moves hand across the window at the place where settings can be changed.*))

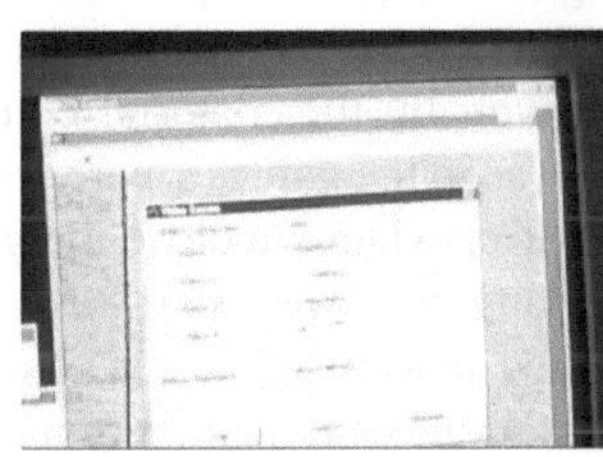

14 <<p>yea, we have more than one to select from it.>
15 increase contrast-
16 (1.84)
17 brighten-
18 (7.39)
19 C: * now (0.20) there is too much noise.
20 T: yea.

Subsequently, and as if he were speaking to himself, Theo also described the present situation as offering too many options to select from (turn 14). He not only made changes but also articulated them in words, "increase contrast" and "brighten," as if were exemplifying the relation that binds descriptive words to the actions so described. His actions were accompanied by changes in brightness and contrast but the image behind the window remained as fuzzy as it was. Craig now described it in terms of there being "too much noise" (turn 19), consistent with his earlier assessment of a poor ("bad") resolution. Theo made a statement of agreement.

In the course of our data collection, there were other moments when there was "nothing" on the screen, when there was nothing but "noise." But we became good at attributing that particular part of noise to the fluid in which the retina is suspended or to some other part of the instrumentation. In the present situation, the "noise" was more like that of a television monitor when there is no signal on the cable or when the channel is slightly mistuned. We did not know the source of the noise. Our exploration, our search for something that both will change and explain the situation, was part of the process of providing a clearing, in and from which entities come to emerge until one of these – hopefully – comes to make the difference.

When there were problems, here as elsewhere during our research, team members never engaged in the kind of searches that psychologists and cognitive scientists talk about, where the problem solver searches a known problem space for specific and specified items – similar to someone searching for a particular knife or pot in his or her kitchen. The present situation was rather different. It was a search for something that we did not even know what it was. We were going through the proverbial haystack without knowing what we were looking for; and we were looking for something without any assurance that we would know what it was until the point that the equipment was working again. As Theo formulated at another instant during that morning, he did not think that they had changed anything since the previous day. Yet the image was different. The process in which we were therefore engaged was more like groping in the dark (we literally did), when one finds oneself in an unfamiliar place in complete darkness. All we can do is move about without knowing into which direction we have to move to change our situation. There is no way that we can judge whether or not a particular movement, a particular direction, or a particular action is more or less successful in getting us out of the situation. But through our actions we clear a field, constitute a clearing, and begin to separate figure from ground. In the laboratory, there was a general sense of what the overall outcome ought to be, and there are certain constraints to action given by the material conditions of the situation, but there is no map or gradient (as in the games where we are told "hotter" and "colder") that could help us identify if or when we close the gap.

The language we used was an integral part of the exploration; it primarily constituted a form of action, with an intentionality of *in-order-to*, rather than a representation, with its intentionality to be *about* something. There were no "meanings" required, no understanding, just ways of using the words to get our work done. Language was a resource for pointing out some things, making them salient, and aligning others to ones own ways of being attuned to the field. When language appeared to be insufficient, we also had available gestures as means to point, encircle, or iconically articulate an entity. The language was part of our being attuned to and integrated with the world rather than a re-representation of this world separate from use while engaged in the search for the causes of the "different field." Language and the whole of the scientific research activity with which it was interwoven – here, a search in the dark to find the reason for the darkness of the monitor we were using – constituted a *language game* (Wittgenstein 1953/1997).

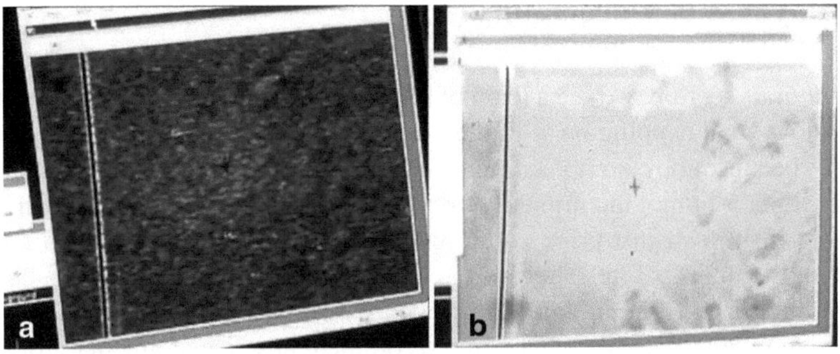

Fig. 2.1 (**a**) The display of the image from the microscope when Theo said, "May be it's the trouble with the large screen." (**b**) Five seconds later, and after another few fiddling moves, the desired image was present and the scientists continue as if nothing had happened

Cleared, Cleared Up, and Cleared Away

In the end, the problem suddenly disappeared as one of the software settings, after being changed, brought about the desired image. Immediately prior to that moment, Theo framed for the first time the real change that had occurred since the previous day, "Maybe it's a trouble with the large screen" (Fig. 2.1a). Here, "large screen" was not a way of saying "large" in absolute terms. The "large screen" stood in contrast to the smaller monitor that we had used on the previous day. The significance of the denotation becomes only clear in this historical perspective. Up to that point in their work, we had used a smaller computer monitor, for we only had displayed the graphs before, whereas one researcher worked searching for suitable cells through the ocular. Craig had decided to replace it with a monitor currently connected to another computer in another lab. Articulating the large screen and thereby bringing it into the clearing, turning what was in a mode of present-to-hand into something in the mode of ready-to-hand, changed the way in which we were attuned to the situation and quickly resolved the issue. Then, 5 s later, the desired image appeared. However, there is no sign visible or audible on the videotape recording the situation that something special had occurred: the equipment was working again. Instead, we apparently continued as if our problem had never existed at all, that is, as if we had not just spent 15 min struggling to find how to get our CCD image to show up on the monitor. But with the disappearance of the problem, therefore, that which had been in the foreground receded, hiding again in its ready-to-hand modality.

Everydayness of Uncertainty

In the previous section, we see a team of scientists at work, in a situation where they do not know what is going on with their equipment other than that it malfunctions. Precisely because they do not know *what* it is that malfunctions or what is at the origin of it, they cannot intentionally orient towards doing something specific about and fixing it. In fact, just prior to the image returning in the way that it had been present on the night before, Theo still announced a possible rather than a definitive cause of the image from the microscope being as it was. In the process, they fiddle with this and they fiddle with that – in the complete absence of any plan for fixing a clearly defined problem. They pull down a window here and then pull down a window there. When something does show up, it may be taken to be significant – such as when Craig asks about the AVIs and 640 × 480 that have shown up on the monitor while Theo fiddled with the display. In such a situation, anything showing up on a pull down window is a potential candidate for further investigation and possible source for the microscope image to be other than intended. But all of this makes sense, for to have a plan at all, we needed to know what the problem was, which could then be the goal in the solution (search) procedure. The problem in this situation did not lie with these scientists (including myself). Rather, the problem, if there is any, lies with inappropriate modeling of problem solving in the real world.

The scientists in the video do know what they want to end up with: something that shows more-or-less clear images of the material on the microscopic slide similar to the way in which they would see them when looking through the ocular (Fig. 2.2a). If they did not have those images, they could not make the two measurements that they required for getting an absorption spectrum. There already was uncertainty as to the nature of the objects because a broken rod looks like a cone, and the different cones may be confused with one another. Sometimes, even though there appeared to be a rod or cone, when a measurement was conducted there was nothing visible in the graphs. But the beginning is the image, because it allows the scientists to line up the objects they find with the light beam from that they use in the light absorption measurement. It was only when the image actually existed that we were able to continue in our day's work. In Chap. 3, I turn to the work associated with this uncertainty that was arising from the double bind relating the two forms of inscription observable in Fig. 2.2b. Uncertainty will be connected with relating visible images with graphs, an uncertainty in part created because of the very decontextualization processes required for creating scientific knowledge independent of particular specimens. That is, *those graphs that we ultimately study and that constitute one of the foci of this book cannot be understood independently of the uncertainty that accompanies the scientific investigation in the course of its realization from the beginning to the end. Any account of graphing as social practice therefore has to account for the inherent and radical uncertainty that separates any natural phenomenon and the inscription (graph) said to stand in for it when the phenomenon is actually absent.* Part of the account that I provide in the following Chap. 3 is concerned with the scientists' efforts to come to grips with the

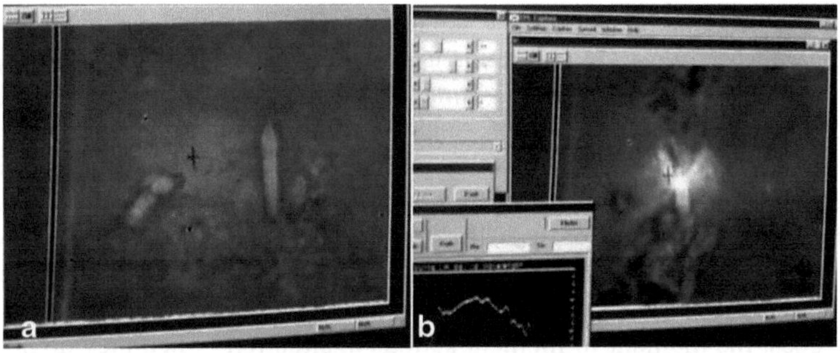

Fig. 2.2 (**a**) Scientists aim at an image that shows clear objects, here some rod-shaped photoreceptors. (**b**) They take a measurement from a cone-shaped receptor

uncertainty and, in part, to remove it by acquiring pertinent information about the conditions in which the fishes have been living prior to being caught and transported to the laboratory. However, even at the university uncertainty entered the conditions of the fishes, as some of these were held in a shared facility, where feeding, light, or temperature might vary without the scientists' knowledge.

In this situation of uncertainty, there was nothing that unsettled these scientists. It is as if the uncertainty was nothing special. Indeed, it is not. As the chapters that follow show, uncertainty was a travel companion of the scientific research team. Uncertainty is nothing extraordinary, but rather part of what it is to do science. It may be that scientists are good at precisely that: coping with the mundane everydayness of uncertainty. It is not that uncertainty does not exist outside the scientific laboratory. It does and pervasively so even though, as a feature of the same everydayness, we do not attend to these uncertainties and subsume them just as we subsume other, more predictable aspects of our lifeworlds. This, too, will be a feature of the account in the following chapters, for the scientists subsume their early results to the common way of modeling the changes in the absorption of light in the course of the life cycle of one salmonid species. These scientists will fit their observations to the canon only to realize in the course of their inquiries that the canon does not describe their data very well.

Uncertainty temporarily ends when scientists arrive at a point where they have a stable and accountable, and, therefore, reportable phenomenon. The remainder of their work then consists in cleaning up the data and getting the study written up and accepted for publication. While waiting for the peer review process to do its thing, scientists are already onto other things, again dealing with uncertainty until these, too, have settled and something else has emerged for publication. Other than in the social sciences, where the peer review process is highly uncertain, the natural scientists in this laboratory were less worried about that part of their work, as if they were certain that they would get published whatever their results. As Craig said at one point when it became apparent that our results were not consistent with the canon, our work and data were highly appropriate for publishing – it was just that our story was changing.

In the present chapter, we see the scientists make do with the situation at hand. *Tinkering* and *bricolage* are two terms used to talk about situations where people make do with the things at hand. Thus, an earlier study from an ethnographic perspective used the metaphor of tinkering for describing scientific research (Knorr 1979). The notion of tinkering invokes images of making do with the materials and tools at hand, sometimes – in situations of adversity – converting and subverting their intended use. But these materials and tools are known and available to the tinkerer. In the present fragments, we see scientists in another way, one in which they are infrequently shown in as far as the ordinariness of their groping goes, the moment-to-moment actions in situations when they do not know what is going on. In groping, they do not even see, know, or have available those entities that ultimately will turn out crucial to their success. As the quotation from these scientists' research paper shows, once the situation was behind them and once the CCD-based camera worked, its unproblematic use could be framed in a single sentence. We had become so familiar with our instrumentation and could make it reliably work. In its everydayness, therefore, the process of groping about also tells us about how everyday familiarity and its associated knowledgeability come about. Our deep practical knowledgeability is the result of our extensive, embodied explorations that brought forth a lifeworld with joints (articulations) separating objects and entities that can be articulated (talked about) and distinguished. I propose this groping in the dark as a metaphor for learning, as an exploration in the unknown, by means of which our known world expands into the clearing cleared in the process. We talk in the process, or rather, talking is part of the process of groping. It has the function to make salient or to ascertain mutual alignment to an aspect of the world. When we do not talk it is because the salient is taken to be self-evident – in which case it does not need to be articulated because "the pointing out of the statement is completed on the ground of what is already disclosed in understanding or rather circumspectively discovered" (Heidegger 1927/1977, p. 156).[8]

In the next chapter, I turn to the uncertainties involved in the relation between the entities on the microscopic slides – i.e., the images that the scientists were after in the present chapter – and the graphs depicting the absorption of light in photoreceptors as a function of the light spectrum. It turns out that especially in the beginning, before the data collection became so much a routine that a masters-level graduate student could do it, the relation between image and graph was tenuous. And again, just as in the present chapter, the tenuous nature of the relationship between two different inscriptions in the chain of translations from the natural world to a scientific fact stated in mathematical and discursive form (see Chap. 1) was part of the everydayness of doing scientific research.

[8] For Heidegger, "understanding [Verstehen]" is to be heard in the way Wittgenstein (2000) would hear it, that is, as knowing one's way around the world. Thus, Heidegger uses the example of a turn signal on a truck to show that we do not interpret it to "understand" it or infer some "meaning." Rather, we see through the turn signal knowing that the truck will turn.

References

Amerine, R., & Bilmes, J. (1990). Following instructions. In M. Lynch & S. Woolgar (Eds.), *Representation in scientific practice* (pp. 323–335). Cambridge, MA: MIT Press.

Bourdieu, P. (1997). *Méditations pascaliennes* [Pascalian meditations]. Paris: Éditions du Seuil.

Collins, H. M. (2001). Tacit knowledge, trust and the Q of Sapphire. *Social Studies of Science, 31*, 71–85.

Garfinkel, H. (2002). *Ethnomethodology's program: Working out Durkheim's aphorism*. Lanham: Rowman & Littlefield.

Hawryshyn, C. W., Haimberger, T. J., & Deutschlander, M. E. (2001). Microspectrophotometric measurements of vertebrate photoreceptors using CCD-based detection technology. *Journal of Experimental Biology, 204*, 2431–2438.

Heidegger, M. (1977). *Sein und Zeit* [Being and time]. Tübingen: Max Niemeyer. (First published in 1927)

Knorr, K. D. (1979). Tinkering toward success: Prelude to a theory of scientific practice. *Theory and Society, 8*, 347–376.

Redfield, P. (1996). Beneath a modern sky: Space technology and its place on the ground. *Science, Technology & Human Values, 21*, 251–274.

Roth, W.-M. (2004). Perceptual gestalts in workplace communication. *Journal of Pragmatics, 36*, 1037–1069.

Roth, W.-M. (2012). Mathematical learning: The unseen and unforeseen. *For the Learning of Mathematics, 32*(3), 15–21.

Roth, W.-M. (2014). On the birth of the intentional orientation to knowledge. *Encyclopaedia – Journal of Phenomenology and Education, 17*, 91–126.

Suzuki, D. (1989). *Inventing the future: Reflections on science, technology, and nature*. Toronto: Stoddart.

van Eijck, M., & Roth, W.-M. (2011). Cultural diversity in science education through novelization: Against the epicization of science and cultural centralization. *Journal of Research in Science Teaching, 48*, 824–847.

Wittgenstein, L. (1997). *Philosophische Untersuchungen/Philosophical investigations* (2nd ed.). Oxford: Blackwell. (First published in 1953)

Wittgenstein, L. (2000). *Bergen text edition: Big typescript*. http://www.wittgensteinsource.org/texts/BTEn/Ts-213. Accessed 30 Nov 2013.

Chapter 3
Uncertainties in/of Data Generation

In Chap. 2, scientists are shown to be working in situations marked by uncertainty. Precisely because they are in a discovery science, they do not know the object of their ultimate finding – much like Christopher Columbus who could not have known that he would *discover* the Americas. The media, however, rarely provide information about the variation in the data of the sciences or medicine, about laboratory contexts, or about what has not been included in the measurements or analysis. This is important because the "details of laboratory work, and of the visible products of such work, are largely organized around the practical task of constituting and 'framing' a phenomenon so that it *can* be measured and mathematically described" (Lynch 1990, p. 170). It may therefore not surprise to find scientists who will critique the kind of mathematical inscriptions with which students are presented in their science courses. Thus, upon seeing a graph of ideal birth rates and death rates to model the temporal dynamics of a population as can be found in any introductory university textbook on ecology (see Chap. 1), an internationally known marine ecologist suggested to me: "You're never gonna find a data set that looks like this. This is a theoretical model, it's based on, you know, nice mathematics and equations, and it's the way we think the world probably works" (Roth 2001, p. 14). He further suggested never having seen a data set that would contain a perfect relation, because "in the real world, [there] is a constant fluctuation" (p. 14). That is, to be able to unpack the claims made in the scientific literature or in the popular media, we need to be familiar with how the laboratory context *might have shaped* the data collection to get a sense of what is included in and what has been excluded from the data mobilized in support of the scientific claims. Without such familiarity, even professors lecturing undergraduate classes may erroneously relate graphs and the phenomena in the world that these are intended to make present. The purpose of this chapter is to report on how scientists deal with uncertainty in and arising from data generation as a possible starting point for rethinking what we might want to do in STEM education to afford student development with respect to using graphs and graphing in the face of uncertainty.

W.-M. Roth, *Uncertainty and Graphing in Discovery Work*,
DOI 10.1007/978-94-007-7009-6_3, © Springer Science+Business Media Dordrecht 2014

An explanation of the process of data generation generally and that of the data that underlie scientific claims specifically should be of interest not only to those educators interested in producing more scientists but also to those who focus on general STEM literacy (e.g. because we want to have engineers who engage in an issue even if they do not know what the issue is rather than technicians who do what they are told to do). Thus, across the media, we are confronted daily with the results of yet another medical study suggesting that eating more rolled oats, kale, or fish (oil) diminishes the incidence of certain medical conditions even though the claims are made based on populations rather than individuals so that any scientific claim may or may not bear out at level of the person. Being able to critically analyze such reporting might be considered an important goal of education generally. For example, while writing these words, I was directed to an "Infographics" with the subtitle "Sitting is Killing You" (Medical Billing and Coding 2010). One of the panels reads: "Sitting increases risk of death up to 40 %," specifying that "Sitting 6+ hours per day makes you up to 40 % likelier to die within 15 years than someone who sits less than 3." Should I be alarmed giving the fact that I am sitting at my desk for 8+ hours, sometimes up to 14 hours? On what kind of data are such claims based? More important is this question: "What was *not considered* as data in stating these claims?" To knowledgeably unpack such claims generally and the limitations thereof more specifically, recipients of such messages actually need to be more familiar with, for example, what data distributions look like.

We can represent the death rates of individuals as a function of the amount of time they sit per day. Many of those sitting less than 3 h a day die before those (all those shaded dark grey, representing the overlap of the two distributions) sitting 6+ hours per day (Fig. 3.1a). If the same graph represented the efficiency of a drug, the distributions might show the relative benefits of a placebo (left) and a drug (right). In this case, there would be many individuals in the sample for whom the placebo worked just as well as the drug. Studies with school students suggest that they better learn to interpret when such distributions are included. They might ask, "What is the percentage of people sitting 6+ hours who have a higher risk?" Thus, participating in the construction of data will "ensure that the students [are] actually analyzing data rather than merely manipulating numerical values" (Cobb and Tzou 2009, p. 167). Other forms of information, too, appear in the media and yet are frequently not knowledgeably employed (e.g., Fig. 3.1b). For example, a psychologist might find that the correlation between *IQ* and science achievement (*Ach*) is given by the equation $Ach = \mathrm{m} \cdot IQ + \mathrm{k}$. How much should a science teacher be concerned? Teachers need to know why a student underperformed so that they could address his/her needs – and these needs are treated in the statistical model as error. That is, what the teacher needs and ought to do to address a student's need *has been deleted* from this scientific model of achievement; and they need to be more familiar with the ways in which scientists deal with the uncertainties inherent in their work. That is, to meet students' needs to know what they are supposed to learn, the teacher has to know how to address the deviations from the norm indicated by the regression curve (Fig. 3.1b).

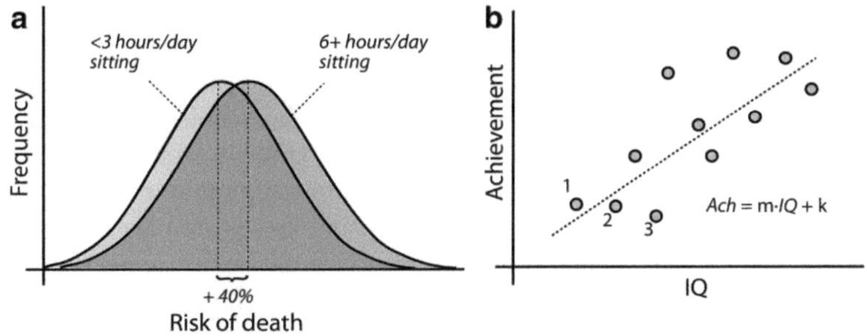

Fig. 3.1 (**a**) Hypothetical distributions of risk of death for people sitting less than 3 h per day risk and for those sitting longer than 6 h per day. Although on average there is a 40 % higher risk for those sitting more than 6 h per day, any given risk level includes many individuals sitting for less than 3 h per day. (**b**) Even though there is a (hypothetical) relationship between IQ and school achievement, higher IQ does not necessarily mean higher achievement, as the relationship between data points 1, 2, and 3 shows

Watching the videotapes I took in the laboratory allows us to observe how scientists decide, at different instants along the trajectory that takes them from living fish to the representation of retinal light absorption in a research article what to include and what to exclude from their data. This selection process begins in the laboratory, where the scientists make a first decision about whether to keep (saving it to the hard disk) or scrap a measurement. This is the place where scientists make first decision about what is in and what is out. In school science, students are presented with tasks and task conditions that they have to address and for which they are held accountable. Like in other everyday life situations but contrary to school settings, scientists also choose to abandon a problematic issue (Lave 1988). That is, under certain conditions, those facing a problem abandon it rather than spending time and resources in the perhaps futile attempt of trying to solve it.

Finding the Ideal Graph

When scientists collect data, there is a lot of uncertainty. Frequently, there is noise; and the signal, that which the scientists are really interested in, is either absent or shows up only tenuously. But the scientists work towards an ideal. This ideal graph hardly ever appears without additional work being done. But on the rare occasions when such an ideal actually does show up, scientists are particularly pleased. They talk about a measurement as exhibiting features that resemble the anticipated ideal. In this section I present the work done in identifying such ideals. Instances of such work can be recognized when the scientists say something like "Rather nice ... pretty 'pretty' if you ask me." These ideal curves allow them to work with other data such that these come to be shaped towards this ideal, often involving many transformations and summations over many cases. The identification of an ideal is shown in Fragment 3.1, which begins with the noises made when Craig opens and closes the shutters that allow the sampling beam to fall onto the microscopic slide. The research team members in the laboratory hear these noises as specific transitions in the events. Here, these noises precede the announcement that a reference measurement is to be taken. Theo responds by formulating that the measurement is under way (turn 003). Once the opening and closing noises are heard again, Theo is familiar with what has happened and formulates that the reference measurement has been taken. Any lab member present is aware of the fact that Craig now is aligning the photoreceptor with the beam. After a long pause, Craig states the name of what he has located, "a single cone" (turn 007), and he then announces the scan, that is, the part of the measurement where the light beam goes through the chosen photoreceptor.

Fragment 3.1

```
  001  C:  ((click)) (0.75) ((cluck)) ref
  002      (0.82)
  003  T:  under way
  004      (1.77) ((click)) (0.83) ((cluck)) (0.40)
  005  T:  reff ´done
  006      (16.49) ((lab members
           know that Craig is
           aligning the photore-
           ceptor with the beam))
→ 007  C:  ((click)) (0.78)
           ((cluck)) ((Craig
           opens the shutter for
           allowing the light to
           come through)) (1.49)
           sINgle cONe (0.62)
           scA:N
```

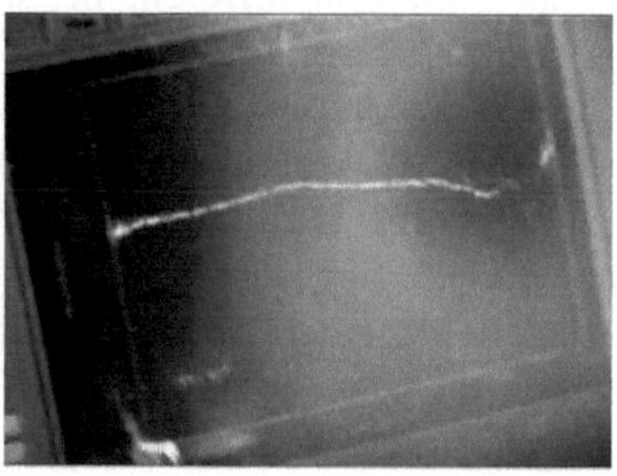

```
  008      (7.27)
→ 009  T:  * looks prETty grEEN to mE; but e:h (1.21) rUZer
           nICE ACtually (0.31) huh
  010      (1.12)
  011  C:  o:kAY (.) could be a
           double cone
           sidewa[ys ].
  012  T:        [yea]
  013      (5.61)
  014  T:  * looks ´prETty `prET-
           ty if you ask me
  015      (0.56)
```

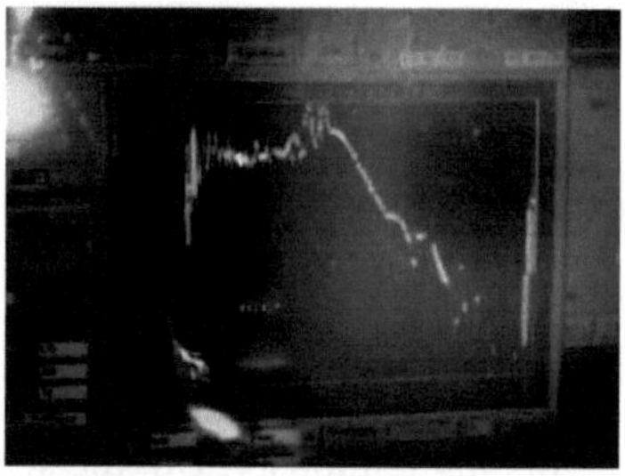

```
  016  C:  o(.)kAY=
→ 017  T:  =<<dim>but i zink it is in the green region.>
  018      (1.86)
  019  C:  okay; save that. (0.27) do you want me to blEACH
           it?
  020      (0.73)
  021  T:  i zINK we dont need thIS one.
```

There is another longish pause, after which the computer monitor displays the difference between the two intensity measurements. Theo is adjusting the scale, as the difference between the intensity distributions next to and through the photoreceptor is very small. “Looks pretty green to me,” Theo says and adds, using the disjunctive conjunction “but,” that it is “rather nice, actually” (turn 009). That is, Theo states the acknowledgment that the curve is a rather nice looking one, but there is a problem. This problem is apparent if one knows that the green photoreceptor cone is actually paired with a red cone. When Craig takes a measurement on one of these, he would announce a “double cone.” That is, we have an opposition here between the single cone that Craig has announced and the curve that Theo sees as resulting from the green member of a double cone.

After some time has passed, Craig acknowledges this possibility by saying that what he is looking at what may be a double cone but seen from the side – in which case the sampling beam would have gone through both the red and the green member. The signal from this member does not show up, however; or, rather, the team members do not articulate the graph as exhibiting this second aspect. Theo acknowledges Craig’s explanation. After a long silence, he describes the curve as something that looks “pretty ‘pretty’.” However, after Craig acknowledges this description (turn 016), Theo again uses the disjunctive “but” to introduce his assessment of the recording as consistent with a green cone – leaving open that it is a contrast to the single cone (blue, UV) that Craig describes as having seen. Craig acknowledges, requests the data to be saved, and then asks whether they should “bleach” the cone. Bleaching means shining light on the photoreceptor for about 2 min until all of the light-sensitive molecules have changed chemically. A subsequent measurement would then no longer exhibit an absorption spectrum (Fig. 3.2a). It constitutes a form of experiment where, after the procedure, the phenomenon as disappeared from hand. This change from presence to absence of the absorption curve is therefore proof that the observed absorption curve “was real” rather than artefactual. In the present instance, there is a proposal | acceptance turn pair sequence (turn 020 | 21, 21 | 22) as a result of which they do not need to bleach this one. This process would have enabled them to establish the absorption more clearly as the difference between the spectra before and after bleaching. It would have allowed them to compare two measurements through the photoreceptor rather than comparing the measurement with the reference, which has been taken next to the photoreceptor and therefore is not taking account of any absorption or effect from the cell walls and within cell fluids. In contrast to the astronomers featured in another study (Garfinkel et al. 1981), however, our team cannot repeatedly vary the phenomenon by shifting the telescope, thereby literally fails to have its phenomenon “*in hand* at all times in the inquiry” (p. 137). As the present example shows, we frequently leave out the bleaching part when we have sufficient evidence to have the phenomenon in hand: “Do you want me to bleach this one?” “I don’t think we need this one.”

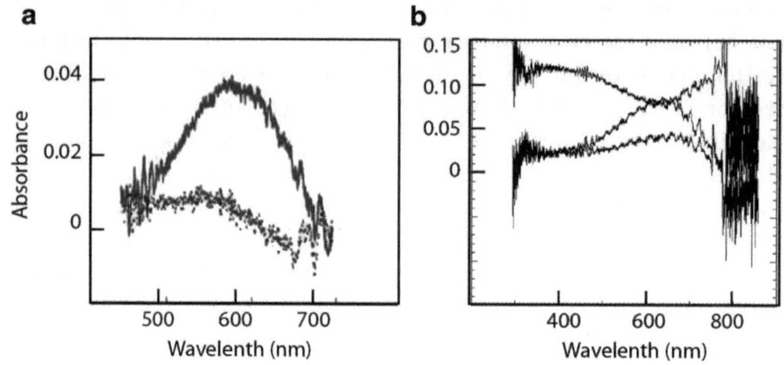

Fig. 3.2 (**a**) "Bleaching" a cone constitutes a form of experiment. First, an absorption spectrum is taken (*upper curve*). Then, the cone is exposed to light thereby changing it chemically. As the cell no longer is alive, it does not regenerate and, therefore, can no longer absorb. This is shown in the *lower curve*. (**b**) Even bleaching curves have to be selected. In this situation, the original signal from the unbleached cone is weaker (*bottom*) than that from the bleached cone (difference between the two is *top curve*)

The raw graphs seldom feature what there is to be seen. In fact, there is nothing such as a raw graph. To decrease the uncertainty about the quality of the current absorption spectrum, our team tends to scale the data so that the phenomenon exposes itself, which allows us to make a rapid decision whether to retain the measurement, which then becomes data, or whether to "chuck" the measurement, so it cannot become data. Sometimes we keep a measurement and decide later whether it should be discarded and not taken into consideration (see below). In this laboratory, certain expressions allowed me to learn when the data belonged to the ideal type, when they looked the way scientists wished the data would present themselves all of the time. The expressions included "pretty," "nice peak," "it's [looks] pretty good," "pretty 'pretty'," and "beauty." It is with reference to these ideal types that our team members excluded other measurements as irrelevant. If these did not look pretty, then measurements were tossed. Characteristic expressions marking the appearance included "I struck out on this one," "quite a bit of absorption," "flat liner," "bleached," "very hard to read," "too much in here that I want to look at," or "photo products." In each case where such a descriptor occurred, the associated measurement was discarded. That is, our team was becoming increasingly familiar with what good measurements were to look like and to be included to become data and what unusable measurements looked like so that these could be disregarded before having to be considered in an explanation of the relevant phenomenon.

Seeing Through the Clutter

In the difference curve seen after a reference and sampling measurements have been done and subtracted one from the other, there tends to be a lot of clutter. It is uncertain whether the absorption graph constitutes (usable) data or not. In this section I show the work scientists do to see through the clutter to make a first decision about whether to retain data measurement as potential data or whether to chuck it. In my field notes and transcriptions, I repeatedly entered comments of the type "there is a potential graph, but C discards it. The novice to the setting would probably not know at all what to do with this graph." As research on concept formation showed, humans learn about the nature of a concept from the contrast of instances and non-instances. In Fragment 3.2, the scientists eventually discard the measurement without providing explicit reasons for doing so. They read the measurement as possibly being consistent with some object that they know. But because it does not fit the phenomenon they are after, the measurement is not retained. The episode begins with Craig's announcement – around the time he formulates the scan – that he is looking at "either a single cone or a broken rod" (turn 011). When the first amplified images of the absorption (difference) spectrum shows up on the monitor, Theo comments, using the disjunctive conjunction "but" that he is looking at (the signal of) what looks like a "bleached rod" (turn 019). This description picks out one of the two possibilities that Craig has formulated, modifying it by the adjective "bleached," which means, a signal in the region where the rod would be expected but that is much weaker. He follows up his description by producing an extended chuckle. After a pause, however, Craig points toward a peak on the left part of the screen, querying, "what is *this,*" and ends the statement with the disjunctive conjunction "though" (turn 023). Theo produces a statement that in fact acknowledges the preceding one concerning the presence of the feature and, following a long pause, comes to name the possible peak: "UV a [alpha]," that is, the peak our research team had been after and would be reporting in the article to which these data contributed. I then ask a question about the location of the baseline (turn 015) – upon which the curve is "grafted" and which would be subtracted by an algorithm that Theo has written – to which Theo responds by providing a more extensive (than normal) reading of the possible things that might have caused the features of the graph.

Fragment 3.2

007 C: scan
008 (0.71)
009 T: <<pp>under way>
010 (2.92)
→ 011 C: now its EIther a sINGle cONE or a brOKen rOD.
012 (0.40)
013 T: * <<pp>alright> ((*modifies graph, magnifies difference*))

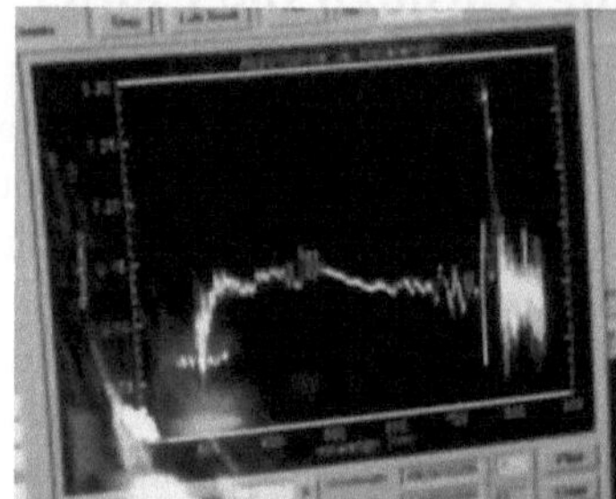

014 (10.29)
015 M: is that something that looks similarly?
016 (1.38)
017 C: yea. (0.18) it cAN look ˘similar.
018 (3.85)
→ 019 T: * but now looks like a bleached rod.
020 (0.56)
021 <<dim>hu hu hu hu hu> .hhfs

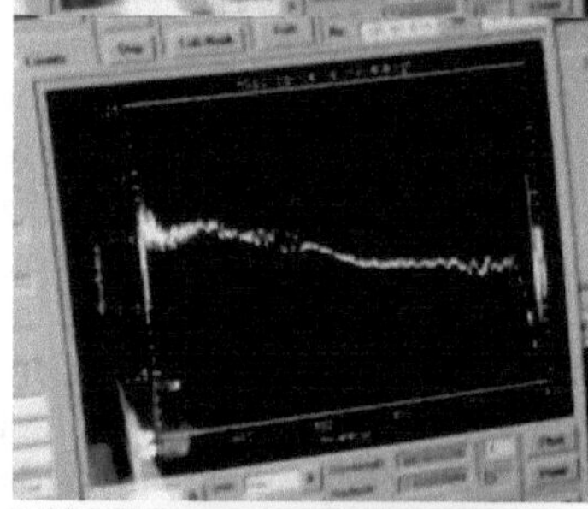

022 (1.26)
→ 023 C: * <<f>well> whats thIS though. ((*points to middle, "fuzzy peak"*))
024 (0.45)
025 T: yea.
026 (5.49)
→ 027 T: yea ze ze [u: vee a]

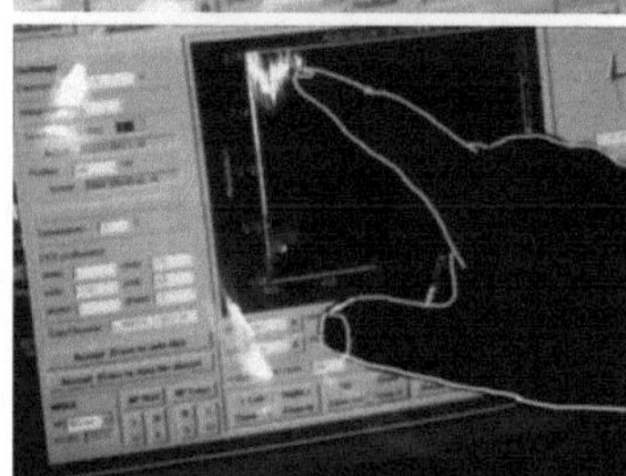

028 M: [would bASe]line be dOWn here * ((*points to "a"*))
029 (0.58)
030 T: yes. baseline would (.) would be * down here. ((*moves back and forth around "b"*))
031 (1.03)
032 T: id be one pOSsibility. (0.53) dis * ((*points to "c"*)) could be a (.) blEACHed <<f>rOD.> (0.19) dere is a little remem remnant with the photoproduct [right]
033 C: [yea]
034 (0.40)
035 T: dats one way of reading it.
036 (2.33)
037 T: ze other one is to read dis ze whole branch from <<dim>here down is something which> ((*from upper left to "a"*)) (1.04) <<f>caused by (.) in ze reference;

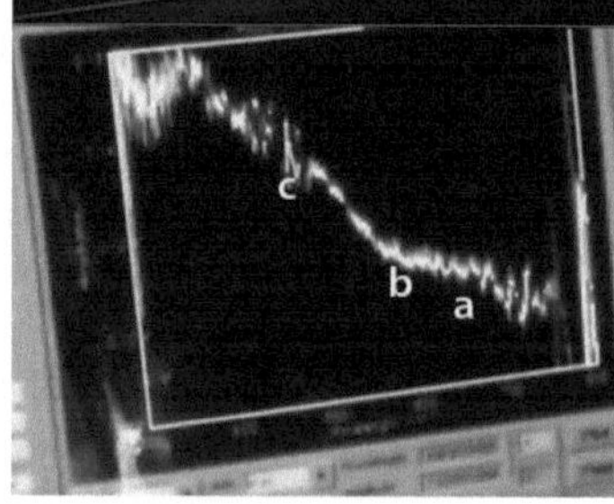

038 (0.65)
039 M: uh hm
040 T: and dat we have somezing else really going on here
((*left "peak"*))
041 (1.57)
042 T: <<p>but dat (??) in the positions here
043 C: <<p>alright i=11, (0.99) venture (2.49) ((*clack*))

He locates the right end of the baseline in the graph marked by the letter "b" and then points to the area marked by the letter "c" suggesting it could have been caused by a broken rod. Finally, he suggests that the entire "branch" could be the result of something in the "reference" (measurement) so that it is not caused by the photo-receptor at all (turn 037). Craig then announces that he is "venturing" on, which concludes the episode and starts the search for a new cone. They have not saved the data. Theo, in not challenging Craig's "decision," and by not taking the initiative to save the data on his own, de facto accepts the decision to discard these data. Subsequent to the transcribed part of the episode and in response to my question, Theo comments: "We don't think we can use it."

In this situation, the scientists "venture on." A spectrum where there is only a faint hint of a Gaussian-shaped absorption curve is removed from the data set. During the discussion of the first several months of my presence in the laboratory – amounting to over 3,000 data points – the lead scientist repeatedly suggested removing some of the measurements, which amounted to removing most measurements on some (unfortunate) days. However, we do not know what the relation is between these 3,000 data points that the scientists discuss and all those instances that they have not included while in the laboratory. In the subsequently published studies from this work, there is no hint about this relation. As a recent article in *Science* suggests, there is a lot of research conducted where scientists highlight a few positive data in the face of a sea of negative data (Couzin-Frankel 2013). Because of this suppression, other scientists may not have an inkling of an idea why their attempts to reproduce a study fail – it failed most of the time in the original study, which only reported the one time that the experiment did in fact work. Returning to our laboratory, in most situations we did actually retain measurements. This may be driven by the need to have sufficient data points for a particular phenomenon. Thus, in some data retained, the peak hardly showed up at all and was not very different from the one that was discarded in the fragment: the signal is of the same order as the noise (Fig. 3.3). As we wanted to extract the location of the maximum of the peak from the data that we had retained, we needed to clean these up – a phenomenon enabled by the nature of inscriptions themselves (e.g., Latour 1987). But the graphs themselves tended to exhibit noise (e.g., "we have some noise on top here that is the problem"), which necessitated extracting the animal (i.e. signal) from the foliage (i.e. perceptual ground). For example, scientists looked at the curves and saw them as approximately Gaussian-shaped that "sit" on an incline. Because the incline is considered an artifact, we had to "subtract" it from

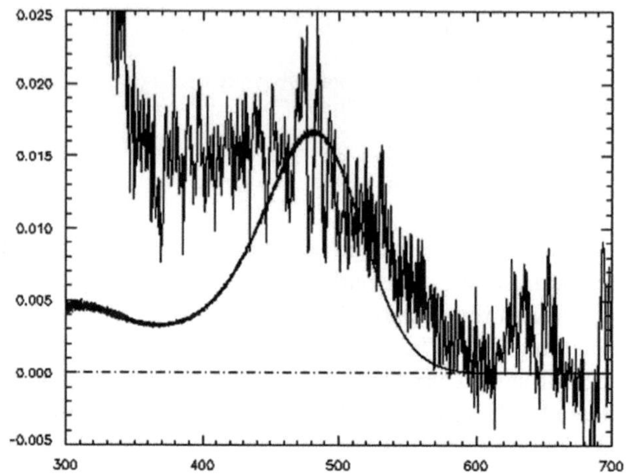

Fig. 3.3 The scientists retained this absorption spectrum although it is barely noticeable, especially for the untrained eye. The noise is of the same order as the signal. Why should the peak be retained, here modeled by a polynomial fit, whereas the other one between 600 and 700 nm would not be retained?

the actual measurement. The resulting curve is "cleaned up" or fitted in one of a number of ways. Thus, we ultimately noted in one of our publications:

> Each record was linear detrended if necessary (Hárosi 1987). A nine-point adjacent averaging function was used for line smoothing, and the smoothed curve was normalized to zero at baseline on the long wavelength arm and to one at the centre of the α-band. The fit of the normalized curve was compared with a nonlinear least-squares routine to the upper 20 % of the weighted A_1/A_2 averaged Govardovskii et al. template (Govardovskii et al. 2000) (based on the centre of the α-peak ±40 nm). (Temple et al. 2008, p. 3880)

It is from the fit that we extracted the wavelength at which the absorption curve has its maximum. In essence, in such processes we get rid of the variation in the measurements to extract what we de facto *take to be* the real data; in getting rid of the variation, we *make* the data what we *take* them to be. Because our team intended to get rid of unwanted detail in its data, we used, for example, a Fourier transformation procedure ("FFT" and "inverse FFT"). The basic idea underlying this procedure is that any mathematical function can be represented as a sum of sine curves.[1] This sum, which may consist of an infinite number of terms, is called a Fourier series. Once represented as a Fourier series, we "lop off" the higher-order

[1] Technically, any function *f*(*x*) can be expressed by means of a Fourier series. A Fourier expansion of a function *f*(*x*) is given by

$$f(x) = \frac{1}{2}a_0 + \sum_{n=1}^{\infty} a_n \cos(nx) + \sum_{n=1}^{\infty} b_n \sin(nx),$$

where the coefficients can be found by

$$a_0 = \frac{1}{\pi}\int_{-\pi}^{\pi} f(x)dx;\; a_n = \frac{1}{\pi}\int_{-\pi}^{\pi} f(x)\cos(nx)dx;\; b_n = \frac{1}{\pi}\int_{-\pi}^{\pi} f(x)\sin(nx)dx.$$

Once such an expansion has been found, scientists eliminate the higher frequencies, for example, by setting all $a_n = 0$ and $b_n = 0$ for $n > 5$ or $n > 7$, whichever leads to a "nice" curve when the Fourier expansion is reversed.

frequency terms, which corresponds to getting rid of the high frequency "noise" in the curve (the many ups and downs in the raw curve visible in Fig. 3.3). We then retransform the series by means of an *inverse Fourier transformation* into a curve that then looks similar to the original but excluding the "noise." That is, we include the measurement but exclude a lot of the variation in it.

Sources of Uncertainty

In the preceding section, we see that for scientists to know that they are seeing a photoreceptor of a specific type, they need to see a particular type of curve; but to know they are seeing a particular type of curve, they need to see a particular kind of photoreceptor. But the receptor depends on what scientists see under the microscope, the preparation they have done, what has happened to the fish since it left the tank and came to the laboratory, their instrumentation, and so on. As a result, they find themselves in a double bind where two forms of knowledgeability – one related to the visual images of the slide contents and the other one related to the graphs – mutually implicate each other. That is, uncertainties related to the object of inquiry arise in and from the praxis of scientific research. In this section, I show that if the graph they obtain and what they see under the microscope actions are not consistent and in this way can be used to stabilize each other, one or more prior actions may be called into question. Scientists realize that they had not done what they were certain about as having done; that is, there is a problem in the "doing" part of the work–gloss pair. These prior actions and the outcomes of observations (observational actions) presuppose each other and therefore stand in a mutually constitutive relationship. In this section, I exhibit the scientists' work related to uncertainty and getting some sense of control over it.

Practical Uncertainty

In discovery work, objects of interest, facts, are distinguished from other objects denoted "artifacts." Scientific research outcomes become discoveries only when the scientific community at large cannot question the nature of the scientists' actions, and thereby turn facts into artifacts – otherwise other scientists would claim that the discoverers "do not know what they are doing." What scientists want to achieve is a high degree of reproducibility. If this is not the case, when other scientists cannot reproduce some data, doubt will reign about the quality of the research and about the reality of the phenomenon reported (e.g., Collins 2001). That is, the ontology of

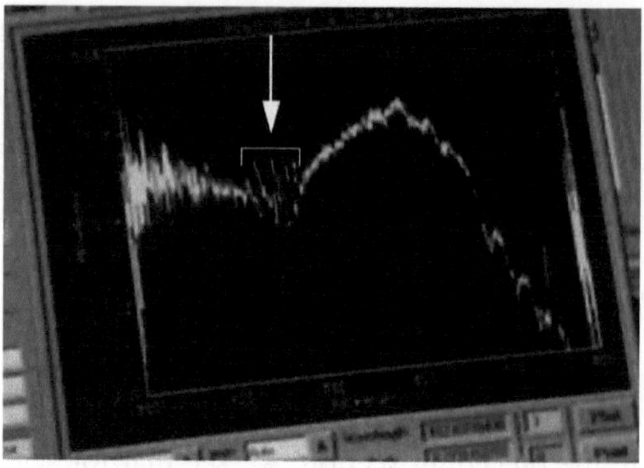

Fig. 3.4 The *arrow* indicates the "ringing" that always showed up in the data despite our efforts to get rid of it by changing the light source and by making sure everything in the path of the light was in order

actions is an emic issue in the scientific community. Artifacts are signals that look as if they refer to the natural object of interest but in fact are due to the mediation of instrumentation or are result of the specific form of the researchers' agency (Lynch 1985). In technical, scientific, and medical usage, artifacts (Lat. *ars*, art and *facere*, to make) are not present in the natural object, though, as shown here, even natural objects only exist in and as of the artful, practical actions of the scientists. If the artifacts occur together with the signal, they are troublesome because they make it appear as if the researchers do not know what they are doing and therefore are unable to get a "clean," that is, reproducible signal. Scientists therefore may devise ways that rid their representations of any additional features of known or unknown origin.

Some features of the graphs in my study of the scientific laboratory were recurrent but undesired: we had repeatedly tried to identify where they originated and attempted to get rid of them. One of the known problems in this laboratory was a feature glossed as "the ringing." The term denotes a particular feature, a strong and periodic signal that consistently showed up on almost every graph, sometimes being of the same order of magnitude as the absorption spectrum itself (Fig. 3.4).[2] We generated different possibilities for how aspects of their apparatus came to be superimposed on the signals of interest. In quote (A), Theo stated the hypothesis that the ringing is an optical effect, which he attributed here to the reflecting objective; in quotation (B), he stated the hypothesis that the ringing was caused by the flickering of the xenon lamp that produced their sampling beam. Craig also

[2] In the natural sciences, implicitly or explicitly, some recorded signal $h(x)$ is taken to be the result of a *convolution* of the true signal $f(x)$ from the phenomenon and the instrument function $g(x)$. The three functions are related according to

$$h(x) = \int_{-\infty}^{+\infty} f(t)g(x-t)dt.$$

When scientists know the instrument function $g(x)$, then they can obtain $f(x)$ by means of process called *deconvolution*. But even if they do not have $g(x)$, they tend to eliminate the influence of their instrumentation by developing better equipment.

figured that the feature was brought about by the spectral characteristics of the xenon lamp.

```
A T: Yea, this one is a feature that we are getting. It
     must be optical. . . . So what you do get is the pat-
     tern from the reflecting objective you saw.
B T: It might actually already be the flickering of the
     xenon lamp. Some people have xenon, feedback circuits
     built in so that it's, it's more stable. But now it
     is basically an arc that is radiating and of course,
     that flutters quite a bit.
```

The ringing was unwanted, particularly in its consistent appearance. Craig feared that the presence of the feature in a published graph may be read as an indication of our incompetent or, at least, of their partially competent laboratory practices. We did not want this feature to show up in our scientific publications without at least knowing where it originated – its presence could throw doubt on the facticity of the phenomenon or the skill of the experimenter. We therefore attempted to rid our inscription of this feature by changing parts of the apparatus, for example, the xenon lamp using feedback circuits or using a lamp with different spectral characteristics. Craig also asked Theo ordering and installing a new lamp. But it, too, turned out to leave what was described as an artifact intact. All of our attempts were of no avail. The ringing remained. "Doing [changing *X*]" did not bring about the desired change, which meant that the problematic feature was due to a different action. With the ringing, uncertainty, too, remained concerning the data itself and the appropriate way of extracting the relevant information – λ_{max} and half-maximum bandwidth – from the recorded curves.

Material Uncertainty

Material things are often considered to be objective because of their factual permanence. Material actions, though realized in and through the bodies of human agents, have, among others, an ephemeral quality in addition to leaving more (e.g., modified material objects) or less durable traces (e.g., sound waves) in the material world. The materiality and durability provides actions with their objective nature. Experienced practitioners may question their observational actions, doubting what they see, but they normally take their material actions for granted in the sense that they take them as aligned with the goals (intentions) that had brought them forth. If an action has not realized its goal, it is reproduced often with some slight modification (researchers "try again," implying that it will work this time). In this subsection, I provide some examples of situations where we measured only to conclude after the fact that what we thought to be the case was in fact not the case and that we had not done what we thought to have done. The measurement process itself introduced uncertainty into our laboratory work.

There were many instances where a team member, after several hours of work, began to question what he had really done at some earlier point, or simply noted that something he had done previously must have been inappropriate. That is, some prior work–gloss pair "doing [*X*]" may have been the culprit such that the work ("doing") was not conform to its gloss *X*. The member questioned the nature of his actions only after confronting some contradiction later on. Such contradictions existed in, for example, the absence of photoreceptors on their monitors or the non-alignment of visual image and graphical representation. The following fragment from the laboratory conversation exemplifies how a sequence of actions, previously taken as having done what it was intended to, was rendered problematic while searching for photoreceptors. A gap thereby was opened between the goal (intention) and the action that followed – much in the same way that there are gaps between any two paired inscriptions that constitute translations of each other (Chap. 1).

After scanning the microscope slide for a while without finding a conical photoreceptor for recording an absorption spectrum, Craig offers up a question about an action that had occurred much earlier during the day. His present observation raised doubts about "dissecting according to plan" as the appropriate gloss of what he had done.

Fragment 3.3a

```
01   C: now, struck out on that one. this dissection real-
        ly bothered me. that dissection really
        bothered me and i dont know why.
02   T: you want to go back to the old way you did it?
        where you just cut and took the lens out.
03   C: yea i am gonna do that.
```

In the morning, Craig had conducted the dissection and extracted eye and retina from it. He had continued to take a retinal piece, "macerated" it and mounted it on the slide, and then started the experiment. That is, the dissection initially remained unquestioned. Craig continued the preparation and measurement as if it had been appropriate. At that time it was a dissection that evidently had achieved what it was supposed to achieve. Now, however, Craig put the previous action into relief – it apparently had not done what it was supposed to do. In this, he changed the nature of "doing [dissecting]" after the object, at some later point, turned out not to be what it was supposed to be. Craig suggested that he did not know why the dissection had bothered him. However, about 1 min later, after I asked what it was that had bothered him, he did articulate an explanation: "Well, I couldn't visualize the fold in the eye. And therefore I couldn't open up the slip and get into the retinal– have really a limited view of what I am supposed to be going after, and that dictates the success" and "I didn't use a razor blade in the macerating process. And I notice that I don't get those fringes of photoreceptors that, that I have grown to like somewhat."

On the day following an episode concerning the problematic status of the "minimal essential medium" (MEM), and after Theo had made a new bottle of it, the action of "doing [preparing the MEM solution]" was questioned.[3] Throughout the episode, Craig used the microscope to look at the slide, whereas Theo and I were focused on the computer monitor. Craig began the episode by questioning the pH value of the solution that they had in the previously prepared "stock." He questioned whether it had been [pH =] 7.2 (turn 04). Theo responded that the pH had been 7.42, which constituted only a two-hundredths difference with what he currently had produced (turn 05). He went on to explain that it had been the stock before the previous one that had had the pH of 7.2 (turn 09). Craig asked about the total quantity that Theo had prepared (turn 11), which Theo confirmed as having been 1 L (turn 12). Responding to my question, Craig made a statement that could be heard as explaining that he was not finding any cone-shaped photoreceptors and seeing rods shrivel up, which meant that there was possibly something wrong with the MEM solution. This rendered "doing [drawing conclusion]" uncertain (turn 21).

Fragment 3.3b

```
04   C: was it seven point two last time?
05   T: no, seven point four two (0.33) and the same time,
        (1.31) not the same only about two hundredths
        difference.
06      (0.58)
07   C: okay.
08      (1.38)
09   T: it was only the time before i had the seven point
        two.
10      (2.12)
11   C: and you made up a full liter; right?
12   T: yea. (0.92) its always one (0.24) of these (0.24)
        jars of the (1.81) em=ee=em.
13      (0.58)
14   C: um.
15      (3.45)
16   M: do you see trouble?
17      (1.20)
18   C: well (2.81) its just that i=m not seeing (0.54)
        cones and i=m seeing rods (0.16) shrivel up pretty
        quickly, which means that the osmolarity of the
        medium its (2.77) possibly suspect.
19   M: um.
20      (10.95)
21   C: its hard to draw conclusion just now. (8.40) just
        not seeing the (0.39) number of photoreceptors i
        normally see.
```

In this situation, we collaborated to find a possible cause for the problem of locating the object of interest, cone-shaped photoreceptors. Not finding cones, or taking a long time to do so, is not problematic itself, though it is indeed annoying. Any number of actions could have not done what they were intended to do and

[3] Minimal essential medium (MEM) is a cell culture solution containing amino acids, vitamins, minerals (inorganic salts), nutrients (energy source), and various other components.

thereby have could have led to the absence of the cone receptors. This had been the case when Craig suspected that his dissection was problematic or that the way in which he prepared the eye prior to taking the retina from it was somehow inappropriate. In the present case, the search for the culprit action took a different direction, because there was an additional perceptual clue provided by the preparation itself – other entities currently not of interest, the rod-shaped photoreceptors, were identified as "shriveled up" (turn 18), which may have been indices pointing to the "MEM solution" and its preparation rather than some other source of trouble. If redoing a suspect part of the preparation leads to a finding, then this is a strong indication for the team that the action identified was the culprit. Redoing here functions to establish causality between the work | gloss pair "doing [making new MEM]" and its intended and anticipated outcome "doing [seeing cones]"; this causality clearly establishes material continuity and contravenes ambiguity thereby lessening the existing degree of uncertainty.

This section therefore shows that there is a gap between plans and intentions, on the one hand, and practical actions, on the other hand. This introduces uncertainty into the research process, for a cause–effect relation can no longer be ascertained between plans and practical actions. The situation is irremediable, as this gap exists in principle (Henry 2000). The different individuals wanted to do something and believed they had done it – such as Craig with his dissection – and continued as if they had done it only to reconsider it sometime later. Even more so than the cases Suchman (1987) discusses, the scientists not only had plans but also completed actions being convinced that the former appropriately described the latter. Sometimes hours, sometimes months later, they are confronted with reasons to reconsider and renounce this relation. Sometimes, for example, it might have been in the late afternoon after having spent hours in the laboratory collecting data or rather doing what turned out to be "trying to collect data." At other times, it took several months until Shelby found out that Sam and he had not extracted the cells that they had intended to extract. One of the things scientists thereby come to be confronted with is how to loose their phenomenon (Garfinkel 2002). They lost it, but not quite in the way that Garfinkel and his colleagues had lost theirs while attempting to redo Galileo's experiments on the inclined plane: they had "no idea of why, or how, or what [they] had done to lose it, or where to look to find it again" (p. 276). Craig had lost his phenomenon without knowing it, but once he came to the conclusion that it *was* indeed lost, he quickly articulated a reason why, when, and where.

Uncertainty of Tools and Instruments

Tools and instruments mediate practical action (Fig. 3.5). Uncertainty therefore does not only pertain to the things created by the scientists' actions, but also to the means of production (tools, instruments) that are used in, enable, and therefore mediate individual actions and the activity as a whole. If scientists are uncertain about the relation between their intended and practical actions, then ambiguity may

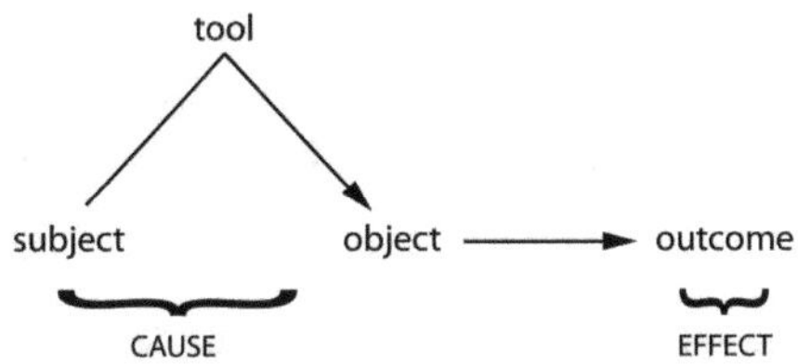

Fig. 3.5 In this activity theoretic model, the subject acts upon the object using tools to bring about some outcome. If the outcome (effect of action) is problematic, then the trouble may have its source in the tool or in tool use (action)

also be arising from a tool or instrument of scientific production. The following two episodes exemplify this point.

Craig noted that the photoreceptor cells appeared in huge clusters under the microscope rather than being spread out; he ascribed this to using a scalpel rather than the previously employed razor blade in the process of maceration. The scalpel therefore was not doing – in his hands – what he thought it was doing. Importantly to the argument of this chapter, this knowledge was not available to him as the action unfolded; it only became available through some contradiction that appeared much later in the sequence of actions, that is, in the outcome (Fig. 3.5). Although Craig formulated to have been thinking at the time as doing something, he now was uncertain whether he actually had done what he had believed to have done. However, the uncertainty here derived from the tool – which interacted with the object at hand in a different way than the tool he usually used – rather than from his action. In the following episode, the (indescribable) way in which the sample looked to Craig's eyes led him to question the solution in which the retina is kept after excision. In this case, the MEM solution had been appropriate on previous days, so that it is not the making that is questioned but the degradation occurring in time. Here, usage of an old MEM solution is the action that is being questioned (turn 22) *after* several hours of investigation and after already having questioned the way he had dissected the first fish.

Fragment 3.3c

```
22   C: is this last weeks em=ee=em?
23      (0.95)
24   T: yes.
25      (1.78)
26   C: ((While he is looking through the microscope.)) i
        think we ought to make new em=ee=em.
27      (0.30)
28   T: yea, i=ll make new. (1.17) today is (0.15) yea.
29      (0.87)
30   C: this is roughly a week away from when it was made.
31   T: yea; it is about a week now.
```

After Theo confirmed that the MEM solution had been prepared the previous week (turn 24), Craig and Theo enacted a suggestion | acceptance pair in offering "ought to make new MEM" (turn 26), "I'll make new . . . today" (turn 28). In the following two turns, the two confirmed their common ground with respect to the fact that the MEM solution was about 1 week old (turns 30, 31), which is a suitable gloss for its deterioration and for preparing a new one. Here, it is uncertain whether MEM was doing what it was supposed to be doing in the scientists' hands – preserving the retina in its container and the sample on the slide – leading to the work involved in "doing [making a new solution]." Production means mediate activity and actions. The results of actions therefore bear the marks of this uncertainty in production means (tools, instruments), though scientists normally make these marks disappear in the same way that they bring about the disappearance of their own actions.

From Scientific Action to Scientific Fact

At various stages in the philosophy of the hard sciences, it has been recognized that natural phenomena are bounded by theory and, particularly relevant in the present context, by human activity. Thus, asserting that the ontology of natural phenomena is circumscribed by human activity is equivalent to denying "the hallowed independence of the world of representations from the world of embodied practices" (Gooding 1992, p. 66). The data presented in the preceding subsections show that human actions are indeed bounded by the material nature of the human body, which belongs to the world of natural objects and objects it creates (artifacts). Actions, though seemingly ephemeral in some respects, have the same material nature as objects and inscriptions, as they are constituted by the materiality of the human bodies that bring them forth. The nature of the action is of the same type as that of the objects, and therefore both are involved in their mutual stabilization and constitution that is taken to reduce the uncertainty in and of each.

This dialectical – i.e., mutually constitutive and implicative – nature of objects (the result of "doing [seeing photoreceptor]") and material actions ("doing [X]") also entails that the contents of the microscopic slide (the outcome of "doing [seeing photoreceptor]") and the graphical representation of the absorption spectrum (result of "doing [seeing spectrum]") stand in a constitutive relationship, that is, they mutually presuppose, implicate, and stabilize one another (Fig. 3.6). What it is that has been observed on the microscopic slide cannot be known until the corresponding spectrum has been produced; but the spectrum cannot exist without something on the slide that causes the two raw spectra to be different. If the two do not stabilize, or if they stabilize but with respect to an entity of artefactual nature, then laboratory members frequently search for, and render uncertain, what one or more actions really rather than presumably have done. They may see something on the slide, but if a corresponding signal in the graph is missing, they remain empty handed. Similarly, if there is something in the graph without a corresponding something in the image from the microscopic slide, they are empty-handed again.

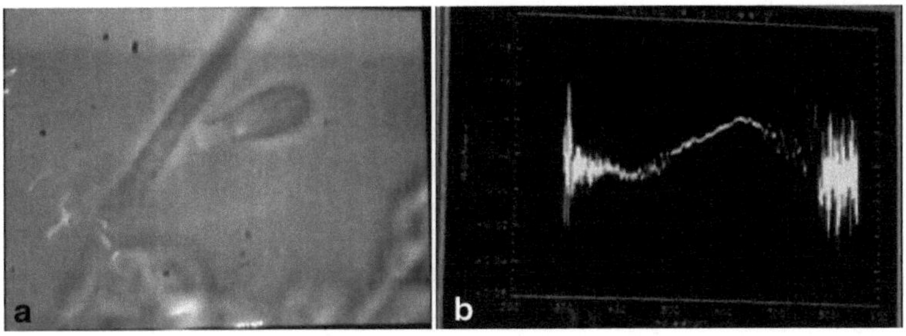

Fig. 3.6 The image of the microscopic slide contents (**a**) and the graph (**b**) mutually presuppose and implicate each other: To know what is on the slide scientists need to know what is in the graph and to know what is in the graph they need to know what is on the slide

As a result, they question the nature of the actions that has led to whatever is on the slide, that is, its precise nature (i.e. its ontology).

These results do in fact explain why scientists have tremendous difficulties interpreting graphs, even if these come from undergraduate textbooks in their own field (Roth and Bowen 2003). When experienced scientists had trouble saying much about what graphs are intended to express, they generally had questions about how the data were collected, whether the graphs presented real data, where the data were collected, what instrumentation was used, and so on. That is, the scientists in that study appeared to lack the knowledge and disappearance of uncertainty that is associated with the mutual stabilization of the actions along the continuum of scientific production.

At issue here is not the fact that actions can be subsumed to the paradigm of the text and therefore are subject to the phenomenon of *interpretive* flexibility of artifacts as the source of uncertainty. Existing studies of the uncertainties and ambiguities associated with the operational deployment of technology and its effects – while good in articulating the social constructivist issue of interpretive flexibility (e.g. Bijker et al. 2012) – *cannot* get at the present issue, because they study *effects* rather than the performative quality of actions unfolding in real time – which I have marked as work associated with "doing X." The real issue is that scientists in a very strong sense themselves do not know what they do all the while the artifacts and tools (technology) they use in the process have quite durable and certain character to them – not until they have the outcomes (effects) of their action in hand (Fig. 3.5), which provides them with some level of assurance that they actually have done what they had intended to do. Interpretive flexibility of technology is the *result* of the fact that practical action is unknowable in principle, so that its outcomes never can be anticipated with certainty. Uncertainty is irremediably inherent in scientific (tool-mediated) action. The art of the scientists lies in making outcomes sufficiently repeatable so that the irremediably inherent uncertainties are minimized. Quantum physics and its uncertainty relation may serve us as analogy here, for it provides a frame for thinking about uncertainty as inherent in any form of action.[4]

When Is Data Legitimately Excluded?

In the preceding section, I describe how scientists exclude data along the way even before they get to the analysis of the location of the peaks of the spectra; we also see how uncertainty arises from the very practical work that they conduct to get the data. To know what they have under their microscope, scientists need the graphs that allow them to distinguish among different materials on the slide, but to know whether there is something in a graph they have to know what is under the microscope. What the nature of the measurements included is therefore depends on the nature of the measurements not included. The nature of the data does not derive from the measurements themselves: even if unacknowledged, there is always an irreducible figure | ground relation. Distinguishing this figure (data) from ground (noise, background) – where ground is necessary for the figure to appear – is part of the data collection process that allows scientists subsequently make a knowledge claim. Here, they do a first selection in the laboratory. If there is too little evidence that the data "meet inclusion criteria" (Craig), then these are not even saved. Later, as discussed in the session analyzed below, further exclusion criteria are made operative. Thus, Craig suggests excluding all those absorption curves data that have a peak at a wavelength below 503 nm ($\lambda_{max} < 503$ nm), that is, less than what previous research has reported to be expected for the vitamin A_1-based chromophore (absorbing chemical) and everything above 527 nm, which is more than the expected vitamin A_2-based chromophore. In this way, the measurements included would be selected based on the results of previous research. However, if the true range of the wavelengths were to be different, then the scientists would have already eliminated data that could have been used for revising the accepted range of the maxima for the absorption curves. In this section, I describe how our scientific research team came to exclude some measurements from the data and how others entered. *By excluding measurements, scientists in fact reduce uncertainty associated with the quality of specific (sets of) measurements that would have to be considered in the constitution of the data presented in an article in support of a knowledge claim.* What will show up in the graph ultimately published – see description in later chapters – will be based on a selection no longer available to the readers of the study that the various author collectives from this laboratory ultimately published.

[4] The uncertainty principle states that the product of the uncertainty of complementary variables such as position x and momentum p always is greater or equal to some constant: $\Delta x \cdot \Delta p \geq \hbar/2$, where $\hbar$ is the reduced Planck's constant.

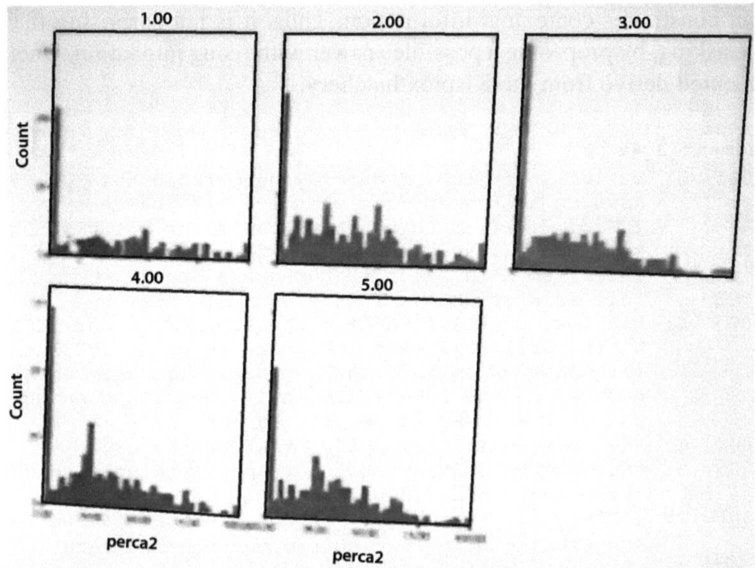

Fig. 3.7 One of the slides projected at the data analysis meeting presents, for coho salmon from five different contexts, the distribution of photoreceptors according to the amount of porphyropsin (%A_2) in the retinal rod-shaped photoreceptor cells. The amount is derived from the position and width of the absorption curves such as the one shown in Fig. 3.3

In the present situation, Theo suggests retaining these data until after they have a better sense about the quality of their data. This would be consistent with an orientation of retaining judgment until a better explanation has been arrived at. It also keeps open the possibility for revising the scientific canon with respect to the range of possible λ_{max} values. At the moment, the decision which data to retain is based on the curves that other researchers have published, whereas Theo proposes getting the quality of the data so high that they themselves can decide which λ_{max} ("lambda max," wavelength where absorption is maximum) to take, and, therefore, to establish their own scale for the A_1/A_2 ratios. This part of the meeting begins when, following a comment about the variability of the data, Shelby presents the results to the other team members (Fig. 3.7). It is a series of five histograms. As Shelby says, these are "batches" of data collected in 2-week intervals from fish that the laboratory received from the Kispiox fish hatchery.

The videotapes show that relevant to the interpretive work of the scientists is their knowledge of where the specimens are coming from, that is, from which river system or hatchery they derive. Without knowing where the data come from and how these were collected, this group of scientists struggles. This is so because, as seen in another study, even the identification of the species of a specimen depends on contextual information: a group of scientists could not distinguish between the young of three or four species unless they knew *where* in the river these young have been caught (Roth 2005). That is, scientists could not classify the fish unless they knew where it was caught, which is a property external to the specimen and, therefore, constitutes contextual information. Thus, it is not surprising that Craig would be asking, by proposing a possible answer with rising intonation, whether the data presented derive from the Kispiox hatchery.

Fragment 3.4a

```
  001  S: so ive just kind of quickly summarIZed this um we
          have um hIStograms for what ive call bATches
          ((Fig. 3.7)) um so you have the first batch which
          came in well and two weeks lAter and two weeks
          lAter, so EAch bAtch is two weeks apart.
  002  C: this is kispiOX?
  003  S: kispiox. um ALl fISh from the kISpiOX so our fIRSt
          batch april thIRteenth this was our distribUTion
          for ALl fish um individual rods s histogram um all
          plotted on the same scALe so by the sECond batch
          and what we sEEM to be gETting here is
→ 004  C: kay so bt whY: thAT spike of A:ONe there?
  005  S: this is where i say that perhaps we=re getting
          these individual [rODs]
  006  T:                  [kkm ] the problem is that we we
          ARe wORking with them on the munznbEAtty dAta
  007  S: oh no; this is; this is not both.
  008  T: this goes from fIVehundredzrEE to
          fIVehundredtwENtysEVen
          nANo[mEters      ]
  009  S:     [<<p>alright>]
  010  T: and so if we have sOMething lets say at say
          fOUrninetyEIght or fourninetySEVen that would be
          in the tALL bAR in there
  011  S: yea
  012  T: so,
  013  C: uh um
  014  T: so what we really have to decIDe is what ze rANge
          is [of the ]
→ 015  C:    [well it] its lOWer than fIVehundredandthrEE
          then it doesnt belONG in there.
```

Shelby confirms the origin of the slide as coming from the Kispiox, that he received the fish on April 13, and that the graph presents the distribution for all individual rods from which measurements were taken. Not articulated because available to all gazing at the graphs is the abscissa bearing the label "percA2" (i.e., % A_2) and the ordinate being labeled "counts." The laboratory members present also take the fact that the second graph entitled "2.00" is the "second batch" to be the result of Shelby's moving the cursor to this graph while naming it (turn 003). Craig asks about the "spike of a = one," which, as can be seen from the plot, is a high count for the first bin of the histogram corresponding to 0 % A_2 (porphyropsin) and, therefore, equivalently to 100 % A_1 (rhodopsin) (turn 004). Shelby and Theo, who had collected and processed the data, respectively, take turns in making statements that explicate the situation. First, Shelby begins by making an attribution to the effect from individual rods, but Theo points out that they are working with the "Munz and Beatty" data, that is, with the algorithm for determining the A_1/A_2 ratios from a given polynomial regression equation with the specific

λ_{max} that they determined from the data. This determination itself requires one of several possible mathematical procedures for approximating the absorption curve – a published seventh-order polynomial, curve smoothing, or a process of removing high-frequency parts of the curve through Fourier and reverse Fourier transformation. Theo points out that the Munz and Beatty regression curve is based on the 503 and 527 nm as the λ_{max} for vitamin A_1 and A_2, respectively. But in our data there are curves with $\lambda_{max} = 498$ nm or $\lambda_{max} = 497$ nm. These data "would be in the tall bar in there" (turn 010). He also suggests that our team has to make a decision about the range of, but he does not succeed in completing his statement as Craig interrupts him with the categorical statement that "if its lower than 503 [nm], then it doesn't belong in there" (turn 015). As the Munz and Beatty (1965) study had been conducted, among others, on the five Pacific salmon species (*Oncorhynchus*),[5] including coho, Craig's statement makes this to be a strong reason for excluding data that appear to suggest maximum absorption lower and higher than the range set in the 35-year-old study that is one of the referents for our work.

Although Shelby and Theo appear to accede, they also articulate further reasons for retaining the measurements under discussion. Theo, who is not a biologist by training, makes a statement that we can hear as wondering whether the range that Munz and Beatty had offered is "exclusive" for the coho, thereby implying that there might actually be a different range for the coho. He states not knowing what the range of the λ_{max} would be for the coho (turn 020) and that they do not yet have sufficient or "sufficiently good" data to "decide [them]selves" (turn 022) and, therefore, where to expect these to lie (turn 024).

Fragment 3.4b

```
016   T:  <<p>yea right>
017   C:  its
018   T:  you see what I dONt know is this is
          fIVehundredthrEE t fivehundredtwentysEVen is the
          um exclUSive rANge which which is pOSSible for ze
          for ze coho
019   S:  yea
020   T:  i dont know what the cOHo curves where from where
          to where they do go,
021   S:  yea; thats right a [good point]
022   T:                     [and our   ] data isnt gOOD
          enOUgh yet to decIDe oursELVes which
023   S:  yea
024   T:  where we expECt it to [be]
```

[5] Atlantic salmon is from the genus *Salmo*, whereas Pacific salmon is from the genus *Oncorhynchus*. There are actually six species from this genus in the Pacific Northwest, five bearing the common name salmon (chinook [*O. tsawytscha*], chum [*O. keta*], coho [*O. kisutch*], pink [*O. gorbuscha*], and sockey [*O. nerka*]) and one with the common name rainbow trout (steelhead [*O. mykiss*]).

025 S: [i:] think i think that
thats a rEElly good pOInt that if we were to have
it; if we were to get rID of thAT say no NO more
zeros ye actually can go below fiveothrEE then
weed start to sEE that the cURves are a bETter
shAPed; ((*gestures an inverse parabola in the
air*))
026 T: <<p>yea>
027 S: becOS that thats a rEALly good pOINt cos we dont
know for sure that the protein in coho is the sAMe
as what <<dim>munznbeatty hve done it for other
fish like rainbow trout and salmon and so it> mAY
BE slightly different.

Shelby's statement supports the one Theo articulated. It suggests that if the team was to plot the measurements below the 503 nm minimum wavelength then the curves (histograms) would exhibit better shapes (turn 025). He accompanies this suggestion by an iconic gesture that outlines a slightly skewed Gaussian curve, as visible in the histograms displayed (Fig. 3.7). His statements further suggest that the coho salmon that we investigate may in fact have some differences in their protein that is part of the absorbing molecule so that the λ_{max} of the associated visual pigment might change with respect to the data that Munz and Beatty (1965) provided and approximated with their regression equation for determining A_1/A_2 ratios. Theo makes a statement to the effect that he does not trust the first and last one or two bins of the histogram (turn 028), and Shelby affirms, "that's a really good point" (turn 029).

My own statement in turn 032 pertains to changes to the curves that might be observed if the bin size were enlarged to 10 % and whether the expected shifts in the A_1/A_2 ratios that are expected over time would be better visible. In response, turning my statement into a question, Shelby's statement asserts that the distributions were already visibly shifting to the left, which means, a shift toward less A_2 (porphyropsin) and more A_1 (rhodopsin) as would be expected from fish at that time of the year just prior to migration.

Fragment 3.4c

028 T: so I wouldnt trust the first ten to five percent
of the and from ninety <<p>five plus or some
[thing like]>
029 S: [yea no thATs] a really good pOINt
030 T: <<p>zats really out of range rEAlly>
031 S: so that thats
032 M: ee if you take a bIN size you have fIVe now?
033 S: uh bin sIZe of yea [fIVe]
034 M: [fIVe] if you took tEN would
the the cURves come out clEARly and shift alONG ah
as we go through tIMe?
035 S: u::m well they they dO kind of now anyways; so
well, the first ones, first ones our very first
day so i=m you know not; this one here maybe not
but; these ones hERe seem to be slOWly shIFting
towards the rIGht u lEFt rather; towards more a.
and thIS one hERe is looking pretty nICe almost
→ 036 C: yea thATs thats nice. i mean i i=m hA:Ppy with
what i:ve see there; thats nICe data.

It is perhaps his status of an outsider to the community of biologists that makes Theo less susceptible to the strong disciplinary constraints of the reigning paradigm in the field of biology. In Craig's case, whose career to that point spanned a 30-year period has been mainly within the field of salmonid fish vision and its paradigm, deviating from the paradigmatic canon may be more difficult. The effect of this canon may have been particularly strong, because its originator, George Wald, had been a co-recipient of the 1967 Nobel Prize in Physiology or Medicine. This same pattern can be observed in the videotapes when our team explains the data of changing A_1/A_2 ratios in the course of the life history of the coho salmon, as these unfolded during our research project.

In the end, we find out what good data are to look like, and measurements that otherwise are excluded, first in the laboratory where our team is assessing the absorption spectra and discard those that do not fit what we want. Here again, Craig makes a statement that suggests that certain data points "do not belong here" and articulates an assertion concerning what should be included because it constitutes "nice data" (turn 036). There is a contradiction not made salient by any one of our group present to the meeting: the fact that the team already has excluded many other measurements that might make the results look even less nice, and that Craig further suggests to remove all those data that lead to the high peak for the bin in which the amount of A_2 in the photoreceptor is between 0 and 5 %. That is, the data look nice because we have made them look nice not because these are inherently nice.

Theo and Shelby can be heard acquiescing: Craig thereby comes to be reified as being categorical about excluding the data points below a cut-off point suggested by other, much older research. Craig then listens to the discussion among other laboratory members. But it is not just an arbitrary decision to drop data. Because our team ultimately has to defend its decision – when attempting to publish the study – we have to be able to articulate a set of reasonable inclusion and exclusion criteria that our reviewer peers can buy into.

The issue about inclusion and exclusion of measurements was not settled but came up repeatedly during the 2-h laboratory meeting. Thus, for example, some 30 min after Fragment 3.4a, the issue became again the topic of our talk. The fragment begins just after Elmar has asked about the last data point for the Kispiox hatchery, something of special interest to him because he has done a lot of research on fish in this geographical area. Shelby notes that the last batch had arrived on June 12 and that he had already shown it prior to Elmar's (late) arrival (turn 037). Shelby states that there are a number of measurements with $\lambda_{max} < 503$ nm and, as before, suggests that these data "should not necessarily [be] discounted" (turn 039). He begins by pointing out that especially with the data we do have so far, one would get some particular result. But Craig insists again on the point that anything that does not meet the criterion of being "within the window of λ_{max}" "ha[s] to be rejected" (turn 42).

Fragment 3.5a

```
037   S:  <<pp>which is, this (.) this is one> batch five
          was the twelfth of june; yes so that was our last
          batch, so the one i showed you was indeed the last
          one i guess
038   E:  and it was?
039   S:  so what; what I would deal with eventually what
          was really interesting was that um this is a
          matter of fact to the fact that there is some
          readings that are below fiveothree; and that we
          shouldnt necessarily discount those. and that they
          may not actually bE fiveothree; but youre gonna
          obviously get some ((  gestures triangular shape   )),
          specially with that kind of anal[ysis]
040   E:                                      [yea ] yea
041   S:  <<dim>you get some of [the]>
042   C:                        [so ] if if they dont if
          they dont meet the cRItErion of bEing within the
          wINdow of a lambda max then they have to be
          rejected
```

Theo responds, making a statement that argues for retaining the data. He formulates not being sure that the range is the correct one and proposes to use a different range for retaining measurements: 495 nm $< \lambda_{max} <$ 535 nm, which in fact extends the heretofore accepted range by eight points above and below (turn 043). Craig articulates twice the oppositive conjunction "but," which can be heard as insisting: had they checked the data for coho salmon in the Munz and Beatty (1965) article? Shelby asks whether this study has in fact coho data, to which Craig responds asking whether they have gone to the literature to find out. That is, Craig does not insist on asserting that Munz and Beatty *actually* have coho data but asks whether they (Shelby, Theo) have looked into the literature more generally. Shelby says that they have been in the lab and therefore "not gone so deep into it yet" (turn 047). Craig also offers the suggestion that they go to the paper by Alexander et al. (1994), which is the one on which our entire research project is based.

Fragment 3.5b

```
043  T: yea, no doubt, a few weeks ago i am not sure that
        fiveothree to fivetwentyseven is the correct
        range. i would say its fournINEtyfive to
        fivezirtyfIVe is the correct range
044  C: but but did you go to mUNznbEAttys coho?
045  S: um not yet do they, do they have coho? munznbeatty
046  C: well i mean did you gO look at the literature and
        see whether
        the[re is ]
047  S:    [we hav]ent been out of the lab; so we havent
        gone thAT deep into it yet.
048  C: um
049  S: but we obviously need to goo
050  C: maybe look at alexANders papers too
```

Shelby notes that "they just go to the standard, five-o-two that could be the technique they are using too and they use a raw average mean" (turn 051). Not only is the number 502 different from the number 503 that the previous speakers had articulated, but also the Alexander et al. paper does not at all mention such a number. The paper refers to the same Munz and Beatty study that already has been discussed in this meeting. Craig then accedes in the sense that he gives a reason why Theo might be correct that there are maximum wavelength peaks below 503 nm (turns 054, 056). He refers to the possibility that there could be variations arising from difference within the chromophore-binding counterion pocket (turn 058). Changes in the amino acid sequences near this pocket may result in changes of λ_{max} of the visual pigment. That is, he articulates a detailed knowledgeability with respect to the chemistry associated with the rhodopsin (A_1)-associated vision processes that has been a central research issue of recent decade. But he then insists on removing those measurements from the present analysis, giving a particular emphasis on the center part of the verb. Theo, in turn, insists on his preference for retaining curves with λ_{max} values being 5 nm above or below the currently accepted range (turn 059), and Shelby – in using the confirmative "yea" followed by the conjunctive "because" that is followed by a reason – apparently supports this position (turn 060). What we do not see here is a discussion of the fact that the Munz and Beatty study provides an algorithm for establishing the A_1/A_2 ratio given that pure A_1 has a $\lambda_{max} = 503$ nm and A_2 pure has a $\lambda_{max} = 527$ nm. What would it mean for A_1/A_2 ratio if $\lambda_{max} < 503$ nm or $\lambda_{max} > 527$ nm? This question cannot be answered unless our group was to establish a different range of values with an adjusted regression equation to estimate the appropriate A_1/A_2 ratio.

Fragment 3.5c

```
  051   S: yea they they just go to the standard fiveotwo.
           but that could be the technique they are using too
           and they use a raw average mean so theyre not
           using the actual is small point for the individual
           rods
→ 052   C: yea which maybe,
  053   S: <<p>which isnt bad>
  054   C: no i i knOW what you are saying
  055   T: yea
  056   C: theodore there maybe that maybe um variance withIN
           the OPsin; the perfORmance within the OPsin. um
           molecule.
  057   S: and it could also be the experiment
  058   C: and its could. also. it could it may have
           something to do with chromophore binding within
           the uh um the counterion pocket.((Throws up hand,
           as if saying "who knows?"))) um. ((As if think-
           ing)) there could be a variety of things
           explaining those short wave length lambda maxes
           but I think for pURposes of analyses that has to
           be remOVed.
→ 059   T: i would prefer to drop five percent and above the
           five percent to fiveninetyfive and the rest.
           <<pp>id prefer that do this>
```

```
060  S: yea; because is the same with the other end too.
        because there
        is [that one point yeah]
061  T:    [where it is higher ]
062  S: its just tapering ((gestures far toward his
        right)) off at either end. yea. if we got rid of
        the if we get rid of thOSe two ends it starts
        looking very normally and distributed. its slight-
        ly skewed.
```

Shelby elaborates that there is a similar issue "at the other end" (turn 060). A closer inspection of the histograms (Fig. 3.7) shows that in each of the five "batches" the very last bin is indeed higher than those to the left. He states that removing the two ends would make the curves "start looking very normally and distributed" though they remain "slightly skewed" (turn 062). Shelby does not say that the measurements should be removed. It remains open whether our team should use an extension and the resulting change in the shape of the histogram, because the data are "tapering off on either end." Extending the acceptable range would produce normal curves that are slightly skewed – toward a higher percent value of A_2 and, therefore, to longer wavelengths, as seen in the histograms.

The publication resulting from this study ultimately will show that our team is going to retain the lower wavelength limit of $\lambda_{max} = 503$ nm but accept longer wavelength maxima on the other end (Temple et al. 2006). In our publication we describe what we have done as using a different than the heretofore-used Munz-and-Beatty algorithm for estimating the A_1/A_2 ratios (i.e., Govardovskii et al. 2000). Our article will state having used the Munz and Beatty algorithm as a second estimate and having used the average of the two for deriving the relative amount of A_2 present (in %). The more recent paper had not done measurements on salmonids but published a general algorithm based on the observation that across a broad range of animal species, the shape of the absorption curves is independent of the λ_{max}/λ ratio. Based on the data Munz and Beatty (1965) had published for coho salmon, our team derived, using a least square regression, a third-order polynomial for the determining the position of λ_{ma} and therefore for the A_1/A_2 ratio. Biologically, the explanation given in Shelby's dissertation and the associated scientific journal article is in terms of the broadening of the spectra towards *longer wavelengths*, consistent with observing $\lambda_{max} > 527$ nm, whereas there are no processes that would explain the observation of $\lambda_{max} < 503 \pm 1$ nm.

In summary, as exemplified in this meeting, our research team addresses the issue of the abnormally high counts in the first and last bin of the histogram. These are the result of the fact that Theo and Shelby have counted all data with $\lambda_{max} < 503$ nm as indicating the presence of 100 % A_1 and counted all data with $\lambda_{max} > 527$ nm as indicating the presence of A_2 even though the previously established curve maxima for the two chromophores are 503 and 527 nm, respectively. The statement Craig made is equivalent to saying that the corresponding absorption curves – even though these might look "nice" and fall into the category of "beauties" – do not meet inclusion criteria and therefore should be excluded.

They should be excluded even though there is a possibility that the maxima shift because of chemical processes or because of some other reason. For the purposes of their present analysis, they should be excluded. The two individuals less enculturated and invested in the canon (Theo, Shelby) oppose this recommendation and express the preference of retaining the data.[6]

Knowledgeability and Familiarity

In this chapter we see how scientists produce the data by including or excluding some but not other measurements. Any scientific claim, therefore, will be the end result of transformations that begin with pieces of natural matter that come to be transformed along a chain consisting of different material | form (see Chap. 1). Eliminated from the data will be many of those measurements that introduce uncertainty. If a real knowledgeable use of the graphs requires familiarity with the original phenomenon and the transformations through which it is turned into a scientific fact, then knowledgeability with respect to graphing means something like being able to make the symbolic ascent from the claim to the original setting in which measurements have been produced. This chapter presents an investigation into the scientific practice of data generation in the face of inevitable uncertainty for the purpose of reflecting on the design of science (and mathematics) education.

In the first part of this chapter, I show how uncertainty arises from the very engagement in the research process. Both actions and tools (instruments) are sources of uncertainty. In the second part of this chapter, I present several episodes that show how scientists retain measurements even when there is a contradiction between the visual assessment (single cone) and the graph (green member of double cone). The scientists do retain measurements when they are more-or-less certain that they can make use of these in an ensuing publication. In the featured episodes, they decide to conduct only part of the measurement because they already have sufficient information with respect to this object. They have their phenomenon in hand even if they do not bleach and thereby destroy the photoreceptor to see whether the signal disappears. Our research team also can be observed discarding a measurement even though there is evidence that we retained elsewhere data of a very similar quality. In discarding it, the measurement does not even enter the consideration of shaping the data used to support the research claims about the phenomenon at hand. In such choices, scientists shape the data as much as receiving them from nature. What we ultimately work with differs from merely dealing with error variance. Therefore, the data ultimately made public – including both true and

[6] Disciplinary vision is a dialectical phenomenon: it is required to seeing certain phenomena and, in so doing, no longer sees (other) phenomena. Foucault (1975) articulated the general relation between discipline (imposition of force) and discipline (form of knowledge), and my own research showed the process in operation in the becoming of ecologists (Roth and Bowen 2001).

error variance – is set against the "non-data" that in fact remain invisible because the associated measurements have already been discarded. The phenomenon that we ultimately report, therefore, rises as figure against the ground in what our research group presents. This itself is set against an invisible ground of all the possible responses that our team obtained when we probed nature.

In the third episode, our team is confronted with the results of the earlier selection. The episode shows how in the face of existing experimental results, the chief scientist requests chucking out all those of the remaining data that do not fit the paradigm – here locations of the peaks below 503 nm and above 527 nm. Although the team members who collected and processed the measurements suggest retaining these until they know more, Craig states what can be heard as arguments in favor of excluding these based on the scientific canon. There is a tension, however, because this very project, in its totality, ultimately will overthrow the canon on the variations in the composition of the photoreceptor molecules (between rhodopsin and porphyropsin). It would eventually turn out that our team loosened the requirement for the upper boundary – without providing information as to how this decision affected their assessments of the A_1/A_2 ratios – but we did not change the lower boundary of acceptable λ_{max} values.

It is evident from the present analyses that our research team made its selections based on both an intimate familiarity with the laboratory equipment and the entire process by means of which retinal tissue came to be transformed into %-A_2 distributions. Our team as a whole exhibited an orientation to the field in attempting to adhere to the canon even when the data themselves appeared to contradict it. Of course, we did adhere to the canon by using legitimate equipment or by extracting A_2 ratios from the λ_{max} determined by templates rather than, for example, by a best-fit polynomial grounded in the measurement points themselves. At various stages in the process, measurements were dropped and thereby became invisible in and to the production of the phenomenon – which is always based on the measurements retained rather than those that are excluded from consideration. Not that this team differed in this practice – scientists talk about this practice in one of their two flagship journals (e.g. Couzin-Frankel 2013). As one well-known geneticist told, they selected among all those experiments and measurements done those that produced a good story (Suzuki 1989). They select from all the measurements conducted. Within the retained measurements, the phenomenon came to stand as figure against the ground (unexplained variation). In this chapter I show that even to the actual transformations of the measurements, scientists do what they can so that "order is not simply constituted" but it is "exposed, seized upon, clarified, extended, coded, compared, measured" so that it can in fact be "subjected to mathematical operations" (Lynch 1990, p. 163). Measurements that can be anticipated to resist the processes of order production are simply excluded as unsuitable because, for one or another reason, "they do not meet criteria for inclusion." Moreover, I show in this chapter that mathematical operations – e.g. curve fitting, FFT, inverse FFT – are used to make the measurements suitable for subsequent modeling.

Much like Lave's (1988) shoppers in the supermarket, our research team was in control over what to do and which measurements to retain for the analyses that it ultimately reports. Our inclusion and exclusion criteria were grounded in our familiarity with all those instances that do not even qualify for entry into the data source. These literally constituted the frame that allowed only some measurements to enter into consideration. This frame therefore reduces the original messiness and uncertainty to allow the phenomenon to appear more clearly against the ground then it would if every measurement had been included. Without an integral knowledge of where the data come from, how they are generated, possible problems in the production of data, and how data differ from non-data, even we – all trained scientists with experience in the collecting and processing of data – would be hard pressed to make conclusions and support claims. Making such decisions is important, for example, in democratic decision-making processes. This became evident to me when the mayor, town council, and town engineers in my former hometown based a decision on constructing a water main to supply people with running water on the report of a particular scientist who only collected data on a single day and in only one-sixth of the homes concerned. They did not take into account, and even omitted from entry into the data sources, more than 30 years of anecdotal information that locals had collected about the water (e.g. Roth 2008). That is, these municipal officials could perceive a phenomenon emerging from their data rather than a different phenomenon that would have emerged if all the information had been considered that was available at the time. Although in many discussions scientific facts are held against and contrasted with anecdotal and subjective information, Bayesian approaches that integrate the two forms of data are used even in the hardest of the hard sciences – physics (e.g. Dose 2004). In my study, some savvy citizens did however point out the problems in the methods of the scientist. These citizens serve me as an example of what I envision as the outcome of STEM education – persons willing to engage in a critical interrogation of data presented in the pubic arena as part of policy debates.

In summary, then, this chapter shows that scientists do not just interpret decontextualized data. They require familiarity with the natural setting and with the measurement process and device criteria that include or exclude some of these measurements. The resulting graphs (inscriptions) are an integral part of the entire research process, and familiarity with it is a requirement for interpreting them. Thus, the graphs have a part–whole function to the research as a whole – or, more technically speaking, they are synecdoches of the research processes, that is, parts of the research process that point to the entirety of the process.

References

Alexander, G., Sweeting, R., & McKeown, B. (1994). The shift in visual pigment dominance in the retinae of juvenile coho salmon (*Oncorhynchus kisutch*): An indicator of smolt status. *Journal of Experimental Biology, 195*, 185–197.

Bijker, W. E., Hughes, T. O., & Pinch, T. (Eds.). (2012). *The social construction of technological systems: New directions in the sociology and history of technology*. Cambridge, MA: MIT Press.

Cobb, P., & Tzou, C. (2009). Supporting students' learning about data generation. In W.-M. Roth (Ed.), *Mathematical representation at the interface of body and culture* (pp. 135–170). Charlotte: Information Age Publishing.

Collins, H. M. (2001). Tacit knowledge, trust and the Q of Sapphire. *Social Studies of Science, 31*, 71–85.

Couzin-Frankel, J. (2013). The power of negative thinking. *Science, 342*, 68–69.

Dose, V. (2004). Data analysis via Bayesian probability theory. http://www.mpg.de/841061/forschungsSchwerpunkt. Accessed 1 Dec 2013.

Foucault, M. (1975). *Surveiller et punir: Naissance de la prison* [Discipline and punish: Birth of the prison]. Paris: Gallimard.

Garfinkel, H. (2002). *Ethnomethodology's program: Working out Durkheim's aphorism*. Lanham: Rowman & Littlefield.

Garfinkel, H., Lynch, M., & Livingston, E. (1981). The work of a discovering science construed with materials from the optically discovered pulsar. *Philosophy of the Social Sciences, 11*, 131–158.

Gooding, D. (1992). Putting agency back into experiment. In A. Pickering (Ed.), *Science as practice and culture* (pp. 65–112). Chicago: University of Chicago Press.

Govardovskii, V. I., Fyhrquist, N., Reuter, T., Kuzmin, D. G., & Donner, K. (2000). In search of the visual pigment template. *Visual Neuroscience, 17*, 509–528.

Hárosi, F. I. (1987). Cynomolgus and Rhesus monkey visual pigment: Application of Fourier transform smoothing and statistical techniques to the determination of spectral parameters. *Journal of General Physiology, 89*, 717–743.

Henry, M. (2000). *Incarnation: Une philosophie de la chair* [Incarnation: A philosophy of the flesh]. Paris: Éditions du Seuil.

Latour, B. (1987). *Science in action: How to follow scientists and engineers through society*. Cambridge, MA: Harvard University Press.

Lave, J. (1988). *Cognition in practice: Mind, mathematics and culture in everyday life*. Cambridge: Cambridge University Press.

Lynch, M. (1985). *Art and artifact in laboratory science: A study of shop work and shop talk in a laboratory*. London: Routledge & Kegan Paul.

Lynch, M. (1990). The externalized retina: Selection and mathematization in the visual documentation of objects in the life sciences. In M. Lynch & S. Woolgar (Eds.), *Representation in scientific practice* (pp. 153–186). Cambridge, MA: MIT Press.

Medical Billing and Coding. (2010). Sitting is killing you. http://lifehacker.com/5800720/the-sitting-is-killing-you-infographic-illustrates-the-stress-of-prolonged-sitting-importance-of-getting-up. Accessed 29 Dec 2013.

Munz, F. W., & Beatty, D. D. (1965). A critical analysis of the visual pigments of salmon and trout. *Vision Research, 5*, 1–17.

Roth, W.-M. (2001). "Authentic science": Enculturation into the conceptual blind spots of a discipline. *British Educational Research Journal, 27*, 5–27.

Roth, W.-M. (2005). Making classifications (at) work: Ordering practices in science. *Social Studies of Science, 35*, 581–621.

Roth, W.-M. (2008). Constructing community health and safety. *Municipal Engineer, 161*, 83–92.

Roth, W.-M., & Bowen, G. M. (2001). Of disciplined minds and disciplined bodies. *Qualitative Sociology, 24*, 459–481.

Roth, W.-M., & Bowen, G. M. (2003). When are graphs ten thousand words worth? An expert/expert study. *Cognition and Instruction, 21*, 429–473.

Suchman, L. A. (1987). *Plans and situated actions: The problem of human-machine communication*. Cambridge: Cambridge University Press.

Suzuki, D. (1989). *Inventing the future: Reflections on science, technology, and nature*. Toronto: Stoddart.

Temple, S. E., Plate, E. M., Ramsden, S., Haimberger, T. J., Roth, W.-M., & Hawryshyn, C. W. (2006). Seasonal cycle in vitamin A1/A2-based visual pigment composition during the life history of coho salmon (*Oncorhynchus kisutch*). *Journal of Comparative Physiology A: Sensory, Neural, and Behavioral Physiology, 192*, 301–313.

Temple, S. E., Ramsden, S. D., Haimberger, T. J., Veldhoen, K. M., Veldhoen, N. J., Carter, N. L., Roth, W.-M., & Hawryshyn, C. W. (2008). Effects of exogenous thyroid hormones on visual pigment composition in coho salmon (*Oncorhynchus kisutch*). *Journal of Experimental Biology, 211*, 2134–2143.

Chapter 4
Coping with Graphical Variability

In Chap. 2 I describe the ways in which scientists deal with situations when they do not know what is going on, that is, I describe the work of coping with uncertainty in scientific research generally. In Chap. 3, uncertainty is shown to arise from the very actions that produce the materials from which scientists take measurements and strategies for eliminating some of these measurements so that these do not become part of the data. In this chapter, I am centrally concerned with the *work* of coping in another type of context, that is, when scientists are confronted with variance in the graphs that they produce from their measurement sets. That is, I am not so much concerned with the variance itself but what scientists do, their work, when confronted with variation that they do not yet have in their firm grip and that, therefore, is not ready to hand. When the absorption spectrum is plotted – often after several manipulations that coax a signal from the data that initially do not reveal the presence of useful data (Fig. 4.1) – a more or less Gaussian absorption curve (referred to as "a beauty," see Chap. 3) may be observed under ideal circumstances (Fig. 4.1b). The wavelength corresponding to the maximum of the absorption spectrum ("lambda max" or λ_{max}), together with the width of the spectrum at half height ("half-max bandwidth" or "HBW") can be used to calculate the proportions – in the laboratory, scientists referred to it as "ratio" – of porphyropsin ("A_2") and rhodopsin ("A_1") ($\%A_1 = 100 - \%A_2$). During the laboratory work, the members of our team tended to make a first assessment of each absorption spectrum rapidly cleaned of some background noise. If it looked promising, they then saved the data for subsequent processing to determine the exact position of the maximum along the wavelength scale (measured in nanometers) and the half-max bandwidth (given as a wavenumber measured in cm^{-1}).[1] Before such information could be extracted, a lot of processing had to be done. As Fig. 4.1b shows, the absorption curve "sits

[1] In the physical sciences, waves with a length of λ are often expressed in terms of wavenumber k, defined either as $k = 1/\lambda$ or $k = 2\pi/\lambda$. The wavenumber therefore corresponds to the number of waves per unit distance or radians per unit distance, respectively.

W.-M. Roth, *Uncertainty and Graphing in Discovery Work*,
DOI 10.1007/978-94-007-7009-6_4,

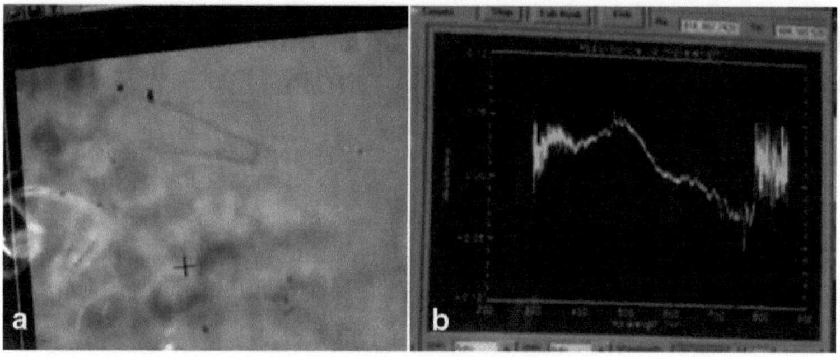

Fig. 4.1 (**a**) Among the many entities seen under the microscope, the scientists pick the bottle-shaped one, which they tentatively identify as a "blue" or "UV" (−absorbing) cone. The resulting absorption curve "sits" on a background, part of which can be thought of as a sloped line descending from *left to right*. (**b**) Considered to be the "baseline," it will be subtracted so that the absorption curve rests on a horizontal line (© Wolff-Michael Roth, used with permission)

on" a background signal, here approximately linearly falling from top left to bottom right. This background has to be subtracted to get something like a Gaussian "sitting on baseline" that allows a better determination of and λ_{max} (nm) HBW (cm^{-1}).[2] To find the wavelength corresponding to the maximum of the absorption spectrum (λ_{max}), our research team used a published polynomial that was fitted to the data. Because this function was fitted to the long-wavelength side of the absorption spectrum only (i.e., on the right in Fig. 4.1b), the determination of half-maximum bandwidth requires a different procedure. To determine it, we produced the Fast Fourier transform (FFT) of the entire Gaussian-shaped absorption spectrum, cut off higher order terms,[3] and conducted an inverse FFT to obtain a "clean(er)" curve from which the half-maximum bandwidths required for determining the A_1/A_2 ratio are more easily extracted.

The data tended to be highly uncertain. As shown in Chap. 3, we decided already in the laboratory whether to retain or scrap some of the measurements. This decision was made based on our sense of whether the data could be cleaned up enough to give a useful result after further processing. The decision also was driven by the correspondence with a possible candidate cell on the microscopic slide. As I show in Chap. 3, when the perceptual assessment (rod or UV, blue, and red/green cone) was consistent with an emerging peak in the spectrum, the data was retained for the moment. But even during further processing, it often was not clear whether an absorption spectrum

[2] Subtracting the background shifts the highest point towards longer wavelengths, which makes intuitive sense when we think about more being taken away from the part of the spectrum, where the line is higher, than from the right, where the line is lower.

[3] As shown in Chap. 3, this corresponds to removing the higher-order frequencies in the Fourier spectrum, i.e., the high frequencies and white noise visible on the curves (Fig. 4.2b) are cut off.

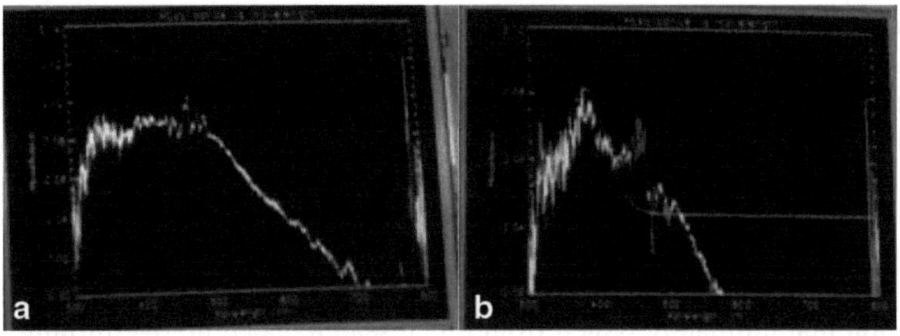

Fig. 4.2 (**a**) The initial difference spectrum often does not reveal one of the expected curves. (**b**) After some manipulations – including "pixel shifting" and "baseline" adjustment – a candidate for a UV (−absorbing) cone has emerged consistent with the theoretical model (smooth curve) (© Wolff-Michael Roth, used with permission)

was real or artefactual (Fig. 4.2a). From the absorption curves, we determined, as central part of our work, the amount of rhodopsin and porphyropsin in the cell. This, in turn, required knowing the location of the peak in terms of the wavelength in the visual spectrum. Thus, for the rods researched in the current experiment, previous research suggested values between 503 and 527 nm. If the peak was outside this range, then it was not of use – even though, as shown in Chap. 3, for the particular species (coho salmon), the values for A_1 and A_2 peaks may not be the same as for other salmonid fishes that were used to arrive at the 503–527 nm range. As a result, any peak "takes 'shape' in and as of the way it is worked, and *from* a place-to-start with *to* an increasingly definite thing" (Garfinkel et al. 1981, p. 137).

Investigating scientists' use of and talk about graphs when they do not yet know what the canonical position will be – and therefore inherently involves uncertainty – constitutes a manner of investigating the real, visible work on which knowledgeable graph use and graphing are based. Because during discovery work, scientists often do not know the "natural object" that they are in the process of identifying – such as the astronomers discovering the first Galilean pulsar (Garfinkel et al. 1981) – and do not know whether the graphical signals they obtain in their data refer to anything, there is a radical uncertainty at work that does not tend to be described in research on graphs and graphing. Thus, as shown in Chap. 3, to know whether a graph on a recording device shows something of interest, scientists need to know the natural object, and, conversely, to know that there is something like a natural object rather than an artifact, they need to know that the graphs exhibit something real rather than an artifact.

The purpose of the present chapter then is captured in and by the following questions: What is the real work underlying graphs and graphing in the discovery sciences when nobody knows in advance what will be the state of affairs after the fact? How do scientists cope with and tame the uncertainty in their graphs that is an integral part of discovery work, where outcomes are not available beforehand?[4]

[4] This was the case in our laboratory, where we thought our work would confirm what Craig had called the "dogma," but what ended up as an overturning of this same dogma after many years of research.

Answers to these questions are not only interesting to those theorizing scientific expertise, which tends to be studied when scientists are on familiar terrain, but also to better situate the phenomenon of uncertainty as a possible characteristic in life generally. I investigate how scientists cope and deal with uncertainty in data and graphs while the latter are produced and talked about *in the course of* discovery work, that is, prior to the point when scientists are more or less certain that they have found a reproducible phenomenon of interest. That our research team was in the course of doing real discovery work can be gauged from the fact that the publication that ultimately issued from this research ended up overturning a long-held scientific canon with respect to the chemical changes in the retinal composition of coho salmon. At the time, we still believed that our work was consistent with "the dogma"; and we did not know yet that we would be debunking it. But before we could investigate the correlations of porphyropsin levels with life stages, we needed to know whether the measurements we made would lend themselves to making supportable claims. In the following, I analyze an extended, 15-min episode from a laboratory meeting where the uncertainties in the data, data collection and analysis procedures, and graphical modeling of the data were especially salient, exemplifying the work of coping with uncertainty. In the course of this episode, the various laboratory members contributing to the discussion draw and project a considerable number of graphs (models, data) and use gestures to overlay these with additional models (e.g., best fit curves) for the purpose of coming to grips with "the template," which Theo formulates that our team ("we") has trouble with. The entire sequence of graphs and gestural models is presented in Fig. 4.3. In the sections that follow, I analyze the unfolding event as it leads from inscription to inscription until our team has finished its deliberations concerning the uncertainty and variance within its data set. The analysis moves through the 13 stages displayed in Fig. 4.3, which exemplifies the prevalence of graphical and gestural signs in the discovery sciences.

When the meeting begins, Shelby has already projected histograms of the "data" that the team has so far collected, that is, the distribution of data points (histogram) for different levels of porphyropsin ("%A_2") in bin sizes of 5 %. But Craig asks for an overview of the "data" they have and the specifics about the origin of the fish that were sacrificed (hatchery or wild for the different geographical areas involved and those fish retained in the university freshwater tanks). Shelby shows Craig a sheet with a listing of seven different batches of coho salmon (*O. kisutch*) and, pointing to the each of the seven entries, elaborates the entries by providing information on where the fishes had been sourced. Shelby notes that the total sample, all seven batches taken together, constitutes a "broad spectrum" of fish. Craig begins a new entry in his laboratory notebook and then begins to ask a question about data and data collection that begins the episode presented in this chapter before our team returns to discuss the graphs that Shelby had projected to begin with. There is therefore a clear, member initiated and defined episode that interjects a clarification period prior to getting to the data analysis proper.

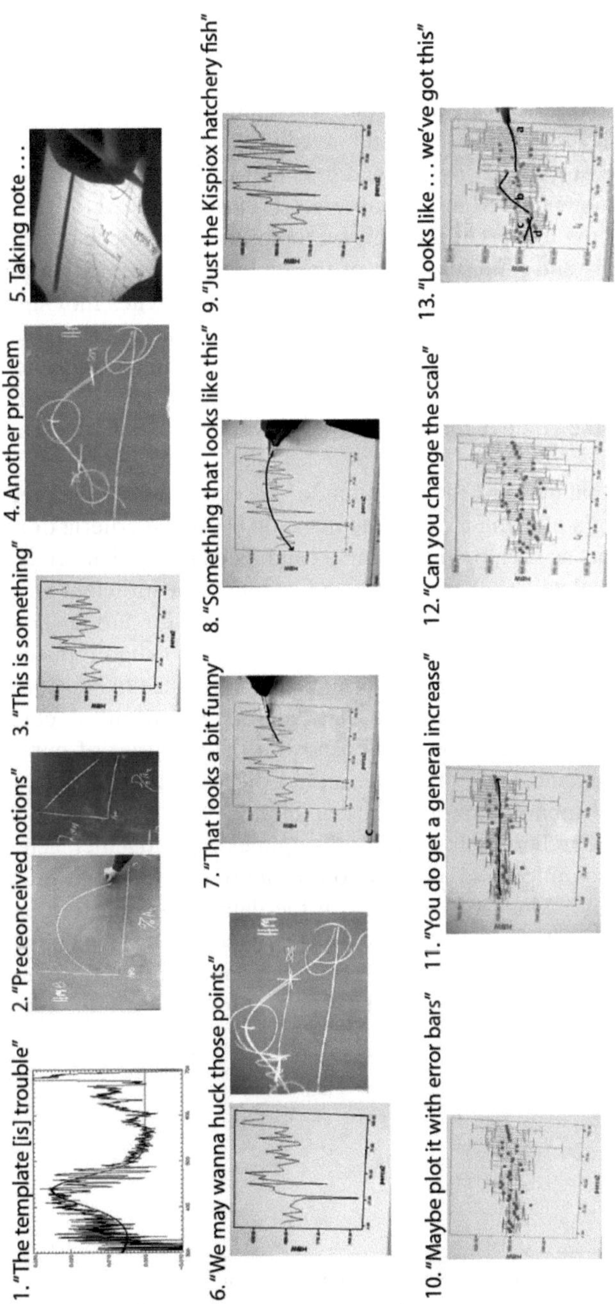

Fig. 4.3 The account of how scientists attempt to "fish for certainty" is provided in 13 sections that are organized around aspects salient to the scientists themselves

Checking Criteria Controlling Variability

"So you are looking at a half-max bandwidth match"

Just as the ultimate discoveries of scientific research themselves, particular topics of talk in data analysis sessions are occasioned productions that members to the setting accomplish in almost unremarkable (and invisible) fashion. The topic of the "wave form" and its measure given by the width at half of the maximum height of the absorption spectrum emerged in a situation itself clearly marked off ("okay, now") when the head of the team (Craig) begins a topic: the acceptance of raw materials for further analysis and representation of the biologically interesting phenomena. The episode is ended as such when Shelby, in turn 131, invites moving to the first summary data slide, an invitation that the team members accept (i.e., by producing an invitation | acceptance turn pair). That is, the beginning and ending of the episode featured here was marked in and by the laboratory talk itself, such that the talk provided the context for the specific conversation about uncertainty in the data.[5]

The fragment that delimits the beginning of the episode from the preceding talk is opened by Craig's offering a candidate question about the criteria of acceptance, which Theo accepts as such by producing a candidate answer that briefly describes that he used a template (turns 002, 004) and as long as a sufficient amount of the template is reproduced by the data, he retains the data for further analysis (turn 010). In the course of answering, Theo makes a statement that introduces what they "are having trouble [with]," which is subsequently suggested to be addressed when the "spectra are becoming cleaner" so that "more stringent rules" could be used (turn 012). The nature of the spectra depends on the activities of our team while extracting fish eyes and from these the retinal pieces that are macerated so that pieces of it can be mounted on a microscopic slide. This therefore is the kind of situation where researchers from multiple disciplines work together, achieving both a distribution of cognition and a disruption of the representational infrastructures typical of each field. Much of the noise in the data is thought to arise from the preparation and the selection of cells (e.g., clearly separating the cells that derive from the ventral versus those that derive from the dorsal part of the eye). It is only now that the data become cleaner that our team "can start to ask waveform questions" (turn 018).

[5] That is, beginnings and endings are differently organized than in school mathematics or science, where students are told to "interpret a graph," whereas in the present instance, the content of the conversation is not pregiven but emerges itself in, and as the result of, the sequentially ordered laboratory talk.

Fragment 4.1

001 C: (:T) okay now when collecting the data and analyzing
the data (1.53) u:m (1.51) ´whAT (0.88) critE:Ria
(0.39) have bin:: (0.27) IMplemented (1.10) for curve
acCEPtance.
001a (2.13)
002 T: de basic; (0.20) `uh:: de tEMplet; (0.35)
003 C: the templ[A:te],
004 T: [the] template we are working with; (1.13)
we are having trou[ble]
005 C: [can] i keep thIS? ((*holds up the sheet with information on origin of coho in sample*))
005a (0.41)
006 S: uh (0.35) <<dim>yea, i have a copy of this one>
007 C: okay;
007a (0.25)
008 S: sorry;
008a (1.55)
009 C: okay so?
010 T: so uhm; (0.73) so i=m using de template and if zeres
sufficient, (0.45) uh (1.07) amount of ze template
<<dim>reproduced by ze curve i keep it.>
010a (0.80)
011 C: sO:: youre looking at halfmax bandwidth () match
012 T: y:es::; dats it um; (1.70) we=re (.) our spECtra are
becoming clEAner and (0.30) s:o we can apply more
stringent rules. (0.26)
013 C: okay. (0.19)
014 T: and den deres a second question; is do we want to
analyze ze wAVeform more strongly or ze position;
(1.39) so and dats anOTher choice which has to be
made at some point (1.35) and i have been uh (1.54)
putting my emphasis on de position; (.) right now;
(0.35)
015 C: lambda [max]
016 T: [yea]; lambda max. (0.90) and so also in my
analysis i have not yet tried to uh (0.99) to sheck
what de what de bandwidz gives us.
016a (0.59)
017 C: O:kay.
017a (0.20)
018 T: so i i dont zink; i zink now we=re getting good
enough zat we can start uh (.) to ask waveform
questions.
018a (0.26)
018b C: right.
019 T: but with ze prelimi (.) ze first data sets it wasnt
good enough. (0.83) <<dim>we wouldnt have any,>
020 C: like forinstance, i mean what you mIght, (1.33) um a
priori what you would expect to sEE is the bandwidth
(0.46) to go from (1.45) u:m::. (0.38) the
porphyrOPSin (0.99) bandwidth which is slightly wider
than the rhodOPsin bandwidth (0.49) but thEN you
would expect to see an INcrease in bandwidth beYONd
the a=two (1.22) representing the a=one=a=two
mixtures

In their sequentially ordered turn taking, our team members together articulate the possibility that they can get the A_1/A_2 ratios from the waveform (i.e. the shape of the spectrum) and the half-maximum bandwidth.[6] Theo explains that as long as a curve reproduces the template, he keeps the (absorption) spectrum (raw data) by saving it to the hard drive. The "template" is a polynomial published in the literature that is fitted to the absorption spectrum derived from the difference between two measurements and adjusted so that the "baseline" is approximately parallel to the wavelength axis (abscissa) and the curve maximized in its appearance by making minute "pixel shifts" between the two measurements to minimize noise (Fig. 4.2). The issue in this episode is in regards to the measurement of HBW from curves such as the one displayed, which is rendered difficult by the variance clearly observable in the difference between the idealized curve – "the template" – that Theo has overlaid on the measurement (Fig. 4.2b). As long as "there's sufficient ... amount of template reproduced by the curve, [Theo] keep[s] it" (turn 010). As his reply indicates, Craig hears this as an indication that he is trying to use the half-maximum bandwidth – which is not really the case, despite Theo's affirmation that they are only now "getting good enough data that [they] can check what the what the bandwidth gives [them]" (turn 016). This will be seen later in the episode when Theo explains that he is interested in the part of the spectrum from λ_{max} to the place on the curve where it has half of its maximum height. In fact, Theo "rectifies" what they have been doing in describing that the question of whether to emphasize the waveform – i.e., shape of the absorption curve – more strongly or the position (i.e. λ_{max}) currently is answered in favor of the latter case, whereas the question of the waveform has to await the point when the "spectra are becoming cleaner" (turn 012) at which point they may "apply more stringent rules" for the selection and identification of the amount of A_2 chromophore in the cells analyzed. He "has been putting [his] emphasis on the position," which is, on the position of the maximum of the spectrum along on the wavelength axis (i.e. λ_{max}).

In turns 014 and 016, what will become the topic of the conversation comes to be re-introduced, and Theo again articulates the trouble of the first data sets that were not good enough (turn 018), and the required quality is reached only at the point of the meeting (turn 019). With turn 020, Craig accepts and reifies the topic: the distribution of the half-maximum bandwidths of the recorded absorption spectra. This conversation is clearly marked and set apart by the introduction of the topic and the subsequent uptake on the part of Craig. Almost exactly 15 min later, our team equally clearly shifts the discussion topic again (turn 131). We return to looking at the biologically interesting data: the distribution and changes in the percentage of porphyropsin (A_2) in the retina of fishes collected at different times, from different hatcheries, from the wild caught in the same river systems as the hatcheries, from the larger river systems that include other tributaries, and in the estuaries of these larger rivers. In the episode featured here in its entirety, the topic is the quality of the included data on which the results, discussed in the subsequent

[6] In the end, there will have never been a follow up when the curve does not pan out in the way that Craig projects it should.

part of the laboratory meeting not featured here, are based.[7] The changes in porphyropsin levels can be gathered directly from the λ_{max} that Theo has been reading off the data by using "the template," a polynomial the maximum of which, after a best-fit procedure, indicates the position of the maximum of the spectrum.

In this first fragment, half-maximum bandwidth as the topic for the subsequent 11.5 min emerges from the sequentially ordered laboratory talk occasioned by Craig's question about the criteria for accepting absorption spectra into the data set from which porphyropsin levels would be calculated. Theo accounts for what he had done by saying that he kept data when enough of the template, the results of previous research by other scientists, has been reproduced. But the second indicator for porphyropsin levels, half-maximum bandwidth, could not be extracted from our spectra because these were not (yet) clean enough. We observe in this first fragment of the episode how the subtopic of the analysis session becomes the members' topic in and through the sequentially ordered turn taking. That talk not only has a topic but also makes the relation in and of which we all are part. The topic therefore also exists *as* the relation. That is, the possibility for this topic exists in, arises from, and constitutes our societal relation. That the possibility exists *as* the relation is significant, because it means that the societal relation itself *is* the higher psychological function rather than merely begin the context for a "co-construction" of mathematical "meaning" in the way researchers with a sociocultural or postmodern bent tend to theorize such instances. And the topic becomes realized as such in, as, and through the subsequent talk until some other turn sequence brings it to a close (turn 131 below). At that point the team members present turn to what will have been, at the end of the day, the main topic for and of this meeting: porphyropsin levels at different hatcheries measured thus far. Following the emergence of the topic of bandwidth, Craig then launches into what we may gloss as a "think-aloud" (turn 020): Craig repeatedly writes and erases drawings and eventually copies down into his notebook what he has produced on the chalkboard.

Preconceived Notions

"Just sort of preconceived notions"

Contrasting their written productions, which are carefully edited, scientists may make statements inconsistent with the current canon of the field while they speak in laboratory meetings, lectures, or think-aloud protocols. Although scientists often do not state it in the way Craig would do in this meeting, they operate with preconceived notions that describe what they see and how they see it.[8] In the process, inconsistencies may arise because data cannot ever be seen in and for themselves but always already

[7] In that part, the quality of the data no longer is at issue but the data are taken as if there were no uncertainty related to their quality.

[8] The relationship between what there is to be seen and talk constitutes the function of speech: "λόγος [logos] lets something be seen (φαίνεσθαι [phainesthai]), namely that what speech is about,

through a language that makes them appear in a particular way. Inconsistencies, however, often go unnoticed and thereby remain irrelevant for following the trajectory of the unfolding conversation. This happens in this meeting, too, when Craig articulates some "pre-conceived notions" about the mathematical relations relating HBW, λ_{max}, and porphyropsin/rhodopsin levels. The fragment furthermore exemplifies the role of graphs in conceptualizing possible yet-to-be realized outcomes of research.

In this fragment from the session concerning the uncertainty in our data, Craig makes a statement that comes to articulate the relationship between the half-maximum bandwidth of the two chromophores and mixtures thereof that "a priori . . . you would expect to see" (turn 020). He describes the porphyropsin bandwidth to be wider than the rhodopsin bandwidth and that for the mixtures of the two chromophores, the bandwidth would be even higher than that of the already higher porphyropsin bandwidth. He does not explain why this ought to be. He suggests that there is a "sort of transition" and then gets up and walks to the chalkboard (turn 024).

Fragment 4.2

020 C: like forinstance, i mean what you mIght, (1.33) um a priori what you would expect to sEE is the bandwidth (0.46) to go from (1.45) u:m::. (0.38) the porphyrOPSin (0.99) bandwidth which is slightly wider than the rhodOPsin bandwidth (0.49) but thEN you would expect to see an INcrease in bandwidth beYONd the a=two (1.22) representing the a=one=a=two mixtures

021 T: uh hm

022 C: to a varying degree;

023 T: yea

024 C: okay so youre gonna see (1.21) ((*pencil in air, begins "drawing" line*)) u::m:; (1.61) this ((*some gesture in air, then gets up*)) (0.81) sort of transition ((*Craig walks towards chalkboard*)) where (1.24) yknow if you plot (2.68) um (3.88) a half ((*writes "1/2 m"*)); (5.93) ((*erases 1/2 m*)) half (0.53) max (1.27) ((*writes "HMB"*)) bandwidth (0.46) ((*draws ordinate*)) versus (14.01) ((*draws abscissa; writes "*A_1/A*" on the left end, then erases it; then writes *A_2/A_1*; then erases it*)) or <<pp>or in percent> ((*writes "%*A_2*" in center of abscissa*) (6.56) ((*looks repeatedly left and right ends of abscissa*)) from lets say an hundred to zero. (2.14) you wouldexpect to sEE (1.61) something like thISS * (7.60) right?

025 (1.59)

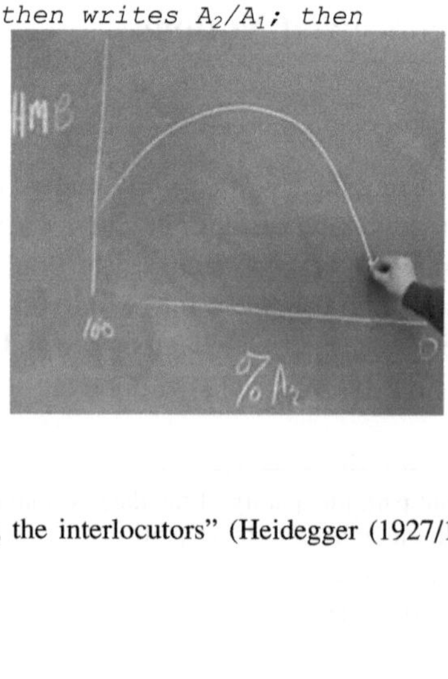

---

and indeed *for* the speaker (the medium), or rather, the interlocutors" (Heidegger (1927/1977, p. 32).

026 S: yea, more or less, maybe a bit hIGher on the one end ((*gesture toward the chalkboard*)), but same idea
027 C: [right]
028 S: [like] more on that end.
029 (5.17)
030 C: and whereas for lambda mAX (11.84), youd expect to see that relationship *
031 (8.63) ((*Craig walks back to table*))
032 C: so jst as sort of, preconceived (1.02) nO:tionS:: ((*Shelby works at computer*))
032a (5.40)

As he arrives at the clean and empty chalkboard, Craig stares at it for 3.0 s before beginning to write "1/2 m." He pauses, produces an interjection, then erases what he has written and then writes "HMB," each letter corresponding to the first letter of the word he utters "half" "max" "bandwidth." Although the normal denotation in the literature is HBW, there is no objection and the alteration in the symbol thereby is accepted. He then produces first the line corresponding to the ordinate next to "HMB" and then draws what will be the abscissa. He first writes "A_1/A" near the origin of the Cartesian grid then erases it. He writes A_2/A_1 in the same place only to erase it again (Fig. 4.4). Craig then writes "$\%A_2$" in the center of the abscissa. There is a pause in the talking. Craig looks right, left, right, and then to the left (origin) of the axes system and writes "100 %"; he turns to the right and writes "0 %" while uttering the corresponding numbers. Just as he says "you'd expect to see" he places the chalk, then moves the hand lower; he pauses, then draws the inverse parabolic line, pauses, gazes to the left toward the beginning of the line, and then adds a little to make the end lower. He then sketches a cross at the endpoint and adds one to the beginning of the curve. There is no reaction at first (1.59 s), and he then states "Right?" with an increasing pitch that Shelby realizes as an invitation to comment. His statement can be heard as suggesting that the situation is "more or less" in the way stated, "maybe a bit higher at one end," but it is the "same idea" (turn 026). Although Craig states an affirmation (turn 027), and although Shelby reiterates that the curve should be higher on one end, Craig does not change the graph. He then turns to the right part of the chalkboard and produces a second Cartesian graph, thereby producing a new segment of the episode.

In this production, there are long pauses, repeated writing and erasure and re-writing (Fig. 4.4). All of these erasures and re-writing are signs that the thought emerged in, with, and being shaped by the communicative production in real time on the board. A further indication that this instant reflects thinking that emerges in and from communicating exists in the fact that Craig subsequently copies what he has produced on the chalkboard into his own research notebook (see below).

Fig. 4.4 The production of the abscissa label takes three attempts until the one consistent with the "preconceived notion" appears on the chalkboard (© Wolff-Michael Roth, used with permission)

A paper that some members of our team had published just a month before this meeting stated that one of the objectives of the group's research was to model the relationship between λ_{max} and HBW, which "is not well understood for paired pigment fishes" (Hawryshyn et al. 2001, p. 2437), a statement that is followed by a reference to a 1994 paper that the first author had published with the leading researcher (Hárosi) in the field. The graph Craig has drawn bears an iconic relation to those that Hárosi (1994) published, though the latter plots HBW against λ_{max}, both measured in wave number (i.e. $1/\lambda_{max}$ [cm^{-1}]). Learning scientists have referred to this phenomenon as iconic confusion, though this confusion here exists in the relation between two graphs rather than between a graph and the phenomenon it models.[9] At the time of this conversation, the relationship still is uncertain consistent with Craig's account of what he had produced as "just as sort of preconceived notions" (turn 032). In its most recent paper, the researchers had three data points for HBW, for A_1 chromophore, A_2, and a mixture of the two, and the relationship was as anticipated, smallest for A_1, then A_2, and largest for the mixture. This is the kind of background that can be found underlying Shelby's statement in turn 26 as a form of "critique" of the fact that Craig's curve appears in this description ("may be a bit higher at the end," turn 26) to be symmetrical or lower on the 100 %-A_2 side, whereas the HBWs of A_1 and A_2 not only are different, expressed in the different ordinate values at the left and right end of the graph, but in fact A_2 (porphyropsin) ought to be higher as per Craig's own opening description (turn 20). That is, Craig articulates one relation between the bandwidth of the two chromophores and draws this but in a figure where the higher percentage is to the left rather than on the right. Shelby intimates that "the one side" ought to be higher. It is unclear whether this statement is about the right end that ought to be higher than the left – as it would be in the case that the A_2 concentrations were going up from left to right or whether "the one end" is in fact the left end of the graph but that

[9] For example, there is research showing that students inappropriately associate undulating speed-time graphs and curves on a race track: the maxima and minima are said to occur when there are curves rather than in the straight parts of the track where the cars can drive the fastest. That is, there is a confusion where the "turns" in the sinusoidal graph and the "turns" on a racing track are said to correspond to each other, when, in fact, an actual measurement would show that the fastest speeds are on the straight lines of the track and the velocity moves from positive to negative values where the sine curve intersects with the abscissa.

Shelby suggested it to be higher (turn 026). Craig does not make changes but marks the two end points by means of crosses – as if ascertaining the correctness of each. In this turn sequence, therefore, we observe an offer for correction and a non-acceptance, as the graph stays precisely as it has been prior to the turn pair.

After the fact we can say that there is an unnoted problem here: If the relation between HBW and amount of porphyropsin were of the kind that Craig shows here, then HBW could not be taken as indicator of porphyropsin levels, as for most values there would be two corresponding A_1–A_2 mixtures. When HBW is approximately parabolic, as in the figure that Craig has drawn, then plotting it against $1/\lambda_{max}$ in fact yields a linear relation as per the studies by Hárosi (1987). The same result is achieved when the absorption curves are modeled as Gaussians of different width and the mixture as a weighted sum of the two Gaussians: the resulting bandwidths, in wave numbers, vary linearly with $1/\lambda_{max}$. As λ_{max} has been shown to vary linearly with the percentage of A_2 in the photoreceptor, the relation between bandwidth and $\%A_2$ can be anticipated to be approximately linear. What the Hárosi (1994) paper does show are parabolic relations of the kind that Craig and Shelby discuss, but these are between HBW and λ_{max} both measured in wave numbers (i.e., cm^{-1}). This conversation does not pick up on this problematic.

Tentative Identification of Patterns

"When we look at something like that . . ."

An important requirement for getting a handle on[10] uncertainty exists in knowing what the exact nature of the data is, what has been included and what has not been included, what the criteria for inclusion and exclusion are, whether what has been done can be defended in terms of such criteria, and what the level of the integrity of those curves is that result after multiple transformations of the original data have been implemented. Because each step in the translation according to Fig. 1.1 involves crossing an ontological gap, a crossing that is inherently grounded in and legitimized by scientific practice (Chap. 1), the scientists can guarantee the quality of their work only if they are (relatively) certain that each step in the chain of transformations from "natural object" to the ultimate claim is defensible. That is, scientists have to ascertain that an inscription replacing another inscription or summarizing the many inscriptions it collects – e.g. each data point represents 20 absorption spectra – retains and is equivalent to the fundamental structure. They have to have assurances that this chain of translation is unbroken – in the face of the uncertainty associated with actions described in Chap. 3 – so that a continuity may

[10] "Getting a handle on" should be read in terms of the things that we encounter in their ready-to-hand modality (Heidegger 1927/1977), that is, when the things do not require our conscious awareness thereof in the way we are not aware of the pen when jotting down some notes – unless there is something wrong with the pen.

indeed be assumed in the process of mapping nature onto the graph ultimately featured in the publication. It does not then come as a surprise, perhaps, to hear Craig repeatedly ask about the criteria that were applied in accepting or rejecting spectra. An important aspect of the work is that it is distributed over team members and that at the time of their emergence nobody has control over or knowledge of all the translations that occur between the living fishes, on the one end, and claims about varying porphyropsin levels displayed later in the meeting, on the other end. An interesting aspect of this stretch of talk is that knowledgeability arises in, for, and from the social relation that individuals subsequently will articulate on their own.[11] In this section, I exhibit the work of identifying patterns when it is not yet known what ultimately will be accepted as the factual pattern that is accepted as corresponding to nature.

Testing the Preconceived Notion

"That's something we could plot up there now"

After Craig had walked back to his place at the table, Shelby proposes to look at the data he has and suggests "that's a something that we could plot up now" (turn 033). As Shelby works on his computer, Theo explains that he has some "disturbances" on the "long wavelength [side]" and there is disturbance on the short wavelength (turn 036). He talks about the problems arising from "the reference," which means, from one of the two measurements of light transmission through the microscopic slide taken next to the retinal cell of interest (here, a rod). He thereby denotes a recurrent problem in the data, where a "ringing" feature (seen in Fig. 4.1b to the left of the peak) was thought to arise from somewhere in the equipment. Our team had already spent over $1,000 to purchase a sampling beam-producing lamp that was hoped to eliminate the problem of oscillations on the low-frequency side of the absorption spectra. But all efforts were to no avail.[12] Theo here makes a statement that suggests that as the team comes to acquire greater control over this variation, by producing a reference beam "with increased quality" "as well as [an increased quality of] the specimen," we will "really have the whole [absorption] curve available for" the determination of λ_{max} and HBW (turn 036).

[11] This is consistent with the claim that any higher order psychological function exists in societal relations first (Vygotsky 2005). Societal relations are the central topic of Chap. 10.

[12] When I arrived in the laboratory, I contributed to the understanding of the phenomenon by producing a mathematical model that explained why there would be such "ringing" given a rectangular characteristic of the lamp in terms of intensity over wavelength and given the fact that the measuring beam had to traverse four interfaces between different mediums when going through a photoreceptor (microscopic slide fluid-cell wall, cell wall-cell inner, cell inner-cell wall, and cell wall-microscopic slide fluid).

Fragment 4.3a.i

```
032   C: so jst as sort of, preconceived (1.02) nO:tionS::
         ((Shelby works at computer))
032a     (5.40)
033   S: <<p>um:;> (4.21) so we can actually; <<p>i mean tHAts
         a something we could plot up there now <<all>if we
         wanetto>>
033a     (0.32)
034   C: kay
034a     (0.49)
035   S: <<pp>um;> (2.52) should i
         take now (0.66) sort of
         (0.95) yea givita try>
         ((mutters to himself))
         ((Craig is writing in his
         notebook))
035a     (2.62)
036   T: (:C) as i say what we do
         get is a, for example, ze
         long wavelengz zeres a
         clear () uh disturbance
         from ze reference where we
         have short wavelengz
         disturbance. (1.16) ‘so
         (2.46) wiz increased
         quality of ze reference as well as de specimen all ze
         zings increase (0.77) and den we
         really have de whole curve available for ze.
036a     (0.55)
037   C: all right (0.19)
038   T: i would really like to determine ((parabolic gesture
         with lH index in direction of chalkboard HBW curve))
         ze curve from our (0.56) data <<dim>but its really
         not quite good enough at the present.>
038a     (0.89)
```

Theo notes that he would like to determine “the curve” from the data collected by this team but that these are “really not good enough” (turn 038). That is, instead of having to work with a template to model the long wavelength side from the research literature to be used to extract λ_{max}, he would like to get the template from our own work and specific to coho salmon. But at this instant in their research, our measurements are “really not quite good enough” (turn 038). Craig, turning towards Theo, asks whether the rejection of the data were determined by the criteria considered (turn 039).

Fragment 4.3a.ii

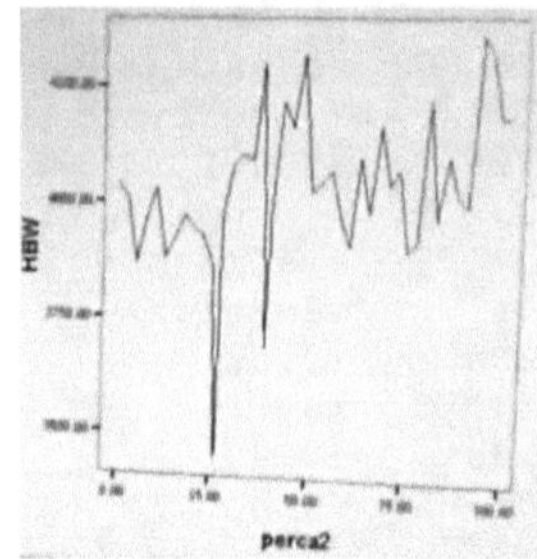

039 C: * nOW; (0.92) whEN: the criteria were being consIDered (0.19) did thAT determine rejECtion on spECtrA?
039a (1.46)
040 T: ̄what (0.24) well; ((8:20))
041 C: (:T) sO in other words when we look at hIstogrAMs ((*points to projected graph*)) like for instance when we look at something like thAT ((*points again*)); (1.07) fOLded into the dAta do you s:EE; (3.37) is:: () is thIS reflECting spECtra of hIgh=intEGrity,
041a (1.11)
041b T: <<p> ̄yea>
041c C: or isthis reflecting the () the whOLe gamut.
041d (0.27)
042 S: right (0.13) for right now is (0.23) its whatever (0.12) theodores accepted; (0.20)
043 T: <<p>yea.>
043a (0.37)

Shelby had succeeded in projecting a graph that plots half-maximum bandwidth (HBW) against the percent of A_2 ("perca2") (turn 039). There is a pause, then markers of hesitancy on Theo's part, which provides others with an opportunity to take the speaking floor again. Craig takes the opportunity to articulate the issue of integrity of spectra (i.e., the data) (turn 041) in terms of the graph that Shelby presents. Craig proffers a candidate question about whether "that," pointing to Shelby's graph, is "folded into the data" or whether "this" (graph) is "reflecting the whole gamut" (turn 041c). Theo, who initially uttered a marker of agreement ("yea"), does not respond to the alternative Craig proposes. Rather, Shelby responds saying that "it" is "whatever Theodore has accepted" (turn 042). Theo accedes.

Here, then, a question has arisen as to the quality of the data and whether or how this quality is reflected in the graph currently displayed on the screen. The displayed graph exhibits considerable variability, and the question | reply sequence establishes that the displayed data reflect whatever Theodore has accepted according to the criteria that the team has to defend. This quality now is the issue that the *conversation* has to address, that is, an issue that has to be conversationally and, therefore, supra-individually managed. This occurs in the articulation of problems that were identified during the data collection.

Articulating Problems During Data Collection

"The problem is that they were really long curves"

Problems do not just exist. Rather, something might become a problem or used to explain a problem at some later point in time. In this subsection we see how data come to be denoted as problematic after the fact, when whatever happened during the day these were collected comes to be brought to the discussion. In the same way that Craig articulated one afternoon that his dissection might not have been what he thought it to have been (Chap. 3), something might be articulated as a problem when issues arise subsequently rather than when what comes to be recognized as the problem originally surfaces. In the present situation, half-maximum bandwidth is at issue; and to explain certain problems, Shelby returns to identify something from an earlier point in the research project as a possible cause. Shelby says that he "can present" what "is in [his] database" (turn 044).

Fragment 4.3a.iii

044 S: its in my database thats what i can prsent to you (0.24) right nOW (0.20) but we::ve ((*turns to Theo, gestures back and forth from Theo to himself*))
045 T: but ze problem really is zat zey were fairly long curves; (7.06) ((*walks to chalkboard, draws a curve*)) * so ze question is, i can look hERe ((*circles "foot" on left of Gaussian*)), i can look hERe

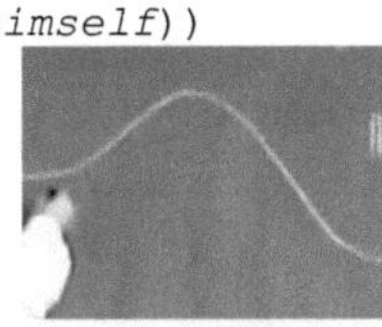

((*circles* λ_{max})) (1.05) and i can look here ((*circles right "foot" of Gaussian*)). * (1.24) de rule of u:m (0.27) of macnichol says (0.25) fifty percent ((*marks line at half-max on right "leg", writes 50%*)), (1.25) dis ((*between the marks on the curves*)) is the important stretch for him. ((*Marks off line at* λ_{max}, *then gestures along graph between the two marks*))

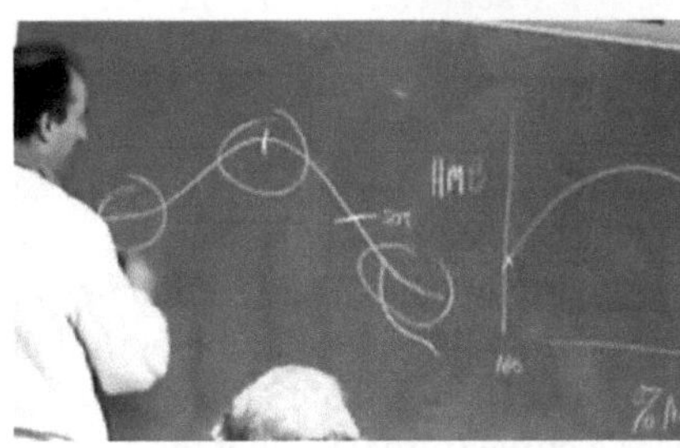

045a (0.63)
046 C: mm hm
046a (1.48)
047 T: but and so dis is basically; USUally (0.26) problems i guess a drop off here, ((*draws change to graph within right-most circle, see screen print in turn 45*)) (0.33)
048 C: yea:
049 T: otherwise a very clean spectrum ((*draws abscissa line at base of curve*)) (0.22)
050 C: thats r[ight]

051 T: [so i] dont zrow it out (0.47). sometimes i get a drop off here, i dont zrow that out eizer coz. (0.65)
052 C: yea
053 T: * as long as i have enough peak in here between, under, above ze fifty percent [line;]
054 C: [right]
055 T: i guess dats how i see it. (1.57) den i (.) den i keep it.

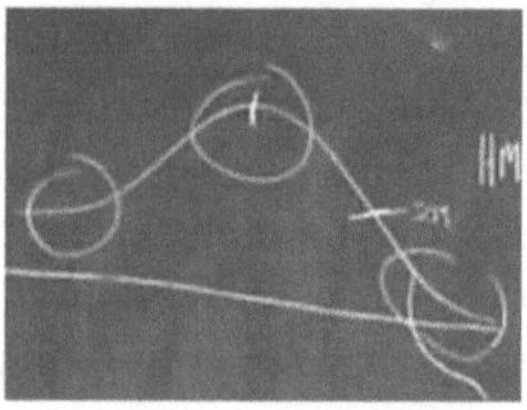

055a (40.11) * *((Craig writes into his notebook copying exactly the diagrams that he has drawn on the board)) ((A "natural" break at this point, This is some indication that the researcher himself considers it worth*while *to keep this, something not already cooked up but the result of thinking in action;*))

Shelby turns to Theo, points to the research associate and to himself, uttering what might be the beginning of a description what they ("we've") have done. As if he were invited to add, Theo formulates the existence of a problem ("the problem is"), and begins with a description of the curves as being "fairly long" (turn 045). As Craig before, Theo gets up and walks to the chalkboard. Upon arriving there, he produces a curve (turn 045) that the other team members present recognize to be an idealized but widened absorption spectrum. Theo marks three areas where he "can look" (turn 045). He refers to the rule by MacNichol, which states that the important region is defined by the right part of the curve above the 50 % of its height. To determine the height, Theo states that he needed to know where the base lies (left and right circle) and what the maximum height of the curve (λ_{max}) is. First for the right side of the curve (turn 047) and then for the right-hand "foot" (turn 051), Theo draws the kinds of features that he observes – and willingly demonstrates and explains to visitors to his office – while analyzing the data that he receives from the wet laboratory (where he participates in some of the data collection sessions operating the software). These features, sudden drop offs in otherwise "clean spectra," make the determination of the baseline points difficult and, therefore, the precise determination of the height of the absorption curve and introduces variance in its location along the wavelength axis. That is, these features introduce uncertainty over and above those that arise during the preparation of the photoreceptor cells. Although an "otherwise clean" graph does contain such problems, Theo explains that he does not reject the data at this point "as long as [he has]

enough peak in here, under, above the 50 % line" (turn 053). Such a curve still allows him to determine the porphyropsin levels because λ_{max} can still be obtained, and, therefore, the amount of porphyropsin present. He *formulates* that this is how he sees the situation, and based on it he keeps the data (turn 055). There is an extended pause in the talk, occasioned by and affording Craig's recording of the graphs (turn 55a).

In this series of sequentially ordered turns, the participants therefore find out[13] that Shelby has included in the plot currently displayed all the data that Theo has provided him with, which are based on what he, in turn, has received from Shelby who collected the curves from the retina in the dark laboratory. Theo states that there indeed are curves that exhibit sudden drop offs at the left and right end of the otherwise clean absorption spectra. He does not throw these data out, at least he does not do so as long as he has "enough peak in here," that is, between the half-maximum point and the top of the peak at λ_{max}. Although not specifically stated as such, what is available *in* and *as* their relation is a possible reason for their HBW data to be noisier than they should be in some ideal case.

This exchange relation actually produces a lot of the background necessary for an appropriate reading of the graph that is now projected on the screen. That is, what does not already exist as the background knowledge for our team as a whole is produced right here and now, and it exists *in* the relation, *for* the relation, and *as* the relation in which we find ourselves and that we make simultaneously with and as part of our discussion of data quality. Craig, whose reading of the data is analyzed below, would not be able to say what he says at that point if it had not been articulated here, in, through, as, and for the societal relation with others.[14]

Theo's presentation may be heard as anticipating possible problems in the interpretation of the data, as he explains possible variations in the data that might wash out the true trends that possibly exist or that make interpretation possible. He presents matters of fact that he, as the person responsible for the preparation and manipulation of the data, is entirely familiar with but that might not be, and in this case are not, known to the other. One important aspect of this team is that Shelby (and other helpers, including Sam) tends to produce the raw measurements – i.e. the experimental absorption spectra. He saves whatever he deems potentially usable to

[13] Any knowledgeable analyst or reader having access to the materials can find out the same.

[14] I emphasize this point of language being from, for, and returning to the other – as do others (e.g. Derrida 1996; Vološinov 1930) – because all too often language use is theorized from the perspective of the speaker when it should be analyzed from the perspective of the recipient for whom a statement is designed (Heidegger 1927/1977). This approach is consistent with the pragmatic argument that language *use* rather than inaccessible "meanings" are pertinent to theorizing language (Wittgenstein 1953/1997).

the hard drive. Shelby passes the raw absorption spectra (measurements) to Theo,[15] who then uses various algorithms and templates to clean up the curves and establish λ_{max} and HBW. Theo then passes the results back to Shelby, who has taken the lead on this experiment and who monitors both sets of values for the coho salmon from different hatcheries and river systems over time. So Shelby, although he has collected the raw data evaluated here, does not know exactly what is represented in the curves he currently projects; he only knows that it represents everything that Theo has provided him with. Theo, in turn, only works with the graphs that Shelby provides him with. The extent of that which is known exists only *in* their societal relation, first, and only subsequent to conversations such as this one can it be known, talked about, and used on the part of individuals. But even if something exists *in* their societal relation, we cannot be certain that it will exist as individual psychological function. A case in point is the difference between what has been published about the HBW including those publication that have made use of Craig's own data (e.g., Hárosi 1987, 1994), the linear relations, and what he presents as the ideal curve for the half-maximum bandwidth to be found in fish retina with varying mixtures of rhodopsin (A_1) and porphyropsin (A_2).

Theo also makes a point that is important for being able to follow the unfolding of this conversation. Earlier he has stated keeping the data if they matched the template. This might be heard as implying a match for the entire absorption spectrum. But in this fragment he notes that what is important, according to MacNichol (1986), is that only a small part of the mathematical curve actually matches the data. This is the match that we can observe exhibited in Fig. 4.2b, where it is also clear that there is a lack of agreement on left part of the spectrum – especially in the "foot" part of it. This also has implications for the determination of the measure currently under discussion: half-maximum bandwidth.

Excluding Recalcitrant Data

"What we may want to do is huck those points"

In Chap. 3, I already describe scientists at work excluding measurements from entry into the data set. In this section, I add to this description to exhibit further criteria for excluding data. As indicated above, the general scientific tenor is that there is a direct equivalence between fundamental structure and mathematical structure captured in the couplet {fundamental structure $\leftrightarrow$ mathematical structure} that establishes the exact equivalence between the two parts. Fundamental structure, however, directly reveals itself only in nice data. That is, in Fig. 4.1b,

[15] This was a simple transfer from one hard drive to another via a local network.

we see a nice and clean peak; in Fig. 4.2b, on the other hand, the noise surrounding the peak is almost of the same order as the peak itself. It is not so nice and does not reveal on first sight the presumed fundamental structure, here the absorption spectrum. That is, we already see above that there are absorption spectra that do not lend themselves to easy classification. In this session, part of the talk concerned the possibility or need to exclude some recalcitrant measurements, that is, measurements that do not fit into the theory – or preconception – and in fact, would be more consistent with some alternative theory, pattern, or preconception. In the present case, Craig proposes: "What we may want to do is huck those points." When scientists articulate trouble reading or interpreting data because of variations of unknown or uncertain origin, observers are provided with opportunities to see what scientists do when the deep structures visible in idealized graphs – such as those that Craig and Theo have drawn on the chalkboard – are not immediately available in real, experimental data. At this point in the meeting, the team is confronted with a plot of the half-maximum bandwidth of the spectra it collected (i.e., HMB) against porphyropsin levels (% A_2). This plot, projected in turn 039, reflects the real data; it constitutes a stark contrast to the idealized curve Craig has sketched on the adjoining wall. This idealized graph presents the expected ("preconceived") deep structure whereas the graph that has arisen from the exchanges between Shelby and Theo exhibits the real case, the deep structure of which is at issue in this subsection. What will scientists do to come to grips with the apparent variability in the currently displayed data plot? In the sciences, the exclusion of "outliers" is a common practice. The concept and associated actions are part of the practices designed to remove what is considered clutter so that nature exhibits its form more easily than it would if all the clutter where retained. However, if a "point" represents many measurements rather than a single one, excluding it based on the notion of "outlier" becomes more problematic (unless, perhaps, if scientists recognize some systematic error in the many measurements underlying the single point). In this third fragment of the episode, the possible exclusion of points is the topic of the talk.

Craig announces that he is done ("okay") (which marks the 40 s of silence as an accepted offer to take the time for taking notes). Shelby then orients the other laboratory members present to the plot he has already projected by explicitly linking it to the one that Craig has "just talked about" (turn 057). Shelby explains that his plot "is in the reverse" and that he could plot it "in the other way around" so that the "bottom axis goes from 0 to 100" (turn 057). But he also formulates that other team members can "get kind of the general impression of what [they] are looking at" (turn 057). Theo, who is still standing next to the diagram that he has earlier drawn immediately follows Shelby by saying that there is "a little noise on here" while pointing to the left side of his absorption graph.

Fragment 4.3b.i

056 C: okay
057 S: thats that plot you just talked about. um; its in the reverse, so i could reverse it the other way around but the bottom axis goes from zero to a hundred instead of, (0.50) so you get QUITE of the general imprESsion of what you were looking at;
057b (0.75)

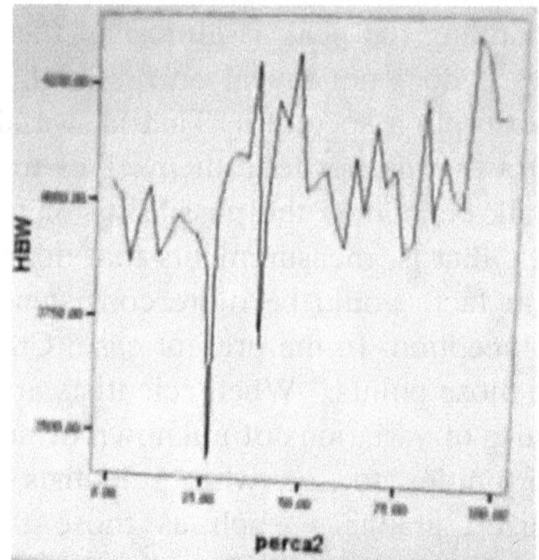

058 T: we have a little noise on here ((*points to "ringing" on left*)) and * so i=m just taking the half bandwidz from hERe to hERe (1.76) ((*points intersection of curve with half-maximum on right, then with noise on left*)). so dats why you see dese dips ((*points to projection, up and down gestural movement*))are when dis little peak would go zrough the fifty percent cuts off my ALgorizm. (1.40)

In Theo's statement appears "a little noise." This noise is, as all team members know, the familiar "ringing" – oscillations overlaying the left branch of the Gaussian – the origin of which is not exactly known. Because of this variation, the half-maximum bandwidth is extracted by the algorithm to the point where the "ring" first crosses half of the maximum height, so that its measurement ends up being smaller than it should be (i.e., when taken at the point where the idealized Gaussian should normally run, which would have been the case had the Fast Fourier Transform be applied). The algorithmic extraction of half-maximum, therefore, introduces uncertainty in the determination of A_1/A_2 ratios.[16] The algorithm that Theo has designed at that instant in the research does not project where the Gaussian *would* lie but takes the half-maximum bandwidth at the location along the wavelength axis where the absorption curve first falls below half of maximum height.

In this situation, team members observe the half-maximum bandwidth twice. First, there is the horizontal chalk line that Theo has drawn in the curve. Moreover, Theo has produced a movement twice, a movement that produces and then makes salient the construct of half-maximum bandwidth as a distance. He has drawn the horizontal line at half the maximum height. The distance that his arm and hand

[16] This is one of those instances discussed in Chap. 3, where part of the instrumental process introduces variations that scientists attempt to get under their control.

cover while pushing the chalk against the board *is* the visible half-maximum bandwidth – though when represented, this bandwidth is measured in cm^{-1}, the inverse of a distance that does not have an embodied equivalent (unless in the form of inverse speed, given in time taken per distance unit, which yields a number that is larger with slower speeds). Second, he reproduces the movement, which now becomes a sign not only indexing the line that everyone else can see, but also repeating the original movement. This movement, therefore, is not a representation of something else: it is an originary movement that is capable of reproducing itself without external (cognitive) mediation by means of signs (mental representations, thoughts). This is a form of mathematical knowing in movement, where physical enactment is an experience of – and corresponds to familiarity with – physical concepts.

Here, then, Theo modifies the original graph for the purposes at hand. He uses the graph to show where uncertainty in half-maximum bandwidth determination enters our work. That is, the original smooth graph was to show one feature and was produced for the purposes of communicating it: the places that are of importance in determining the height of the curve and, with it, the half-maximum bandwidth. Now that the data exhibit a lot of variation, he modifies the *ideal* and idealized graph to introduce one of the aspects that can produce variation. That is, the variation and uncertainty visible in Shelby's graph does not arise from the source, the retinal cells, but from the transformations that the measured absorption curves undergo towards producing the porphyropsin content (determined by means of λ_{max}) and the half-maximum bandwidth, which the team hopes to turn into a second reliable measure of A_2 levels. It allows Theo to articulate where variation comes into the determination of the half-maximum bandwidth as existing in the database at the present point in time. Theo proposes something, though the speech volume becomes so low that it is no longer possible to retrieve what he is saying while Craig overlaps uttering the tentative proposal to "huck those points" (turn 061), which Theo accepts ("yea").

Fragment 4.3b.ii

```
059   C:  yea. (0.75)
060   T:  so we have to <<p>[make (??)>]
061   C:                    [so what we] may wanna do is huck
          those points.
062   T:  yea.
062a      (1.29)
063   S:  so we could;
          <<all>yea jst so we
          could go back and
          look at those data
          point; and
          actually find out
          what data point that
          is> ((points to one,
          then the other tip
          of bottom peak));
          (0.78) <<all>so if
          we could just kinda
          keep off
```

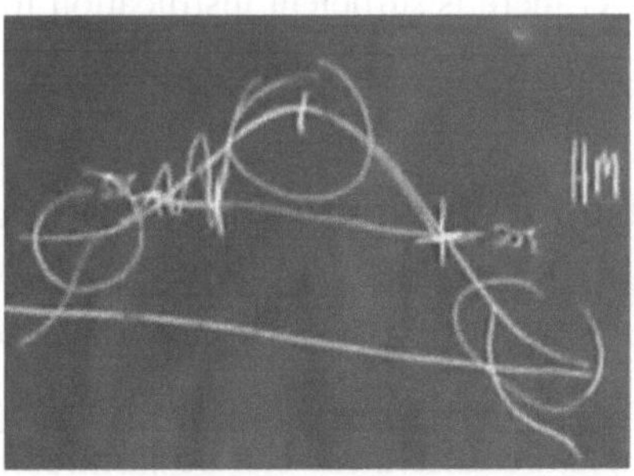

```
064  T: and what we=re doing
        now is we=re keeping
        de image of every
        analyzed curve
        (0.50) dats in ze
        database; so you can
        look at whats
        hAPpnd. (0.99) so
        you can go
        specIFically to dose
        two low points
        ((points toward the
        screen, wiggles in-
        dex finger up and
        down)) and see what
        the curve looked
        like.
```

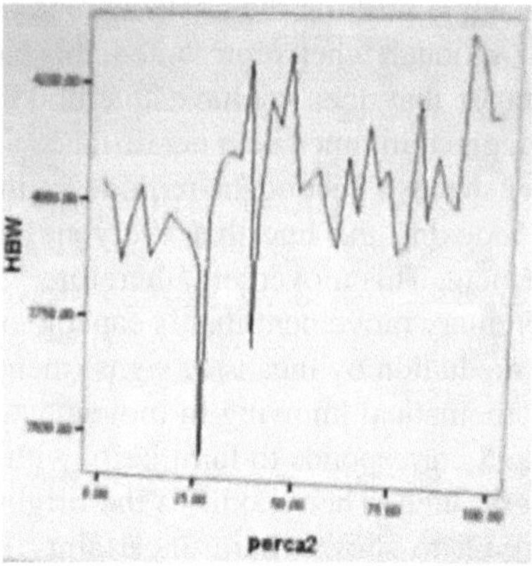

Here readers may note the sequence of events in which the scientists produce an accepted curve. First, Craig diagrammed what he formulated as prediction of what the curves should look like (which readers may want to attribute to knowledge gained from previous research). Because Shelby's curves did not fit this template (curve), due to large variation, Craig justifies the "hucking" of those two points. He does so even though each point plotted represents the calculated mean of a series of measurements, and therefore are not simple outliers (singularities). Shelby's curves become malleable and "noise" becomes the subject of scrutiny.

In this section, the elimination of some data points – or, in fact, an average of a collection of data points – is articulated as a possible way of dealing with variation and the uncertainty deriving from it. Whether this is a feasible action, however, is not being established in this conversation – possibly because it takes some detailed investigations of the kind that is named in Shelby's statement (i.e. that the background to the data points require further study). This statement is elaborated in and by the following turn, as part of which Theo describes what they have been doing: keeping the image of every curve analyzed, which allows the team to "look at what happened." That is, hucking data points is not a lightly taken statement but is only the beginning of what requires a lengthy process of investigating the data to see whether there is sufficient justification for including or excluding these.

Testing Hypothetical Exclusion of Data Point

"If we just ignore that for the time being, it looks like . . ."

Once scientists know that they can legitimately ignore some data points, for example, because these can be justified to be outliers, then they are in a position to read the data in new ways. They may, as happens here, *provisionally* ignore some data points to see what it yields, much in the way painters do after making a stroke on the canvas and then step back to evaluate the effect of their move and then to correct or continue with what they have done. That is, it would be typical to articulate a move and thereby to objectify it to be in a better position for evaluating its implications. The advantage of such moves is that new insights thereby arise from an external, epistemic action that reveals its possibilities by inspection and further possibilities for theorizing through reflection.

Getting a Grip on Difficult Data

"I don't know what's going on here . . . that looks a bit funny"

In the discovery sciences, patterns do not just present themselves in the way that school science and mathematics appear to assume. Patterns do not just exist; but they are not merely "constructed" (i.e., ad lib) by agential subjects of activity. What can be said about data is a function of what the scientists *find* to be displayed; and what they find displayed is a function of their preconceived notions and what is materially there. If the data require a different way of looking, a rupture first has to occur with the habitual ways of looking. Following Theo's explanation of a possible source of the variation in the data (turn 064), Craig provides a first reading – which might be heard as a first attempt in articulating the deep structure in the data. He points to the right-most peak and marks it as something that "he does not know what's going on" (turn 065) and, prior to articulating a trend, he suggests "ignoring it for the time being."

Fragment 4.3b.iii

```
064   T:  and what we=re doing now is we=re keeping de image of
          every analyzed curve (0.50) dats in ze database; so
          you can look at whats hAPpnd. (0.99) so you can go
          specIFically to dose two low points ((points toward
          the screen, wiggles index finger up and down)) and
          see what the curve looked like.
064b      (1.52)
065   C:  s:O::; (0.82) if wE:
          ((gets up and walks to
          projected graph)) (0.28)
          can get, <<all>i i dont
          know whats going on
          here.> * (1.26) that
          that looks a bit funny;
          but if we just ignore
          thAT <<p>fer (1.14) the
          time bEing, it looks
          like> (1.46) thIS
          ((points to chalkboard))
```

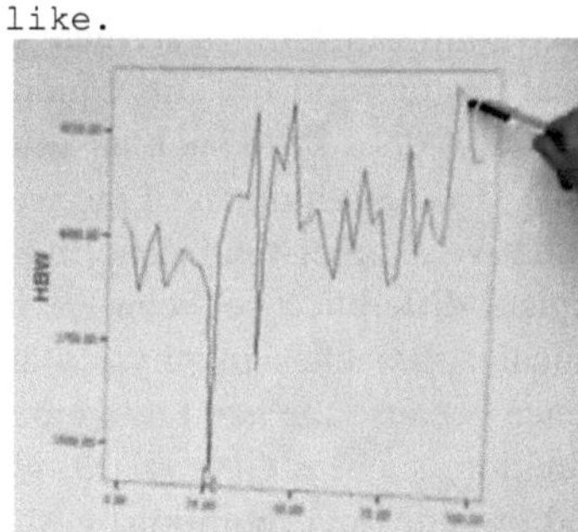

is where it; ((*hand moves left and down*)) * (2.09) the (0.78) if we rUN a regrESsion on thiss an (0.58) we igNORe (0.20) some ((*point to right most peak*)) of the ((*center peaks*)) extrE:mes, like here ((*2 downward peaks*)), here ((*2 top peaks in center*)) (0.41)

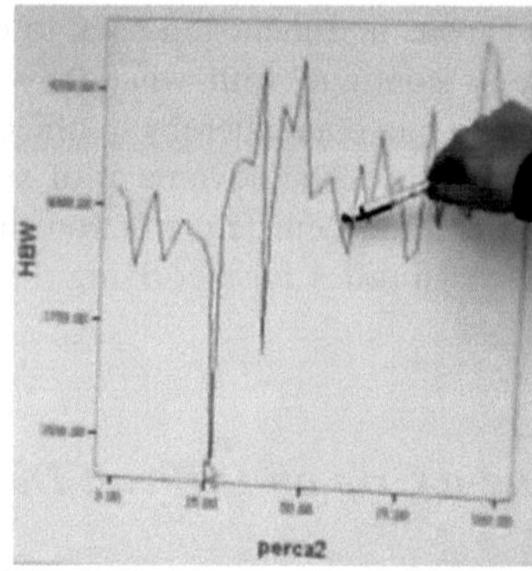

066 S: this is [thrEE] thousand data points, so that it is [

067 C: [here] ((*right-most peak*)) [yea

068 S: a pretty bI:g dataset, this is NOT jsta couple points, this is the whole shebANg so your bound to get some (0.69) extrEMe points.

He also points to the two center peaks upward and to the two downward peaks, marks them verbally as "extremes," and suggests "ignoring them." He finally proposes "running a regression on this." At this point, Shelby brings out a contrast in suggesting that "this is not just a couple [of] points," but in fact "this is the whole shebang" (turn 068). That is, we see and hear a proposal for "hucking" or "ignoring" certain "extremes," but this proposal is not accepted but rather confronted with the statement that there are 3,000 data points. Shelby contrasts this with the description of not just being "a couple of points." This large number, as he states in an elaborating manner, implies that there are *some* extremes. But because this is a large number, any trend would be rather stable with respect to a small number ("some") extreme points.

The extreme points are not just there. To exist qua orderly object in the hands of the research team and therefore, to exist in the modality of being ready-to-hand, any "point" has to be demonstrably "extreme." Here, this is achieved in the sequentially produced contrast Craig's statement produces between the point made salient by the indexical gesture and the subsequent hand movement that not only bypasses the point but also does so to such an extent that it becomes an instructable instance of "extreme." Although much of the work on gestures does not treat the two gestures as distinctive – other than denoting them as "deictic" and "iconic" – there is an ontological difference between them. This difference constitutes the heart of the dynamical approaches in physics and mathematics (Châtelet 1993). There is a difference because the hand movement enacts a curve, represented in mathematics by a relation, such as $G(x, y) = 0$, whereas the individual points on such a curve, denoted by indexical (pointing) gestures, represent a parametric or function of the kind $y = f(x)$ where the separation of data and result is already accomplished.[17] It is

[17] The equation $x^2 + y^2 - 1 = G(x, y) = 0$ expresses a relation between x's and y's, here, a circle with radius 1; it includes all values x and y that satisfied the equation. This relation therefore

precisely in the gesture that a virtual relation becomes actual and material, in and through the body of the scientist. That is, "the gesture envelops before grasping and sketches its deployment before denoting or exemplifying" (p. 33).

Tentative Identification of a Pattern

"It looks like to me what we have is something that looks like ... *this*"

In the midst of recalcitrant data, imposing a pattern is the reverse side of having patterns emerge. There is therefore a dialectical process at work, in which imposition and emergence of pattern play off each other. Hand gestures do not just show what pattern there is but in fact allow the eyes to see what there might be. This is but another instantiation of the relationship between articulation (joints between things we encounter in the world) and articulation (signs that we use to tell the things apart) that Heidegger (1927/1997) originally stated for the relation between things (beings) and language. That is, the eye movement required to see the trend in the data emerges together with or follows earlier hand movements that have been tracing out a space of possibilities (Roth 2012a). Once the eyes or hands knowledgeably reproduce a movement, a bifurcation may occur so that this movement comes to be accompanied by ways that enable making the movement present again in its absence. Thus, a random movement with the hand may allow a pattern to become visible and, subsequently, become an iconic gesture denoting this pattern. In the following, we see how the hand movements outline such possible patterns in the data, which will be seen when the eyes are entrained into the same movement.

Craig has suggested "huck[ing] those points" that show a lot of variation (outliers). This variation exists in a demonstrable manner such that when required, the spiky nature of the "downward spikes" can be exhibited for anyone not yet convinced. He again suggests ignoring some of the outlier (turn 065), but Shelby has responded saying that the dataset is large and that in this case it is likely to have some that reflect large variations. Craig now suggests running a "polynomial on it" to "see what sort of correlation we can come up with," but adds that they first have to "*recursively* go back through the data" (turn 069) to see whether "those points [outliers]" are "due to errant spectra" (turn 069). He further suggests that these could "have provided considerably more variance to that point in the transition," thereby referring to data reflecting – according to the "dogma" that framed their work at the time – a changeover from high porphyropsin levels thought to be typical of salmon in freshwater to lower porphyropsin levels thought to be typical of salmon in saltwater.

corresponds to a moving gesture. In a function such as $y = f(x)$, each x from the domain is mapped onto a corresponding value y from the codomain. It therefore correspond to an indexical gesture that points to some target.

Fragment 4.3b.iv

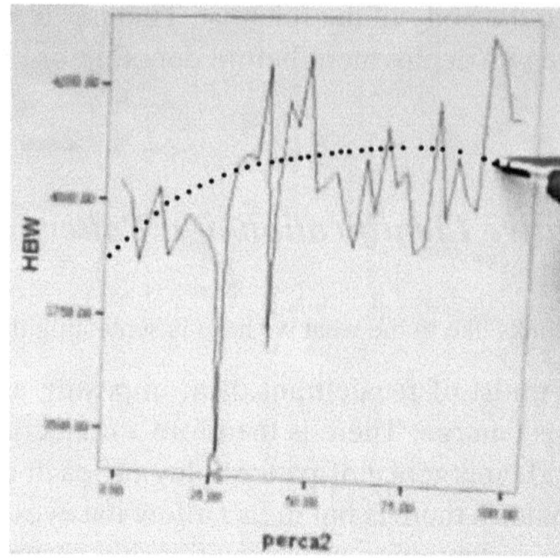

068a (1.22)
069 C: yea; (0.43) it lOOKs like to mE what we hAve is sOMething that looks like (0.38) thIS ((* *draws imaginary curve from right to left*)) (0.31) looks ((*points to his HBW graph on the chalkboard*)) (1.81) like thAT. (2.63) a::n:d (3.64) we could run a polynOMial on it and sEE:; (2.39) ((*walks back*)) what sort of CORrelation coefficient we can come UP with. (1.91) but the FIRst thing that one would have to dO is (0.88) recURSively go bACK thrOUgh the dAda; (0.72) and mAKe sure that thOSe ((*points to screen*)) (0.41) lOW points: (0.65) are not dUE to (0.21) ERrant (0.99) <<f>or> (1.41) <<f>s::pECtRA:> which:: () are NOt meeting critERia (0.91)
070 T: <<pp>uh hm>
071 C: right, or the criterion; (2.08) A:N:d (0.37) have provided (1.95) u::m: (0.74) conSIDerably more vARiance to thAT partICular point in (0.46) the trANsition; ((*horizontal gesture toward screen and parallel to the ordinate*))
072 (4.36)
073 C: uh becOS; (0.66) i would say with the excEPtion of; i mean EVen if we inclUDe thAT (1.76) ALL the data except for those tWO:: (1.13) dOWNward spIkes::. ((*Points to screen*)) (2.70) now those twO: downward spIKes::; (2.54) when you say youre including ALl the dada does that mEAN that you=re; ((*leans forward points to sheet with record of data included*))
073a (2.47)
074 S: <<all>((*open hand gesture to sheet*)) tsee yguys i jst, i jst did it everything quick>; (0.29) all of these
075 C: all means everything?

As he continues, Craig first makes a statement that acknowledges what Shelby has said ("yea") and then provides an elaboration of the data by showing the trend that they display (turn 069). Placing his pen against the projection screen, Craig "draws" a parabolic curve that looks like the one he has earlier drawn on the chalkboard (offprint in turn 069) – even though he has just prior to this instant exhibited the opposite relation. There is a high degree of similarity, even in the relative position of the HBW, which is higher for 100 % A_2 than for 0 % A_2.[18] That

[18] Shelby has drawn the abscissa from left to right, and, therefore, in the reverse of what Craig has done on the chalkboard; see turn 057.

is, in this situation, Craig's statement explicitly notes that the trend his hand has just traced in and through its movement, "this," "looks like that," the earlier graph. That is, the same movement in the reverse that earlier had produced the line on the chalkboard now, ephemerally, exhibits the trend in the data. In other words, what we might alternatively see as a parabolic trend, visible if a regression were to be performed on the data displayed, becomes an idealized inverted parabolic trend (fundamental structure) represented in the trajectory of the pen over the data.

In the course of this talk, Craig has actually changed what he brought out as the trend underlying the data. Initially, he gestures – indexically/iconically in his movement, iconically with the hand inclined – to show the linear relation from lower left to upper right at the right end of the plot. He then turns and points to the graph on the chalkboard (turn 065), which in fact runs differently than what he has shown just now. His movement then exhibits a curvilinear relationship with a maximum in the center (turn 069). That is, the hand movement both exhibits what is required to demonstrate the curvilinear relationship and serves as an instruction for seeing (again) this very relationship. It therefore exists twice, in the form of "two technologies": as orderly phenomenon that everyone in the room can see once he has learned how to look and as embodied practice, which, when enacted, *gives* immediate access to the so articulated phenomenon of the curvilinear relationship.

In this situation, the HBW versus %A_2 graph provides an indication of the variability and uncertainty in the data, perhaps more sensitive than other indicators because of the reasons that Theo earlier elaborated. Craig insists on checking the data that these all meet the criteria, especially those "low points" (turn 069). He expresses concern about the fact that "in the transition, that is, the transition of the coho salmon from their presupposed freshwater rhodopsin-related A_1 chromophore to the porphyropsin-related A_2 chromophore thought to be dominant in the saltwater environment. Data such as those in the "downward spikes" would introduce unwanted variation in the determination of the actual proportion of A_1 and A_2 in the visual pigment of the fish. Craig brings the conversation back to the issue of what has been included in the plot to constitute the data ("When you say you're including all the data, does that mean that you're . . ." turn 073), to which Shelby responds that he did something quick. Craig insists, "does that mean everything?" and Shelby affirms ("Yea"). But again, he insists that there are "a lot of data points" (turn 076), as if he wanted to warn the participant to "huck those points" all too quickly.

Gestures Foster Seeing

"Actually I'm surprised that that's so spiky"

Gestures afford seeing not in the least because the movement involved projects possible movements for the eyes, which are required to see anything at all.[19] At the

[19] Experiments in which an image is fixed on the retina show that within a few seconds, the person only sees grey. Without eye movements parallel and transversal to a line, the line could not be perceived (Roth 2012b).

same time, however, these possible movements allow features of the background to protrude into consciousness not in the least because these stand against those patterns of movement that are the movement required for patterns to emerge. Thus, as soon as the gesture imposes and makes emerge a possible pattern, certain features of the ground come to stand out, such as points far from the trajectory of the hand.

Craig begins making a statement about being "pretty . . ." but he is interrupted by my question about how many data points there are in each of the spikes (turn 077). This statement can be heard as a question co-articulating uncertainty about what exactly each data point reflects: individual or collated measurements. Shelby ascertains having heard my utterance as a "good question" but says that he does "not know how to tease this apart" (turn 078). His statement articulates surprise about the "spiky" nature of the graph and that he may have "chosen the wrong type of graph" (turn 078). His statement elaborates in reiterating that there are 3,000 data points, which would mean that it "takes a lot" for the curve "to go down that far" (turn 078). And then he says again that he is "kind of surprised." We can hear him express surprise about the variation and, therefore, co-express his anticipation to have a much smoother graph. Craig then makes a statement the content of which is the hypothesis that Shelby has "mix[ed] quite a broad spectrum of fish" and "that could have something to do with it" (turns 079, 081). In fact, the fishes that have contributed to the data were from different hatcheries and from different river systems. This "could have something to do with [the variation]" (turn 081). Here, only the familiarity with the background to the data collection allows Craig to articulate this hypothesis.

Fragment 4.3b.v

```
075a S: yea.
075b C: oh okay;
076  S: and so there may be it has a lot of data points;
        (1.06) <<dim>so i=m pretty i=ll just=>
077  M: =for a spIKe, how many data points are we talking
        about; (0.16) for each of these spIKes.
078  S: well thats a good question. (0.67) <<p>um (0.89) i
        dont, i dont know how to tEAse that apart on this
        particular;> (0.41) a=ACtually i=m surprISed that
        thats so spIky, maybe i chose the wrong type of graph
        to dO. (0.62) on this. <<pp>i dont know> (1.95) um
        (1.36) cozifthats a whole bunch of data points; i
        mean theres three thOUSand youd think itd take a lot
        to make it go down thAT far. (0.79) so i=m kind of
        surprISed. ((13:46))
078a    (1.31)
079  C: well youre ALso mixing aquite a broad spectrum of
        fishhere ((gestures over data source sheet)) so
        [that]
080  S: [yea ]
081  C: cOUld have something to do with it;
082     (0.74)
```

Graphs not only stand in as expressions of thought – as in the earlier presentation when Theo displays what he has been working with and seeing while processing the data – but also are used as resources and occasions to think with in a public forum. What we observe in this situation therefore is thought unfolding with expression, new expressions giving rise to new thoughts. We witness changes in what is being articulated and thought. Thought and expression are manifestations of one and the same process, not because expression follows thought but because the laboratory members present can find their own thought in their expressions.[20] It is not surprising, from such a perspective, to see Craig return to his seat and then note down what he has just expressed. In the course of his talk, Craig articulates for everyone else present what trends he can see in the data. These trends, however, do not correspond to the one that he has outlined earlier on the chalkboard. In fact, the trend is opposite to it with respect to its slope. He suggests that the bandwidth *increases* as the percentage of A_2 moves to 0 (as displayed in the graph), whereas in the diagram, the maximum bandwidth exists in the center of the A_2 concentrations from 0 to 100 %. It will turn out that some of the data points that he *now* determines to be suspect are not the ones he subsequently marks as suspect.

The graphs not only are the topic of talk but also become ground for articulating relations that would be much more difficult to present in verbal form. Verbal indices ("here," "there," "this") are used to "tie" the indexical (Fig. 4.5a), indexical/iconic (Fig. 4.5b), and iconic (movement) expressions (Fig. 4.5c) to the graph. In the deictic gestures, the graph as a whole or individual points are the topic, whereas the graph is the ground against the *dynamic* hand/arm movement that expresses something not itself in the display. It is layered over and above the display but represents something "underneath" or underlying the data, the deep structure. The erratic data points are but concrete expressions of extraneous variation of some true signal that is made visible in and by Craig's hand movements.

This part of the meeting throws further interesting light on graphing because Shelby plotted half-maximum bandwidth with a connected line graph. None of the other team members presented commented on this matter, thereby making it an acceptable way of representing the data at hand. Yet the scientists do not apparently engage in the *critical* evaluation of trends that go against "some preconceived ideas" about the relationship between HBW and $\%A_2$.

Delimiting the Data by Sets

"I could chose just one set"

Random variations that obscure true relationships underlying data are potential trouble spots. This is so because random errors may exhibit a linear relationship when the true relationship is exponential. This has to be particularly the case when

[20] There is a long history of scholarship on the relation between thinking and speaking and the presence of thinking to itself (e.g. Hegel 1807/1979; Merleau-Ponty 1945; Vygotskij 2005).

the size of the error that comes with each data point is unknown but subsequently stated to be systematic. Attempting to grasp the source of variation or limiting the variation by controlling the nature of the data, therefore, may be an important step to being able to model what are taken to be the true relations. Looking at what scientists actually do might further elucidate this question. The data that Shelby currently displays in this meeting were taken from coho salmon sourced both at the Kispiox Hatchery and Robertson Creek Hatchery. There might be a possibility that the outlier data come from one or the other of these batches of fish. One way of pursuing where the variation comes from is to engage in a disaggregation of the data.

In this fragment of the episode we can see that when scientists do not have a ready answer, they do not and cannot know whether some action will produce the anticipated changes in the data displayed. But any orderliness that after the fact will be accounted for arises from the practical aspects of the work, which leads to "transcendental orderliness" (Garfinkel et al. 1981, p. 141) of the natural objects in the scientists' hands. In the preceding subsection, it is the "transcendental" disorderliness, the properties of *this* display that the work of the scientists comes to provide. In fact, orderliness and disorderliness reflexively make each other possible as ground against which the other emerges as figure. They do not exist independently but require each other for their definition and support. There is therefore always uncertainty, arising from the production of one or the other of the inscription (see Chap. 3), which threatens the stability of the pair of inscription. It has been noted that inscriptions may in fact hide certain aspects of mathematics in the workplace – these become black boxes (Williams and Wake 2007). Work has to be done to unpack that which is hidden and invisible in a black box – such as the number of data underlying each point plotted invisible to Craig but pointed to by Shelby. In the present instance, the particular inscriptional form Shelby has chosen exhibits and hides variation simultaneously. It exhibits variation that our team intends to get under control or knows the source thereof; it is a variation that muddies the anticipated or real trend in the phenomenon. But the inscription also hides the variability of each data point, itself the mean of many measurements, which total over 3,000. In the following, our team enacts the work that it hopes will unpack and make visible the source of the variability at hand.

Immediately after Craig's articulation of the hypothesis that the variation might be caused by pooling all the fish in their study, Shelby suggests that he "could just choose one set" (turn 082). His statement can be heard as a proposition to use fish that had come from the Kispiox River in northern British Columbia (turn 087); Craig accepts by restating that possibility (turn 088).

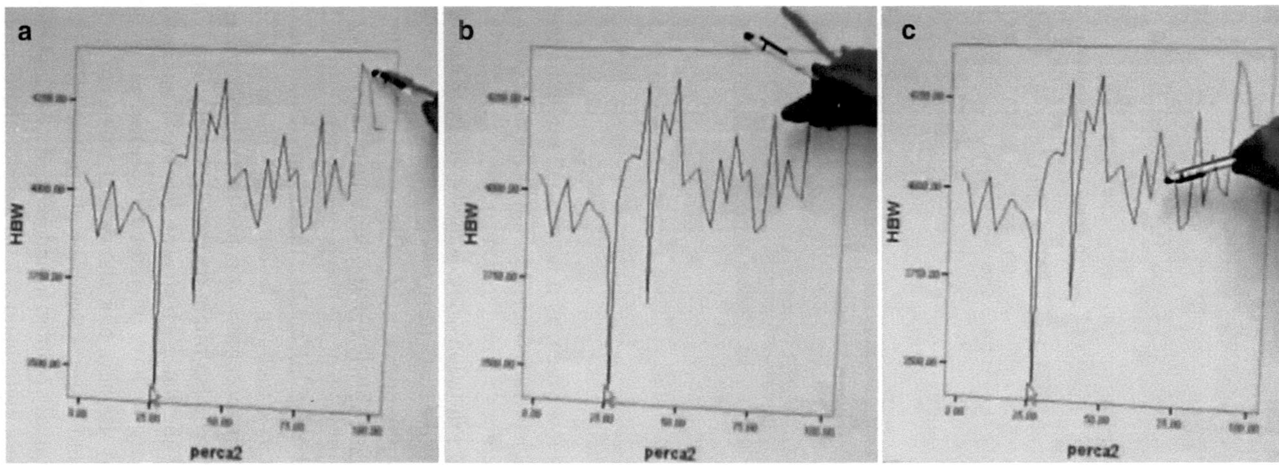

Fig. 4.5 Craig marks and makes salient specific features that he articulates to be suspicious and gestures – deictically in (**a**), deictic-iconically in (**b**), and iconically in the way the pen parallels the earlier produce hand/arm movement in (**c**) – the trends that are exhibited in and by the data

Fragment 4.3c

083 S: i could just choose ONE set. (0.15) <<p> if ye [wan]t>

084 C: [yea].

085 S: yea sure; (0.19) thats easy to do.

085a (0.52)

086 C: lets, (0.30) d (0.16) d=do we have the mO:st for the kispiox [fish]

087 S: [yeah]; <<p>i=ll select the kispiox.>

087a (1.23)

088 C: just the kispiox hATchery fish. (0.22)

089 S: yea; (2.65) ((*works computer*)) and <<dim, p, to him-self>ah i wonder if i should choose; maybe an error bar plot would be more appropriate, no thats not gonna give us. the wrong pointa. (1.31) h:: which way. we=ll try the line again, just leave it the way it is, see if we can get this a bit cleaner>. * (1.11) so were still getting those sAMe; (1.85) so those are actually in the kispiohhx, those two points ((*"touches" each with cursor*)) that are going out like that () <<p>interesting>

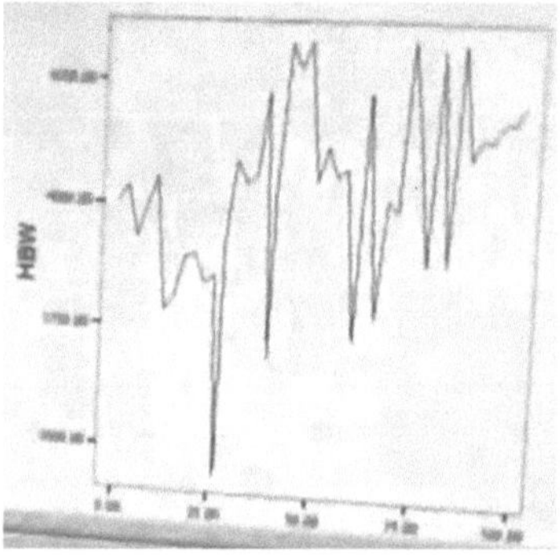

090 (8.92)

091 C: hm

092 (12.16)

Shelby speaks with a low voice, as if he was talking to himself while selecting the data and form of analysis (turn 089). As soon as the new graph representing only the Kispiox River fish comes up on the screen, he notes: “So we are still getting those same … so those [variations] are actually in the Kispiox [fish] those two points that are going out like that” (turn 089).

The result of these actions appear to be baffling: As soon as the new graph for the “Kispiox fish” is displayed, Shelby notes that these “are still getting those same [points],” while “touching” each of the two peaks with the cursor. He then notes that “those two points,” “that are going out like that,” “are actually in the Kispiox” (turn 089). He marks this fact as something “interesting.” There is a long, 21-s pause, as if the graph had left us all speechless, a pause that is only interrupted by Craig’s interjection (“hm”) – which also might be heard as indicative of puzzlement. The selection of a subset of the data, which was introduced as a way of decreasing variation actually appears to have increased it. However, a direct comparison between the former and the new plot shows that there are in fact new features that have appeared and old features that have disappeared (Fig. 4.6). Thus, the one peak that Craig had repeatedly pointed to and marked as a candidate for elimination has disappeared. But on the right-hand side of the plot, there not only is a new peak but the existing peaks are as high as those in the middle earlier designated for elimination. What is important to this meeting is what arises in the relation between Craig and Shelby: the two lower peaks that are still present. Shelby characterizes

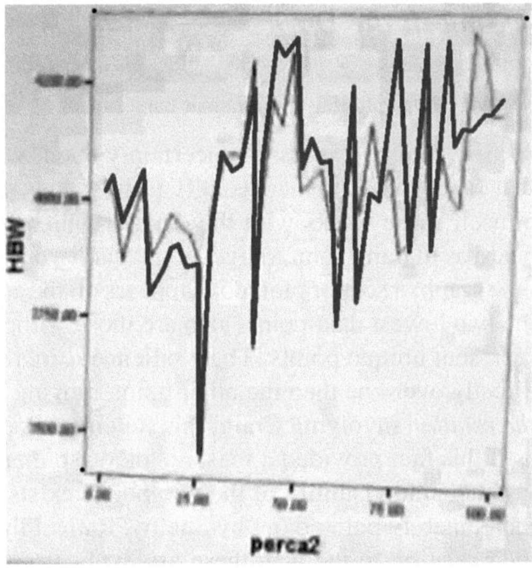

Fig. 4.6 Direct comparison between the total data set and the set deriving only from the "Kispiox fish"

them, therefore, as particular to the Kispiox fish. After and thereby breaking the long pause, Craig tentatively ("maybe") proposes a plot with error bars.

In this part of the episode, readers therefore observe how uncertainty comes to be proliferated in the very attempt to get it under control. At this point, the team does not talk about where the variation comes from and, therefore, how to get it into grip. Much in the same way as some of the team members were fiddling with the computer monitor in Chap. 2, the team is fiddling here with different ways that might promise but does not guarantee the control it is aiming for.

Using Graphs to Delimit Errors in Graphs

The preceding move has not let to the intended reduction in variability of the data and, therefore, in the intended reduction of uncertainty. In such a situation, especially when scientists are not yet familiar with the domain, any action is as good as another (Chap. 2). It is only after the fact that a value "good" or "poor" can be attributed to action and its ascribed intention or, in other words, that the quality of fit can be determined between situated actions and the plans (see Chap. 3). It may not come as a surprise then that Craig makes the proposal to try another avenue: plotting the data with error bars. The uncertainty of what such a move may bring about is expressed in the adverb "maybe," which modalizes the verb "to try" as a possible but not necessary next move. The verb "to try" implies decision between alternatives and finding out about something that is doubtful or obscure. Perhaps unsurprisingly, cognitive scientists have denoted this problem-solving method as the trial-and-error approach. This section, therefore, exhibits more of the tentative moves to bring order to the situation and, in and through the work made visible to everyone present, make this order socially available.

A Shot in the Dark

"Maybe try plotting it with error bars"

At first, Shelby expresses uncertainty about whether "that" is "going to work with that" and, speaking increasingly faster, states that he will give it a try. He mutters to himself as he works with the data, at which point I begin a statement that can be heard as making some suggestion ("can't you put this," turn 096) at which point the new graph, a scatter plot, now appears on the screen. At this point, Shelby notes that the two lowest data points also are those without error bars, and that they therefore represent unique points. Their salience is increased by the cursor, which he moves directly over one then the other point, moving back and forth three times. *In and as the relation* involving Craig, this statement is confirmed and marked as salient. It is as if this fact provided a reason: "okay, so there you go" (turn 101). That is, in this instance, the visibility of the two points exists in and as their relation. It is not just there, merely pointed out by Shelby. Rather, the distinction of these data points and their relation to the way these are explicated is a result of the relation. "They *are* just two unique points," that can be pointed to (by means of the cursor) and that can be seen and confirmed on the part of the listener ("yea," and "here you go").

Fragment 4.4a

```
093   C: kay hu, (1.33) maybe try it plOTtin:g: with (0.30)
         error bars
093a     (0.95)
094   S: yea, but i=m i=m not sure if that is going to work
         with that(?) <<all>i=ll give it a try> (3.63)
         <<pp>a:nd (1.57) see? (0.51) twentytwoand (4.49)
         whats there thats the question now
095      (0.67)
096   M: cant you put this;
         (0.27) * oh okay
097      (3.32)
098   S: so you get these tWO
         points; yea look at that
         those tWO points that
         have no error bars,
         ((points to two points,
         moving cursor back and
         forth)) theyre just two:
         (.) unIQue points
099   C: yea; (0.36)
100   S: <<p>yea.>
100a     (0.76)
101   C: yes there you GO;
102   S: yea its quite different;
         so you do get a general
         INcrease (0.52) from zero to ahundred; ((moves cursor
         in straight line from left edge to right edge of
         graph)) (0.64) very vERy subtle ((gestures with cur-
         sor))
102a     (0.53)
```

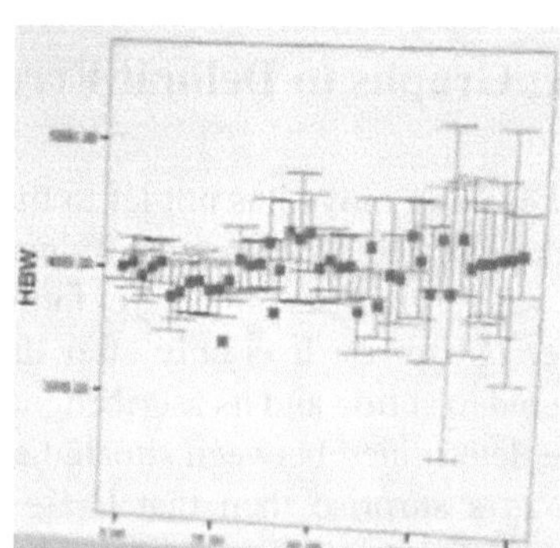

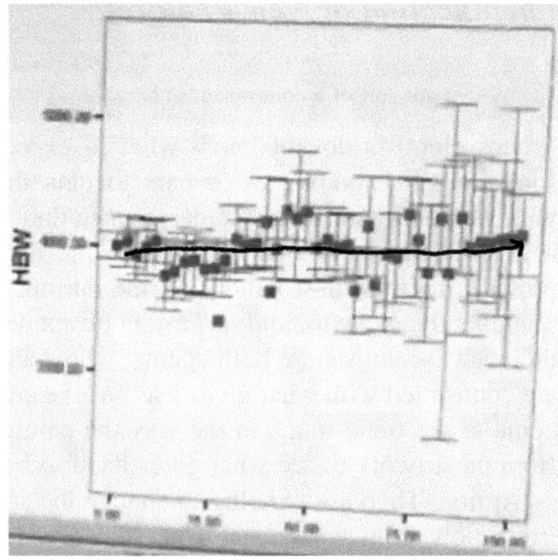

Fig. 4.7 Shelby moves the cursor across the data set to show a "general increase from 0 to 100" (turn 102)

In this instance, the discovery of the two points as singular and therefore as negligible to the trend is itself an occasioned production that emerges in the course of the team's inquiry. Shelby first draws a comparison to what has been: "yea, it's quite different" (turn 102). He then states an implication for the trend of the data: "you do get a general increase from zero to a hundred, very, very subtle" (turn 102). As he makes the statement, the other members to the setting can see the cursor move from the cluster of points at the left margin of the graph to the aligned cluster of points on the right (Fig. 4.7). The line, therefore, not only is pointed to but also iconically produced by and available in the cursor movement. Although the relationship is different from that which one of Craig's statements has articulated before, other team members present do not confront the two different hypothetical deep structures in the data and the trends that they express. At this point, nobody in the room comments on the fact that Shelby's statement can be heard as a proposal of a linear correlation whereas Craig has made visible a parabolic relationship: one of these is still visible in the line on the chalkboard, the ephemeral gesture perhaps present in perceptual memory.

In this instant, therefore, plotting the error bars yields an image of a funnel opening from left to right. Any acceptable regression curve will lie within that funnel. Even a cursory look allows following the more-or-less linear cursor movement from the left to the right that goes with the statement "you get a general increase from zero to a hundred" (turn 102). To some extent, the mystery about the underlying pattern has been resolved through the move to use error bars that provide indications of the variations within the data. The variation present has been disciplined by the disciplinary and disciplining practices accepted within the field.

Identification of New Features

"We've got sort of a component in here"

When scientists do not know what to expect when they follow a hunch or do something like adding error bars to data that they have considered heretofore without, new features may emerge once they are confronted with the results of an action (Roth 2012a). "Shooting into the dark" might just yield something that provides them with a handle on the current problem. "*Fishing* for certainty" or "Finding the straw to hold on" might be apt descriptions for the process at work. In this view, scientists are both agents, who decide to take actions, and patients, who are confronted with what gives itself as the unintended result of their actions. They come to see trend much in the way the painter comes to see when she steps back from the artwork to see what gives itself to be seen.

At first, Theo asks Shelby to change the scale on the "y-axis" (turn 103). Their irreducibly joint work establishes the new limits on the plot. Craig affirms and names what is happening "a higher resolution there" (turn 107). There is a pause (3.90 s), then the new graph appears, and then another 6-s pause unfolds (turn 109a). Craig is the first to speak. He gets up, walks to the screen, and then moves his hand in front of it to exhibit three trends in three parts of the plot.

Fragment 4.4b

103 T: can you change de scale from saying (0.20) on (0.64)
the y=axis
103a (0.48)
104 S: ah::: yep. (2.00) yea what would you prefer it to be.
104a (0.35)
105 T: maybe from zree zOUsand to (0.44) five zOUsand
((*00:15:58, Shelby changes scale on axis*))
105 (0.65)
106 S: <<p>okay. so a minimum>
107 C: yea; higher resolution there; yeah
107a (0.61)
108 S: <<p>i was wondering.> (0.82) maximum five thousand?
109 T: yea ((*00:16:14*))
109a (3.90) * ((*new graph appears*)) (6.00)

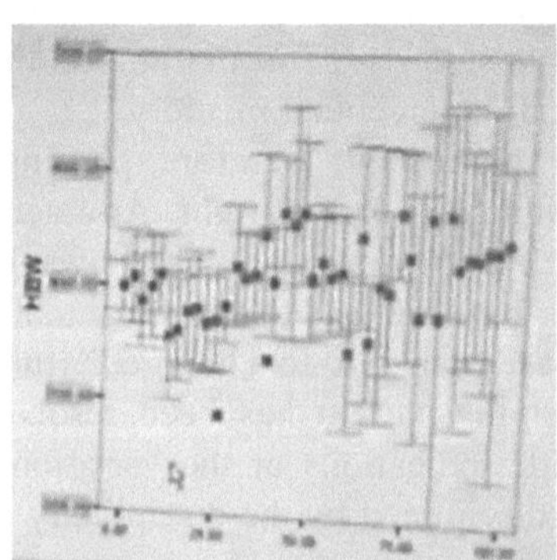

110 C: so what it lOOKs like is
((*gets up and walks to the projection screen*))
that weve got thIS:::
(1.00) HEre ((*draws imaginary straight line on the rightmost of graph, Fig. 4.8a*)) and then
weve got sort of a comPOnent in hERe ((*makes arch gesture through middle of graph, Fig. 4.8b*)); and then it goes
(0.68) ((*slightly upward movement, Fig. 4.8c*))
111 T: ah (0.23)
111a C: down again ((*gestures downwards at leftmost of graph, Fig. 4.8d*))

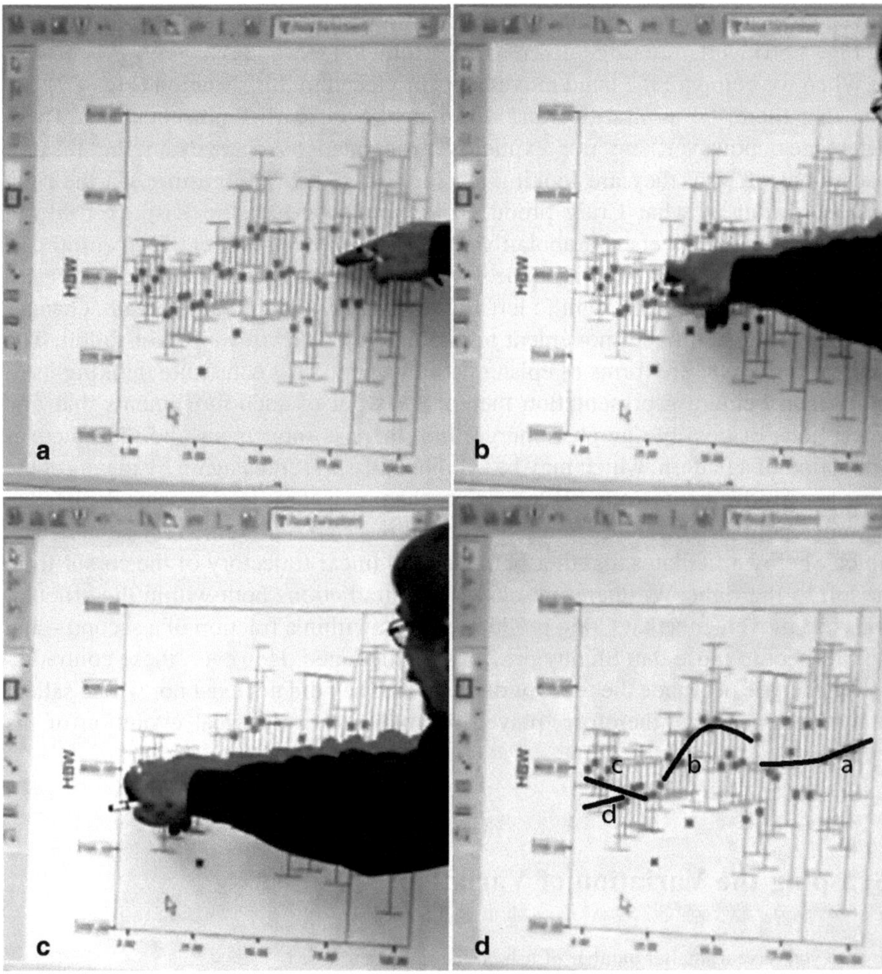

Fig. 4.8 Craig gestures the trend in the data. (**a**) There is a linear relationship as per his iconic hand shape and the movement of the hand from *upper right* toward *lower left*. (**b**) There is a hump in the center, which has, following the gesture, an approximately parabolic shape. (**c**) There is a linear relation on the *left* end of the display, as per hand gesture from *right* to *left* that moves in an approximately horizontal manner (though he says "down"). (**d**) Craig moves his hand in front of the data points, exhibiting in these movements the trends in the data

Craig's statement brings out – not unlike in his first analysis – that there is a linear trend on the right-hand side of the plot (a, Fig. 4.8. He then makes salient a "component," which, as the shape of the hand trajectory shows, is approximately inverse parabolic (Figs. 4.7b: b and 4.8d). He repeats the gesture first backward and then forward again. In the 0.68-s pause that follows, his index finger first makes a slight upward movement following what might be seen as the trend of the first 11 data points (c, Fig. 4.8d), then moves back toward the point where the parabolic movement ended and produces a second movement now slightly downward (d, Fig. 4.8d) while uttering "down again" (turn 111a).

When we compare the hand movements produced in this situation (Fig. 4.7), we note that there are similarities and differences with earlier presentations. These differences, however, are not explicitly articulated as changes in the idea or assessment of what they are looking at. The first part of the gesture, on the right end, is similar to what Craig produced in the context of the Kispiox fish, but different from the inverse parabola that he produced even earlier. The central part does have the inverse parabolic shape. Finally, on the right end, Craig first produces a movement upward from right to left and then, almost in the same breath, changes the description and hand movement to one that goes down from right to left. The hand gestures here are forms of epistemic movements that constitute thinking itself rather than being a *re*presentation thereof. It is out of such movements that first perceptions of possible graphs emerge: first there is movement and then there is perception of a pattern, which may be symbolically presented again by means of the same hand movement (Roth 2012a). Again, the trends that are seen to be depicted in the movement are different from the "general increase," a "very, very subtle" one, which Shelby articulates together with a nearly linear trajectory of the cursor from the left to the right. We therefore observe contradictions both within the articulations and movements that Craig produces – even within a fraction of a second – and with respect to those that Shelby previously articulated. However, these contradictions again are not made the topic of discussion. They did not exist nor where salient at that instance and, therefore, played no role in the historical evolution of the meeting.

Grasping the Variation of Variation

> "Do you have a smaller number of fish on the right hand side?
> Because the overall variability increases there"

Grasping what is underlying the variation, that is, literally getting a hand on and making handy, is crucial to the possibility of explaining the biological phenomena said to be represented in the inscriptions. Variation in the data can have multiple sources, some of which are related to the phenomenon itself, that is, turn up in the measurement. Others are related to the statistics involved in calculating the extent of the variation. Thus, both standard deviation and standard error are weighted measures of the total variance as measured by the sum of the squared deviation from the mean. This "sum squared," as it is called in some sciences, is expressed as

$$SS = \sum_{i=1}^{n} \left(X_i - \overline{X}\right)^2 \tag{4.1}$$

where n is the total number of data points, the x_i are the data points, and $\bar{x}$ is the mean.[21] The sum squares, that is, the total sum squares SS averaged over all the data points decrease with increasing number of data points. A scientist familiar with statistics, large variations, especially in the co-presence of means with smaller variations, might – but does not have to – see this as another possibility for attributing the varied variances to specific sources.

Shelby then makes a statement that can be heard as an evaluation of Craig's presentation saying that he finds it "not surprising" and continues to articulate a reason. He talks about the data representing not individual fish but rods that are grouped. He ends by saying that he is skeptical of those data near 0 % and 100 % (A_2). Craig follows the comment about being skeptical with by querying the nature of the error bar – whether they represent ± 1 standard deviation. By asking the question in this way, the normative character of this value over other possible values comes to the fore. Shelby responds by saying that he had plotted using 95 % confidence intervals. There is a long pause and then begins some elaboration about wanting to use 1.96 standard deviations[22] and just as he was beginning to suggest what this would lead to "then you would get" (turn 114), Theo interrupts him making salient the points "up there." Shelby continues by suggesting, "one" (1 SD) "would be a bit cleaner" (turn 116). In this, he articulates what we can hear as a reference to Craig's question about the error bars corresponding to $SD = 1$.

Fragment 4.4c

112 S: <<f>wELL> and thats not surprising because in actual
fACt you (0.41) i i find it hard to belIEve that
therere; well when you mitigate individual rODs into
groups ((*circles 5 left-most data points*)). this is;
this is not indiVIdual fish, this is (0.50) each rOD
grouped right. (1.41) so de () de rODs that are one
hUNDred ((*circles right-most 4 data points*)) () or
ZEro ((*circles left-most 5 data points*)) (0.80) might
be: (0.25) u:m (0.25) uh ones i=d be skEPtical of
perhaps (0.59) <<p>if there ?>

113 C: this is plus or mINus one standard deviation?

113a (0.39)

114 S: ah this is ninetyfive percent confid=dence intervals.
(5.54)
i mean really youd wanna use one point nine six
standard deviations (0.37) and then you could get

[21] In Chap. 12, I show how even eleventh-grade students come to be familiar with the principles underlying variance and use these principles for identifying best-fit functions through the data that they have collected in investigations of their own design.

[22] The two statements, standard deviation (SD) = 1.96 and 95 % confidence intervals are equivalent.

115 T: <<p>once you get up here> <<p>hu hu hu>
116 S: yea, one one would be () a bit cleaner; (0.58) certainly that is
117 T: yea hu
118 S: yes, but for,
119 T: ninetyfive, dats too much [hh hu]
120 S: [yea, no] definitely, yea much, yea. (3.37) bt i think youre right; i think the general trend is there from from low ((*points to the left-most group of data points*)) to
121 C: [yea]
122 S: [some]thing going on up in here perhaps ((*points to the right-most group*)) to down here; ((*points to left-most group*)) () yea.
122a (5.59)
122b S: so didyou wanna see the other stuff that ive already plotted (0.39) that is kind of a summary of what weve got to date
123 C: yea yea
124 M: do you have a smaller number of fish on the right hand s=side because the the the overall variabILity INcreases there; as you go to the right;
125 S: to the right?
126 M: yea, [dont you] dont you think?
126a C: [uh hm]
127 S: um, ACTually in thIS case yeah you=re right there have been fEWer fish in the hUNdred percent (0.21) porphyrOPsin
128 M: uh hm
129 S: so that makes sense, you get that
130 M: uh hm
131 S: more (?) variability. most of our fish have been more in the a=ONe range; (0.32) most of them (0.36) and in fact a hundred percent a=ONe has been our probably our number ONe (0.44) recording so. (0.53) so i=ll, i can show you that data right now really quickly,

Theo suggests that the 95 % confidence intervals are a bit much, thereby adding to the preceding question about the standard deviation being 1 and the assertion that "one would be a bit cleaner." Shelby affirms that Theo is right. He then says, all the while looking straight at his computer monitor that someone ("you") is right and points to "trend from low" "to something going on up in here perhaps" (turns 120, 122). At this point, Craig utters a confirming "yes" (turn 121). Shelby then offers to move on and look at "the other stuff that [he has] already plotted," which is "kind of a summary of what [they] have got to date" (turn 122). Although Craig affirms, my own question is an invitation to stay on the topic and constitutes a refusal of moving on. The statement I am making asks about the number of fish corresponding to the right side of the plot: Is there a smaller number of fish? A reason for the query follows immediately, "because the overall variability increases there, as you go to the right?" (turn 124). Shelby seeks to confirm the region at issue ("to the right?" and then states that they have had fewer fish "in the hundred percent porphyropsin," which has the retinal (chromophore) produced from vitamin A_2. Shelby ascertains that it makes sense to have greater variability when there is a smaller number of fish; and he adds that most of their fish have been in the 100 % A_1 range, the chromophore in rhodopsin; this range corresponds to 0 % porphyropsin and the corresponding vitamin A_2-based chromophore. Shelby then makes another offer to move on and look at the other graphs he had prepared. This time, the offer is accepted and the team begins to look at the measures A_2 distributions for fish from different hatcheries and river systems.

At this point, this discussion of the half-maximum bandwidth has ended. There is no clear conclusive statement on the part of any one member but a clear proposal and acceptance to move to some other topic. This new beginning inherently constitutes the closure of what will become the preceding topic. And this opening | closing is itself available in the sequentially ordered turn taking that produces the unfolding communicative situation. As in many other instances, rather than making the problem the core topic of the discussion (e.g. the causes for the monitor that is not working in Chap. 2), the scientists go on to some other point much in the same way they continued after the monitor displayed what they wanted it to display. In many instances, they do not wonder why something does not work. As soon as it works, as we see in Chap. 2, they go on with their business never wondering what it was that did not allow them to make sense or do their work. In the present instance, Shelby initially offers to go on and show another graph (turn 122b), which is in fact the one the meeting had started out with. My question, however, returns to the immediately preceding issue and therefore also constitutes a non-acceptance of the offer to move on. In and as the exchange with Shelby, the greater variability visible on the right-hand side of the graph comes to be attributed to the lower number of measurements collated in each point. The end of the episode concerning the variability of the data and the discussion about half-maximum bandwidth is achieved following the second attempt, when Shelby offers to "show . . . that data right now really quickly" (turn 131). Crag affirms, the meeting turns to that issue. That turning therefore has been the outcome of an interactive achievement of the group as a whole, as the affirmation offer was not further mediated by another turn at talk.

Consequences for the Control of Variance

As a result of the preceding discussions and discussions during subsequent meetings, Shelby, first alone and then in his collaboration with Sam, began an attempt of getting the variance into his grip by controlling the location in the fish eye from where he was getting the photoreceptors to be studied for their absorption of light. In the past, this location had not been an issue. In all earlier work in the field, scientists had ground up retinal tissue and took measurements from a sample that included many photoreceptors. The new technology that Craig had developed allowed taking light absorption measurements from individual photoreceptor cells. Initially, that is, just prior to the collection of data for the investigation researched in this Part B, our team was indiscriminately taking photoreceptors from anywhere in the retina. It was in response to the great variations reported in this chapter that Shelby started taking only photoreceptors from the dorsal part of the retina. That is, the method evolved in the course of the investigation. It evolved even further when Shelby and Sam began noticing something, which they later conceptualized as an artifact of their method.

> There was a problem with our methodology over the summer-time when Sam started with me we started dissecting differently and that was ... it had some serious implications, for the results, unfortunately, ended up ... well I can show you a graph of what happened but where we took the retina from yea. And so we were starting to take a strip um because at the same time we were doing um where is it here (looking for graph). So instead of just getting the dorsal and then taking a small corner of the dorsal we started to take a strip because we were hoping to get more cones and that was particularly applicable to the stuff that we were doing with the T3–T4 [hormone] treatment because we wanted to record ah the change in porphyropsin rhodopsin in rods and in cones and so by changing the methodology we didn't really we didn't think that it would, well we didn't really think about it I guess enough to think that it would have much of an impact and because we don't do our analysis until a lot later we didn't find out until too late. And it's really a shame because it meant that I lost a whole segment of my analysis, of my dataset, which is really crappy. So basically what happened was– so we were originally taking the eye and cutting it in half and then cutting this in half, this dorsal section and taking this little corner and that's all we'd take. But in order to get more cones we started taking a strip all the way across here, which took us out of just the dorsal [part], it took us into the central area of the eye. And so what that did, it caused a change cause there's variation across the retina in the percent porphyropsin. I didn't I didn't really suspect that there would be that much of a change across this diff– across that difference. But what I ended up doing is– there's the data set there so what we have is this summer's data these four data points are sitting much higher than they should've been so the regular trajectory would've been to come down here, increase in the winter time, and then decrease again towards the summer. But because of this increase porphyropsin um from the sampling technique it's caused an increase. So if we look at there: it's spread out more, you can see more clearly there, that's much higher than we've had any time before in Robertson Creek. I don't think that's environmentally significant, I think it's method– methodologically changed. Yea exactly like right around um February March we started looking for um cones in our other samples the T3–T4 treatment, but because the just because of the way it was like Sam was relatively new and you know this technique we were using, he just used the same technique for all of the dissections and didn't vary it between Robertson Creek and T3–T4 treatment or I didn't. I mean, whatever, whoever was doing the dissections, we just didn't change it when we put on a new technique. That's what we started using and we didn't think about the implications until I analyzed the data at the end of the summer and then we realized and the problem is that we're just too busy in there all the time that we don't have time to analyze as we go. And that was really was really too bad. I tested that so I've got um corner versus strip, this one here I think. After I saw that I went back and I and I couldn't go back and check the data set that we already had. So what I had to do is I actually had to do another experiment. So I actually did an experiment where I took a corner and a strip from the same fish and compared the percent porphyropsin for each. And I only did those on three fish, just quickly to see if there was an obvious trend, because I mean a real test you'd have to do a lot a lot like ten fish at least. But with three I can get a sense of whether or not it's in the same direction and I think, "Well, this will show that in fact it is that there's more porphyropsin when you take the strip relative to the corner piece." So what I have to do is, I have to take that data and say it's unreliable because the methodology was different even if even if there wasn't a difference I can't um say that I'm confident in that dataset because we changed the methodology. So I would have to take those data points and remove them from my data set

> and start again. So basically I would say that you know I've got, I had a full annual cycle up to that point anyways pretty much, I would take that out and say that we stopped for this period of time and restarted it again at this point when we– so that I know when I'm back to the regular so that was uh my last sample that I did the proper way again was I think um September

The meeting can be said to have had consequences, which led the team to a different trajectory and practice during its data collection. This did not eliminate problems, however, but rather created new sets of problems and uncertainties that became salient only long after the fact. I exhibit in this chapter the tremendous amount of work scientists mobilize to grasp and thereby get control over uncertainty in their data. As they are not in a position to know where the uncertainties and variability derives from they can only hope that from their work greater control will emerge without having any assurance that this will actually occur. Or rather, scientists may be confident in their competencies of producing a coherent account after the fact when out of the apparent uncertainty and variability some sense of order eventually emerges. David Suzuki, a well-known Canadian geneticist become broadcaster and environmentalist that the writing up of scientific research results, while producing a coherent story, communicated "nothing of the excitement, hard work, frustration disappointment and exhilaration of the search" (Suzuki 1989, p. 192). Nothing, too, remains after the fact of the uncertainties that mark pretty well every step until some insight revealed itself. Thus, to write up the results, Suzuki suggests they "riffled through all our records, selected the ones that said what we wanted and then wrote the experiment up in the proper way: purpose, methods and materials, results and so forth" (p. 192). A recent article in *Science* suggests that such preferential selecting from the data is common in the natural sciences (Couzin-Frankel 2013). That is, uncertainty does not only disappear because it is not part of the logic of the write-up but also because it is so much part of the everyday work of science that it has become unremarkable and, therefore, goes without saying.

References

Châtelet, G. (1993). *Les enjeux du mobile: Mathématique, physique, philosophie*. Paris: Éditions du Seuil.

Couzin-Frankel, J. (2013). The power of negative thinking. *Science, 342*, 68–69.

Derrida, J. (1996). *Le monolinguisme de l'autre ou prothèse d'origine* [Monolingualism of the other or prosthesis of origin]. Paris: Galilée.

Garfinkel, H., Lynch, M., & Livingston, E. (1981). The work of a discovering science construed with materials from the optically discovered pulsar. *Philosophy of the Social Sciences, 11*, 131–158.

Hárosi, F. I. (1987). Cynomolgus and Rhesus monkey visual pigment: Application of Fourier transform smoothing and statistical techniques to the determination of spectral parameters. *Journal of General Physiology, 89*, 717–743.

Hárosi, F. I. (1994). An analysis of two spectral properties of vertebrate visual pigments. *Vision Research, 11*, 1359–1367.

Hawryshyn, C. W., Haimberger, T. J., & Deutschlander, M. E. (2001). Microspectrophotometric measurements of vertebrate photoreceptors using CCD-based detection technology. *Journal of Experimental Biology, 204*, 2431–2438.

Hegel, G. W. F. (1979). *Phänomenologie des Geistes* [Phenomenology of spirit]. Frankfurt/M: Suhrkamp. (First published in 1807)

Heidegger, M. (1977). *Sein und Zeit* [Being and time]. Tübingen: Max Niemeyer. (First published in 1927)

MacNichol, E. F., Jr. (1986). A unifying presentation of photopigment spectra. *Vision Research, 26*, 1543–1556.

Merleau-Ponty, M. (1945). *Phénoménologie de la perception* [Phenomenology of perception]. Paris: Gallimard.

Roth, W.-M. (2012a). Tracking the origins of signs in mathematical activity: A material phenomenological approach. In M. Bockarova, M. Danesi, & R. Núñez (Eds.), *Cognitive science and interdisciplinary approaches to mathematical cognition* (pp. 182–215). Munich: Lincom Europa.

Roth, W.-M. (2012b). *First person methods: Towards an empirical phenomenology of experience*. Rotterdam: Sense Publishers.

Suzuki, D. (1989). *Inventing the future: Reflections on science, technology, and nature*. Toronto: Stoddart.

Vološinov, V. N. (1930). *Marksizm i folosofija jazyka: osnovye problemy sociologičeskogo metoda b nauke o jazyke* [Marxism and the philosophy of language: Main problems of the sociological method in linguistics]. Leningrad: Priboj.

Vygotskij, L. S. (2005). *Psychologija razvitija cheloveka* [Pyschology of human development]. Moscow: Eksmo.

Williams, J., & Wake, G. (2007). Black boxes in workplace mathematics. *Educational Studies in Mathematics, 64*, 317–343.

Wittgenstein, L. (1997). *Philosophische Untersuchungen/Philosophical investigations* (2nd ed.). Oxford: Blackwell. (First published in 1953)

Chapter 5
Undoing Decontextualization

> [A] *concrete conception* of commodity . . . *coincides* with the *theoretical understanding* of the entire totality of the interacting forms of economic life. (Il'enkov 1982, p. 105, emphasis added)

In this chapter, I provide a case study of the work that goes into undoing the decontextualization that occurs as a normal and integral aspect of scientific research work in the course of the translations from the natural phenomenon towards the knowledge claims stated in language (see Chap. 1). At the heart of the definition of science lies its objectivity, that is, the fact that some study can be repeated at any time and anywhere in the world and still give the same results (Latour 1988). When reproducibility cannot be achieved, doubts are raised about the quality of the knowledge claims, and the text submitted to peer review may not be accepted for publication in recognized research journals (Collins 2001). To reproduce results, the conditions have to be the same or comparable, or, in other words, contingencies have to be stripped from the specimens and the remaining conditions have to be reproducible in other places so that the specimens that are used in different places and at different times are in fact comparable. In the context of the present study, for example, indicating the water temperatures in which the coho salmon from the hatcheries were held in the laboratory until the instant that they were processed and used in the data collection would be one specification towards such reproducibility. In field studies, the scientists would have to make modifications to the local context so that the conditions again become comparable to those found or achievable in any scientific laboratory (Latour 1988). On the other hand, not stating which measurements have not been admitted into the data leads to difficulties explaining why experiments are difficult to reproduce (Couzin-Frankel 2013).

As a result of the preparatory work that takes scientists through the chain of translations to increasingly inclusive inscriptions, context and contingencies have been stripped from the samples used in scientific research to the point that nothing but numbers and graphs are left. These numbers and graphs in themselves, however, tell scientists very little if anything. Scientists do not just "interpret" them, as if there was a way that would take them from the world of the text to the wild or the

W.-M. Roth, *Uncertainty and Graphing in Discovery Work*,
DOI 10.1007/978-94-007-7009-6_5, © Springer Science+Business Media Dordrecht 2014

hatcheries were the coho specimens had originated. Previous work suggests that scientists have difficulties reading graphs when they are unfamiliar with the context and research methods that led from the raw phenomenon along the chain of translations to the graphs (Roth and Bowen 2003). When scientists are familiar with the generation of data and the translation process – described in Chap. 1 – then they treat these as transparent means that allow them access to the natural phenomena that they are so familiar with (Roth 2003). What I show in this chapter that has not been described in the literature is that scientists engage in long and exacting work to undo the decontextualization that they have produced to be able to read the results of their work in terms of the real context from which the specimens have been sourced. Decontextualization that comes with collating an increasing number of data from different contexts is part of a historical trend that began during the Renaissance (Edgerton 1985). At that time, graphs and graphing appeared as a new means of production in the sciences,[1] their power deriving from the fact that many measurements from the same or across context can be displayed visually at the same time. Graphs became a quintessential and defining characteristic of the sciences such that these, as we know them today, would not even exist. In the production of scientific knowledge, graphs play an important role because they depict, at a sufficiently abstract level, general tendencies in the relation of two or more variables. When scientists talk about their own graphs, they do not just know graphs but the graphs deriving from their work also appear to bear metonymical relations to their work. For graph readers unfamiliar with the context that gave rise to a graph, this inscription constitutes something like a black box, which requires as much work to be reopened as it has taken to produce. As a black box and metonymy, the graph stands for the production as a whole. The practical sense related to a graph – and, therefore, its use – has the possibility to be articulated in and structured by an explanation because of the producing scientists' practical knowledgeability and situational familiarity that precedes, accompanies, and concludes the explicatory process.

On Being Familiar with Black Boxes

One important problem with existing studies of graphing is that these do not tease apart familiarity and processes that function independent of it. For example, in Chap. 1 I refer to a major study of graph interpretation that had asked H. Simon, a Nobel Prize winning economist and a cognitive scientist, to read/explain a supply/demand graph that economics students tend to encounter in one of their first classes. His response to the task is very similar to the ones that those biologists provide who

[1] Now referred to as Cartesian graphs simultaneously arose from the work of René Descartes and Pierre de Fermat, and were subsequently developed following the translation of Descartes' *La Gé ometrie* into the scholarly Latin by the Dutch mathematician Frans van Schooten.

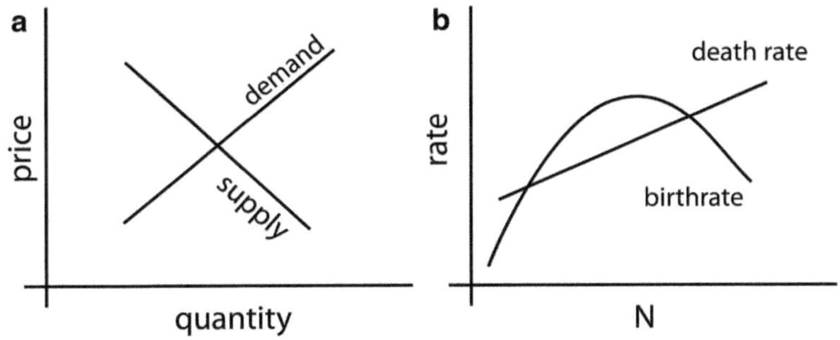

Fig. 5.1 (**a**) This supply-demand graph was used in a study of how an expert economist reads graphs. (**b**) This birthrate–death rate graph from undergraduate ecology was used in studies with biologists and physicists, many of whom had trouble reading or interpreting it in the way that course instructors of introductory ecology courses would have accepted as correct

teach undergraduate courses and who are asked to explain birthrate/death rate graphs (Fig. 5.1b) typical of undergraduate courses (see Chap. 1).[2] These graphs featured two intersections one of which should be read as giving rise to a stable equilibrium the other to an unstable equilibrium. But non-university research scientists were less successful at a statistically significant level than their university-based peers, who are familiar with the kind of graphs used in the study.

In Chap. 1, I describe two related investigations in which a total of 33 scientists (16 biologists, 17 physicists) were asked to talk about graphs from introductory courses of biology shows that even with tremendous levels of training, only 9 (27 %) of the scientists gave correct answers on a graph that bears structural similarity to the oxygen-shrimp frequency graph. This graph features the distribution of three types of plants – distinguished by their photosynthetic mechanism (C3, C4, and CAM) – along an elevation gradient. Once I excluded university-based biologists who were teaching at the undergraduate levels, only 2 of 25 (8 %) scientists correctly answered the question. Because all of these individuals were successful scientists, standard psychological approaches that use "cognitive deficit" to explain such results surely are inappropriate.[3] Clearly, something else is at work. It is precisely this something else that is at issue in the present chapter. I suggest that the familiarity with the natural phenomenon and the entire translation machinery that connects graphs with phenomena allows scientists to read the graphs rather than some feature attributed a priori and without further investigation to cognitive abilities. That is, I use an anthropological approach to study how scientists come to use and become familiar with data and graphs during exchanges with each other rather than attributing graphing competencies to any special feature of the mind.

[2] In biology, such graphical models constitute a rather recent phenomenon developing especially during the twentieth century (Kingsland 1995).

[3] When research on graphs and graphing is conducted in schools, then cognitive deficiencies are mobilized for explaining why students fail to provide standard answers (e.g. Berg and Smith 1994).

This investigation shows, as reported here, that it is familiarity that is at the heart of scientists' graphing competencies. That familiarity is at issue arises from the fact that the scientists in the present study did not and, apparently, could not interpret their graphs until after undoing the decontextualization that they had enacted to extract the retinal tissue samples that they were working with.

The difficulties people experience with explaining what a graph is supposed to express may parallel the issue of analogies. This is so because in the acknowledgment of an analogy, a person articulates a relation between two situations – which may be because of perceptual characteristics or because of characteristics ascribed to "deep structure." The important part is that the person has to be familiar with the two situations related, especially when the relation is to be at the "deep structural" level. In a similar way, a graph is said to express patterns in some natural phenomenon according to the {fundamental structure ↔ mathematical structure} couplet accepted in the sciences. That is, knowledgeably "interpreting" the (graphical) mathematical structure is equivalent to stating that a person talks about something in the natural world that is expressed and denoted by the mathematical structure. It is apparent that this would require familiarity with both domains as well as with acceptable forms of mapping aspects of one domain onto aspects of another.

Graphing difficulties are often attributed to the "abstract" nature of. One possible way of modeling the relation between abstract and concrete representations (inscriptions) is by means of the black box metaphor (Latour 1987). Thus, an increasing level of contextual detail is deleted in discovery scientific work until the scientists arrive at their stated claims (e.g. Latour 1993). The result is a chain of inscriptions, each of which is separated by an ontological gap from all other inscriptions; these gaps are bridged only in and through scientific practice (Chap. 1). There is therefore no natural and necessary relationship between soil samples placed in stacked drawers and line graphs representing the horizontal and vertical distribution of soil types: practices are responsible for the equivalence of the two. Scientific research thereby moves from the concrete natural world, where specimens are taken (abstracted), and conducts measurements that are presented again in an equally concrete world of sign systems (graphs, equations); and these are used to substantiate verbally articulated claims. In talking about graphs, the scientists move in the opposite direction, which is made difficult because each gap between two representations in the chain constitutes a black box that has to be reopened before the gap is bridged. For those familiar with the entire process of abstraction – scientists, technicians, and students alike – it appears to be easy to go all the way back from the representation to the natural world from which the abstract inscriptions were derived. I heed Latour's (1987) advice to seek purely mental-cognitive explanations once everything related to observable practices has been examined. An opportunity for studying the appearance of graph-related competencies exists when scientists *first* come to be familiar with phenomena and their mathematical substitutes, that is, during a process of mathematization and *before* they individuals can rationalize the process of their learning.

In this chapter, I describe and analyze the first of a series of meetings that occurred over a 2-year period, with about 3,000 usable measures from seven sampling sites already available. During these meetings, our team attempted to come to grips with the trends that might be observable in the data. This is the same meeting where they also discussed the selection criteria and variation in the raw data (Chap. 4). By selecting this meeting rather than all the others that subsequently occurred (e.g. when the study was almost completed), my analysis pre-empts any possibility that the scientists' interpretations would "re-write" the history of their work in the face of the results they did obtain.[4] Historians of the natural sciences have described this tendency to re-write their own discovery process; and this tendency was observed here when Shelby later began to explain to me that he knew all along there were problems with the canon that had framed our team's early analyses. However, during this first meeting, there is no evidence that the scientific canon would eventually be overthrown (see also Chap. 7).

At this point in the research project, our team was convinced to be reproducing the results that Alexander et al. (1994) had published 8 years before. Shelby regularly referred to that study whenever he gave a talk somewhere – e.g. in one of the participating fish hatcheries, in the graduate seminar of the biology department, at conferences – and has prepared a PowerPoint slide featuring the results in which he also provided additional information, including the precise dates of data collection that he had reconstructed from other information (Fig. 5.2). Throughout the scientific study, from its very first conception, this graph served us as a referent for designing the research to the interpretation of the data (see Chap. 7). It turns out that the originally published data and the graph that Shelby reproduced for us, as well as the related knowledge, were important for following the unfolding talk that our team generated in the course of the various meetings. Even my personal knowledge – gathered during the ethnographies of the two participating hatcheries that supplied the wild and hatchery-raised coho (see the introduction to Part B of the book) made available to other team members during these meetings – became an important resource for articulating the conditions under which the young salmon were raised and, therefore, some of the possible reasons for similarities and differences that would explain the variation (or lack thereof) in the data. In fact, the information I provided constituted an affordance in the sense that without having this background, our team would have been at a loss in constituting the significance of their data. At the same time, hindsight shows that knowing about the Alexander et al. (1994) study turned out to be a constraint, for it became a lens through which our team initially saw its data; and this, ultimately, led us up a garden

[4] On a number of occasions, scientist members on our research team "re-wrote" the history of our project, when in conversations and email exchanges, they provided a posteriori explanations that differed what they and others had said initially. That is, the re-writing of history is endemic to science, just as Kuhn (1970) described it half a century ago.

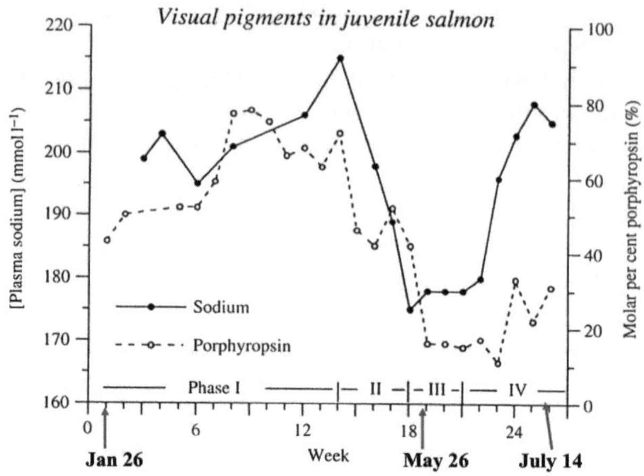

Fig. 5.2 For the purpose of presentations, Shelby has reconstructed the results of another study, which our research has used as a paradigm. It becomes the referent for the findings from our research (© Wolff-Michael Roth, used with permission)

path. It took us a considerable amount of time to realize that the data we had collected actually undid the canon established by the Nobel Prize winning research done some 60 years earlier (see Chap. 7).

The Work of Familiarity with the Data-at-Hand

In this chapter, I investigate what scientists do with their data once translated into graphs when they do not yet know what they ultimately will have found out when everything has been said and done. As the following analyses show, scientists are out on a limb with their ways of modeling phenomena and their conjectures; and it is precisely because their struggle occurs in and as part of sequentially ordered interactions that their *interpretation processes* can be studied. In the middle of their game at science, they cannot draw on later results and familiarity. These are available only after a research project has been completed, in other words, after everything has been said and done, and when they are sure they can get their work through the peer review process. What is relevant information to familiarity *emerges* unpredictably from the (societal) relations that are characteristic of laboratory meetings. This is especially important because these ultimate findings were very different from what they initially had thought to find. In this section, I present scientists' initial work at becoming familiar with their data, after the first few data points have become available. In the following four subsections, I exhibit scientists' (a) initial, positive impressions and acceptance of data, (b) need to see the data in context with other data, (c) expressions of uncertainty, and (d) postponement of more definitive interpretation of their data. I summarize the scientists' efforts to become familiar with their data by discussing how context, which has been stripped and eliminated in the movement of the research process up to that point in the process from going left to right in Fig. 1.1, is creeping back into it. In fact, going left to right in the chain of translations (Fig. 1.1) and stripping context are synonymous, mutually implicate and constitute each other.

First Impressions and Acceptance

"This is the means of each batch, so you can get a sense" (Shelby)

In this first data analysis meeting, Shelby used his laptop and a projector to present the results from the first five batches of fish that Theo and he have processed in the laboratory. Together they have obtained the absorption spectra, which Theo has subsequently cleaned up and transformed to extract λ_{max} and half-maximum bandwidth, the information needed to determine the relative amount of porphyropsin (i.e., "perc[ent] A_2") present in the photoreceptors (Fig. 5.3). In Chap. 3, I analyze the discussion we had about the exclusion of data on the left end of the histograms. Here I return to that meeting at a point where our team is concerned with relating the histograms to the natural phenomenon they are interested in: the life history of coho salmon. Following the measurement in the laboratory and Theo's extraction of relevant information, he had returned the results to Shelby, who analyzed them using the then current version of the SPSS software package. The data collected from the coho salmon sourced from the Kispiox Hatchery and the nearby creeks and river repeatedly became the topic in this 2-h meeting.

The video shows Shelby presenting the distribution of porphyropsin from each batch of fish. He provides an overall assessment: "these ones here seem to be slowly shifting toward the right," and then he self-corrects, "to the left rather" (turn 001). That is, he expresses seeing the distribution maxima to be shifting to lower values of A_2, even though we might hear him not to be all too confident ("I'm, you know, not . . .") about the data in the first panel when he says that it represents their "first day." He describes the data ("this") as "looking pretty nice," and then adds the hedge, "almost" as if he were not quite sure or as if he wanted to keep the statement tentative for a little longer (turn 001). Craig not only agrees that the data are nice but also provides an emotionally laden assessment: he formulates being happy about what he sees (turn 002). This is the kind of data that they often denote – in the laboratory and during meetings – by the noun "beauty."

Fragment 5.1

```
001  S: so well the first ones first ones our very first day
        so i:m you know not (2) this one here maybe not but;
        these ones here ((Fig. 5.2)) seem to be slowly
        shifting towards the right uh left rather; towards
        more a. and this one here is looking pretty nice
        al[most]
002  C:                                         [yea ] yea
        its eh its eh nice; its eh mean its eh i=m happy with
        what its i see there; its eh n[ice] data=
```

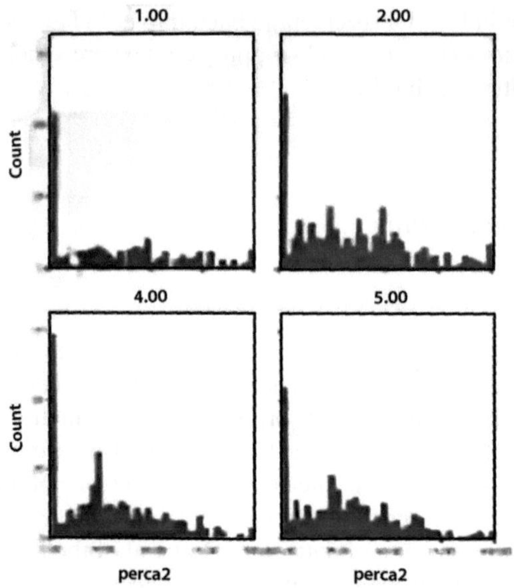

Fig. 5.3 Shelby projects onto the screen the cell counts for different amounts of porphyropsin content (%A_2) for five different sampling locations (numbered above the graphs) (© Wolff-Michael Roth, used with permission)

```
003  S:                              [yea]    =yea it is it
        is. so then the next one that ive done then ive just
        kind of eh, quickly just run means for these ((shows
        SPSS-generated list of means)) so this is the mean
        for each batch of fish; <<acc>jst so you can get a
        sense of the standard deviation its eh going on.> its
        pretty wide theres no question here. u:m and this
        gives you a general sense of that data you just
        looked at so for batch one um its eh got a range from
        anywhere from seventyeight in a fish to sixpercent
        o:f a:two. So pretty pretty wide range; and ive done
        that for each batch.
```

Shelby then continues to provide descriptive statistics about the means and standard deviations, on the basis of which he notes that the distributions are wide and that "there is no question here." In any other context, this might have been an innocuous statement. But in the present meeting, where we work on establishing a method for determining release dates based on the A_1/A_2 ratio, having wide distributions is far less than ideal. This is an important question as the within-fish range is from 75 to 6 % A_2. The scientific canon at the time states that the A_1/A_2 ratio is a function of the physiological changes in the fish during the process of getting ready for migration into the saltwater environment. If the *within-fish* variations are so large, this might be used to question the entire model. But the team members present do not question the model at this point. In not questioning, we implicitly accept the data – even though in another part of the same meeting, Craig requested the high peak in the first bin (left most frequency) and the data in the right-most bin be removed because they do not "meet the criteria" (see Chap. 4).

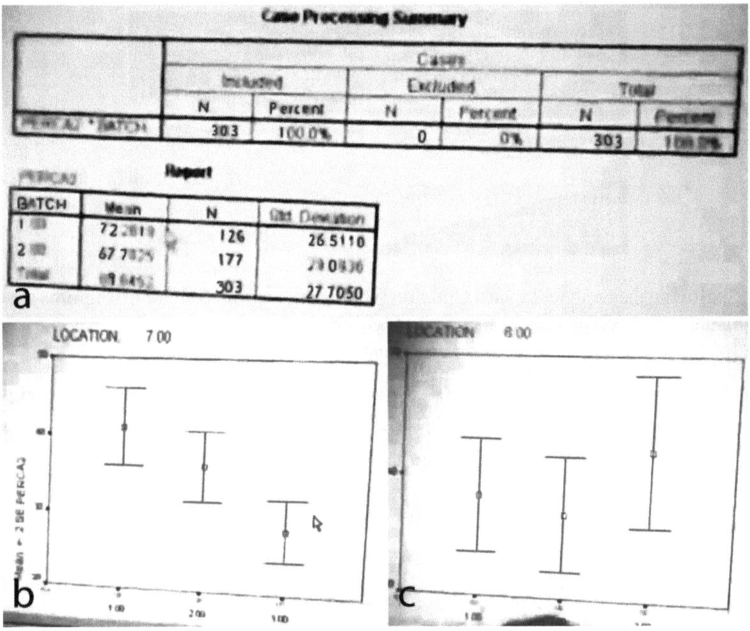

Fig. 5.4 Shelby projects onto the screen (**a**) case summaries, (**b**) the percent porphyropsin (%A_2) measures for Location 7, and (**c**) the percent porphyropsin (%A_2) measures for Location 7 (© Wolff-Michael Roth, used with permission)

Putting Data in the Context of Other Data

"Do you have comparisons?"

Biologically of interest are the changes in percent A_2 as the season progresses. At the time, the scientists are collecting data in 2-week intervals to capture what they believe, based on the scientific canon, to be the physiologically driven change from one to another form of photosensitive material: rhodopsin (vitamin A_1 analogue) and porphyropsin (vitamin A_2 analogue). A few minutes after having considered the data from our main participant hatchery (Robertson Creek, Fig. 5.4), Craig requests to look at the second hatchery. The Robertson Creek data show, on the one hand, a decreasing porphyropsin level in the fish that migrated to the sea (Fig. 5.4b) but a renewed increase in those fish that the laboratory retained for research purposes in freshwater containers (Fig. 5.4c). Shelby pushes a sheet of paper toward Craig, saying that he "got both of [them] there in small histograms" (Fig. 5.5). But, as he does not have the means as numbers available, he requests the batch numbers and then produces, while seemingly talking to himself, the requested information so that everyone can see it on the projection screen. Shelby says that he has plotted the distributions in the same manner as the preceding figures he presented and now refers to a table that presents the means and standard deviations for two measurement episodes (batches) and the processing summary (offprint in turn 012). He reads the relative amount of A_2 in each of the two batches from the table he has

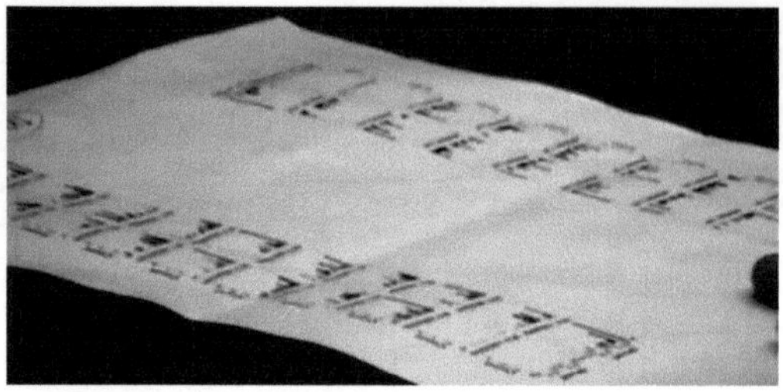

Fig. 5.5 Shelby pushes a sheet toward Craig on which he has printed small histograms to which he refers in his talk (© Wolff-Michael Roth, used with permission)

generated just now: 72 and 67 %. He immediately provides a commentary: “They are pretty high in porphyropsin” (turn 014). He elaborates: “these are wild fish that are on their way to the wild, these are migrating to the sea” (turn 014).

Fragment 5.2

```
008  S: its eh got its eh got um both of there in small
        histograms * ((shows Gregg a sheet)) um these are
        most important observations in the fish its eh just
        quickly plotted out the means very quickly here um
        its eh choose the um data so the data I wanna choose
        um can you give me the number that you want me to
        look at? Location number?
009  C: u:::m one versus two
010  S: okay u::m what its eh showed you; its eh showed you
        one its eh just choose two then and we will go from
        there
011  C: okay
012  S: <<pp>and so, we want to () and get the means for
        those ((Shelby is talking very low volume, while
        looking for the data)) so the means and we=ll look at
        uh at batch percent a=two and see up there ((thinking
        expression)) we have three> ((long pauses)) the first
        batch was seventytwo percent a=two, ((the group looks
        at Fig. 5.4a))
013  C: uh hm
014  S: the second batch was sixtyseven. um. so theyre theyre
        pretty high porphyropsin; these these wild fish
        thatare ON their way to the wild; these are
        mI:grating to sea um
015  E: <<p>what about eighty percent>
016  S: um but theyre high up the river; so [you]
017  C: (:E)                                [no ] ((to
        Elmar))
018  S: see what im saying they its eh seem to be shifting
        thAT much in advance
019  C: yea
020  S: its eh what surpris[ed me].
021  C:                    [yea   ]
022  S: which is unexpected; cause the the suggestion from ()
        the previous work was that they seemed to shift a lot
        in advance
023  C: <<f>okay>
```

To follow this conversation in the way that the research team members do, the statements need to be considered with respect to the established canon, confirmed by the Alexander et al. (1994) study in reference to which the present research is conducted. Elmar makes reference to that study, where maximum porphyropsin levels were around 80 % (Fig. 5.2). Shelby responds using the adversative conjunction "but" followed by the statement that the fish "are high up the river," which he then elaborates: "they do not seem to be shifting *that* much in advance" (turn 018). He formulates having found this a surprising result given "previous work," which proposed that the fish "seemed to shift a lot *in* advance" (turn 022). He then modalizes the comment by adding that there could be another cause, which would make the instant of changes shift from year to year. Craig subsequently makes a statement that proposes looking at the data from the other hatchery in the study (i.e., those presented earlier in the meeting and reproduced in Fig. 5.4b, c), the wild fish caught near the hatchery, and those that have already migrated to the estuary (about 30 km away). He makes a comment to the effect that these data (hatchery and estuary) would not make a difference.

The laboratory members indicate surprise about their specimens' high porphyropsin amount. This is in contrast to the coho from the Kispiox Hatchery itself, where porphyropsin levels are at about 30 %. Based on the data from the second hatchery and based on the published data (Fig. 5.2), this value would mean that the fishes are ready to go out to the ocean.[5] But the data had been collected when the fishes still were in the hatchery. On the other hand, the wild fish from the same geographical location have very high levels of porphyropsin, much higher than the fishes from the Robertson Creek Hatchery. These, as the latter, also started to migrate. Shelby articulates this as a surprise, which Craig confirms ("yea").

In this situation, the scientists do not just interpret and make sense of the data. They do not just inductively get from the data to some kind of result. Instead, they already begin to mobilize their familiarity with the geographical contexts where the fish were caught to situate and interpret the numbers. Their familiarity is a lens through which they see the data. This lens is made from what scientists currently know about other hatcheries, published data, and the scientific canon they are adhering to. Through this lens, the data are seen as *not fitting* with other things. It is precisely because of their background familiarity with the life cycle of the coho salmon and with the literature that the data do not fit into the picture. The porphyropsin (A_2) levels are high given that fish on the way to the ocean should have much lower porphyropsin levels. One possibility is, as Shelby suggests (turn 018), that the shifting does not occur much in advance of the migration, especially not "a lot in advance" in the way this has been suggested in the scientific literature. To put this into context, we return to Fig. 5.1. Thus, the present results suggest that reading the economics or biology graph requires a priori familiarity with economics

[5] As pointed out in the introduction to Part B, the canon suggested that the porphyropsin levels (% A_2) would drop from somewhere between 80–90 % down to around 5–10 % similar to the graph in Fig. 5.2. Thus, if the retina of coho has 30 % porphyropsin, the fishes are closer to what was thought to be the saltwater composition than to the freshwater composition.

or biology to say anything useful and intelligible about the graphs. It is precisely because H. Simon already knows economy, about the phenomena denoted by the terms supply and demand and is familiar with using these terms that he is in the position to provide a reading that exhibits a high degree of knowledgeability. On the other hand, the children asked about the shrimp and oxygen graph in Fig. 1.2c may not have known about and have familiarity with the relation between oxygen in water and the frequency of certain organisms. But I conjecture that the seventh-grade students appearing in one of my studies (Roth and Barton 2004), who investigated the frequencies of different organisms in the different parts of a creek and correlated these with the speed of the water, would succeed at a much higher frequency. This is so because they were explicitly using the conjecture of oxygen availability with their frequency counts.

Uncertainty – Again

"Okay, this is reversion data?"

One may note that the laboratory members, though initially looking at the large variations in the retina from the individual fish – which, according to Shelby, fall between 6 and 78 % A_2 – now are looking at the mean values only, no longer considering possible explanations related to the large variations or differences in variations. The team members present already have seen several plots, including the one that they refer to in the opening of Fragment 5.3. The dogma states that when the coho salmon get ready to head for the ocean, their porphyropsin levels drop, as in the April–May period of Fig. 5.2, whereas these levels are thought to increase again if the coho were held back in freshwater beyond the normal time of migration. The presented data are consistent with this information (Fig. 5.4b, c). The "reversion data" Craig refers to (turn 001) are from the Kispiox fish that the team keeps in freshwater tanks at the university for subsequent testing.

Fragment 5.3

```
001  C:  okay this is revERsion data
002  S:  yea i would suspect because we had held the fish back
         for a few weeks, well three fOUR weeks past the time
         that they release them; so we are into the reversion
         with the kispiox as well
003  E:  <<p>from the hatchery>
004  C:  yea
005  E:  ze ones from the hatchery zey dont have the stimulus
         eether ze shift to do ze wEIGht to do the shift
         probably. ((Craig nods)) they probably need something
         to get off when they are exISting in a hatchery
         envIRonment.
006  S:  well well theyre still in fresh water too; so i mean
         ver they drop off to twentyseven which is pretty
         close to those other ones
```

While looking at the data, Craig does not make a statement that could be heard as articulating a visible trend to offer his interpretation subsequently. Rather, he states seeing the mean values of A_2 decrease and then increase again as "reversion data" (turn 001). This expresses considerable familiarity – based on his work in the field and reading of the literature – that reversion data look like this. No special "interpretation" is necessary just as we do not need to engage in interpretation our neighbor's "What a sunny day today!" when the sun shines. Shelby confirms, articulating why "reversion" is a reasonable descriptor: the fishes have been held in freshwater past the date when these would normally be released to begin their migration. That is, he no longer looks at the data on their own but contributes to the reading through the lens of what has been done to the fish. This therefore can become part of a common explanation for the actual values of the measurements.

Elmar, who had done his PhD work on olfactory imprinting of a salmon species, suggests that the hatchery fish may not have the stimulus for beginning the physiological changes and that they need something "to get off in the hatchery environment" (turn 005). Shelby adds that the fishes currently are in fresh water and that the mean level of A_2 has dropped to 27 %, which is similar to the values that have been measured in the fishes from the other hatchery (turn 006). He suggests that this level "is not bad" given the environment in which the fish are raised in the Kispiox hatchery. As everyone in the room knows from having visited the place, the containers are in an enclosed hall with low light conditions and with a water temperature that is constant throughout the year rather than varying as it would in nature and in the comparison hatchery where the fishes are kept outside. Moreover, Shelby adds, this value is good given that other salmon have a higher value at the point when they are "going into sea"; the Kispiox hatchery fishes are "already at the bottom," that is, at a point that corresponds to the lower parts of the published graph. Shelby continues by elaborating the reference for his assessment of the Kispiox data: the Robertson Creek hatchery releases its fish at a point so that when it arrives in the ocean, its A_2 values are 44 % or 47 %. Craig and Shelby produce a confirmation that indeed the release value lies around a mean porphyropsin level of A_2 equal to 44 %. Elmar comments that this is "pretty high," and adds – using the oppositive conjunction "but" – that "they [fish, hatchery] are close to the ocean." That is, he can be heard saying that the value of A_2 is high compared to some implicit, non-articulated norm. But his statement is intelligible given that the hatchery is close to the ocean.

Postponement

"Any kind of ability for us to predict comes from watching [that] curve"

Shelby, in a long turn at talk without interruption (Fragment 5.4), then elaborates for the benefit of the remaining team members present the relationship of the data to the existing theoretical canon and published research. The fish do not appear to "shift entirely [from A_2 to A_1] *before* they go," that is, leave the breeding grounds and hatchery to head downriver and toward the ocean. Our research team's ability to predict the proper release time – i.e., the objective of this research – comes from "watching the curve" and making a decision to release the fish from a given hatchery when the curve is at a yet-to-be-determined point. The problem that our team currently faces is that we "don't have the initial stuff," that is, the measurements that constitute the early part of the graph supposed to lie somewhere near 80–90 %. This would have meant to start "back in March," when we might have been able to observe "that really nice eighty [or] seventy percent line," that is, the early part in the Alexander et al. graph (Fig. 5.2) where measured A_2 values are high and nearly constant.

Fragment 5.4

```
001  S: well, and yea, i dont know, i () its seem to me that
        they dont shift, they certainly don't shift entirely
        before they go, and there is no question about that
        um () wild or hatchery, theyre not they couldnt be
        shifted before they go, and there is no question
        about that, so any kind of any kind of projections
        comes from watching a curve for a while, and starting
        to see some kind of but the problem is that we dont
        have that initial stuff, so that maybe there is some
        you know, from back in march, there maybe in that
        really nice eighty or seventy percent line ((gestures
        horizontal line, then dipping of the curve)) and then
        dip dip dip tu tu tu and it just starts dip dip dip
        and then bang down to go so if we can say from the
        beginning of that curve just start and then you know
        that we had four weeks or three weeks and that might
        be more useful and predictable.
```

Everyone "in the know" can hear that Shelby is taking the lens of the Alexander et al. graph. This is so because, while talking, he enacts features of the graph by means of gestures (Fig. 5.6) and statements that bring into this meeting the currently absent graph from the Alexander et al. paper. Everyone in the room can recognize this graph in Shelby's gestures because of their familiarity with it. The gestures are instructions, which direct the recipients' eyes to move in the same general way that the graph also instructs them to move (Roth 2012). In a first iteration, Shelby describes what they might have observed had they begun taking measurements in March. He produces the baseline, which is actually high in A_2. His right hand, which heretofore participated in gesturing the level of the baseline, moves backward and then takes a pointer configuration. The left hand iconically shows the level in reference to which the subsequent measurements exhibit change. That he is talking about the measurements can be seen from the fact that he points to imaginary locations on the graph, then moves and makes another forward pointing gesture as if he were plotting data points. Each time his hand moves forward pointing to some location on the virtual graph, he utters "dip," as if he was making a dot. Each "dip" and the forward movement of the stretched finger makes present, for those ready to hear and see, a data point. But at the same time, each point

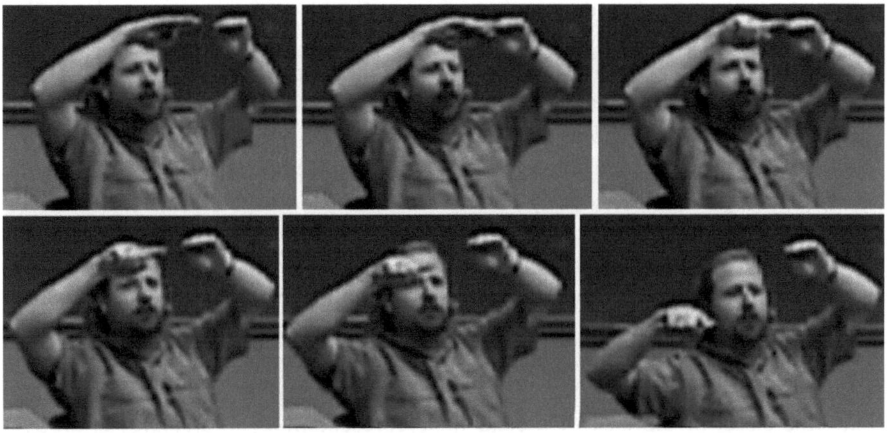

Fig. 5.6 Shelby uses hand gestures to show how the data points should fall; the points that he plots ephemerally in the air relate to each other as the data points in the Alexander et al. graph (Fig. 5.2) (© Wolff-Michael Roth, used with permission)

corresponds to a particular state of the real living fish. The gestures, iconic in their relation to the Alexander et al. graph, thereby bridges between a representational world somewhere along the chain of translations (Fig. 1.1) and the phenomenon of interest in the material world. This means that the scientists have to be familiar in and with both worlds, as well as with the translation process, so that their body can combine these in the way a mathematical function relates domain and range.

Following the first presentation, Shelby reproduces what we expected at the time, though now in terms of a continuous graph, whereby the right hand moves along a continuous trajectory from the reference point toward the bottom. In these two presentations, one can thereby first see the individual measurements that would be expected and then something like a best-fitting curve drawn through all the measurements – in the way it would be done in the resulting publication. When Shelby produces the gestures for a third time, he verbally co-articulates the first several points by making repeated "t's" and then marks the three points associated with large downward distances saying "dip, dip, dip." The right hand then disappears completely below the table, becoming invisible just at the instant when he voices the words "bang down." Finally, he directly names what he has been talking about: "the curve." His statement suggests that our team has to start by looking at the beginning of the curve, which leaves about 3–4 weeks for predicting when the downturn will commence.

From Decontextualization to Recontextualization

Science is a successful endeavor because its concepts are applicable to a range of phenomena. The range becomes wider the lesser the inscriptions are tied to the specifics of any context; and inscriptions become more inclusive and abstract the less contextual particulars they include. It may therefore not surprise to learn that birdwatchers more easily (learn to) classify specimens when using field guides that employ drawings than those including photographic images (Law and Lynch 1990). The former contain less detail and, therefore, are more easily applied to a range of varying birds than photographs that are highly specific because of contextual detail included. But decontextualization, as the abstraction it implies, while increasing the power of thought, simultaneously comes with a loss: "Every abstraction is nothing other than a sublation of certain clear ideas/representations [*Vorstellungen*], which is generally done to more clearly imagine/represent [*vorstellen*] that which remains" (Kant 1763/1956, p. 803). Learning by abstraction, however, does not mean that we should do away with context but implies "*negative attention*, that is, a real doing and acting that is opposed to that action by means of which an idea/representation [*Vorstellung*] becomes clear" (p. 803, original emphasis). Out of this negative attention, "the Zero, or the lack of a clear idea/representation [*Vorstellung*] is brought about" (p. 803). Otherwise – e.g., "simply a negation or lack" – there would be no difference between the intention to know and the intention not to know.

In the meeting fragments analyzed in this section, it is evident that our research team does not see the data points abstracted from everything else. Rather than looking at the data independent of context, simply articulating relationships and mathematical patterns, they "look through" the measurement points seeing, on the one hand, the graphical inscription, and, on the other hand, the real events in the hatchery, where our team got the fishes, and in the river and estuary, where we captured additional specimens for sampling purposes. In the past, I proposed the notion of *transparency* to talk about the fact that scientists use graphs as if these provided direct access to the phenomena (Roth 2003). Although some researchers have critiqued this notion (Noss et al. 2007), I continue to find it useful in the present context because in knowledgeable use graphs do precisely what signs are intended to do: efface themselves so that the thing it stands in for becomes visible. Thus, we generally do not attend to the features of a voice but to the thing that a word orients us to. If a person points to a tree outside the window and says "look at the apples," we orient towards the fruit on the tree and not on the sound that linguists would transcribe, using the international phonetics alphabet, as /æp(ə)l/. We normally do not attend to the qualities of the sound, on the accent or dialect of the person, or the characteristics of her speech. Instead, it is something about the apples out there on the tree that is to stand out. It is precisely when the sound disappears and nothing but the apples in front of the window are salient that the speech sign has done its work.[6] In the same way, graphs do their ultimate work when they disappear from consciousness and orient the user to the phenomenon itself.

Returning to the setting of the data discussion, the team members see their currently available measurements through a lens that works precisely when it is

[6] In other words, signs embody an inner contradiction. They do so because to stand in for something else, that is, to make something else present that currently is absent, signs have to efface themselves. When we stare at a graph or wonder about some sound or ink trace for which there is good reason to consider it as a word, then the sign, the sound or ink trace, is considered in its own right and no longer provides us with transparent access to that other thing that it is supposed to help us make present.

transparent. That is, without having to make these thematic, scientists see their data in terms of (a) the theoretical canon, which relates the A_1/A_2 ratio and other physiological aspects (e.g., response to "salt water challenge") to the freshwater and saltwater habitat and (b) the graph that has the Alexander et al. published data on A_2 ratios during a 12-week period covering the onset of migration plus some weeks before and after. In a way, the team's preconceptions shape what we see and how we talk about the graphs and what we make present with them. Assessments and evaluative commentaries are provided *in terms of the graph they anticipate to obtain as their outcome*. An explanation is sought for the deviations by constructing possible reasons that are based on their familiarity with the context from which the fish derive. Thus, the Kispiox measurements do not tell us something inherently; rather, these data are seen through the lens of our team's familiarity with the conditions in which these fishes are raised and the differences between these fish and those in the nearby wild fishes living under very different conditions. One possible reason for the deviation is that the timing may not be quite as reported in the scientific literature that we are familiar with at the time or that the timing might differ from year to year. What is left untouched here is the nature of the graph itself: It constitutes the paradigm within which this group of scientists operates and looks at the data.

At this early stage of the research, then, our team does not "let the data speak for themselves." Moreover, it is not just the reigning paradigm (theory, empirical work) that shapes what there is to be seen. There already is a lot of embodied experience with the particulars of where the specimens were caught and under which conditions they have grown up that our team brings to the effort of employing and deploying the inscriptions that we are looking at. That is, what can be learned from the data does not derive from the abstract properties and relations between the data points but depends on contextual particulars of the original context from which the fishes have been sourced. If this is the case, then one might hypothesize those individuals to have difficulties reading data who are not intimately familiar with the data sources and the associated practices by means of which the data are acquired. This includes, as the present research project shows, experienced research scientists.

Unpremeditated Reconstruction of Context

> Or, perhaps, while sleeping, I returned without effort to an age of my primitive life, forever gone, found again such as my childhood terrors . . . I had forgotten this event in my sleep; I remembered it again as soon as I had managed to wake up. (Proust 1913/1946–47, p. 13)

In his celebrated seven-volume novel *In Search of Lost Time*, Marcel Proust discovers near the end that only *involuntary* memory is capable of resuscitating what time has made it lose. It is in the seventh volume that time is found again. In analogy to this novel, scientists lose much of the context surrounding their specimens as they produce their data – where data refers to anything that is used in support of the knowledge claims made in scientific publications. I show in this section that in their process of explaining the data, our team reconstitutes precisely this context that we have left behind earlier as part of abstracting the photoreceptor cells from their contextual particulars. That is, returning to the earlier quotation from Kant's work, our team first has applied negative attention to the contingent aspects of our fish specimens, and then has had to spend a lot of effort to bring back into focus that which we had left behind. That is, only when we return the arduous road to the original sites – equivalent to opening up the black box – that we regain the lost context so that we can make the relation between graphs and natural phenomenon that "graph interpretation" implies. We had to undo and reverse the decontextualization that our scientific method had produced. In the following three subsections, I exemplify the reproduction of context in team's discussion of (a) an ensemble of environmental factors, (b) the age of hatchery-released fish, and (c) the weight-porphyropsin link they had previously constituted.

An Ensemble of Environmental Factors

Our team seemingly struggled with the fact that in one setting (Robertson Creek), the wild and hatchery-raised fish exhibit similar mean levels of porphyropsin; in the other setting (Kispiox Hatchery), there are vast differences between the wild fish and those in captivity. In the present fragment from the data analysis session, those who conducted the measurements (Shelby, Theo) make mention of size differences. Our lead scientist links these differences to differences in "life history strategies" (turn 007), that is, Craig draws on a particular concept in biology to explain why the data might differ. It is not that the data tell him about differences in life history strategies, it is the strategies that tell him about differences – much in the way data coders did not arrive at hospital practices from the hospital data, as they were tasked to do, but used their knowledge of hospital practices to code the data (Garfinkel 1967). It is the practical familiarity with the world that allows us to use signs that make absent things present rather than the other way around. This practical familiarity with the world of the coho salmon and our world of research envelops the team's reading of the data and graphs, preceding, accompanying, and concluding it. Such a statement is consistent with evolutionary and individual-developmental considerations: the human species was familiar with its lifeworld prior to the arrival of (self-) conscious language use, and small children are familiar with their world prior to their use of language, particularly their conscious awareness of language use. This language comes to be integrated into, and indistinguishable from, the world we inhabit and that in a way inhabits us.[7]

The session continues with Craig then offering another concept: the fish that they are dealing with represent two different age classes – lower modal (i.e., early)

[7] In this regard, Bourdieu (1997) writes that "what is comprehended in the world is a body for which there is a world, which is included in the world but according to a mode of inclusion that is irreducible to a simple material and spatial inclusion" (p. 162).

second year coho versus upper modal (i.e., late) first year – which may be due to differences in life history strategies (turn 007).[8]

Fragment 5.5a

```
001  C:  and () it would be nice to () it doesnt look like
         theres any dIFference between the hATchery and the
         wI:L:d at relEASe time
002  E:  in robertson
003  C:  in rOBertson. now there dOES appear to be a
         dIFference between the hATchery and the wIL:d at
004  S:  fifty percent
005  C:  attum
006  S:  kispiox
007  C:  kISpiox becoz theres appARently life hIStory strAtegy
         dIFferences in the populATion. () kay youve got () in
         one case; () i mean reAlly what you wanna knOW is
         whether you lOOK at a: UPper mODal fIRst year and a
         lOWer mODal sECond year () okay thATs what i think
         you guys sAW
008  S:  (:E) another quick point; where were the brOOd stock
         for the kISpiox HAtchery collected inITially. ()
         <<p>do you know?>
009  E:  uh; yea i wrote it down, thOSe two rivers where we
         got the fish from
010  S:  oh it is? maybe it would make a big dIFference if the
         brood stock originally came from closer to sEA then
         you would expect
011  E:  no
```

Shelby introduces another "quick point" by asking Elmar whether he knew where the fishes originally were sourced. Elmar responds that these are from two rivers. This appears to surprise Shelby ("Oh it is?" [turn 010]). Rather than pursuing the idea of the two rivers, Shelby then makes a statement that can be heard as suggesting a big difference if the brood stock used in the hatchery were from a river closer to the ocean. That is, at that moment our team does not appear to know where the Kispiox coho had been caught as the source of the roe (eggs) and milt (sperm) that produced the current brood. If these coho were from a river closer to the ocean then these would be more similar to the fishes from Robertson Creek, which lies only 30 km upstream from the estuary to which it is linked via the Stamp and Somass rivers, whereas Kispiox Hatchery is nearly 300 km upstream from the ocean. Here, Shelby draws on geographical factors that might distinguish the fishes and their physiology. This is consistent with the observations I made in a study of fish biologists and hatchery workers, who had trouble and ultimately could not categorize a fish specimen caught in this estuary but, in trying to come up with criteria, drew on geographical factors that would differently affect the sheen on the two species considered (Roth 2005a).

Craig asks Shelby to write down "the question," and then restates that we might be dealing with two age classes of fish based on the fact that the hatchery coho

[8] Life history theory suggests that the events in the life of an organism are of a temporal nature such that it reproduces the largest number of offspring. "Strategies" are those behaviors that are thought to maximize offspring and offspring survival.

salmon were of different size.[9] Elmar says that we would have to look at the [fish] scales, which, because these have annual growth rings, are a means for our team to read the age of a fish. He states that this would be important given that we received on that very day a batch of fish from the Kispiox Hatchery with very differently sized specimens. Craig picks up on the size differences, for which he offers two hypotheses: either the fish are exposed to an ensemble of environmental factors or are being held over and therefore represent an older class of fish.

Elmar then makes a statement that tells us where the fish had come from. This provides possible answers to both Shelby's and Craig's contentions (turn 023). The fishes are from different creeks and these creeks constitute very different ecological niche conditions. It turns out that Elmar not only knows that the fish came from different river systems but also is greatly familiar with each of these systems (turn 023). In one instance, it is a quick flowing creek, whereas the other creek flows through many ponds and even a lake. It is a slow-flowing creek. At this point, he also co-articulates a typical disease that comes from living in such a river system, which is the disease that research team members had detected in some of the specimens (turn 023). Our team members can hear his response to Craig's question concerning the origin of the big fish as pertaining to the Statsnat River, where the productivity is higher and therefore leads to bigger fishes. These fishes are also diseased.

```
Fragment 5.5b
023  E: the first batch was from eh a river clifford creek
        that doesnt thats basically a flowing river all the
        way through. zsecond batch come which i found out
        afterwards they just said well theyre wild fish the
        fence blew out of clifford creek so dey just gave us
        fish from statsnat () statsnat is one beaver pond
        after another the whole river down we had to go over
        it by helicopter. theres even one lake in it; all the
        rest r beaver ponds. sO: quite dIFferent and dats
        where zee uh () um disease occurred too and zis
        disEAse is typical for slow slowflowing water anOZer
        indicator;
024  C: is that where the big fish came from?
025  E: yes zats where the big fish came from so dey must
        have probably way higher productivity in
        skats[nat river   ]
026  C:      [did you guys] keep the morts?
027  S: ((Elmar turns to Shelby, they look at each other)) no
028  E: no
029  S: we didnt we thought about looking at them bcoze i
        noticed that one of them had a fish in its stomach;
        that was unusual
030  C: hO:ly shIT that does sound like ah
031  S: seemed a bit unus[ual for the fish]
032  C:                  [lower modal     ] year two
```

[9] "Age class" refers to fish born in different years. Thus, in any river, there may be young coho salmon in their first year (referred to as "zero plus [0+]" a specified number of months) or in their second year (one plus [1+] some specified number of months).

In this exchange with Craig about the mortality, Shelby articulates yet another concrete observation that is relevant to the age problem: "one of them had a fish in its stomach" and he adds an assessment, "that was unusual" (turn 029). Thus, Shelby provides an explanation why he did not keep the dead fish; he and Elmar then look at each other, each responding with a subdued "no," almost as if they were children caught doing something inappropriate. Shelby says that he found a fish in the stomach of one of his specimens, which he formulates as having found unusual. Craig comments even more strongly: "Holy shit" (turn 030). While Shelby repeats the unusual nature of this observation, Craig states that this "does sound like a lower modal year two" (turn 032). Elmar confirms: "they are definitely year-two, different release class" and therefore different that those that were "supplied from the hatchery" (turn 032).

Later, Elmar also suggests that the specimens are from different age classes and adds that "they had an auxiliary clip at that time," which Shelby confirms. "The clip" refers to the fish hatchery practice of removing the adipose fin – a fin on the back of a fish, which is believed not to be necessary for efficient swimming (Fig. 5.7). Fish found without adipose fin definitively are hatchery-raised coho, though, in contrast to other marking practices, this does not reveal the specific hatchery where the fish has been raised.[10] Craig then notes that the other fishes, the smaller ones, to be coho released during the present year from the hatchery. Elmar responds negatively, suggesting that the small fish are wild based on the fact that they do not lack their adipose fin ("they have no clips, no ad[ipose fin]"). Craig *formulates* what he is in the process of doing: he is trying to figure out whether the smaller fish are 1 year and the larger fish 2 years old. Shelby offers two possible explanations: (a) the stream where these small, 9-g coho have been caught has a low productivity (which does not allow the fishes to put on weight as these normally would) or (b) these fishes migrate toward the ocean during their first year.

At this point our team is grappling with identifying the age class of our specimens. The coho salmon we analyzed were of different size, which could be because of low productivity environment or because of different age classes. The wild fishes have come from different streams and are of different size. An additional piece of information are the gut contents, which assists the team in narrowing down the possible age class. In the course of grappling with the problem, the team members bring up and discuss various possibilities. These possibilities derive from our concrete familiarity with the river and creek system in the area and our

[10] During my ethnographic research in the hatchery, I observed several such practices. For example, by increasing and decreasing the water temperature of the developing fishes immediately prior to or following hatching, fish culturists change the thickness of the growth rings observable in the otoliths (ear bones). Each hatchery has a specific sequence giving rise to a specific pattern that some of the personnel extracted from the ear bones of dead adults that had returned to the hatchery and surrounding rivers. The second practice of marking salmon involves the injection of a coded wire tag into the "nose" of the fishes. When adults are caught or found dead, a metal detector will indicate the presence of such a tag, which, once removed from the carcass, will reveal the origin of the specimen.

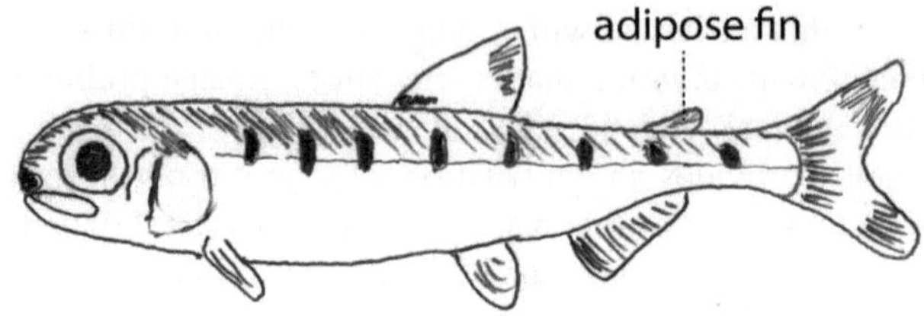

Fig. 5.7 Young salmon with adipose fin. Hatcheries frequently remove this fin, marking the fish as hatchery raised (© Wolff-Michael Roth, used with permission)

familiarity with fish according to which only year-two coho would eat smaller fish, diseases that exist in slower rather than faster flowing waters, and so on. Following the conversation, one can see these possibilities arise as something new. This contingent articulation of new information gives the learning that occurs an emergent, unpremeditated quality, as certain knowledge of the specifics of the origin of the fishes or their state is articulated in some discursive context only to become significant to another aspect.

How Old Are Hatchery-Released Fish?

The age of the fishes has been introduced as a *possible* confounding factor. Whether it is a confounding factor is neither certain nor confirmed at this stage. Knowing the biology of the species involved is an integral aspect of interpreting this data. Our team's members draw on their familiarity with the source of the data to lay out possible explanatory scenarios. They do not just interpret the numbers to make some claim by inferential reasoning. In this meeting, it is suggested that the data should not be presented in the way they currently are, as there are differences that make the data from the different rivers or age classes incompatible. In effect, our team has found reasons for not explaining the data as is, but some of those present have introduced information that allows us to defer any definitive statement.

In the preceding excerpts, an undetected confusion about the fish may be noticeable, a confusion that would become evident in the subsequent discussion. The hatchery-supplied coho are less than 20 months old – because fertilization occurs in October, the fish are about 8 months old – whereas the coho migrating toward the sea are older than 20 months (i.e., the release age). The weight and size of the fishes are functions of the living conditions and temperature; in wintertime the fish hardly feed and therefore hardly gain weight (and, as my ethnographic work in the hatchery showed, they even may lose weight), whereas the fishes increase substantially in weight during the summer months (hatchery and wild) (Fig. 5.8). This is why hatchery workers, who model the growth of the fish using graphs, feed less in the winter than they do in the summer months leading to characteristic growth curves (e.g. Roth 2005b).

The following segment from the data analysis session shows that some members of our team articulate presuppositions, whereas others know, from the time we have

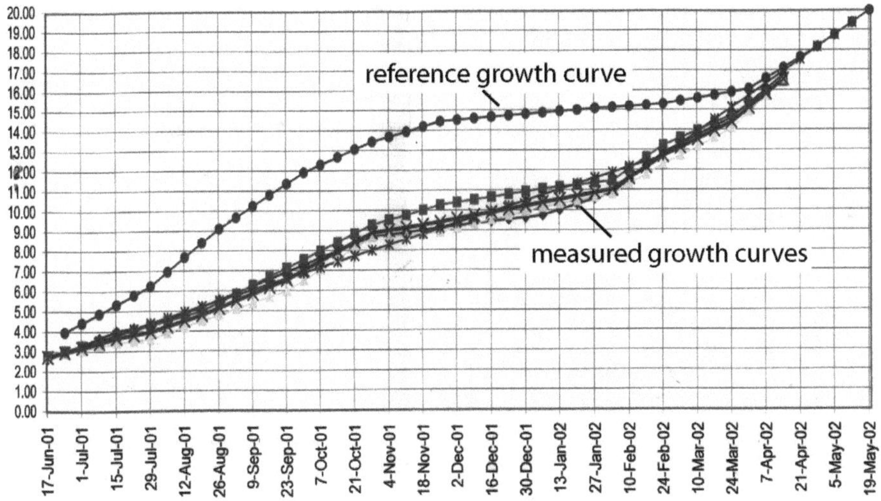

Fig. 5.8 Growth curves for coho salmon at Robertson Creek, including a reference curve for desired growth rate as a biologist had established it and actual growth rates of coho salmon in the different ponds (© Wolff-Michael Roth, used with permission)

spent in the hatcheries, that the truth is different – hatcheries release fish when they are "one plus" (i.e., fish that is more than 1 year old) rather than "zero plus" (i.e., specimens in their first year). In fact, the lead scientist is unfamiliar with the precise release date and with the age of the wild coho salmon when these appear to be leaving the river. My ethnography in the fish hatchery shows that there are differences between the different species of salmonids that are raised in the hatcheries. For example, three species are raised at Robertson Creek: coho (*O. kisutch*), steelhead (*O. mykiss*), and chinook salmon (*O. tshawytscha*). The hatchery releases these at different stages in their life history and at their respective optimum weight: at 20–25 g (about 20 months following fertilization), 60 g (~18 months), and 6 g (~6 months). In the following fragment, two individuals turn out to know the timing of the release because of their extensive presence in fish hatcheries: Elmar and I. Craig states that the coho salmon are released during their first year and that the wild fish are of the same age (turns 074, 076). Elmar, however, makes a statement that contradicts what Craig has said: the fish are "one plus (1 year + [unspecified months]") that is, they are released during their second year of life. Shelby concludes that the "little guys" "are the abnormal ones" that might go out early; and he offers a causal connection ("that's why") between the similarities of "everything" and "them" and for "getting more of them" (turn 084, 086).[11] He states to have "two comparisons of fish that are quite similar," that is, the wild and

[11] Some hatchery workers conduct experiments to find out whether different release times yield higher return rates. Thus, for example, super smolts are fishes retained for up to an extra year prior to release into the rivers and the beginning of the ocean migration.

hatchery-raised coho from Robertson Creek and the wild and hatchery-raised fish from Kispiox Hatchery (turn 086).

Fragment 5.5c

074	C:	[the hatchery] released fish er er zero plus
075	S:	correct
076	C:	the other guys that you looked at i bet you that are zero plus
077	E:	no zey are not; ze hatchery released fish are not zero plus
078	S:	<<p>yea one plus.>
079	E:	zey are one and a half year old zey are one plus
080	S:	<<p>one plus>
081	E:	yea one and a half years more almost two years so deyre holding zem zat long.
082	S:	the wild and hatchery fish.
083	E:	one year and ten monzs
084	S:	so its the little guys that are the abnormal ones they i:t might that might be going earlier the bIG ones is the same as the other
085	C:	oh okay
086	S:	thats why they are in similar size; thats they are similar, everything is similar so thats more similar thats why we are getting more of them so that is really good. so we have tWO comparisons of fish that are quite similar

Craig then suggests that they needed to document "that scenario" and adds that they also needed to document the Robertson Creek scenario, which he formulates as assuming to be a release of zero-plus. Elmar disagrees and proposes that it is "the same thing." There are several brief exchanges at the end of which I contribute the fact that the coho eggs are fertilized in October and the fish are released in May (19 months later). Elmar later (incorrectly) suggests "[a year and] nine months," which I confirm in a constative utterance. It is only at this point that Craig appears to realize that the coho are released in their second year in both hatcheries and he explicitly formulates what he just has learned: "Okay, now then, that's consistent."

Shelby then has a long turn during which he summarizes the results presented and raises the central question: Why are the wild coho near the Kispiox Hatchery so different from the ones raised in the facility? He offers up the possibility that the ones with the lower values are actually released too late "so that they travel for another 50 kilometers with the wrong pigment," that is, with the rhodopsin pigment typical of the marine environment. In this case, then, the wild coho from Kispiox would be the normal ones – because they have retina with porphyropsin amounts typical of freshwater coho – and all the other fishes are abnormal. He ends by saying that this "is possible, but speculation probably."

Elmar offers a change of topic, which, while not allowing us to assess the relevance of what Shelby has articulated, introduces a new piece of information relevant to the question of age. He formulates having had a thought: Two years prior to the meeting, there was a "terrible run" and only 80 returning coho were counted that year in that river (turn 100). As a result, very few offspring – one-plus coho smolt – would be in the river and that most of the specimens that they received from that system would therefore be "zero-plus." He adds that only the scale analysis would tell them the age of the fish and that this kind of analysis is not easy to do (turn 104).

```
Fragment 5.5d
100  E: do you know what i am just thinking. clifford creek
        two years ago had ze most terrible run since ever
        recorded eighty fishes of () almost seven zousands
        down; so dere is probably very little two yearold
        fish very very few two year one plus smolt they are
        probably zero plus smolt but we have to go to ze
        scales to try to
101  S: fishes?
102  E: well i hope we got boze of dem zats why I was hoping
        so we can probably do the scale analysis side by side
        then we can see the zero plus ones or have more of ze
        ozer ones
103  S: yea
104  E: scale analyzes is pretty is not so easy if you dont
        () unless you boze specimen side by side is not so
        easy
```

In this instance, Elmar knows about the biology and that there could not have been many young fish from the age class when there were only 80 adults that had returned to the river. As a consultant to fish hatcheries in that geographical area – and based on what he has learned about salmon during his dissertation – he is very familiar with the hatchery practices and with the life cycle of Pacific salmon (*Oncorhynchus*) species. Because of my ethnographic study of fish hatcheries conducted simultaneously, I, too, am very familiar with the practices in these institutions and the parts of the life history that the salmon spent in these institutions (beginning and end of their life cycle). What Elmar and I know from our extensive experience in the fish hatcheries does not come to be mobilized up front, as a starting point for the process of making sense of the data. Rather, it is in the course of the meeting and *as* relations with others that relevant aspects of our familiarity with the concrete settings of the hatcheries – their practices, the natural environment, and climatic and geographical conditions – emerge as possibly important pieces of information that bear on the situation at hand. Our team's (collective) learning process is the result of the constitutive relation *in*, *as*, and *from* which the new familiarity emerges.

What has emerged, then, is the fact that small fishes other than those attributed to Clifford Creek also have a likely age of zero [years] plus [months], as the run that would have given rise to one-plus coho smolts was nearly wiped out. Craig apparently notes some consistency, and Shelby articulates his sense that there are two good comparisons (wild vs. hatchery-raised, and geographical location) that our data afford. The mystery, then, as Shelby's statement suggests, arises from the difference between the fish sourced at Robertson Creek Hatchery and those sourced at Kispiox Hatchery. In the former situation, wild and hatchery-raised fish exhibit similar porphyropsin levels but in the latter, the porphyropsin levels are very different. Something other than weight has to give rise to the difference. Not articulated here – but which would become salient only years later – is the fact that there is little difference between the zero-plus and the one-plus wild coho. This could have led our team to pursue alternatives to the pattern exhibited in the Alexander et al. graph much earlier (Fig. 5.2). But already in this meeting, Shelby presents evidence for the possible independence of porphyropsin levels from weight or age class.

Undoing the Weight–Porphyropsin Link?

In search of a possible explanation for the differences in porphyropsin (A_2) levels between the hatchery-raised and wild fish from the area our team has spent a considerable amount of time on the question of the age class and weight of the coho fishes that we had received from the Kispiox Hatchery. Some 15 min later in this meeting, Shelby actually projects the results of an analysis that is consistent with an independence of porphyropsin levels and weight. Following a discussion of the conditions under which the fish are kept in the university aquarium facility, Shelby changes topic when suggesting: "there is a quick analysis on the size" that he has conducted; it shows that "there seems to be absolutely zero trend." He projects a graph (Fig. 5.9) – to which Elmar immediately responds by shaking his head in what we might see as an expression of puzzlement and then utters, with rising intonation and using an interrogative: "What is this?"

Fragment 5.6

```
001  S: there is a there is a quick analysis on the size
        there seems to be absolutely zero trend
002  E: what is dis?
003  S: thatssize; fishweight versus percent porphyropsin. it
        shows absolutely no trend whatsoever; whichis true.
        () which is good, <<acc>jst because you mentioned
        something about size> thats why we do that
004  C: yea
005  M: what was it again?
006  S: its just fish weight on the [eh y:eh:x]
007  M:                             [oh okay  ]
008  S: axis versus percent of porphyropsin
009  E: all locations in one graph hu .hh
010  S: in all locations, it is just all three thousands
        <<acc>its a good way to get a sense of some trend>
011  E: yea you are right; you are right.
012  S: and i did in different colors just for the whole
        population but clearly you got nice, you fill the
        whole blank, the whole box as [far]
013  C: [yea]
014  S: we are getting sizeunrelated data
015  C: yea
016  S: to porphyropsin but that i actually expected that
        completely expected that thats why i brought it up
```

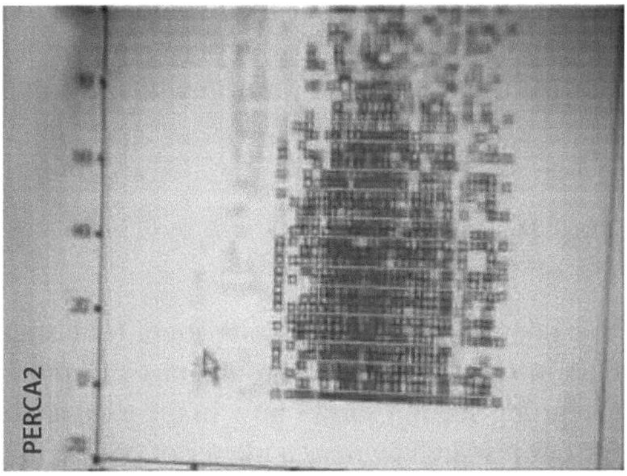

Fig. 5.9 Shelby projects a graph that plots porphyropsin levels against fish weight (© Wolff-Michael Roth, used with permission)

Although one might have assumed that this additional analysis should have settled the issue, our team does not discuss it in this manner. None of the members present picks up on the potential relevance of this analysis. In fact, Elmar appears to be a little incredulous, which can be taken from the fact that he laughs after stating that all data from all locations have been plotted (turn 009). Shelby states that he wanted "to get a sense of some trend," and this is why he plotted "all three thousand [data points]" using "different colors [for the different locations where the coho had been captured]" (turn 012). He concludes that the data fill "the whole blank," "the whole box." This can be read to mean that they "are getting size-unrelated data to porphyropsin" and that is something he "actually expected." He expected this result and that is why he brought the plot into the discussion. There is a long pause, which ends when Craig begins to summarize what they have learned during this meeting.

Here, Shelby introduces a graph. It apparently is consistent with the hypothesis or claim that the porphyropsin levels are independent of fish size. Although this information appears to contradict what our team has been discussing earlier, which is a relationship between weight/age class of the fish – at least at Kispiox Hatchery – and the porphyropsin levels, we do not actually bring these two aspects of the meeting together and do not discuss the implications thereof. With hindsight, the similarity in the data from zero-plus and one-plus coho provided us with a possibility for recognizing that it is not smoltification and the associated physiological changes that drive the difference in the visual system but some other factor. The publication of the paper in which we ultimately would report our results suggests high correlations between porphyropsin levels and temperature or length of day. The claim that there is a relation between these variables and that it may be periodic emerges over the course of the following 2 years, with the massive accumulation of relevant data. At this stage of the analysis, during that meeting, however, what is later recognized as evidence – weight, age-class independence – is not yet seen and described as such. What Shelby just has presented stands on its own and, because Craig continues with a summary of what he sees as emerging, this fact has no implications for the present discussion of the data. For months to come, our team would continue to pursue the hypothesis of a correlation between the saltwater- and migration-related changes in porphyropsin levels rather than the seasonal changes (for the timeline of the ways in which the data were explained see Chap. 7).

The Past Regained

In analogy with Proust's last sequel (one translation of which is entitled *The Past Recaptured*), where the protagonist recaptures the past in finding time again, the foregoing section shows how in the explanatory effort, the lost context re-emerges for our research team as a matter of course in the process of talking about a relation of the graphs and the locations where the fishes had been sourced. But there is nothing in our meeting that would suggest that we *consciously*, deliberately, or in a premeditated fashion sought to recontextualize the decontextualized data. It is in and through our effort to develop an explanatory discourse that we "wake [ourselves] up" to remember what we have left behind. During that meeting, no individual member of the research group has had all the information required or had a grasp on what it took to become familiar with it, or even realized what the relevant information might be. All of this, the nature of *what* constitutes information and what is required to be familiar with the phenomenon *emerged* unforeseeably from our meeting-constitutive societal relation. Emergence here is equivalent to saying that the end result could not have been predicted on the basis of adding up all pre-existing knowledge available across the group but (parts of) what emerges exceeds, and is in surplus of, the sum total of what existed before. In the process, the ultimate graph is constituted as a synecdoche (part) of the research process (whole), as the research team learns to connect the data with the original settings (contexts) from which it had extracted these. Confronted with the data that have been decontextualized in the process of research, our team now can be observed in the process of *rebuilding* the context that it had stripped in the course of producing the graphs.

This chapter contributes to a better way of modeling the relationship between inscriptions on the right-hand side of the chain of translations (Fig. 1.1) – the abstract inscriptions that make the phenomenon present again – and the concrete contexts from which the former emerge in the course of scientific discovery work. It focuses on how our team came to be familiar with the data and graphs. The analyses show that in the data analysis meeting – the present one being documentary evidence of what happened in our early meetings generally – our team struggles integrating the data with what we already know, data that are the result of many steps in the research process that has taken us downstream and away from the data sources. In this process, context literally was lost as pieces of material were extracted from their surroundings and then subjected to measurements, which themselves underwent extraction and abstraction. These steps were part of the chain of translations that took a natural object and transformed it into scientific knowledge claims, where any two chain links were separated by an ontological gap that was bridged by means of our concrete research practices (Fig. 1.1). The work that linked any two inscriptions disappeared, leaving behind only a trace in the form of a graph. To unpack the graph again would require work that opened up the black box and recovered the work that has gone into the linking. This recovery work here had to be conducted by those scientists who had produced the graph in the first place. How much more work might we have to anticipate from those unfamiliar with the habitat of the fishes and the process by means of which we got to the absorption curves? In the process, we, as other scientists, did not just "construct" data-related knowledge. Our becoming familiar with data and graphs was a process of undoing the decontextualization that we had previously enacted. Integration of the graphs with our existing knowledgeability is equivalent to moving upstream toward the original settings where our fishes had originated. That is, we had to move out of the laboratory and back into the wild to be able to articulate the relevance of the graphs we had produced. What we could do with a graph was related to our intimate knowledge (a) with the situations where the fishes had been sourced and (b) with the transformations that the things in our hands underwent in the laboratory and subsequent computer models. Without this (often tacit) intimate familiarity (ground), the data (figure) would say little to nothing – a statement justified by the research results showing that scientists have difficulties "interpreting" graphs that feature in undergraduate textbooks of their own domain.

References

Alexander, G., Sweeting, R., & McKeown, B. (1994). The shift in visual pigment dominance in the retinae of juvenile coho salmon (*Oncorhynchus kisutch*): An indicator of smolt status. *Journal of Experimental Biology, 195*, 185–197.

Berg, C. A., & Smith, P. (1994). Assessing students' abilities to construct and interpret line graphs: Disparities between multiple-choice and free-response instruments. *Science Education, 78*, 527–554.

Bourdieu, P. (1997). *Méditations pascaliennes* [Pascalian meditations]. Paris: Éditions du Seuil.

Collins, H. M. (2001). Tacit knowledge, trust and the Q of Sapphire. *Social Studies of Science, 31*, 71–85.

Couzin-Frankel, J. (2013). The power of negative thinking. *Science, 342*, 68–69.

Edgerton, S. (1985). The renaissance development of the scientific illustration. In J. Shirley & D. Hoeniger (Eds.), *Science and the arts in the renaissance* (pp. 168–197). Washington, DC: Folger Shakespeare Library.

Garfinkel, H. (1967). *Studies in ethnomethodology*. Englewood Cliffs: Prentice-Hall.

Il'enkov, E. (1982). *Dialectics of the abstract and the concrete in Marx's Capital*. Moscow: Progress Publishers.

Kant, I. (1956). *Werke I: Vorkritische Schriften bis 1768* [Works I: Pre-critical writings until 1768]. Wiesbaden: Insel. (First published in 1763)

Kingsland, S. E. (1995). *Modeling nature: Episodes in the history of population ecology*. Chicago: University of Chicago Press.

Kuhn, T. S. (1970). *The structure of scientific revolutions* (2nd ed.). Chicago: University of Chicago Press.

Latour, B. (1987). *Science in action: How to follow scientists and engineers through society*. Cambridge, MA: Harvard University Press.

Latour, B. (1988). *The pasteurization of France*. Cambridge, MA: Harvard University Press.
Latour, B. (1993). *La clef de Berlin et d'autres leçons d'un amateur de sciences* [The key to Berlin and other lessons from a science lover]. Paris: Éditions de la Découverte.
Law, J., & Lynch, M. (1990). Lists, field guides, and the descriptive organization of seeing: Birdwatching as an exemplary observational activity. In M. Lynch & S. Woolgar (Eds.), *Representation in scientific practice* (pp. 267–299). Cambridge, MA: MIT Press.
Noss, R., Bakker, A., Hoyles, C., & Kent, P. (2007). Situating graphs as workplace knowledge. *Educational Studies in Mathematics, 65*, 367–384.
Proust, M. (1946–47). *À la recherche du temps perdu: Du coté de chez Swan* [In search of lost time: Swann's way]. Paris: Gallimard. (First published 1913)
Roth, W.-M. (2003). Competent workplace mathematics: How signs become transparent in use. *International Journal of Computers for Mathematical Learning, 8*, 161–189.
Roth, W.-M. (2005a). Making classifications (at) work: Ordering practices in science. *Social Studies of Science, 35*, 581–621.
Roth, W.-M. (2005b). Mathematical inscriptions and the reflexive elaboration of understanding: An ethnography of graphing and numeracy in a fish hatchery. *Mathematical Thinking and Learning, 7*, 75–109.
Roth, W.-M. (2012). Tracking the origins of signs in mathematical activity: A material phenomenological approach. In M. Bockarova, M. Danesi, & R. Núñez (Eds.), *Cognitive science and interdisciplinary approaches to mathematical cognition* (pp. 182–215). Munich: LINCOM EUROPA.
Roth, W.-M., & Barton, A. C. (2004). *Rethinking scientific literacy*. New York: Routledge.
Roth, W.-M., & Bowen, G. M. (2003). When are graphs ten thousand words worth? An expert/expert study. *Cognition and Instruction, 21*, 429–473.

Chapter 6
On Contradictions in Data Interpretation

Contradictions as integral moments of learning processes have been of interest to learning scientists and STEM educators under various guises: as discrepant events that promote cognitive conflict and conceptual change, as a means to "push" learners from one conjecture to another when there are tensions between results and findings in geometrical proof, as drivers of reasoning when examples and counterexamples differ or when argument and counter-argument come into conflict, as sources of new mathematical ideas arising from the contradictions between perception and beliefs, as a resource of proofing by contradiction when they face dilemmas such as that regarding the parity of zero, or as the driver for increasing students' meta-level awareness, especially features of their thinking. Contradictions are said to promote learning because a "discrepant event *constrains the construction of the new model as well as creating dissonance with the old model*" (Clement and Steinberg 2008, p. 110). The same may be said about examples and counter-examples, argument and counter-argument, that is, whenever two positions on the same issue appear to differ. But the conceptualization of contradiction generally and discrepant events specifically tends to be approached from the perspective of the person who already knows, who design such events beforehand without knowing what a student *actually* perceives and considers. For example, learning scientists may write or talk about "*built-in* contradictions." But such contradictions tend to be recognizable only a posteriori. In gesture studies, for example, it is known that a person may express the same phenomenon in conceptually contradictory way, one in verbal the other in gestural form (e.g. Alibali and Goldin-Meadow 1993). But the person does not know that her conceptions or perceptions contain a contradiction until after she fully develops the more advanced conception in both modalities. Other researchers may think that there are contradictions in statements without also stating *for whom* there was a contradiction and under what guises did it appear. Thus, to take a historical case, Galileo did not see that his way of adding speeds is a limit case that leads to contradictions as soon as the speeds reach certain values, about one-tenths of the speed of light, as shown in special relativity theory. Scientists do not necessarily recognize contradictions as such or know how to remove them until after they have found a way of removing them. This is why,

W.-M. Roth, *Uncertainty and Graphing in Discovery Work*, 179
DOI 10.1007/978-94-007-7009-6_6, © Springer Science+Business Media Dordrecht 2014

for a period of time, the language they use may subsequently be referred to as "muddle" (Rorty 1989). This is so because in contrast to the master craftsperson, who tends to know what kinds of jobs he needs to do, "someone like Galileo . . . is typically unable to make clear exactly what it is that he wants to do before developing the language in which he succeeds doing it. His new vocabulary makes possible, for the first time, a formulation of its own purpose" (pp. 12–13). But, as Rorty highlights, this language, this formulation, is possible only *after the fact*. The new language makes possible to talk about *why* a new language was necessary and *how* it was developed. Only through the final outcome of the evolutionary development do reasons become available so that causal accounts are completely inadequate to explain conceptual development. In the course of the research that will lead to new knowledge, the descriptions and explanations do not yet exist in the way they will be once everything has been said and done. For this reason, the criteria required to make any decisions or interpretations come to exist only with hindsight.

The purpose of this chapter is to provide a detailed analysis of what after the fact can be recognized to have been unresolved contradictions in our team's considerations of the data that we had collected. In the work of our research team, there were contradictions in the way that the data were modeled graphically and algebraically that went undiscovered and remained even in the journal article where we reported our work. In the process, we did not hear or attend to the information that could have given us clues to contradictions and we maintained our new interpretation, which was to become a new conceptual theoretical lens right through the publication of the work.

On Logical and Inner Contradictions

Contradictions come in different kinds: logical contradictions and inner contradictions (e.g. Il'enkov 1982). Logical contradictions – as the adjective that denotes reason (Gr. *logos*) and language (Gr. *logos*) indicates – are contradictions between the contents of two statements pertaining to the same phenomenon. Thus, "[a] logical contradiction arises within reason itself, disrupting it, breaking up the very form of thinking in general" (Il'enkov 1977, p. 105). The contents of the two statements, "This bird is black" and "This bird is white (not black)" stand in a logical contradiction, for the same bird cannot be black and not-black simultaneously. In formal logical terms, this is expressed as $p = \neg(\neg p)$, where p is a predicate (statement) and "¬" is a logical negation. In words the same is stated as "p is not not-p," "p *is-not* not-p," that is, p and not-p are different. As a Venn diagram shows (Fig. 6.1), the two are mutually exclusive. There is no other solution, as a third option is not given (in Lat., *tertium non datur*). Only this kind tends to enter and be discussed in the learning sciences. The statement "$p = p$" also is called identity (Heidegger 2006). In classical philosophy and in the sciences, the statement is the highest law. Yet it already "presupposes what identity means and where it

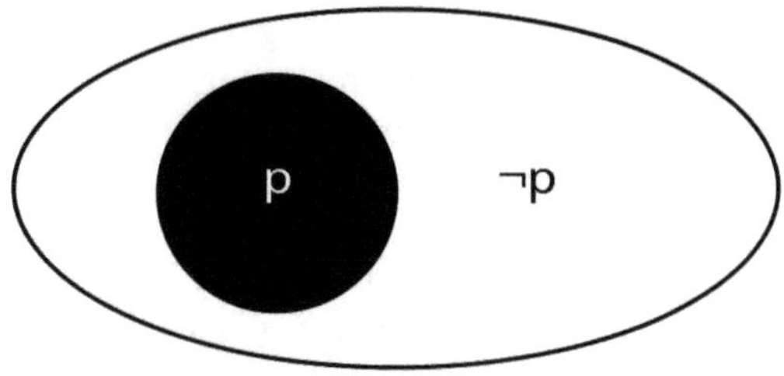

Fig. 6.1 This Venn diagram illustrates the mutually exclusive nature of two statements about the same object, such as "This bird is black" (p) and "This bird is white (not black)" ($\neg p$)

belongs" (p. 35). The second statement about p not being not-p is already implied in the definition of identity, that is, the difference between p and $\neg$p is grounded in the self-identity of p, from which $\neg p$ precisely differs. Logical contradictions of this sort may exist in the natural sciences, and can "be explained by the division of labor," "which more or less restrict[s] each person to his special sphere, there being only a few whom it did not rob of a comprehensive view" (Marx/Engels 1962, p. 318).

It is apparent that a logical contradiction is salient in the conscious awareness of the subjects of activity, who then act upon it in one or another way. It is only when K. Leibniz noted that Descartes' measure of motion (impulse in modern physics) contradicted Galileo's law of motion that a discussion could evolve in which physicists were teasing apart what were to become the concepts of *velocity* (v), *linear momentum* ($p = \mathrm{m} \cdot v$), and *energy* of an object ($E = ½\,\mathrm{m} \cdot v^2$). If something is a logical contradiction in the eyes of an observer but not in those of the subjects themselves, then no consequences should be expected for the thinking and acting of the observed subject. Frequently in learning science research, this distinction is not made. Thus, applied researchers often write about offering learners "discrepant events" without ascertaining that from the perspective of the learner, whatever is being offered is in fact seen as a contradiction. Indeed, contradictions have emerged as a topic of study from two perspectives. On the one hand, contradictions, in the form of discrepant events or states (e.g., between an infinite 0.9999... or $0.\overline{9}$ and the finite 1), are offered to or created for students such as to encourage them to change their current ways of seeing or conceptions. On the other hand, contradictions have been observed between gestural and verbal representations while students provide explanations for mathematical (or scientific) phenomena (e.g. Roth 2002). From an instructional point of view, educators can use this as an indicator of the readiness to develop a more advanced concept – even though the students themselves may not yet recognize the contradiction. From a conceptual change perspective, the presence of discrepant data are supposed constitute anomalies, which set in motion a process that leads from a state of crisis via the generation of new hypotheses, through testing and elimination, the acceptance of a new theory. Associated with a change in conceptual language are structural shifts within a hierarchy of concepts, when the change is within ontological categories, and are changes in the overall structure, when the change are between ontological categories (Chi 1992). The mechanisms associated with such structural changes, and the theoretical discourses that come with them, appear to suggest a sudden qualitative change such as associated with insight learning. Cognitive perspectives on contradictions focus on the latter as drivers of reasoning and changes therein that come about when examples and counterexamples differ – which assumes, as discussed above, that the problem solver actually has some theoretical grasp of the contradiction rather than experiencing something as not working right. The latter situation may not, and frequently does not, lead to greater knowledgeability, even among scientists. The conceptual change discourse about initial conceptions (structure) and change to new conceptions (structure) may reflect a focus on entire discourses. It can give the impression that there are switch phenomena, or conceptual changes based on rational decisions. A language perspective on conceptual change, however, would not make us anticipate radical changes to occur. Rather, languages tend to be learned incrementally rather than wholesale. If we look at two points in language development that are sufficiently far apart, changes become apparent. However, one has to ask, why does language change in the first place? Whereas the response could be that continuous addition of words and the use of grammatical rules incrementally changes language, the *why* of this change is thereby not yet answered. The circle of scholars around the Russian intellectual Mikhail Bakhtin has developed a suitable theory. But to employ this theory consistently requires an introduction to inner contradictions, which I do below.

Second, cognitive development and conceptual change theorists sometimes suggest the co-presence of multiple ways of talking/thinking about pertinent issues. For example, when psycholinguists study mathematical problem solving, they note that at times there are differences in cognitive levels between what children say, generally at a lower level, and what they gesture, generally at a higher level without nevertheless being noticed by the children as such (e.g. Alibali and Goldin-Meadow 1993). In other studies, science students are observed in natural school settings where gesturally presented conceptions preceded the corresponding verbal conceptions on the order of several weeks. The period of co-presence of the different conceptual articulations can be characterized as one of conceptual change. This change is not sudden but appears to occur by forms of communication that is, in hindsight, confused, inconclusive, and comes closest to a "muddle" (Roth 2008). However, confusion and muddle are not apparent to the person and become apparent only subsequently, after the new forms of conceptual talk have evolved and in view of the deeper familiarity with the phenomena that has emerged. That is, during the process of development, the problem solvers do not have a (theoretical) grasp of the contradiction, which becomes apparent as such only with hindsight. A theoretical analysis that explains this situation already had been provided during the 19th century. Thus, from within an event, causality remains invisible; cause–effect relations operating with events – such as a problem solving *process* in the course of its unfolding – can inherently be provided only after the event has come to a conclusion and when the outcomes of the event as a whole are available for analysis (Nietzsche 1954). Precisely the same conclusion was reached more recently following the analysis of the struggles engineers experienced while attempting to operate "intelligent" photocopiers (Suchman 1987) and scientists trying to match visual images and graphical representations as I report them in Chap. 3. In such studies, the relationship between what the engineers and scientists were doing and what they intended doing could be established only after the fact.

Inner contradictions are of a different kind.[1] An inner contradiction exists when a thing does not coincide with itself, thereby offering a radical alternative to traditional logic: dialectical logic. In dialectical logic, a thing is recognized as not being identical with itself. This non-self identity has its source in the fact that the universe is continuously evolving so that a thing never *is* but always *becomes*. Thus, no thing is ever constant and therefore identical with itself but is in continuous flux. Dialectical logic was articulated to generate categories – minimum units of analysis and thought – to think this non-constancy of the world generally and the human life form, including cognition, specifically. As shown in Fig. 6.2, to consider change in and for itself requires categories – i.e. minimal units of analysis – that stand for change. Such categories are basic figures of thought and cannot be further reduced. There is therefore an inner contradiction, because the minimal unit can manifest itself in different ways, depending on when, where, or how we look experimentally (e.g. as the differently shaped and shaded parallelograms in Fig. 6.2). Whereas this might sound strange to some readers, there is actually precedence in the sciences for this kind of thinking: In the thought experiment from quantum physics referred to as *Schrödinger's cat.*

In this thought experiment, a cat is placed in a box together with an atom from a decaying substance. As long as we do not look into the box, we do not know whether the cat is either alive or dead. We can describe with a *formalism* in which the cat is both alive and dead simultaneously – just as on the inside of our category, there is either a rectangle or a parallelogram depending on when we care to look. In the case of the cat, the problem arises that physicists model the situation as a system, which is described by a wave function ψ that is an entanglement of temporally unfolding states of being alive and being dead:

$$/\psi\rangle = \frac{1}{\sqrt{2}}(\ /alive\rangle + \ /dead\rangle\ ) \tag{6.1}$$

This does not mean, however, that what is in the box is alive and dead simultaneously. A cat that is both alive and dead simultaneously is a *logical* contradiction. This contradiction is removed when we think of the system holistically as a unit that *manifests* itself in one (e.g., cat = alive) or another way (e.g., cat = dead). Physicists have learned that the observation itself makes the difference: What we observe depends on when, where, or how we look. Thus, in the case of light, how we look changes whether light manifests itself as a wave (e.g. which explains the refraction of light in prisms and the reflection of light in a mirror) or as a particle (e.g., as in the photoelectric effect on which the light meter in a camera is based). Similarly, in the case of the shearing of a rectangle considered in terms of a minimal unit of change,

[1] Even many scholars who avow adherence to cultural-historical activity theory – where inner contradiction is a central aspect – often confuse these with the logical contradictions.

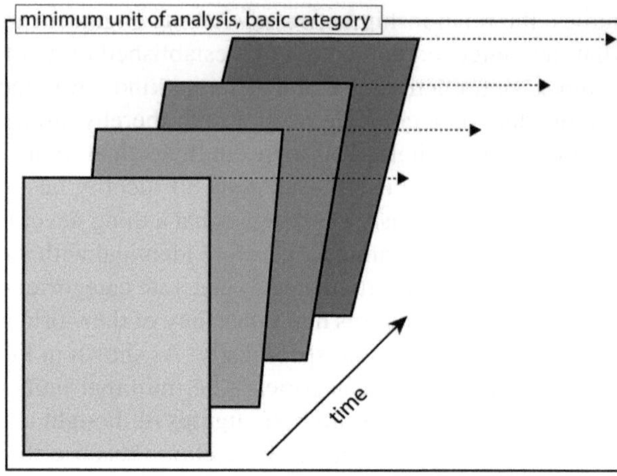

Fig. 6.2 A minimum unit (category) of change describes a system over a period of time, such as the shearing depicted, which transforms the rectangle into a parallelogram. As can be seen, time and forces no longer are external to the system but are internal to it (© Wolff-Michael Roth, used with permission)

what we observe depends on when we look. At one point, we observe a rectangle, and at another point we observe a parallelogram. The *logical* contradiction exists when we think about the system in the ways in which it manifests itself, between the different manifestations: "*Is* light a particle or *is* it a wave?" or "*Is* it a rectangle or *is* it a parallelogram?" The contradiction disappears when the system description and the observation descriptions are considered to be different things.

Applying these ideas to the case of scientific research, in modeling it there is always an inner contradiction, because research, as an event, inherently changes the world (Il'enkov 1977). The minimum unit of a change is shown in the analogy of the shearing of a rectangle (Fig. 6.2). This inner contradiction is different, however, from what we frequently find in the literature, where the terms *breakdown, antinomy, clash, rupture, problem, tension,* or *logical contradiction* falsely are used as synonyms for inner contradictions. Breakdowns, antinomies, clashes, ruptures, problems, tensions, and logical contradictions are (outer) manifestations of inner contradictions. Thus, in the often-used example of a patient-doctor relation – patients are to be helped by and are revenue sources for doctors – there is an apparent contradiction between, for example, recommending a (costly) treatment versus helping the patient by advising to opt for a non-revenue generating treatment.[2] This is only the outer manifestation of an inner contradiction that is systemic to medical care driven by the rules of a market economy.

Using this framework, we can talk about inner contradictions as an expression of the approach modeling discovery science dynamically, with a minimum unit of analysis that is a unit of change (Chap. 1). On the other hand, in any form of activity, statements (interpretations) may co-exist that the scientists themselves do not (initially) perceive. These contradictions are of the logical type, because they pertain to two or more statements about states of affair that are incompatible. But as

[2] The example can be found in the context of studies that draw on or make reference to cultural-historical activity theory.

long as these remain undetected, these do not *drive* the research activity. These situations may be sources of unease, of bewilderment, confusion, and the likes without driving scientists to seek resolutions. In Chap. 2, we see an example of this, where scientists continue with their work once their apparatus performs again. They do not seek to figure out why the apparatus did not work before – though they might try comprehending the source of the phenomenon if it were to be recurrent. An example of this is the consistent "ringing" signal (oscillation) that showed up in the absorption curves we recorded (see introduction to Part II). At one time, Craig thought that it was due to a "flickering" of the light source, which led him to purchase a very expensive one; at another time, he thought that the characteristics of the light were not quite as rectangular as it should be (Fig. 6.3b). But when he installed the new light source, the flicker remained. When I met Craig, he had already spent a great deal of other efforts and financial resources to try and eliminate the ringing; but all efforts were to no avail. In this case, I contributed to his perspective on the phenomenon when I presented him with a mathematical model of multiple reflections on the four interfaces that a beam of light traverses on its trajectory through a photoreceptor cell (Fig. 6.3a). These reflections are superposed over the incoming beam with a box-like wavelength characteristic, which, when modeled by means of a Fourier series (Fig. 6.3b), would generate the ringing. This model showed that the ringing is endemic to the experiment itself and, therefore, cannot be removed by means of different equipment. Craig then stopped pursing the effort of removing this signal by means of changes to the equipment.

The differences in thinking about the sciences in dialectical terms versus thinking about them in terms of the logical contradictions are highlighted in a seminal philosophical text about the nature of dialectical logic (Il'enkov 1977):

> Marx and Engels showed that science and practice, quite independently of consciously acquired logical notions, developed in accordance with the universal laws that had been described by the dialectical tradition in philosophy. It can (and in fact does) happen, even in situations when each separate representative of science involved in its general progress is consciously guided by undialectical ideas about thought. Science as a whole, through the clash of undialectical opinions mutually provoking and correcting one another, develops for all that in accordance with a logic of a higher type and order. (p. 290)

The same results emerged from a recent interdisciplinary symposium involving scientists and philosophers on emergence, complexity, and dialectics (Sève 2005). It was recognized that contradiction is the fundamental aspect common to all sciences not because these resolve apparent paradoxes, which dissipate once a more rigorous formalism is used but because underlying these paradoxes are effective contradictions that no logical treatment in the traditional sense of the word can resolve. A classical example – classical because Hegel used it, and, following him, a range of other dialectical philosophers – concerns the way motion comes to be represented. To resolve the paradox that in movement, an object is both here and not here or it is not in motion, displacement ends up being represented in a series of immobile positions – the contrary of movement. The second way in which the requirement of formal non-contradiction expresses itself is in the form of *insolvable real contradictions* – contradictions between *our representations* and *the real*.

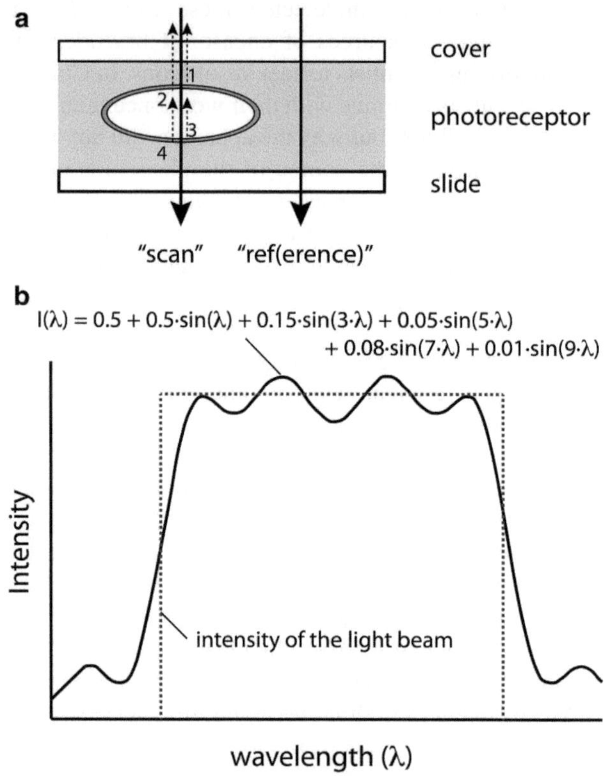

Fig. 6.3 (**a**) The sampling light beam has to transit four additional interfaces between different media during a "scan" when compared to a "reference." (**b**) The sampling beam has approximately a rectangular characteristic with respect to wavelength. This can be modeled as a composite of light of different wavelength using Fourier analysis (© Wolff-Michael Roth, used with permission)

In the present chapter I focus on what I recognized with hindsight to be logical contradictions in the way our research team was talking about and modeling the data. The inner contradictions at work derive from the historical perspective on the changes described in Chap. 7 and modeled in Chap. 9.

Contradictions in the Modeling of Data

In this section, I present our research team's talk about those data that we were in the process of collecting. What became apparent to me only several years after the research was completed that there were (undetected) contradictions that revealed themselves only as dilemmas, uncertainties, (minor) troubles, situations and things that did not totally rhyme or that we could not "put their finger on," had difficulties of "wrapping the head around," were unexplained uneasiness, confusion, riddles, things not working, failure, or more general breakdowns of the ordinary ways of going about things. With hindsight, the problematic interpretations of the data are especially salient during a meeting where we presented preliminary results of our research to the fish culturists at Robertson Creek Hatchery. In the following, I analyze seven fragments from that meeting and the data talk that we provided in the ultimate journal publication. Today one can identify logical contradictions that were not salient as such to our research team or audiences of our presentations at any moment during the research process. Our talk at the time could be characterized as *conceptual muddle*. In the following, I analyze what Shelby was saying during the presentation at the hatchery. However, he did speak for the team as a whole, as we had had a laboratory meeting just days before in which all details of the presentation had been discussed.

A Contradiction That Is Not

> "I'm getting wild fish, hatchery fish, from different populations, and they're doing the same thing"

As pointed out above, the scientific canon that the team followed at the time stated that the level of vitamin-A_2-based porphyropsin levels would decrease substantially at the time that (anadromous) salmon physiologically prepare for their seaward migration. This implies that salmon at different stages in their life history (juvenile fish [parr] vs. migrating fish [smolt]),[3] from different locations along the migration route, and from different ecological systems (where the onset of migration differs), ought to have different amounts of porphyropsin in the retinal photoreceptors. In the meeting fragment analyzed here, Shelby reports that all fish "are doing the same thing." This, therefore, constitutes a contradiction between our data and the canon. But, as Fragment 6.1 shows, the contradiction is not made thematic in such a way that it would have forced us to change our explanations.

During the presentation of results to the staff at Robertson Creek, the important aspect of the graph concerning parr and smolt to fall onto the same curve is explicitly presented and subsequently raised and talked about. In his presentation at the hatchery, Shelby, speaking on behalf of our research team as a whole, points out that all the data points follow the same line, including one "ocean capture point." He overlays the data sets from different origins one after another: hatchery smolt at release, hatchery smolt in estuary, wild smolts in estuary, hatchery smolts in the tanks at the university (fresh water), hatchery smolts in the tanks at the university (salt water), hatchery juveniles in ocean, wild coho juveniles in the ocean, hatchery parr in hatchery ponds, wild parr in Robertson Creek, wild coho juveniles in ocean in winter, and hatchery alevins in hatchery ponds (Fig. 6.4). According to the canon, there should have been considerable differences ranging from the parr in the hatchery with about 90 % porphyropsin (10 % rhodopsin) to the smolt in the ocean with about 5 % porphyropsin (95 % rhodopsin) of the pigment make-up of the receptors. Shelby notes during the meeting that out of all these data,

[3] The life history of salmon includes several stages: (a) fertilized eggs hatch into (b) alevin or sac fry that quickly evolve into (c) parr that become, after staying from 6 months to several years in fresh water, (d) smolt at the time when the fishes migrate to the ocean.

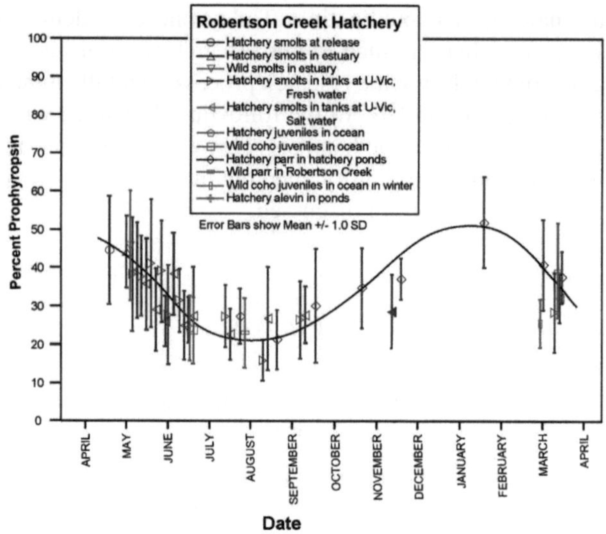

Fig. 6.4 All data points from all sources related to the Robertson Creek Hatchery are plotted on the same graph and modeled by a single sinusoidal curve – which Shelby projected during a meeting with the hatchery (© Wolff-Michael Roth, used with permission)

there is only one point that he characterizes as not following the line (the only filled triangle in Fig. 6.4, recorded during the month of November), which is exactly the one that he had talked about earlier in the laboratory, the point caught from a research vessel of the Department of Fishery and Oceans during the month of February. In the presentation, he says to the hatchery personnel: "I can't tell you exactly why they're not on the same line." He expands by saying that he would like to collect some of the "missed bunch of points" to fill in the picture concerning the coho adapted to ocean life. In this first fragment from the presentation to the hatchery folks, Shelby states that there is a difference between the results presented in this meeting and "what other people had suspected" (line 03), that is, that going to seawater has an effect on fish (lines 05–06).

Fragment 6.1

```
→ so what i can say from that is that salinity ITself
  doesnt change the visual pigments. which is interesting
  because thats not what other people had suspected.
  because thats why theyre called marine pigment and
→ freshwater pigment. bcause people thought that well if
→ you go to marine water then that has some effect on the
  fish and stuff like that. so thats kind of interesting.
  now it dOESnt mean that going to the SEA doesnt have an
  effect cause the sea is a different light environment.
  and i didnt change that environ all i changed was the
→ salinity. so all i can say from this for sure is that
→ salINIty itSELF doesnt have an effect. so going back and
  so going back and forth between brackish and fresh water
  itself dOESnt have an effect.
```

He articulates a conclusion: "salinity itself does not change the visual pigment" (line 01).[4] He describes what others anticipate or have observed. This does not exclude "going to sea" (lines 07–08) and "salinity" as factors driving the change between the two forms of light absorbing opsin (lines 11–12) because he changed salinity but not the light environment that comes with travel to the ocean habitat (lines 09–10).

At this point in the presentation, therefore, a clear picture appears to emerge according to which all data points but one from the Robertson Creek hatchery or its natural environment – which includes samples from very different contexts (see legend of Fig. 6.4) – follow the same pattern. This pattern, for the data as a whole, is circannual and approximately takes the shape of a sine curve. Most importantly, also contributing to the plot are those fish sourced at the same hatchery and kept for nearly a year in the university laboratories (triangles pointing left or right). Half of the fish were kept in saltwater tanks, the other half in freshwater tanks. All tanks were at a constant temperature of 15 ± 1 °C. Although Shelby explains some of the differences between the different environments, he does not make salient the one that exists between them in terms of temperature. Yet it is clear – a clarity that has become apparent to me, a participant, only after the fact – that our team makes the claim of an effect of temperature and simultaneously the data are plotted as if there were no temperature effect. This is apparent, for example, in the comparison between hatchery and wild coho from the river near Robertson Creek Hatchery. The two types of fishes live in water kept at different temperatures for at least part of the year, as shown in the subsequent part of the unfolding presentation and associated conversation.

Noting Something Without Seeing the Contradiction

"You can see, there's a very tight correlation"

At the time of the meeting, our research team has as its main (new) hypothesis that temperature drives the process of smoltification, the beginning of migration, and the adaptation of the photoreceptors for the extended stay in a marine environment. However, during the presentation at Robertson Creek, evidence emerges that the wild coho in the river near the hatchery and the hatchery-raised coho do not live at the same temperature. The hatchery draws its normal water supply for the coho ponds from the creek that gives the hatchery its name and in which at least some wild coho live. However, in the summer time, the temperature of the water is so high that the hatchery mixes the warmer creek water with cooler water from the bottom of the large nearby Great Central Lake before allowing the mix to flow into the coho ponds. As a result, the hatchery-raised coho live in much cooler water during the summer time than do the wild coho that serve as a direct comparison case.

---

[4] By saying pigment, Shelby uses a synonym for the rhodopsin/porphyropsin composition of the photoreceptors, that is, the two "opsin" forms that together absorb the light.

During the meeting at the hatchery, Shelby first presents the tight correlation between the water temperature and the "freshwater fish" (line 02). He describes the correlation and then – suggesting that someone pointed this out to him at a recent conference (line 06) – adds that there was a jump in porphyropsin (line 12) that correlated with drop in temperature (line 09). While talking, he circles the two data points from the month of July with the cursor (marked in Fig. 6.5). He articulates the hope that our team would be able to conduct something like a natural experiment during the next month, where, if the temperature were not to drop (lines 17–18) with a corresponding lower value of porphyropsin level, he would expect the data to "follow the curve a little bit more closely" (lines 18–19).

Fragment 6.2

and this gets really interesting because you can see
→ there theres a very tight correlation ((*Fig. 6.5*)) its a
negative correlation; its the reverse. but its a vERy
tight correlation that what appears to be hAPpening is
that the visual pigments are following the temperature
and whats neat about this is; i took this data down to
florida and i hadnt really had a chance to look at it
closely and someone said to me wow this is following
→ really closely because in twothousandandone you just
→ happen to have a cool spell in late july ((*circles the*
temperature dip in July)) and sure enough look at that
→ the pigments ((*circles 2 data points in July*)) jumped up
→ a little bit just like youd expect if the temperature
dropped a little bit. which is really neat. now i dont
have a lot of data points around that to follow up and
down. but theyre sitting on the same line. so its its
→ pretty nice. and what i=ll see this year hopefully is
→ twothousandandtwo we probably wont have that little cool
→ spell and i=ll see whether they follow the curve a
little bit more closely.

A closer inspection of the graph shows, however, that there are several logical contradictions in this description. First, most of the data points displayed (Fig. 6.5) derive from coho smolts kept in freshwater tanks at the university, where the water is kept steady at 15 ± 1 °C. The temperature, as shown in the next section, is that of the ponds in the fish hatchery. That is, the porphyropsin levels are said to follow temperature even though the temperatures are constant for a large segment of the data displayed. Second, the water temperatures in the estuary are not the same as those of Robertson Creek, the hatchery, or the Stamp and Somass rivers that connect into the estuary, where the briny water already corresponds to a mixture of river and ocean waters. Whereas the river water temperature fluctuates considerably with the time of year, the ocean water temperature remains fairly steady around 6–7 °C. Third, if there is the kind of relation between porphyropsin levels and temperature, then the rise in the former *does indeed* follow the expected pattern – which requires a deviation from the curve that does not take into account the natural variations in temperature. Fourth, in the preceding presentation, our team had shown the same curve as modeling all fish, which means, the curve describes a pattern that is independent of temperature as it provides a trend for fish living in

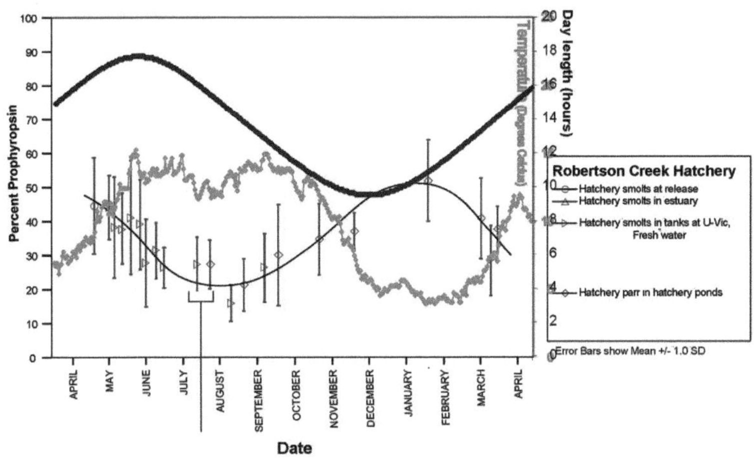

Fig. 6.5 PowerPoint slide of porphyropsin levels in coho salmon at different stages in their life cycle (smolt, parr) and from different settings (hatchery, estuary, university tank) are plotted together with temperature and day length (© Wolff-Michael Roth, used with permission)

very different ambient temperature regimes.[5] In other words, our team has modeled the data by abstracting it from the context and, therefore, from the particular temperature regimes that characterizes the different geographical locations and settings. Shelby will say a little later about the added photoperiod information – i.e. day length (Fig. 6.5, fat solid sine curve) – that it is also correlated with the porphyropsin levels. But this correlation is "not as tight so that when day length starts to go up so does temperature, but temperature lags for longer after [length of day] reaches its peak." On the other hand, his statistical analysis shows that "the correlation for temperature and visual pigments is highly significant."

Not Hearing Contradictory Evidence

"We were actually cooling our water during that period"

About 2 min after the beginning of the preceding meeting fragment, a discussion unfolded in which temperature differences were articulated between the hatchery ponds, normally directly fed by water from the creek but cooled during the summer period, and the creek. This would imply that there should be differences in porphyropsin levels between the wild fish and those in the hatchery if these levels were indeed a function of the temperature. There are also temperature differences

[5] Although separated by only 250 km, the climates at the hatchery and the university laboratory are considerably different with several degrees differences in summer and winter average temperatures.

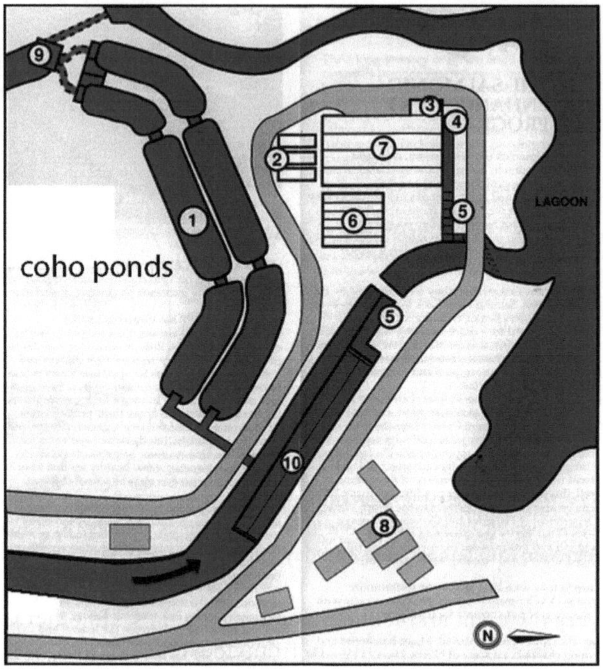

Fig. 6.6 The coho ponds (*1*) are fed with water (*9*) from the creek, to which, at the head, water from the nearby lake can be added. The water from the coho pond flows back into Robertson Creek at (*10*). The creek runs into a lagoon (*right*) that is part of the Somass River (© Wolff-Michael Roth, used with permission)

between the fish kept in tanks at the university versus those that are kept in outdoor earthen ponds in the hatchery. In the video, the differences come to be marked in the interaction with hatchery personnel, but do not enter the considerations of the scientists, who, having shifted to this new model of the variations of porphyropsin levels in the life cycle of the coho salmon, do not seem to hear evidence to the contrary. That is, evidence for a possible logical contradiction is articulated in the discussion, but it is as if the members of our team did not hear what the fish culturists were saying.

The episode begins with Ray asking about the nature of the information plotted in the graph featuring temperature and porphyropsin levels and the graphical model thereof (Fig. 6.5). Shelby does not appear to know exactly what Ray is saying, as he asks Erica, from whom he has received this information (turn 04). Ray then points out, the emphases stressing the contrast of his statement with what appears to be the case assumed, that they are cooling the water (turn 07). As a result, the temperatures in the earthen ponds where the coho are raised (Fig. 6.6) do not follow the photoperiod. In the exchange with Erica, who says that the pond temperature is between 13 and 14 °C, Ray states that otherwise the temperature would be somewhere between 20 and 25 °C in the ponds, which draw their water directly from the creek (turn 12) (⑧, Fig. 6.6). My study in the hatchery shows that the water is channeled from the creek to the top of the earthen ponds (⑧, Fig. 6.6) and released from the bottom back into the creek at a lower point (⑨, Fig. 6.6). Shelby appears to hear Ray saying that "this [graph] looks close to what you guys are doing then" (turn 13), to which Ray responds by reiterating that it "naturally would be up around 25 [°C]" (turn 14). Erica confirms this with a constative statement repeating the temperature value (turn 15).

Fragment 6.3a

```
01   R: whats the temperature of
02   S: whats the temperature of?
03   R: yea
04   S: ah thats a good question; Erica its the creek?
05   E: Robertson Creek
06   S: Robertson Creek?
07   R: so we were ACtually cOOLing our water during that
        period so it doesnt really correlate with
        [photo]period [
08   S: [well ]        [so youre cooling when this is at its
        peak? ((Circles the temperature in the April through
        October period, Fig. 6.5.))
09   R: yea
10   S: whAT do you cool to.
11   E: between thirteen and fourteen degrees generally
        right?
12   R: yea we we=re from from twenty to twentyfive degrees
13   S: okay so this is not hitting that high though because
        this is sitting at ten fifteen ((moves cursor from
        curve to ordinate on right)) so this looks like its
        pretty close to what you guys are doing then. thats
        ten and thats fifteen so you guys would be sitting
        right around there anyways; right?
14   R: yea but in naturally it would be up around
        twentytwentyfive
15   E: twentyfive.
16   S: right. huh thats interesting so is that because last
        year it didnt get as hot? or is this?
```

Shelby visibly is baffled, perhaps confused, and suggests that during the preceding year, the one displayed on the graph, "It didn't get as hot?" (turn 16). The intonation toward the end of this utterance is rising, as it typically would in a question. The intonation also is rising in the fragment of a sentence that Shelby then produces, a fragment that also has the grammatical structure of a beginning question ("or is it this?," turn 16). Erica then makes a statement that explains how they – she and her mentor Mike – attempt to keep the pond temperature between 12 and 13 °C. My ethnographic work revealed that at higher temperatures, certain diseases, such as bacterial kidney disease (BKD) increase exponentially and, with them, the mortality rates. Moreover, coho salmon grow best between 9 and 15 °C. The exchange between Shelby, Erica, and Ray then ascertains the source of the temperature readings as coming or not from the ponds (turns 20–26), which both Ray and Erica state to be from the ponds. Both assert that the water is "artificially cooled" (turns 29, 30).

Fragment 6.3b

```
16   S: right. huh thats interesting so is that because last
        year it didnt get as hot? or is this?
17   E: we we always try tmaintain between twelve and
        thirteen degrees mike likes it thIRteen I like it
        twelve we have a big fight no hh
18   S: right.
19   E: jokin. but like ray had said like it could get up as
        high as twentyfive and we sort of turn ON and shut
        dOWn pumps in the creek to maintain that
        [temper]ature.
20   S: [right ][yeah. so this data you sent me then]
21   E:         [throughout the summer months        ]
22   S: is not from your ponds its from
23   R: its frOM the ponds
24   E: its frOM [the ponds]
26   S:          [oh it is ] from the ponds
27   E: yea but
28   R: but the what i=m saying is the
29   E: artificially cooled
30   R: its artifi[cially cooled]
31   S:           [oh I see okay] yea yea alright
32   R: so it doesnt really correlate with phOTOperiod then
        like like
33   S: not necessarily
34   R: normally that temperatures [r much higher]
35   S:                            [except that  ] it would
        go even higher; like your temperatures are higher in
        july and august.
```

In this exchange, the difference in temperature between the hatchery and wild fishes for the summer months comes to be established. The temperature in the creek could get "up to twenty-five," which is the place where the wild coho are sampled near Robertson Creek Hatchery. That is, there are the same kind or even more radical differences in temperature between hatchery and wild coho in and around Robertson Creek than between all other fish samples used in the research; and yet our team continues collating all the data and modeling them by means of the same curve rather than by different curves. On the other hand, as shown below, at that time our team clearly separates out all of those data from the Kispiox hatchery that are "off the chart" and therefore would introduce very large variations into the picture.

In the present instant, the fish culturists have raised a question about the source of the temperature plotted in the graph. They are very familiar with the hatchery setting, its conditions compared to the conditions outside of the hatchery. They know about the temperatures of the creek and the Stamp River into which it feeds (Fig. 6.6), and the surface and bottom temperature of the nearby lake, where the cool water is pumped from to the channel that feeds the fishponds. It is this familiarity that became a resource for their questioning of the role of the temperature in our data. Although not articulated explicitly, the normal temperatures of

Robertson Creek are co-articulated, and thereby, the different temperature regimes between the "natural" (turn 14) or "normal" (turn 34) and the artificial temperature environment produced by means of "artificial cooling" (turns 29, 30). Shelby initially marks to be seeing what is going on (turn 31), then explicitly states that the temperatures "would go even higher" (turn 35), and finally adds that the "temperatures are higher in July and August" (turn 35), where it remains uncertain whether the temperatures are those in the ponds or creek.

In this instance, a possibility existed to revisit the earlier presented data, or to discuss the earlier articulated drops in the water temperatures and the corresponding increases in porphyropsin levels. It might have been an opportunity to revisit the question of the role of temperature and to ascertain those that constitute the natural environment of the wild coho salmon that led to the data on the graphs. But such opportunities can be considered only with hindsight, because at the time, none of the scientists or hatchery personnel present actually raised the question. The possibility existed perhaps only from the after-the-fact perspective, because if it had been noted as a possibility *in this situation*, one or the other participant (there were 10 individuals attending the meeting) might have brought it out. In this meeting fragment, we therefore see how a member of the hatchery personnel asked a critical and insightful question and how other members to the setting also provided detailed information about the temperature regimes of creek, ponds, and, elsewhere during the meeting, of the lake. This is in fact not the first time that the keen observations and deep contextual familiarity of the hatchery personnel – themselves conducting experimental studies on a regular basis even though they tend to have no more than high school training – provided them with a perspective for looking at the data and experiments not accessible to the scientists unfamiliar with the setting. In other experiments with the fish involving other scientists (from a nearby federal research center), experiments failed or turned out negative results because the scientists acted against better advice from the hatchery personnel, who anticipated the problems based on their local knowledge (e.g. Roth et al. 2008).

Confusion Is Not Contradiction

"This confuses me a little bit"

Following the presentation of the data derived from coho salmon sourced at Robertson Creek Hatchery where the presentation took place, our team's PowerPoint presentation moves on to share the results from the fish sourced at the second hatchery (Kispiox) participating in this study of changes in porphyropsin levels during the early parts of the coho salmon life cycle (Fig. 6.7). This hatchery, as an associated slide has shown earlier during the meeting, geographically is located much further north in the same province with a very different climate,

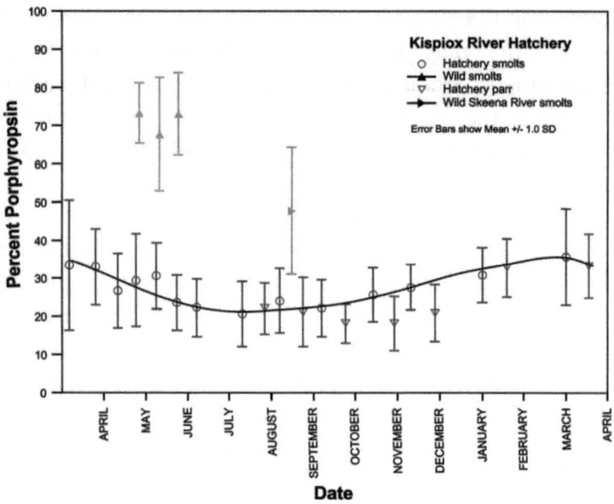

Fig. 6.7 In this PowerPoint slide, Shelby presents for the team the data from the Kispiox data, with only the hatchery-raised data modeled by the curve and the other data treated as "off the chart." (© Wolff-Michael Roth, used with permission)

geography, and ecological context.[6] Whereas the January and July/August daytime temperatures are 5 and 25 °C, respectively, for the Robertson Creek hatchery, the corresponding temperatures at the Kispiox hatchery are 9 and 22 °C. There tends to be no snow in the winter in the geographical area of the former hatchery, whereas there may be snow accumulations up to 200 cm in the area of the latter.

Shelby describes the pertinent features of the hatchery and the graph, and especially points out that it exhibits an annual cycle despite the fish being kept at constant temperature. He suggests to be "confused" by some aspect of the graph (line 02). The confusion is described to arise from the fact that one could expect, if temperature were linked to porphyropsin, colder temperatures at the Kispiox hatchery should lead to higher porphyropsin levels. Whereas during the "colder times of the year," Robertson Creek water temperatures are indeed lower than those in Kispiox facility and, as anticipated, the porphyropsin levels are higher, the same relation is not the case in the summer, when the Kispiox temperatures "are overall colder" (line 09). In this situation, the Kispiox best fit and the Robertson Creek best fit both have a value between 20 and 21 °C. On the other hand, the "wild [coho] are off the chart" (line 12), which is interesting and could be anticipated, "because those wild fish came from very cold, glacial creeks" (lines 13–14).

[6] In the hatchery, the slide was actually not necessary as everyone was familiar with the location of the other participating hatchery and its location despite its distance (about 700 km by air and 1,400 km on the road).

Fragment 6.4

```
  01 ((Presenting the slide in which only the left graph
→ 02 in Fig. 6.8 was presented)) thIS confuses me a little
→ 03 bit because if their temperatures are generally cOLDer
  04 then i wouldve expected if its linked to porphyr, linked
  05 to temperature that the percent porphyropsin would be
  06 hIGHer. i would expect that if theyre cOLder then
  07 ((pause)) in the colder times of the year you guys show
  08 a higher percent porphyropsin. so i wouldve thought
→ 09 these guys would be higher when because theyre overall
  10 colder. but thats not the case. and so there may be
  11 something else at play here. . .
→ 12 the wild are off the chart. and thats really
  13 interesting too because those wild fish came from ah
→ 14 very cold ah glacial creeks with low nutrient levels,
  15 where they believe the coho overwinter for two years
  16 rather than one.
```

Shelby then elaborates on the statement of confusion stating that it is "really hard for [him] to interpret these data (circles the data from the wild coho) because their life history strategies are so different." That is, there are big differences between the wild and hatchery coho related to the Kispiox hatchery. He adds, "I can tell you for sure that the temperatures (points to the data from the wild) are cooler." It can be "correlated that *these* (pointing to the wild coho data) are higher up"; but Shelby insists that our team cannot make predictions about the relationship and differences between wild and hatchery fish from that setting because the conditions at the hatchery are so "removed from the wild conditions."

In this meeting fragment, one observes a contradiction between the two graphs even though these are displayed together. On the left-hand side (Fig. 6.8), only the data from the hatchery, both parr and smolt stages, are aggregated and modeled by the sinusoidal curve – later, in the scientific publication, modeled as a sine curve and correlated with the photoperiod (length of day), with respect to which it was shifted by about 3 months. In the Robertson Creek-related part of the slide (right, Fig. 6.8), however, the data from fish at very different temperatures regimes are compiled and modeled by means of one and the same curve. That is, the scientific practices underlying the two parts of this PowerPoint slide are different, because the equivalent temperature differences in the Kispiox situation are not modeled simultaneously with the same graph, because "the wild [coho] are off the chart" (line 12). There is therefore a logical contradiction, which, however, neither our scientific research team nor the hatchery personnel in attendance makes a note of. Such a logical contradiction always is the (outer) expression of an inner contradiction, such as irreducible contradiction between material phenomena and the theories that model them on the ideal level (Il'enkov 1977). That is, there is what now can be recognized to be a blatant logical contradiction, which, despite all of its salience once noticed, is not apparent to those discussing the data in this setting and within our team during our laboratory-based data analysis meetings.[7]

The final point discussed is the fourth outlier, which comes from the main river into which the glacial creeks feed. This point may reflect data from coho that already are in their second year.[8] Shelby then talks about a study conducted by scientists from the federal Department of Fisheries and Oceans (DFO), who, as he says, believe that the fishes in this river system stop migrating into the bigger river, overwinter there and only leave the river for the ocean during a second year of migration. This would mean that there is "a two-stage system." He summarizes: "so it's really hard for me to interpret this data because their life history strategy is considerably different." He ascertains knowing for sure that the temperatures are cooler than at Robertson Creek and that outside the Kispiox hatchery, the fish are "higher up" in porphyropsin. But then he concludes, "outside of that I can't really make a lot of predictions to what their wild fish are doing relative to the hatchery'cause it seems their hatchery is so far much farther removed from the wild conditions."

Contradictions Undetected in Transaction Sequences

> "It would be even more dramatic if it was our actual if it was in our wilds in the river"

At Robertson Creek Hatchery, our research team presents a comparison between the coho salmon from the present hatchery and another one (Kispiox) that also has supplied fish to the team. The amplitudes of the two graphs are rather different so that the variations in temperature do appear to make a difference in the sense that when the temperature is held constant, there are still annual variations but with lesser amplitude, as the comparison between the two hatcheries shows (Fig. 6.8). Moreover, there are temperature-related differences possible between the wild fish at Kispiox – i.e., the three first outliers in Fig. 6.8 not modeled by the sinusoidal curve – living in glacial rivers and temperatures near 0 °C for 5 months of the year, and Kispiox Hatchery, where the temperature is held at a constant 7 ± 1 °C. When Shelby talks about the two graphs, he suggests that "temperature" "seems to be what's mostly manipulating this" (turn 01).

[7] Teachers might notice or set up such a "contradiction" with students in the hope that they will recognize it as such and, therefore, will be encouraged to learn.

[8] In particular cold conditions, salmon have been reported to remain in the freshwater environment until their eighth year following emergence from the egg.

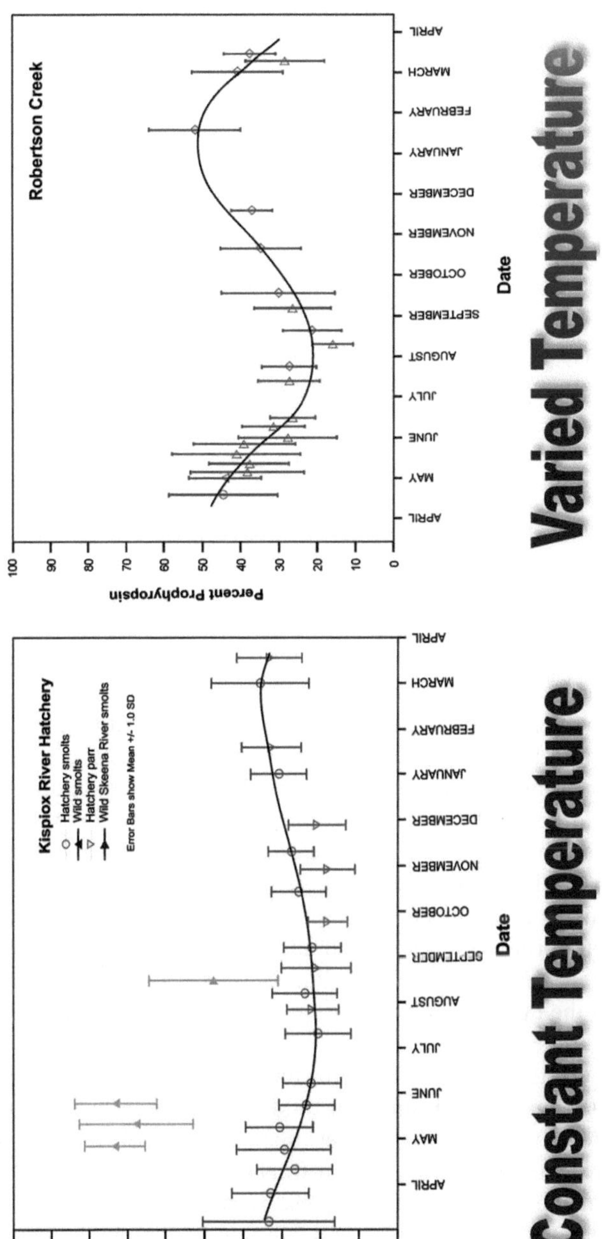

Fig. 6.8 In this PowerPoint slide, Shelby presents for the team the data from the two hatcheries for both wild and hatchery-raised coho salmon (© Wolff-Michael Roth, used with permission)

Fragment 6.5

01 S: so HERe i=ve just compared the two for you ((*shows the two graphs in Fig. 6.8*)) thats the constant temperature ((*points to left graph from Kispiox*)); its a very subtle curve and theres ((*points to the right graph pertaining to the Rodney fish*)) the varied temperature and you can see its higher here. so i=m pRETty sure that we=re talking about a temperature effect here. that seems to be what whats MOStly manipulating this.
02 (0.34)
03 E: and it would be even more drAmATic if it was our actual if it was in our wilds in the river
04 (0.46)
05 E: <<p>that would be [dramatic.]>
06 S: [yE:A:] yEA if temperature is [chang]ing
07 E: [yea]
08 S: more thats right. now the thing though tOO is that (0.57) they seem to pEAK bottom out at twenty percent (0.34) and i hAVEnt seen a fish go much below about fifteen percent. (0.33) so if the temperature gets up to twentyfive (0.32) it would be cOOl to see maybe they do drop down to zero. i doubt they do.

The stark differences between the temperatures within Robertson Creek Hatchery-raised and Robertson Creek wild coho salmon become the topic again in the context of this PowerPoint slide (Fig. 6.8). Immediately following the brief presentation of the slide as exhibiting the difference between the "very subtle curve" for the hatchery with "constant temperature" and the "higher [curve]" when the temperature is "varied" (turn 01), one of the fish culturists suggests: "it would be even more dramatic ifit was . . . in our wilds in the river" (turn 03). That is, the variation would be more dramatic if the curve modeling the wild coho were shown. Erica repeats that this would be more dramatic (turn 05), and Shelby agrees, restating that this would be the case "if the temperature is changing more" (turns 06, 08). That is, there is a collective agreement jointly articulated here whereby greater variation in porphyropsin levels should be observed when the temperature variations are larger. Yet, in a preceding slide, *all* fish from *all* contexts, including river, estuary, and the open ocean were lumped together and modeled by the same curve (Fig. 6.4).

Shelby makes a statement that elaborates the issue: He has not seen porphyropsin levels below 20 % in this study and "it would be cool to see" whether they would "drop down to zero" "if the temperature gets up to 25 [°C]" (turn 08). A little while later during the presentation, Shelby also instructs the fish culturists at the hatchery to keep the fish at the temperature that these were at when caught. That is, he instructs them to keep the coho salmon in a fish trap floating in the creek or in the river that the creek flows into. This would keep the fishes at the temperature of the river until these would be shipped in Styrofoam containers to the university laboratory. There, the fishes normally were processed within 48 h. This procedure would prevent phenomena to occur, which, as the following subsection shows, led fishes in captivity to exhibit changes in their porphyropsin levels with the increased temperature that these were held at in the university tanks.

In this meeting fragment, therefore, the expectation that different temperature regimes should lead to different porphyropsin levels is articulated again, both by a member of the research team and, as an inference, by one of the fish culturists. An all-knowing god-like theorist might suggest that this should have rendered problematic the aggregation of data from different temperature regimes in the Robertson Creek case. But it has not. The logical contradiction has not been or become salient and, therefore, has not brought about a change in our team's research process. There is therefore a logical contradiction in the data modeling that – seen now after the fact – has gone unnoticed during this meeting and even later in this scientific research project right into the final paper that our team will publish. But this logical contradiction exists only with hindsight rather than existing for those in the research project involved at the time the presentation occurred.

Ambient Temperature

"Because our water was warmer than yours, they started to drop"

After having presented the results of the study on the annual changes in the porphyropsin (A_2) levels of hatchery-related coho salmon, our team talks about extensions to the research. We report during the meeting at the Robertson Creek Hatchery some relatively sudden changes that had occurred while conducting an experiment on the effect of hormones on the visual system together with another member of the wider team. These included the test that exogenous hormones – hormones applied to the water and on the outside of the fish – would shift the photoreceptor optimal absorption (λ_{max}) to longer wavelengths, for rods (associated with increased porphyropsin levels) and for red and green members of the double cones in fish eyes.[9] Contrary to what had been expected, the porphyropsin λ_{max} levels increased rather than decreased, as the dogma at the time suggested. As part of this presentation, the team showed results that clearly are evidence for the effect of temperature on the fishes when these are kept for a while in the university-based fish tanks that are at a higher temperature than the ponds in the fish hatchery where these fish were captured and transported to the university. We had sourced the fish from the Robertson Creek Hatchery and had kept these in the university tanks for

[9] In the eye, the two classic photoreceptor cells are rods and cones. Rods are very sensitive and work at very low light levels, but do not allow color distinction. Cone cells have their maximum sensitivity at different wavelengths. The research concerning the changes in porphyropsin (A_2) was done with rods. However, some of the research in this laboratory focused on cones, which, in the fishes investigated, are sensitive in the red and green (red/green double cone), blue, and ultraviolet parts of the spectrum.

2 weeks (line 01). As a consequence, “because our water was warmer than [the hatchery] they started to drop in percent porphyropsin” (lines 02–04). On an accompanying slide that exhibited the different trends for hormone-treated and untreated fish, one can see the drop in porphyropsin of the untreated fish, which Shelby’s statement attributes to the differences in the ambient water temperature (line 08). It is clearly visible especially because the treated fish significantly increased in porphyropsin levels (Fig. 6.9).

Fragment 6.6

```
  01 we kept them for two weeks ((cursor moves from first
→ 02 point [red] down to second point [green])) and because
→ 03 our water was warmer than yours they started to drop in
→ 04 percent porphyropsin . . . so the visual pigments are a
  05 gradual shift. they dont change overnight. so i dont
  06 have to worry about keeping fish in a different
  07 one night. but after two weeks i know that it can have
  08 temperature of water for an
→ 09 effect, cause it happened here ((cursor moves from first
  10 point [red] down to second point [green])) in two weeks.
```

Shelby then emphasizes that the fish “do not change overnight,” which means, he does not “have to worry about keeping fish in different temperatures [for a short period of time]” (lines 05–06). But there will be an effect – as exhibited in the plots and highlighted by the laser pointer moving sharply down from the first to the second data point plotted – “after two weeks” (line 07). That is, in this episode, a temperature dependence of the porphyropsin levels is explicitly acknowledged and substantiated by the data on display. This is so not only for the type of photoreceptors used in the study pursuing the life history changes in porphyropsin levels – i.e., the rods – but also for other types of photoreceptors – i.e., the red-green double cones – studied as part of other research simultaneously conducted in this laboratory.

The porphyropsin changes as a function of temperature observed here would be consistent with the one data point where a dip in the ambient temperature of the ponds (July) appeared to be a corresponding increase in the levels of porphyropsin (see above). It would also be consistent with the data from the Kispiox hatchery, where the wild coho from glacial waters (for 5 months about 0 °C) exhibit much higher porphyropsin levels than the hatchery-raised salmon in parr and smolt stages kept at 7 ± 1 °C year-round. But the data are inconsistent with the indiscriminate aggregation of the data in the two main graphs displayed during the meeting. These two graphs had summarized the findings of our research (Figs. 6.4 and 6.8 (right)).

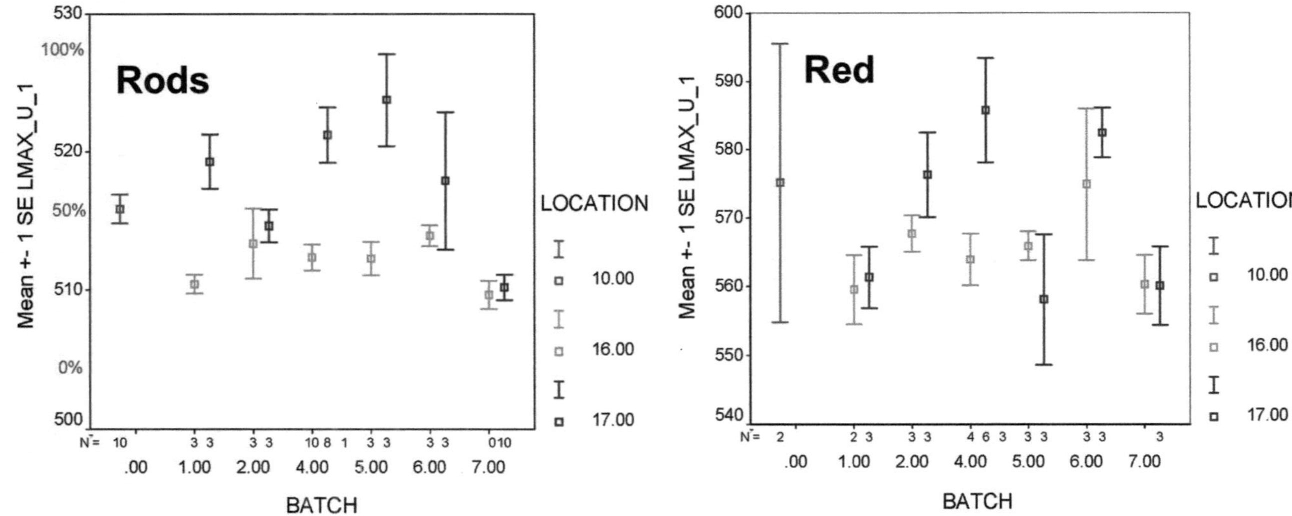

Fig. 6.9 Two graphs, one for rods (*left*) the other one for red cones (*right*), from the PowerPoint presentations showing the results of porphyropsin and λ_{max} measurements in the thyroxin experiment (© Wolff-Michael Roth, used with permission)

Contradictions to the End

> For fish collected from Kispiox Hatchery, a cross correlation with temperature was not possible since this hatchery maintained a constant rearing temperature year round (Temple et al. 2006, p. 308)

What turned out to be logical contradictions in this after-the-fact analysis remained unarticulated within the presentations of the research results right to the point of their appearance in scientific article that reported the results of the study. This was already apparent in a preliminary reporting of the findings from the study presented 2 years after the end of the data collection. This reporting occurred during yet another session at the Robertson Creek Hatchery, because we, as part of the initial research agreement, had committed to providing it with any findings even prior to an official publication. During this presentation, the data for the wild coho salmon caught near the Kispiox facility now were completely absent from the talk and the graph projected during the talk (Fig. 6.10). There are two graphs in the same panel, one presenting the Robertson Creek the other the Kispiox data. The two graphs look very similar, with the Kispiox data shifted towards the later parts of the year and with lower overall porphyropsin levels (%A_2) but with similar amplitude. The variance in temperature is emphasized alongside the absence of temperature variations in the Kispiox situation (lines 04–07). Referring to the graph that displayed the Robertson Creek and Kispiox data simultaneously (Fig. 6.10), Shelby says during the meeting:

Fragment 6.7

the key thing we notice um was that the kispiox river
hatchery uses gets the water from a well and the well
stays the same temperature all year round; seven degrees
→ celsius all year round. this hatchery the temperature
→ varies sharply, four degrees to about sixteen or
→ seventeen celsius. so there is quite a variance here in
→ temperature. but there is no variance in the kispiox.
and that seems to be reflected in the that ((*moves*
cursor along Robertson graph, Fig. 6.10)) cyclical
variation and the amount of freshwater pigment that they
have. the robertson creek hatchery through these grey
points they go from a much higher to a lower ((*follows*
graph)) value thats still on the same cycle; whereas the
kispiox ((*follows graph from beginning, Fig. 6.10*)) that
cycle is compressed; its not as dramatic, which let me
to think that may be i=m looking at some kind of
temperature effect; but not solely temperature.

In this description, temperature is both a cause and not a cause simultaneously. It is not a cause because there are variations in the course of the year in the Kispiox coho salmon, which are raised at constant temperature. Simultaneously there is an effect because the Robertson Creek salmon experience temperature differences in the course of the year. Shelby also presented the porphyropsin levels in coho salmon for the Robertson Creek and Kispiox hatcheries together with

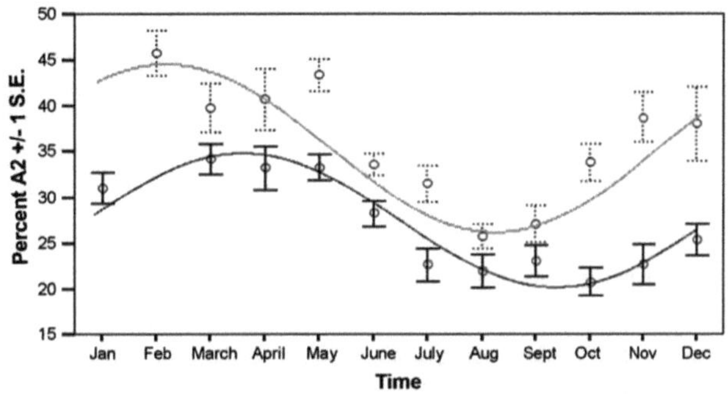

Fig. 6.10 In this PowerPoint slide, Shelby presents for our team the data from the Robertson Creek Hatchery (*grey*) and the Kispiox hatchery data (*black*, with only the hatchery-raised data included) (© Wolff-Michael Roth, used with permission)

environmental data, temperature and length of day (Fig. 6.11). These precise data reappeared in the final publication as Fig. 4a, 4b, and 4c – see the excerpt below – with the difference that the temperature and length-of-day scales increased from the bottom to the top rather than in the inverse way, as this had been done for the presentations.

In the final journal publication, our research team presents the data reflecting seasonal changes of coho salmon from different sources side-by-side without explicitly addressing the absence of variations in the temperature-controlled hatchery other than suggesting that photoperiod may drive the system in this case. Omitted in our journal article are the obvious changes that had been observed between wild and hatchery-raised coho from the Kispiox setting or the changes observed during the thyroxin-treatment experiment.[10] In its first graph presenting the results, our team collates all data from all locations *except* the wild coho fishes from the Kispiox area into the same graph, that is, independent of the hypothesis that temperature makes a difference. We then present a graph in which the data for Robertson Creek Hatchery and Kispiox Hatchery are displayed separately. Our journal article then displays three graph panels together in one plate – a typical practice in the natural sciences (Roth et al. 1999). The first panel shows temperature and porphyropsin levels for Robertson Creek Hatchery versus time of year, the second panel shows day length and porphyropsin levels versus time of year, and the third panel displays day length and porphyropsin levels versus time of year for the Kispiox Hatchery. The main text reads:

> Monthly mean percent A_2 values were more closely correlated in time to temperature than day length at [Rodney] Creek hatchery (Fig. 4a, b). For fish collected from Robertson Creek hatchery, temperature and percent A_2 were negatively correlated (Fig. 4a). Percent A_2 peaked in late February; 1 month after temperature reached its annual low. Cross correlation between percent A_2 and inverse temperature was highest ($r^2 = 0.821$) when the cosine function was phase lagged by 1 month relative to temperature (Fig. 4a). Cross correlation between percent A_2 and day length was highest ($r^2 = 0.855$) when there was a 2-month phase lag between percent A_2 and inverse day length (Fig. 4b). For fish collected from

[10] Dropping data because these do not fit the overall narrative appears to be quite common in the sciences, as a recent article in *Science* suggests (Couzin-Frankel 2013) and as some scientists themselves acknowledge in their autobiographical narratives (e.g. Suzuki 1989).

> Kispiox Hatchery, a cross correlation with temperature was not possible since this hatchery maintained a constant rearing temperature year round. However, a cross correlation between percent A_2 and inverse day length peaked ($r^2 = 0.977$) when the curve fit to percent A_2 data was advanced by 3 months relative to day length (Fig. 4c). The variation in percent A_2 found in ocean-going coho also showed an inverse relationship to ocean-water temperatures (see Fig. 5 in following section). (Temple et al. 2006, p. 308)

The shift of the cosine curves – similar to the equivalent curves in Fig. 6.11 with temperature and day length scales inversed – seems to be temperature dependent. It may surprise that the Robertson Creek wild and hatchery data were collated and modeled together, much like all the data from the experiment were added in one of the figures of the ultimate journal publication. The published study suggests a correlation with temperature – even though, as shown in this chapter, the temperatures are not the same for hatchery and wild salmon in the summer – but indicates that a correlation could not be calculated because the temperature was held constant in the second hatchery.

Contradictions in and of Scientific Research

In this chapter, I provide a description of logical contradictions in scientific research that I became aware of only after the research had ended. During the research itself, what now appears as a logical contradiction was not apparent to any one of the members. In fact, the contradiction, as I show, entered it into the scientific publication and, therefore, had made it through the peer review process. The most salient aspect of the contradiction can be stated in this way: In the context of the Robertson Creek data, the research article reports a correlation with the temperature of the water, but the Kispiox data exhibited the same kind of pattern – a sine curve – even though the temperature was held constant. That is, the fluctuations could not have been due to temperature. With respect to theory building, the logical conclusion therefore would be: (a) either temperature does not play a role in the seasonal variation of porphyropsin, as evidenced by the Kispiox data or (b) the theory differs for the same kind of fishes but from different geographical locations.

There are two aspects to contradiction. Here, I point out the undiscovered logical contradictions in the determination of the data. This form of contradiction is the result of multiple statements that exist simultaneously within a research group or research community but that are mutually exclusive. Although such contradictions may not be apparent at the time, after the fact, they come to be understood as logical contradiction – such as those that Marx had identified in the economical theory of David Ricardo (Il'enkov 1982) – or as inconclusive (conceptual) muddle, from which "the Europeans" emerged when they "found themselves speaking in a way which took [a new set of] interlocked theses for granted" (Rorty 1989, p. 6). The logical contradictions, as hinted at above, are but an expression of a deeper inner contradiction – that between the material world and its ideal reflection in consciousness, as captured in the statements that are made about the material world (Il'enkov

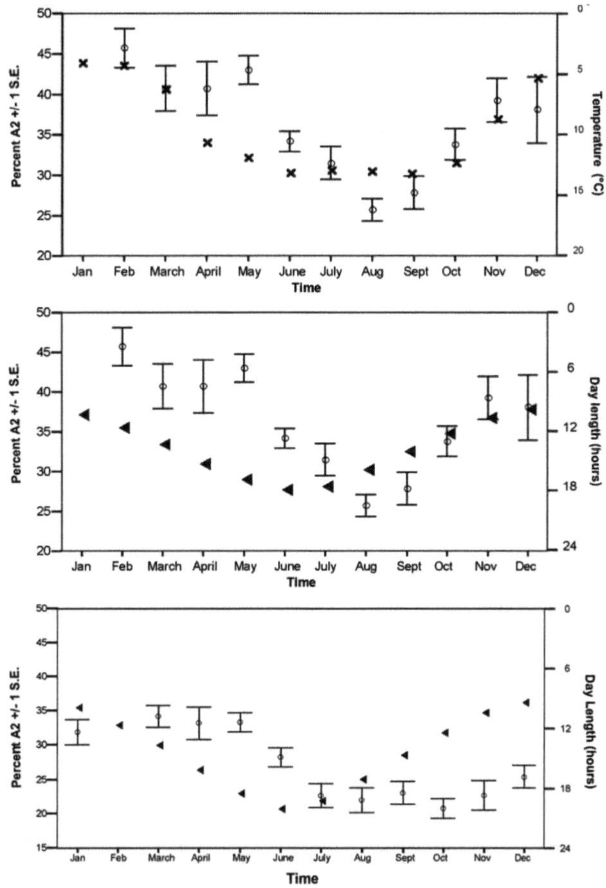

Fig. 6.11 These graphs from a PowerPoint presentation that the research team gives at the Robertson Creek Hatchery to interested personnel feature the porphyropsin data (**a**) from Robertson Creek and temperature of the fish pond (○) (*upper panel*), (**b**) from Robertson Creek (◂) plotted together with day length (○) (*center*), and (**c**) from Kispiox (◂) plotted against day length (○) (*bottom*)

1982; Leont'ev 1983). The inner contradiction also exists because of the inherent movement of life generally and societal-historical research activity more specifically, including the research that I am writing about in the pages of this book.

The present chapter underscores the concurrent possibility of multiple descriptions contradicting each other without that a team of highly trained scientists or their representation- and context-savvy audience recognizes the contradiction as such. Here I show how depending on the discursive context, our research team claimed the existence of a correlation with temperature and the non-existence of such a correlation. We simultaneously claimed that temperature makes a difference – as when we pointed to the differences in porphyropsin between fish from glacial creeks or rivers and a nearby hatchery – and acted as if temperature did not make a difference – when we aggregated the coho smolt from within Rodney Creek Hatchery, where the smolt live in artificially cooled water as compared to the creek from which the water with the normal temperature is taken.

Analysts with information available only a posteriori might exhibit indignation about the fact that what I have shown to be contradictions in our models and interpretation of data had not been recognized at the time. In fact, historical analyses of scientific research suggest that contradictions may remain present and be accepted for some time within a scientific community until a point where a certain threshold is reached. At that point some "scientific revolution" will introduce a theory that allows viewing data in a new way (e.g. Kuhn 1970). Other pragmatic descriptions of the conceptual revolutions in a number of fields – poetry, science, or politics – suggest gradual changes rather than changes based on decision making. Thus, "Europe did not *decide* to accept the idiom of Romantic poetry, or of socialist politics, or of Galilean mechanics. . . . Rather, Europe gradually lost the habit of using certain words and gradually acquired the habit of using others" (Rorty 1989, p. 6, original emphasis, underline added). That is, prior to and during the change process to a new conceptual description, the subjects of activity (i.e. researchers) cannot have the criteria for making decisions between the relative usefulness of two conceptions because the second one is only emerging. Moreover, the entire endeavor of change is rooted in the prior discourse (conception). Transitional discourse inherently contains contradictions – especially when viewed from the dialectical theory of change articulated in Chap. 1 and at the beginning of this chapter. We may, for example, liken scientists' initial discourse to the rectangle in Fig. 6.2, and some temporary final discourse as the parallelogram in the back. If the change in discourse is continuous rather than changing over night or in an instant, then there will be forms of discourse between the two extremes considered within the *same unit of analysis*. Inherently there will be different forms of discourse within the unit. Even if the transition was instantaneous, we would still be required to seek a theory of change that operates with categories of change rather than with categories of identity. In the next chapter, I show that catastrophe theory is one approach in which radical, qualitative transitions are modeled mathematically. There are therefore two ways in which we may look at inner contradictions. On the one hand, there is an inner contradiction because the same unit contains different form of discourse; and there is an inner contradiction arising from the relation between matter and consciousness. The inner contradiction *expresses* itself in logical contradictions when the discourses before and after a conceptual change are compared, because different discourses describe and explain the same situations differently.

The interrelation between the two forms of contradiction was at the heart of *Dialectical Logic* (Il'enkov 1977). The very nature of the sciences as disciplines that develop collectively available knowledge is conditioned by the fact that "a generalization justified for experience as a whole has been drawn on the basis of partial experience" (p. 105). New experiences, therefore, allow the emergence of contradictions between the older and the newer form. More importantly, however, for the emergence of logical contradictions in the sciences is that experience itself is antinomic and, therefore, laden with contradiction. Thus, "in trying fully to synthesize all the theoretical concepts and judgments drawn from past experience, it is immediately discovered that the *experience already past* was itself internally antinomic" (p. 105) as long as it "was taken as a whole and not some arbitrarily limited aspect or fragment of it in which . . . contradictions may be avoided" (pp. 105–106). Logical contradictions may turn out to be the result of an error or mistake. It becomes apparent that logical contradiction is possible as "antinomies that developed in thought as a result of the most formally 'correct' and faultless argumentation" (p. 189). Dialectical logic differs from classical logic in that it considers as inner contradiction what the latter attempts to attribute to "contradiction[s] 'in different relations'" (p. 335). That is, classical logic attributes a contradiction to different statements made simultaneously or across time – e.g., because of development. In dialectical logic, logical contradictions are external expressions of inner contradictions, which reflect the fact that a changing system is thought in terms of categories that capture *change* as internal to the system.

As a result of the two interrelated contradictions, the period of discursive change therefore may be characterized as "muddle" as a positive concept to denote the fact that multiple different discourses co-exist simultaneously and the contradictions that come with this co-existence. There are multiple theses, each of which may develop independently, and only after some time scientists take the interlocking theses for granted. From this perspective, qualitatively new forms (e.g., of talking) emerge from quantitative changes in old forms. Both dialectical materialist logic and catastrophe theory have evolved discourses that handle qualitative changes as the result of quantitative change of a number of different dimensions simultaneously.[11] I show both of these approaches at work in Chap. 7.

Important in the present context is that something recognized at some later point as a contradiction cannot be an aspect of scientific decision making before that point precisely because it is not noted as such. It cannot therefore drive any conceptual change, revisions in the argument, or contribute to the development of a new theory. It does not, in other words, lead to the kind of learning that someone viewing the situation from a god's eye perspective might say that ought to happen. As far as our research team was concerned, there were some confusing aspects in the data that constituted a dilemma but not a major problem that should have prevented publication, and a little uneasiness about the way in which the existing data were modeled. We had some good explanations for excluding the measurements taken from the Kispiox wild coho salmon, because there were only a few data points. But there was no salient contradiction as such. At the time, we did not act and talk as if we were dealing with a contradiction. There was therefore also no need for changing what we developed as an alternative to the current "dogma," which had predicted, as shown in the work of the Alexander et al. study, approximately logistic (sigmoid) decrease of porphyropsin levels around the time that the coho salmon would migrate to sea and an approximately logistic increase of porphyropsin levels at the time that the coho salmon migrate up-river to their spawning grounds. Our research team had found that the porphyropsin levels varied in sinusoidal form with season highly correlated to temperature.

[11] Constraint satisfaction networks, too, can be used to model categorization and choice, where qualitative distinctions are made, based on quantitative, continuously changing relations (e.g. Hutchins 1995; Roth 2001).

I was a member of the research team and an author of the final study. At the time, the contradictions were not apparent to me, which may have been so because of multiple reasons that I can only speculate on today from a perspective with hindsight. What I can say today is that the porphyropsin level P that we published in the end might have been modeled by a function of the type

$$A(T,t) = A_1(T) \cdot \cos(\omega t + l(T)) + A_0(T)$$

where T = temperature, t = time of year in months, $A_1(T)$ = amplitude of cycle, $\omega = 2\pi/12$, $l(T)$ = lag time, and $A_0(T)$ = vertical offset. Such an equation not only accounts for much of the present data but also for those that other studies, such as the one apparently supporting the canon, provided.

The onset of the changes may be temperature related, as the minimum is earliest at a hatchery with approximately the same conditions as Robertson Creek – where the previous study supporting the dogma had been conducted. Here, the temperature hits 9 °C in March, whereas the same temperature was reached at Robertson Creek Hatchery only in the latter part of April. This temperature was never attained in the Kispiox hatchery, where the temperature is at a constant 7 °C. The $\cos(\omega t + l(T))$, shifted as it is by $l(T)$, would account for the position of the maximum of the cosine curve along the abscissa. The factor $A_1(T)$ would provide us with a temperature dependent amplitude, which would account for the fact that the Kispiox annual cycle is depressed with respect to that at Robertson Creek. In fact, the Kispiox data would be modeled by an amplitude $A_1(T = 7\ °C)$. The additional factor $A_0(T)$ would shift the cosine functions vertically as a function of temperature. This, therefore, would allow us to model data of the kind that the research team had from the Kispiox wild coho. To verify this part of the equation, however, data would be required from these fishes for an entire year or at least for an extended part thereof. Because of the different ecological systems, different life history strategies may be at work, too. A Japanese study, as Shelby pointed out 1 day in the laboratory, suggested small amounts of porphyropsin in another salmon species at precisely the time that Sheldon had found in the coho compared to the 0 % levels during the remainder of the year. That is, the levels turned out to be lower on an absolute scale in the open ocean where the temperatures are higher on average and more constant than at Robertson Creek Hatchery.

We might even conjecture that different curves modeling the different sources would have become suggestive if the standard errors of the means rather than the standard deviations had been plotted. Each data point was endeavored to consist of 20 photoreceptors from 10 coho. In this case, for example, the error bars would have decreased by a factor equal to the square root of number of data points, that is, by a factor of about 14 if $N = 200$. The parr and smolt from the Kispiox would no longer follow the same curve but, as can be seen from Fig. 6.7, the parr might be seen to dip in October rather than in the July–September period, when the smolt appear to be at their lowest. However, given that the within-fish variability was very high, our team chose to plot error bars that indicate ± 1 standard deviation.

References

Alibali, M. W., & Goldin-Meadow, S. (1993). Gesture-speech mismatch and mechanisms of learning: What the hands reveal about a child's state of mind. *Cognitive Psychology, 25*, 468–523.

Chi, M. T. H. (1992). Conceptual change within and across ontological categories: Examples from learning and discovery in science. In R. Giere (Ed.), *Cognitive models of science: Minnesota studies in the philosophy of science* (pp. 129–186). Minneapolis: University of Minnesota Press.

Clement, J., & Steinberg, M. S. (2008). Case study of model evolution in electricity: Learning from both observations and analogies. In J. J. Clement & M. A. Rea-Ramirez (Eds.), *Model based learning and instruction in science* (pp. 103–116). Dordrecht: Springer.

Couzin-Frankel, J. (2013). The power of negative thinking. *Science, 342*, 68–69.

Heidegger, M. (2006). *Gesamtausgabe I. Abteilung: Veröffentlichte Schriften 1910–1976 Band 11. Identität und Differenz* [Collected works, Division I: Published writings 1910–1976 vol 11. Identity and difference]. Frankfurt/M: Vittorio Klostermann.

Hutchins, E. (1995). *Cognition in the wild.* Cambridge, MA: MIT Press.

Il'enkov, E. (1977). *Dialectical logic: Essays in its history and theory.* Moscow: Progress Publishers.

Il'enkov, E. (1982). *Dialectics of the abstract and the concrete in Marx's Capital.* Moscow: Progress Publishers.

Kuhn, T. S. (1970). *The structure of scientific revolutions* (2nd ed.). Chicago: University of Chicago Press.

Leont'ev, A. N. (1983). Dejatel'nost'. Soznanie. Ličnost' [Activity, consciousness, personality]. In *Izbrannye psixhologičeskie proizvedenija vol. 2* (pp. 94–231). Moscow: Pedagogika.

Marx, K./Engels, F. (1962). Werke Band 20 [Works vol. 20]. Berlin: Karl Dietz.

Nietzsche, F. (1954). *Werke in drei Bänden* [Works in 3 volumes]. Munich: Carl Hanser.

Rorty, R. (1989). *Contingency, irony, and solidarity.* Cambridge: Cambridge University Press.

Roth, W.-M. (2001). Designing as distributed process. *Learning and Instruction, 11*, 211–239.

Roth, W.-M. (2002). Gestures: Their role in teaching and learning. *Review of Educational Research, 71*, 365–392.

Roth, W.-M. (2008). The nature of scientific conceptions: A discursive psychological perspective. *Educational Research Review, 3*, 30–50.

Roth, W.-M., Bowen, G. M., & McGinn, M. K. (1999). Differences in graph-related practices between high school biology textbooks and scientific ecology journals. *Journal of Research in Science Teaching, 36*, 977–1019.

Roth, W.-M., van Eijck, M., Reis, G., & Hsu, P.-L. (2008). *Authentic science revisited: In praise of diversity, heterogeneity, hybridity.* Rotterdam: Sense Publishers.

Sève, L. (2005). *Émergence, complexité et dialectique* [Emergence, complexity and dialectics]. Paris: Odile Jacob.

Suchman, L. A. (1987). *Plans and situated actions: The problem of human-machine communication.* Cambridge: Cambridge University Press.

Suzuki, D. (1989). *Inventing the future: Reflections on science, technology, and nature.* Toronto: Stoddart.

Temple, S. E., Plate, E. M., Ramsden, S., Haimberger, T. J., Roth, W.-M., & Hawryshyn, C. W. (2006). Seasonal cycle in vitamin A1/A2-based visual pigment composition during the life history of coho salmon (*Oncorhynchus kisutch*). *Journal of Comparative Physiology A: Sensory, Neural, and Behavioral Physiology, 192*, 301–313.

Chapter 7
A Scientific Revolution That Was Not

> Needless to say there may be some controversy now with this information that I've been collecting: it goes against some theories that were proposed by Nobel Prize winner George Wald back about fifty years ago so it'd be quite interesting to see how this work unfolds over the next couple of years. (February 21, 2002)
>
> When I set out to do this I thought I would be writing this great paper that was going to be published in *Nature*, you know, or *Science* ... and when it comes down to it, you're just going to publish that little piece that you worked on and it's going to go to a reputable journal as best as you can and that's that. (January 22, 2003)

The two introductory quotations are statements that the doctoral student (Shelby) on our team made as part of a scientific research project in which it appeared that the apparently reigning scientific canon was undone. In the first introductory quotation, Shelby, nearing the end of the data collection of the research project, expressed the sense of what the data collected so far were showing: These were inconsistent with ("go against") the theories of George Wald, who had received a Nobel Prize in physiology for his work, including that concerning the changing composition of photoreceptors in fishes that migrate between freshwater and salt-water environments. Because the results might contradict the dogma, our research team collected additional data to be able to rule out every alternative explanation that we could think of. Eventually, our research team became convinced that our early interpretation of data, through the lens of the existing dogma that had inspired our research, was incorrect and that the data collected in the study we had designed were more consistent with a different, yet-to-be-established theory. In the process, our team underwent what might be termed a major conceptual change, which did not just involve a different way of looking at the same data but in fact constituted a revision of a perspective on the physiology of euryhaline[1] fishes. However, as the second introductory quotation shows, recorded some 11 months after the first statement, Shelby described that he had to rescind the idea of getting to write a paper in *Nature* or *Science*, the two scientific journals in which major discoveries tend to be reported. The work that our research team (of which he was an integral

[1] Euryhaline = able to tolerate a wide range of salinity; anadromous = ascending rivers to spawn.

W.-M. Roth, *Uncertainty and Graphing in Discovery Work*,
DOI 10.1007/978-94-007-7009-6_7, © Springer Science+Business Media Dordrecht 2014

part) had done was not at all a scientific revolution but an important extension of already existing ideas about changes in the visual system of salmonid fishes.

When Craig, the senior scientist heading our research team, and I originally planned our collaborative work, he explained his proposal. He wanted us to help salmon hatcheries in increasing the rate of salmon returning to their birthplace[2] by providing them with a tool that would allow them the identification of the best time of release of the juvenile salmon they had raised to optimize the return rates of this anadromous fish. Using hand gestures and sketches on pieces of paper he explained that the current "scientific dogma" suggested a change of photoreceptor pigments (a) from high levels of porphyropsin (vitamin A_2-based) chromophores when the salmon are in freshwater (b) to low levels of this opsin when they are in saltwater (Fig. 7.1a). When the salmon return to the river to spawn and die, the reverse change should occur. The six-week changeover periods were thought to be due to morphological and physiological changes of euryhaline, anadromous (coho) salmon in preparation for migration between the two different environments. Five years later, as shown near the end of Chap. 6, we reported in the resulting scientific paper that the changes in porphyropsin levels followed annual cycles independent of the salinity and independent of the life-history stage of the fish (Fig. 7.1b).

> Monthly mean percent A_2 values were more closely correlated in time to temperature than day length at Robertson Creek hatchery (Fig. 4a, b). For fish collected from Robertson Creek hatchery, temperature and percent A_2 were negatively correlated (Fig. 4a). Percent A_2 peaked in late February; 1 month after temperature reached its annual low. Cross correlation between percent A_2 and inverse temperature was highest ($r^2 = 0.821$) when the cosine function was phase lagged by 1 month relative to temperature (Fig. 4a). Cross correlation between percent A_2 and day length was highest ($r^2 = 0.855$) when there was a 2-month phase lag between percent A_2 and inverse day length (Fig. 4b). For fish collected from Kispiox Hatchery, a cross correlation with temperature was not possible since this hatchery maintained a constant rearing temperature year round. However, a cross correlation between percent A_2 and inverse day length peaked ($r^2 = 0.977$) when the curve fit to percent A_2 data was advanced by 3 months relative to day length (Fig. 4c). The variation in percent A_2 found in ocean-going coho also showed an inverse relationship to ocean-water temperatures (see Fig. 5 in following section). (Temple et al. 2006, p. 308)

Initially, the changeover process – from talking about the data along the lines of the "dogma" to talking about it in a new way – did not come easy in part because our team could see its data only through the lens of the dogma. This was so in part because we had decontextualized the data to such an extent that months had to be spent to reestablish the context that would allow our team to grasp the significance of each data point, and in part because the team was developing familiarity with the source field of the data to be able to model and control the natural variation in their

[2] Although there are exceptions to the rule, salmon tend to return to the precise spot where they were born. I had opportunities to verify this at Robertson Creek Hatchery, where the coho salmon were "marked" in two ways that distinguished them from coho raised in all other hatcheries: (a) changing the water temperature during the early growth – either prior to or succeeding hatching – leaves growth rings in the otolith (inner ear structure) or (b) using tiny coded wire tags that are inserted into the nose of juvenile coho salmon.

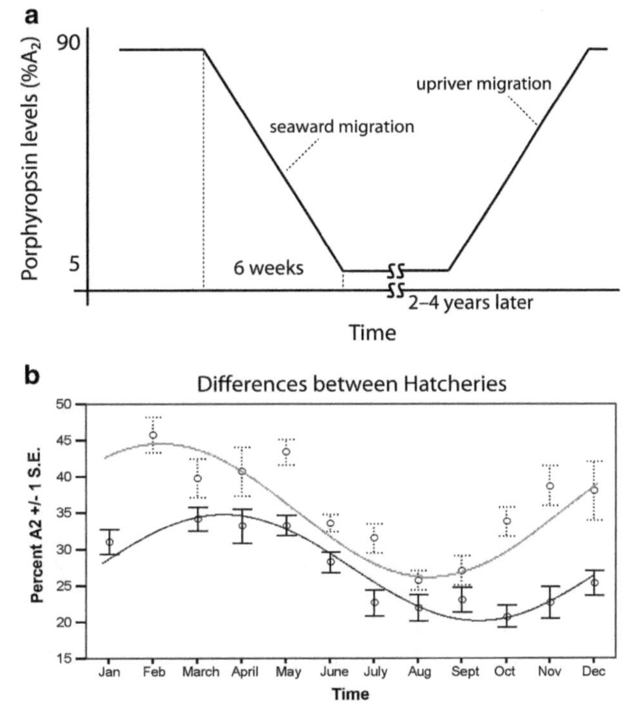

Fig. 7.1 (**a**) The going theory at the time of the research predicted radical changes in the composition of retinal photoreceptors of coho salmon during seaward and upriver migrations. (**b**) At the end of the research, our research team modeled the data as annually varying cosine

source material (Chap. 5). In the preceding chapter, I provide examples of the contradictions in the explanations of the data that existed within our team during and following the data collection period. In part, these logical contradictions were the result of the unrecognized inconsistencies between the existing dominant theory and the data that our team was collecting. After the fact it has become apparent that appropriately modeling the data required a shift in thinking about the kinds of changes that coho salmon actually undergo. This turned out to be related to the annual cycle of the photoperiod and temperature rather than to the migration between water of different salinity. The purpose of this chapter is to provide a microgenetic historical account of the conceptual change in our discourse that happened concerning an aspect of fish biology expressed in the temporal variations of photoreceptor composition, migration, geography, and fish physiology. The conceptual change is that from a step model of changes expressed in Fig. 7.1a to the annual cycles express in Fig. 7.1b. Because the relevant data became available to those undergoing change only over an extended period of time, special opportunities arose to study the conceptual change and resistance to it drawing on cognitive anthropological methods.

Conceptual Change

Over the past 30 years, conceptual change research has led to an impressive array of research on learning in domains that are counter-intuitive to learners; it also provided analyses of apparent contradictions between conceptual change studies focusing on the same topic. Most of the studies have been conducted in physics and the physical sciences, though there are also studies in the domain of biology. Conceptual change research has not gone without challenges: Serious questions have been raised about the usefulness of conceptions and conceptual change approach – i.e., internal cognitive structure – rather than in terms of discourse and other culturally and historically available semiotic (sense-making) means (e.g. Roth 2008; Roth and Duit 2003). The philosophical discussions about conceptual change really are concerned with the contention that "explanations of how language works will also help us see how 'language hooks onto the world' and thus that truth and knowledge are possible" (Rorty 1979, p. 265). The philosopher, in his critique of the mirror view of scientific knowledge, goes on asserting the correctness "in saying that the notion of 'conceptual change' is itself incoherent, and that we need to see through it in order to recognize why debates about it seem both so important and so unlikely to be resolved" (p. 265). An important fact is that conceptual change theorists in the learning sciences seldom use to critique the incoherence of their own theory in the face of the persistence of conceptions among students even when taught under the most propitious conditions – well designed, tested, and researched curriculum taught by a conceptual change researcher with considerable experience in the field of the topic. There is also an incoherence on a societal-historical level in the sense that conceptual change theory does not explain why Aristotelian and other theories outdated in the sciences for hundreds of years continue to exist in the general culture.

In classical conceptual change theory, two types of changes have been proposed, one occurring when there is a rearrangement of concepts within an ontological category (e.g., matter or field, state or process), the other occurring when there is a change from one ontological category to another (e.g., from a material to a field conception) (Chi 1992). There is still an ongoing debate about whether such a distinction is actually helpful, because even scientists freely move between the two, which suggests that a dynamic approach may better reflect observations. The change process involves a reassignment of some fact to a new conceptual category. As a result, a conceptual discourse that initially belonged to one hierarchical organization of concepts subsequently, after the radical conceptual change has occurred, belongs to a very different organization of conceptual discourse. This suggests that it might "be inaccurate to think of conceptual change as a *movement* or *evolution* of a concept from one tree to another in a gradual or abrupt manner" (p. 134). Instead, the concepts may develop independently on the different trees. As a consequence, there would be no shifting even though before–after comparisons make it look as if a shift had occurred. In the present chapter, the before and after conceptions often are co-present: These are co-deployed in some graphical representations featured below so that the decline of one and the rise of the other conceptions are co-implicated. What one observes, therefore, is a shift in the relative salience of certain facts and phenomena leading to what we might describe as a slow gestalt change. Here I propose a theoretical model for describing the emergence of new conceptions and the changeover between two concurrent conceptions based on catastrophe theory.

In the classical conceptual change theory, change involves more than a simple rearrangement of concepts but a "radical" rethinking of metamorphosis and physiology as related to migration. Thus, a "radical conceptual change . . . is to learn to differentiate the two ontological categories, and not to borrow predicates and properties, of [one] category to interpret events in the alternative category" (Chi 1992, p. 137). In the present instance, the conceptual change our research team underwent might be termed radical because it involved a dramatic shift in thinking about the phenomena involved. Natural phenomena were described that did not and could not exist in the older theory. What were thought to be endogenously driven events – porphyropsin levels as a function of hormonal changes – came to be exogenously driven events – porphyropsin levels as a function of temperature or photoperiod. The associated ideas about fish physiology had to be changed as well, because the changes in the physiology had been used to explain the changes in the retina. If the retinal changes no longer correlate with migration periods, the physiological changes that drive other aspects – including response to salt water and silvering/darkening of the scales during ocean-bound/freshwater-bound migration – no longer can account for the retinal changes that occur with seasons. Initially, our research team attempted to fit the incoming data into the existing paradigm, literally plotting our own data together with those of a previous study from the same geographical area (even though, as shown below, the extrapolation from the data collected were inconsistent with a previous study).

For a radical change to be initiated, some new category must be learned and seen in relation to other statements. The person or team then has to realize that their data do not belong to the categorical descriptions they initially used but to the new category. For conceptual change in scientific research, a five-step procedure has been proposed for the changes that lead to scientific revolutions, new scientific thinking, and discoveries:

1. A set of anomalies exists (or an unexplained phenomenon is encountered, or an impasse is reached);
2. their abundance reaches a crisis state (or the individual is overwhelmed by the conflicts or confrontations so that he/she realizes that there is a problem, sometimes referred to as the state of problem finding);
3. a new hypothesis is made or induced (i.e., abduction);
4. experiments are designed to test the new hypothesis, and a new theory is formulated on the basis of the results of the experiments;
5. the new theory is accepted (old theory is abandoned, overthrown, or replaced). (Chi 1992, p. 144)

Even if the learning and conceptual change observed in the present study were considered to be a within-a-tree change, such change may be difficult because it requires shifts of the type that occur while working on "insight problems." But insight learning requires the solution pieces to be "in sight," apparent in the eye of the beholder such as the logical contradictions that I exhibit and discuss in Chap. 6 that have to be seen to be driving discursive change and learning.

Changes that look like qualitative conceptual changes as described above may actually be the result of continuous quantitative changes. How this may be is the subject matter of catastrophe theory, which, in mathematical form, is the equivalent of a dialectical framing of evolution with the emergence of new forms (morphogenesis) (Fig. 7.2). In general terms, a system may be found in one state only (#1, Fig. 7.2). However, quantitatively changing parameters bring the system to a point where a bifurcation occurs (#2, Fig. 7.2). Two states become possible just as in the heating of water to the point of 100 °C, the system can be in liquid or gaseous form. The parameter changes may push the system along path 3 (Fig. 7.2) until minor fluctuations cause a change in the dominant form of talking about and explaining a phenomenon (#4, Fig. 7.2).[3] The system then continues its evolution in the new dominant state that exhibits a different evolutionary trajectory (#5, Fig. 7.2) than it would have had it remained in the original state. Precisely the same description of a qualitative change as a result of quantitative changes is obtained from a dialectical approach (Roth 2009; Sève 2005). A descriptive analysis of the morphogenetic changes has to provide evidence for each of the five instances in the evolution of the system. It is immediately apparent that there is a difference between (a) this catastrophe theoretic and dialectical approach and (b) the classical conceptual approach. In the latter, changes occur because of qualitative changes in the development of the cognitive system. In the former, quantitative changes lead to qualitative changes; and qualitative changes entail subsequent quantitative changes.

The description of developmental changes according to the catastrophe theoretic model requires analysts to provide evidence for each of the five parts of the overall trajectory (Holzkamp 1983a, b; Roth 2009). First, there has to be a demonstration of the real-historical conditions of the preceding developmental level (#1, Fig. 7.2) within and upon which the qualitative functional change develops. This requires a description of all those moments that are relevant to the subsequent stage (#5, Fig. 7.2). From the catastrophe theoretic perspective, this is precisely the articulation of the conditions that are "negated" in the qualitative turnover (#4, Fig. 7.2). Second, it has to be shown that there were objective changes in the external conditions that allow the internal contradictions, which will give rise to the qualitatively new function, that is, a bifurcation point (#2, Fig. 7.2). Those conditions need to be shown to exist that constitute a form of "pressure" in the direction of the

[3] Natural systems, too, may be in one of two states. For example, the phenomenon of supercooling refers to the fact that a substance may be in its liquid form below the freezing point. That is, a substance at a particular temperature may be in either one of two *qualitatively different* states: the liquid or solid, gaseous or liquid, solid or gaseous. Supersaturation refers to another phenomenon of this kind.

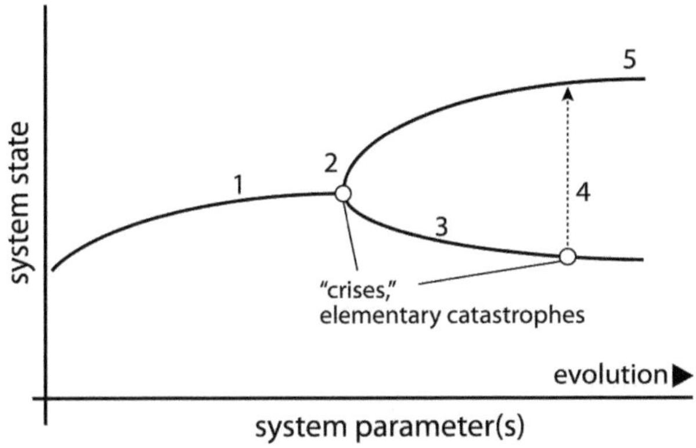

Fig. 7.2 In catastrophe theory, the emergence of qualitatively new states (*2*) and the turnover in dominance from one state to another state (*4*) arise from quantitative changes in system parameters and constitute the elementary catastrophes or crises in the evolution of the system. The descriptive analysis of the evolution requires demonstrating the steps indicated (© Wolff-Michael Roth, used with permission)

new conception. Third, the analyst needs to provide evidence of a functional turnover that relates the pre-existing dimensions in a new way (#3, Fig. 7.2). This is a change at the subject pole of the system undergoing change. Fourth, a change in dominance must be demonstrated, whereby the previously dominant conception is negated and the co-existing new conception becomes the dominant one (#4, Fig. 7.2). This is a second qualitative change, for a qualitatively specific function becomes the dominant function for the system as a whole. As a result, the evolutionary trajectory has changed (#5, Fig. 7.2). Fifth, the analyst must demonstrate of the restructuring process that gives the evolutionary trajectory of the system as a whole a new direction following the becoming dominant of the specific conception. It also requires the demonstration of what happens to previously dominant conceptions, how they have no longer or different functions under the new conditions in the continuing evolutionary history of the system.

In this way, evolutionary change in conceptions does not have to be radical. Rather, it always happens on the grounds of already existing qualitatively different conceptions that are selected under changing conditions. New relevant conceptions do not have to emerge suddenly but rather, they emerge and co-exist with the original conceptions. During the functional turnover one of the new conceptions becomes the dominant one, whereas the previously dominant one becomes a subordinate one. This approach allows us to consider as possible the co-presence of mutually contradictory discourses even on the part of scientists. For example, an astronomer may talk about a beautiful *sunrise* even though in her research or teaching she would use Copernican discourse according to which the earth rotates around its axis, which leads to the experience of the sun moving across the sky.

In the present chapter, I present evidence for the historical evolution of how our research team modeled and talked about the data it collected. This evolution might be characterized as a radical conceptual change – even though the kinds of exemplary (physical) categories are not involved. A "radical" conceptual change might be involved because of a substantial rethinking required in the modeling of the phenomena. According to the previously existing "dogma," porphyropsin (a vitamin-A_2-based chromophore) is dominant in freshwater fish, whereas rhodopsin (a vitamin-A_1-based chromophore) is dominant in saltwater fish. The dogma proposed that in euryhaline anadromous fish, a chemical change occurs from one to the other chromophore "in preparation for" the anticipated changes with change in salinity (e.g., Wald 1957) such that "the biochemical changes in the eye anticipate the migrations" (p. 909). Biologically, the change in the photoreceptor pigment is related to a metamorphosis that "involves a fundamental anatomical reconstruction" (p. 908) or has its "essential character" (p. 909), such as when, in the sea lamprey, indistinct gonads develop into "special external structures are formed for depositing eggs and sperm" (p. 909). Our research team began with the discourse that "the biochemical changes in the eye anticipate the migrations" also occur in coho salmon – as a previous study in the same general geographical area suggested (Alexander et al. 1994) – but, in the course of our research, our discourse changed. In the end, we concluded that even though salmon undergo metamorphosis and fundamental physiological change during the transition from parr to smolt stages (some of which are visible to the trained eye), the change in the porphyropsin/rhodopsin (A_2/A_1) ratio, derived from the shift of the light absorption curve along the wavelength axis, is (a) circannual, (b) independent of the physiological changes and migration, (c) independent of salinity, and (d) co-varies to different extents with photoperiod (length of day light), temperature, and geographical location.

The Big Historical Picture

An account of the evolution of the way in which a research group thinks about a phenomenon presents a series of challenges. First, the extended time period and the multiplicity of events involved renders data reduction difficult because the very detail required for providing an *ethnographically adequate* description extents the limitations of journal articles. Second, the logic of the discovery requires a microgenetic historical (developmental) account (explanation), which traces the evolution from a point prior to the emergence of alternative explanations within the group of scientists to the point of the recognition and acceptance of a discovery. That is, the account has to contain all the stages marked in Fig. 7.2. Third, the matters of interest, the phenomenon, the "animal in the foliage," may actually appear across time and space as they appear in the historical account. Historical and microgenetic accounts have the advantage over causal accounts that they deepen our views on the historical dimensions of a phenomenon. In the following, I provide (a) an overview of the stages of the learning episode, together with the major stages in the present account and the laboratory events that are used as exemplary data and (b) summarize crosscutting patterns in the assertions to which

I refer in the subsequent historical account of the lived conceptual change in scientific research work.

In this chapter I report on an episode of conceptual change in an advanced scientific laboratory that underwent a major conceptual change. Although our team initially talked about the research as overthrowing the "dogma," our research results "only" supported an already existing pattern, though we were doing so at a measurement accuracy that no other team had achieved to that point. Figure 7.3 maps the main stages of the change against the relevant data collected during my ethnographic study and mobilized in the present account. The figure shows that during the stages of the planning and proposal writing, recruitment of the fish hatcheries, and right up to the first eight months of the data collection, our team subscribed to the dogma (see introduction) (Fig. 7.3, ①–④). This constitutes the real-historical conditions and identifiable moments that are negated in the later occurring conceptual overturn (#1, Fig. 7.2). Despite the dogma, team members did articulate alternative hypotheses, which were quickly abandoned, however (e.g., Fig. 7.3, ②, ③). These alternative hypotheses emerged as a result to the objective changes in the conditions (#2, Fig. 7.2), that is, empirical data that did not quite fit the dogma. Eight months into the data collection (Nov. 7, 2001), there was mounting evidence that the dogma might not hold up. Our team was becoming increasingly familiar (quantitative change) with the conditions under which the coho salmon were raised in different hatcheries or grew up in the wild. In one and the same meeting, (a) the chief scientist provided a reading of the data consistent with the dogma and (b) the doctoral student provided a challenge to the dogma but denoted it to be a "complete speculation." This speculatively advanced explanation related the empirical data in a new way: sinusoidal annual variations rather than stepwise changes related to the migration-related changes in salinity (#3, Fig. 7.2). Over the following months, additional data collected in 2-week intervals supported the speculatively articulated alternative hypothesis (Fig. 7.3, ⑤). It is during this period that our team abandoned the dogma and adopted the new conception (#4, Fig. 7.2). As our team wanted to be sure that the pattern that we appeared to be observing of an annual pattern would hold up scrutiny, we decided to collect data for a two-year period. During the remaining 14 months, the additional data collected were consistent with the new conceptual talk (Fig. 7.3, ⑥). Among others, our team conducted experiments reported in other papers that were consistent with the conceptually new forms of talk, which, therefore, was providing a new direction to the work in the laboratory (#5, Fig. 7.2). However, as shown in Chap. 6, there were some unresolved contradictions within the new conceptual model that remained right into the final paper ultimately published (Fig. 7.3, ⑦). A brief, after-the-fact account of the learning episode on the part of the main protagonist (the doctoral student Shelby) is provided (Fig. 7.3, ⑧).

This historical but microgenetic account suggest that the learning episode does not represent merely a gestalt-switch-like conceptual change but an incremental (quantitative) process that that constitutes the basis for a qualitative reversal. In the meantime, there existed a multitude of inconsistent statements simultaneously. It is a period of unsettled time, where, after the fact, the discourse had resemblance with

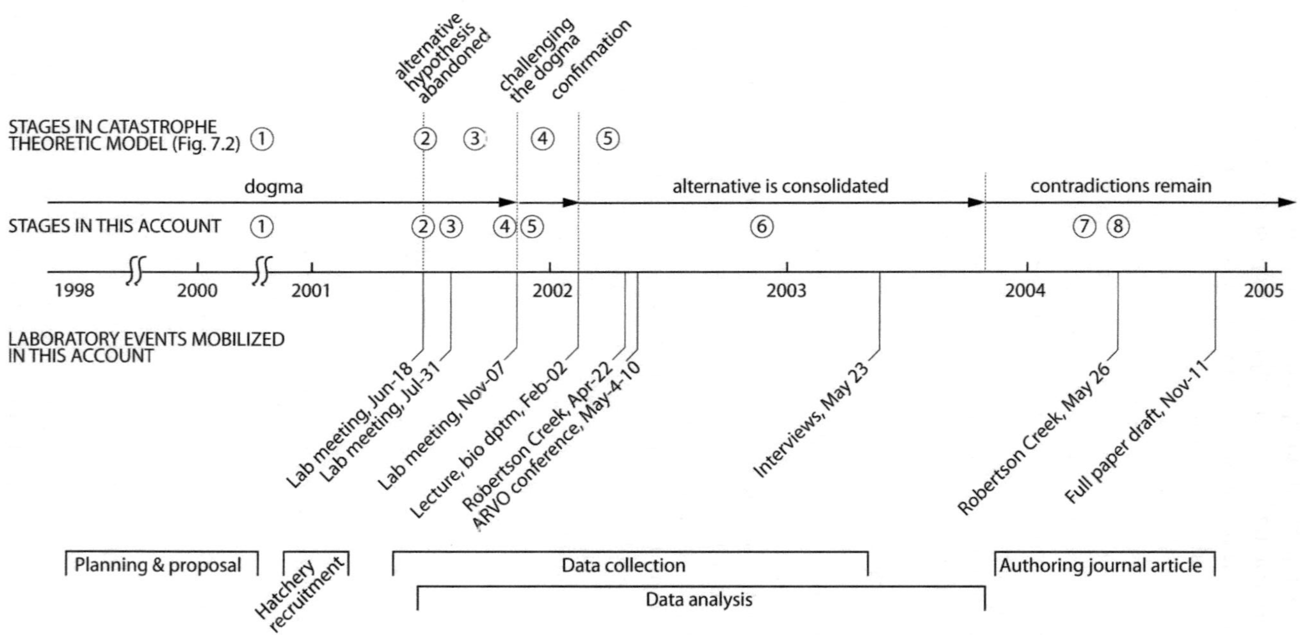

Fig. 7.3 Microgenetic map of the historical evolution of the discourse about the physiological changes coho salmon undergo in the course of their lives, especially as marked by the changes in their visual systems in terms of the amount of porphyropsin present

the "muddle" that some pragmatic philosophers of scientific revolutions describe. We may capture essential aspects of the historical changes that occurred in our research team, which had been implicit in the preceding chapters, in the following six assertions:

Assertion 1 Scientists do not actually interpret the measurements in themselves, which are the result of a process of decontextualization. This is so because the scientists do not actually analyze the mathematical relations, but they are concerned with becoming familiar with the biology of the animal in its natural setting. Thus, even though the data might be fitted best by a particular curve – e.g., in the present context by a cosine function with amount of daylight as the independent variable phase shifted by a certain period with respect to the longest day of the year – the scientists do not report this function or base their explanation on it. Their concern is to find a way of explaining the data based on their largely intuitive, practical sense of how the world works rather than a mathematically exact description of the phenomenon.

Assertion 2 As part of their exploratory data analysis, scientists talk about particular effects and variables that they initially do not follow up upon – perhaps because other things are more salient, because they may not appear plausible. Yet some (considerable) time later, the same ideas originally not salient as figure become the dominant feature (explanation) that the scientists pursue.

Assertion 3 The scientists initially read their data solely through the lens of the particular theory, the "canon" "dogma" (Wald), as apparently concretized in one recent study (Alexander et al. 1994). The scientists fit their own data to this curve and develop data collection strategies in terms of that original curve.

Assertion 4 With increasing familiarity concerning the system as a whole – context of the specimens, instrumentation, and environmental conditions – information previously not salient now becomes a new figure against the ground.

Assertion 5 Increasing familiarity with the concrete details of the source of the fish allows scientist to attribute sources of variability to specific causes. (Weight of fish, no variability parr/smolt, variability across temperature, variability or not across different conditions [salt/no-salt], where to take the photoreceptors from [quick dirty experiment, strip, side].)

Assertion 6 There are (perhaps surprising) contradictions in the sense that the scientists make contradictory claims in a matter of minutes concerning the effects of the same variable. *Muddle*, a descriptive term for such discourse, appears to be justified during this period of conceptual change.

Concerning conceptual change among scientists, it has been stated that "we will not be able to fully uncover the mystery of the discovery process itself" "unless we can clearly lay out what the scientist's initial representation is" (Chi 1992, p. 153). In this chapter, I follow the emergence of a radical change in scientists' conceptual talk, including the inscriptions they use, about the changes anadromous coho salmon undergo during a life cycle that that begins in freshwater (where they hatch and develop through parr and smolt stages), to saltwater (where they develop into mature adults), and back generally into the same (freshwater) rivers where they were born. There is clear evidence for the novelty of the representational talk that they used to model certain aspects of the fish physiology and the representations and theory that emerged in the course of their work.

Beginning Representation

As described in the introduction to Part B, the scientific research project studied here began as part of the larger, nationally funded *Coast Under Stress* project designed to assist coastal communities in coping with economic hardship when their single main source of income came under the threat of extinction. Being a source of income for commercial fishermen, sport fishery tourism, and the First Nations, five of the Pacific salmon species – chinook, chum, coho, pink, and sockeye – are of considerable economic importance to British Columbia. Because of overfishing since the 1900s, hatcheries have been used as a technology to artificially produce young salmon (from the eggs and milt harvested from returning specimens) to be released into the wild once the young are ready for migration to their ocean habitats. Our research was premised by the contention that improving the survival rates of salmon would provide more stable incomes to communities that depend on this natural resource.

During one of our first meetings, Craig told me that he was aware of a study on coho salmon that showed, consistent with the current scientific canon ("the dogma"), a correlation between plasma sodium levels, a measure of coping ability with saltwater environments, and the level of porphyropsin in the eyes of salmon. Shelby, the doctoral student in the project, frequently presented the starting point – the initial conception – using a PowerPoint slide with a graph adapted from the literature (Fig. 7.4). He used the slide to explain the physiological changes that the coho salmon undergo just prior to migration in terms of plasma sodium and how these are correlated with the changes in the visual pigment composition (i.e. the A_1/A_2 or rhodopsin/porphyropsin ratio). For those coho salmon that the hatchery retained in freshwater – he tended to place the cursor near the three right-most measures (around July on the graph) – the plasma sodium levels and the porphyropsin levels increase, that is, "revert back to the freshwater levels." Throughout our project, this phenomenon was referred to as "the reversion." If the trends in the graph (Fig. 7.4) held up, so we reasoned, the porphyropsin levels would indeed be correlates of the physiological changes. The levels of

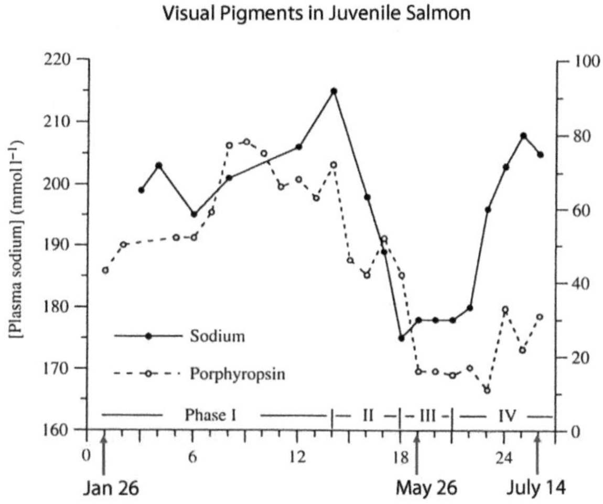

Fig. 7.4 This PowerPoint slide represents the "dogma" in the design and early stages of the research. It was frequently part of the presentations and discourse that the research team used. The presenters had modified an original graph culled from the research literature (© Wolff-Michael Roth, used with permission)

porphyropsin on the date that the salmon are released from a hatchery could be correlated with the number of salmon returning to the hatchery, and, therefore, with the survival rates. Craig also was very familiar with the Alexander et al. (1994) study that had confirmed the canon, as he had provided feedback to its authors prior to publication. The purpose of our team's research was to see whether the pattern would hold up (a) for different hatcheries in different geographical locations in British Columbia, (b) with the number of measurements increase by orders of magnitude, and (c) at a much finer grain especially around the period of changeover from high to low levels of porphyropsin (between about weeks 14 and 19 in Fig. 7.4). Given that in 1999 Craig was about to test new equipment that would increase the speed of data acquisition for measuring porphyropsin levels by several orders of magnitude, there existed the promise of a viable means of assisting hatcheries in optimizing their heretofore-unpredictable and often disastrously low return rates.

This graph, in our team conversationally referred to as "the Alexander study," together with the general claims of the "dogma" (as represented in Fig. 7.1a), was the initial lens through which our team saw the data concurrently generated in the wet laboratory. We knew that the study was relevant as we had in our hands the proposal of an American group of scientists, which had received nearly US $1 m for finding out about the relationship between the visual system of salmon, the readiness of the fishes to migrate, and their return success rates (in the Columbia River basin of Washington State). Craig, however, heavily criticized this other project ("unless they know something that we don't and the rest of the vision science community doesn't know, this is just bullshit, really, it's just not a sound technique"). But we recognized that there was something like a race to find a means of assessing the best time point for the readiness of salmon to migrate and which point during smoltification – i.e. which date – would lead to the highest possible return rates.

Early Explanations and the Need for Context

The *Coast Under Stress* project was funded and, during 2000, our laboratory tested and wrote an article on the new instrumentation to be used. The first data points for the present study were collected in April of 2001. A first data analysis meeting, after the first three bi-weekly data collection episodes had been collected, took place on June 18, 2001. Excerpts from this meeting already feature in Chaps. 5 and 6. Here, I discuss some of the contents of this meeting in their relation to the historical development of scientific discourse in this research team. I present evidence that shows that our team did not simply interpret their measurements but that they reconstructed the settings from which the fish originally derived to be able to make sense (Assertion 1, p. 213). That is, the decontextualization movement that extracted the material from their natural settings for standardized data collection purposes was complemented by a re-contextualization of the data in the sense-making process. During this early stage, the conceptions related to the dogma were dominant, which means, we are at Stage 1 (Fig. 7.2) of our historical trajectory.

First Explanations

> "Okay, everything so far has been predictable."

In the following fragment from Episode 7.1, the members of the team attributed the data fully to the effect of migration. At that point, we had seen two sets of data, each containing three points taken at several week intervals. One set of coho salmon, originally sourced from Robertson Creek Hatchery, had been kept in a saltwater tank at the university from which we regularly took a small sample for measurement (Fig. 7.5a). In the transcript fragment, Shelby suggests that the porphyropsin (A_2) levels were decreasing as expected. He had done an analysis of variance (ANOVA) and had detected statistically significant differences between porphyropsin levels at the three time points when sampling took place. Craig accepts the description as fine but as meaningless until held against the measurements from the group that was placed in a freshwater environment ("that curve there doesn't mean anything until you put against the freshwater fish"). Shelby searches through his data, then projects a second plot (Fig. 7.5b), saying that the ANOVA does not show significant differences between the three means. At this point, Craig summarizes that everything so far has been predictable (turn 01). Shelby both agrees and elaborates that it is not quite so predictable because he had not been convinced that those fishes placed in the saltwater environment necessarily should "go to A_1 [rhodopsin]" (turn 02). What those present are seeing at that point is that the fish kept in saltwater does indeed shift, as the canonical theory would suggest, whereas the fishes kept in fresh water does not shift but rather appears to exhibit a reverse movement (turn 04).

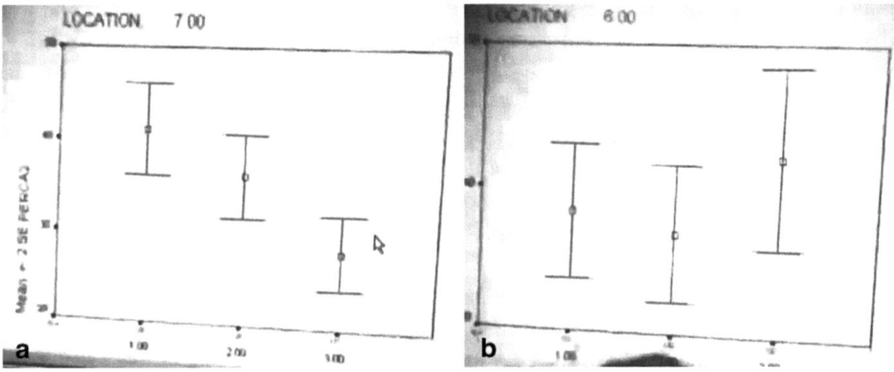

Fig. 7.5 (**a**) The first three data points for coho salmon retained in a saltwater tank at the university. (**b**) Percent porphyropsin (%A_2) from coho salmon that had lived in freshwater (© Wolff-Michael Roth, used with permission)

Episode 7.1

```
01 C: kay so everything so far has been predictable
02 S: yep. well this one actually is not so predictable.
      this is the one where elmar and i talked about it the
      other day no matter what happens its publishable. and
      its very interesting because it wasnt; i wasnt
      convinced; cause no ones really done marine fish yet;
      so I wasnt convinced that you take two sets of fish in
      a tank um mari:ne fish and freshwater fish ((draws a
      line))) that these ones should necessarily um go to a
      [() one]
03 C: [a:=one]
04 S: it wasnt necessarily true because the implications in
      the wild is that um first off that one would say well
      theyre shifting because of hormones um and so thats
      what initiated the shift; thats why it starts in the
      freshwater; thats fine. but in actual fact what were
      seeing so far what we were seeing in freshwater its
      not shifting that much in the wild in the streams yet
      so
```

Shelby later provides a summary: although we had anticipated some sort of shift, we had not actually seen it. This, as he subsequently states, would have given rise to the possibility that other variables "might be controlling the shift [from A_2 to A_1] to that last little bit, to get rid of all that porphyropsin [A_2]." But at the point of that meeting, the measurements collected thus far do in fact suggest that the saltwater environment alone is responsible for the shift from one to the other opsin. Elmar adds while pointing to the currently projected graph (Fig. 7.5b) that it is "nice," because it first drops and then increases again, that is, that the fishes appear to revert, just as the "dogma" suggests.

This part of the meeting is an expression of an orientation that our team has taken up to that point and that dominates during the early part of the data collection. This was so even though, as the later part of this meeting shows, there are unexplained discrepancies with respect to the theory that might have directed us to look for other ways of explaining its data. We are oriented to the "dogma," represented here by the graph from the Alexander et al. (1994) study (Fig. 7.4), which our team takes as an index to what the data ought to look like. It frames what we can see in the data. Time and again, team members present in the meeting refer to the existing dogma using the names of the three authors of this paper ("Alexander-Sweeting-McKeown"), that of the lead author ("Alexander"), or, because Craig knew the author, sometimes by his first name ("George").

These data and the ways of reading these appear to be confirmed by the fishes caught in the wild but near another hatchery (i.e. Kispiox), where the means for the two sets of fishes analyzed were 72.5 and 67.7 $\%A_2$ (porphyropsin). The difficulty that emerges for our team is that there are large differences between the wild and hatchery fishes from the Kispiox location, whereas for the main participant hatchery, Robertson Creek, the wild and hatchery fishes exhibited almost identical porphyropsin levels (48 % and dropping). Much of this meeting on June 18, 2001 was concerned with the attempt to identify the origin of those differences. At that instant, as throughout the research, these scientists discount data because of an "entrenched theory" (Chinn and Brewer 2010) that we referred to as "the dogma." Even though, there are contradictions. Craig says in the meeting that "everything so far has been predictable," whereas Shelby proposes that "this one actually is not so predictable." In the ongoing work process, we, as further shown below, are not quite as systematic and categorical as Chinn and Brewer's cognitive models-of-data theory appears to suggest. In fact, we were in the same position as another Canadian scientist, who critiqued the classical notions of what science is (Suzuki 1989). He suggests that if there were something like a right or wrong answer, "if we knew the right answer beforehand, we wouldn't bother doing research" (p. 191). He continues by saying that "even when we repeat a test that has been well documented, the data we get are not 'wrong' if they fail to conform to expectations. We may have duplicated the experiment poorly, *but the data is all one has*" (p. 191, emphasis added).

An Alternative Hypothesis Emerges But Is Not Retained

With hindsight it is easy to say that the idea for a particular interpretation of the data had been there all along and that the team just did not have sufficient data required to make a decision this or that way. But this would be a reconstruction of the events, that is, it would be Whig history (Rorty 1979). The fact is that what previously was seen as the beginning of a curve consistent with the Alexander et al. study subsequently turned out to be part of a curve that was inconsistent with it. First, the extrapolation of the data backward in time does not coincide with the Alexander et al. graph. Second, the critiqued "reversal" does appear to coincide with the data *and* is consistent with an annual cycle of porphyropsin levels. In both situations, our team had to reverse its earlier statements, a form of conceptual change from the predictions of the canon and the reality of the data as these were produced in the course of our research project.

One of the core competencies ascribed to scientists is their ability to project conclusions toward situations for which direct data do not exist. The present study throws light on this aspect as it occurs during a conceptual change process in scientific research. Our team talked about similarities and differences within and between the two hatchery locations, the differences between the "systems," the differences between different creeks and rivers around the Kispiox hatchery location, which contributed to widely varying sizes in the specimens caught. During the meeting, we are not even certain whether the differences in fish size are due to drastically different environmental conditions (high vs. low feed) or due to different age classes (zero-plus, one-plus) (see also Chap. 6). As part of this discussion, Craig then summarizes that he is "not bothered" by the differences between the hatchery and wild fishes from the Kispiox location but rather by the similarity between the wild and hatchery fishes at the Robertson Creek location. Shelby offers a challenge to him by asking for the evidence on which such an assessment is based. Craig suggests that the hatchery and wild fishes might be on different growth trajectories because the hatchery fish have more food. He expects them to be of different body size. But Shelby responds by saying that the wild and hatchery fishes are about the same size. In fact, earlier in the meeting Shelby had already shown a porphyropsin level versus weight plot to show that there was no apparent correlation between the two variables (Fig. 5.7). And, the associated ethnographic research shows that fish culturists do not feed the same amount but decrease feeding amounts and rates in the winter to match normal feeding habits to avoid oversized fishes at release date (see Chap. 5).[4]

In this meeting, Theo explicitly proposes temperature as a significant variable that plays a role in creating differences in porphyropsin levels (turns 02, 04). This would, therefore, constitute a possible bifurcation point (#2, Fig. 7.2), where in addition to the original theory that describes porphyropsin levels in terms of physiological changes and sigmoid variation of porphyropsin (A_2) levels, there would be an alternative theory that introduces the environmental temperature as a mediating factor in the A_2 levels of coho salmon. The proposal is made following a turn at talk in which Craig suggests that hatchery fish are more synchronized than the wild fishes, which have a greater variance in their body condition. Theo

[4] This is not always the case, however. During my time in the fish hatchery, some fish culturists experimented with "super smolts," that is, they retained the population of salmon to sizes much larger than these might normally develop prior to migration. These fish culturists tested the hypothesis that such supersized smolts have higher survival rates – especially because these would not be subject to predation by smaller fish species.

suggests that the temperature in the Robertson Creek hatchery "was so much warmer" (turn 04), but Elmar responds by stating, "it's pretty similar" (turn 05). He continues to elaborate, following a querying "is it similar?," by saying that the personnel takes lake water to match the outside temperature (turns 07, 09). Yet this assertion is in turn rendered problematic by a statement that Shelby makes (turn 10), who asks Elmar to remember that the same temperature was used throughout the year by changing the depth of the water intake from the water in the nearby Great Central Lake. Elmar and Shelby then introduce uncertainty, allowing subsequent revision of what they have said.

Episode 7.2a

01 G: yea, but the difference=is that the hatchery fish are sYNchronized more so than the wILd=caught fish. the wild=caught fish have this variance= associated with body condition.
→ 02 T: what I was zinking is zat whezer the water temperature is playing a role.
03 S: yea.
04 T: the robertson was so much warmer dan dee, dee wild ones were also faster.
05 E: no its pretty similar.
06 T: is it similar? oh
07 E: because they take um [lake water
08 T: <<p>[i zought because> (??)]
09 E: sey actually try to match i zink outside temperature
10 S: at robertson?
11 E: yea.

Following the featured fragment, Theo returns to the question of the temperatures: at Kispiox, the temperature is about 5 or 6 °C, whereas he formulates thinking that it is 12 °C at Robertson Creek. Shelby notes that this is a big difference, and Theo follows up making an implication ("so") about thinking something (not actually specified), which Shelby acknowledges to be "the good point."

This was the first episode on record during which the temperature was articulated as a variable that somehow might regulate the porphyropsin levels in coho salmon. Although uncertainty is introduced in the contradictory descriptions that Shelby and Elmar state, Theo proposes two differences: between Kispiox wild and hatchery fish, the latter kept at 5 or 6 °C, and between the hatchery fish at Kispiox and Robertson, the latter being kept at 12 °C. In fact, the situation is much more complex because these temperatures correspond to those of the creek in the winter; but the temperatures are artificially cooled with water from the bottom of nearby Great Central Lake during the summer months (see Chap. 6). It is only as our team becomes familiar with the detailed local conditions that we get a better handle on modeling the data produced in the laboratory and relating the models to the real-world context from which the fishes were sourced (Assertion 4, p. 213).

In the following, one can observe the distributed nature of knowledge, which leads to the fact that someone proposes a hypothesis that soon after is abandoned because someone else on the team already has the answer. Many of the descriptions articulated in this first data analysis meeting relate to the environmental specifics of the river systems, hatchery conditions, and geographical differences (distance to ocean, climate) – that is, to context. Without this context, scientists are unable to say much about the data that they have collected so far: why there are certain relationships or why expected relationships do not show up (Assertion 4, p. 213, Assertion 5, p. 213). Scientists, as I show in Chap. 5, find it difficult to say much about data and graphs when they are not very familiar with the system from which the representations have been derived. Our team meeting covers again much of the same terrain, almost as if we had not already talked about these issues. This is evidence that although someone says something in a meeting, the content of the said is not inherently salient to the group as a whole. It is precisely when scientists cover the same ground that some issue then also becomes salient for the person, to whom it may not have been before. Already before, Shelby proposed the "need to know the temperature of maintenance." Yet, as the following episode shows, there were team members who had noted what was happening with the temperatures in the different hatcheries and systems at the time.

Take-Up of an Earlier Hypothesis

"The temperature is so different, y'think this is really important here."

The episode begins with Shelby's description of Theo's statement "pretty good" and that this is really important (turn 01). He continues by elaborating that he means the Kispiox to be a glacial river, with a temperature somewhere between 4 and 6 °C, whereas "the hatchery ... probably maintain[s] at something like 10 or 12 °C" (turn 03). Elmar, who (as I) knows the hatchery, disagrees, suggesting that the fishes are kept at 6 °C. Theo adds that he cannot explain the difference (turns 09, 11), to which Craig adds that he can give a "pretty good scientific discussion" as to the difference between the Kispiox (wild, hatchery) fish but that he "cannot come up with a very good scientific explanation" as to the similarity of the Robertson (wild, hatchery) fishes.

Episode 7.2b

```
→ 01   S: i i=d say theos suggestion yea was pretty good too
          the temperature is so different ythink thats
          really important here then probably too what
  02   G: between?
  03   S: yea i mean obviously the kispiox a glacial river
          theyre running at four or five degrees celsius
          maybe six tops. and the hatchery they probably
          maintain at something like ten or twelve degrees
```

. . .
17 S: we dont have all the data thats thats the problem;
18 G: yea. ((*nods*))
→ 19 S: it could be, it could be the kispiox temperatures are highly different and that robertson arent. and that alONE might explain that
20 G: well i mean thats easy enough to figre out
21 S: yea. exactly. the point is that we dont have all the data yet, i was saying [the difference is the]
→ 22 G: [okay so temperatures]
23 S: yea
→ 24 G: temperatures for kISpiox HAtchery wILd ((*hand gesture: one, other*))
25 E: and robertson
26 G: robertson creek ((*repeat hand gesture*)) wild an hatchery ((*hand gesture, reverse*))

Shelby provides an answer in relating the differences in the Kispiox fish to the difference in river and hatchery temperatures and the similarity of the Robertson (wild, hatchery) fish to the similarity in temperatures (turn 19). Craig's response begins with the statement that this is easy to figure out (turn 20), which he then elaborates verbally and gesturally in the double contrast between the geographical locations of the hatcheries and between wild and hatchery-raised fish (turns 22, 24, 26). That is, he provides an answer to the problem he articulated earlier. Elmar and I provide further information about the two locations. He knows that the Department of Fisheries and Oceans (DFO) collects the temperatures as part of one of its programs. I add my observation that at the Robertson Creek location, water is diverted from a creek into the coho basins during the winter whereas in the summertime, cold water is added to the creek water to limit the maximum temperature in the fishponds. Shelby indicates that this is what he has said and elaborates: "they are controlling the temperature in some way." Craig makes a statement that constitutes a different proposal: to assume conditions are the same between hatchery and wild – "more or less the same temperature" – and that the difference comes from the different access to forage. He then summarizes what they have so far: "In the Kispiox, we've got a discrepancy between outgoing hatchery and wild; at Robertson we have more or less comparable values in wild. Given a limit of data set, the difference between the two scenarios is proximity to the marine environment and that suggest to me that they are maybe within the stream or river physiological preparation for the marine environment." He concludes by stating that this is the "major hypothesis" for the team to pursue and that it would "be interesting to see what the Kispiox wild fish look like in the estuary."

Discourse in Transition

In this situation, our team does not just look at the data. The statistical comparison is only one part of a broader attempt to become familiar with and model the biology involved (Assertion 1, p. 213). As scientists, we always want or need more information. That is, rather than working toward abstraction or intermediate level abstraction, we work toward concretization and total contextualization of our data. The abstraction has occurred together with the movement of the fish from their natural or hatchery setting to the laboratory into the laboratory where these would be euthanized, and eyes extracted and hemisected in preparation for the removal of the retinal pieces (see Chap. 5). Once placed under the microscope, the individual cells were further extracted from their environment as these constituted the only perceptual structures on monitor to which the scientists oriented. As shown in Chap. 5, our explanatory effort required us to engage in the reverse trajectory by taking our measurements back into the concrete settings where the retina, eye, and fish had originally come from and in the context of which the porphyropsin levels have to be explained. It should be noted that our team used temperature differences as a potential resource for explaining the difference in the porphyropsin levels between the Kispiox wild (around 80 %) and the hatchery-raised fish (around 25 %) and the similarity in the Robertson Creek wild and hatchery-raised coho because of the similarity in temperatures. This aspect of our findings would be deemphasized and completely disappear in the ultimately published study (Chap. 6). This is the case even though at the time there are repeated instances when we talk in laboratory conversations and in presentations about what happens to the porphyropsin levels following changes in the temperature of the water. Thus, as shown in the data published in a second paper (Chap. 6) – but not discussed as such because the experiment focused on the effects of hormone treatment – the porphyropsin levels drop significantly within the first two weeks when the fish were kept in the university tanks, where the temperature was 11 °C higher than in the hatchery during the first experiment (January 2002) and 3 °C higher in the second experiment (June 2002). There was further evidence of lower overall porphyropsin levels in marine fish and of the depressed cycles. But this information is not further pursued in the current context.

Much of our struggle in modeling the data during this early stage is related to the lack of context, which we are in the process of constructing before we can describe and explain the data. Later, on June 19, 2002, Shelby would explain this situation while talking to the hatchery personal at Robertson Creek where he had given an update of the team's work:

> What I didn't realize before I started was that the techniques your hatchery uses compared to theirs are considerable different in the way that you raise your fish and because I hadn't traveled up there myself to see the hatchery I didn't find these things out until afterwards so there's some dramatic differences in the results that you'll see as a result of the fact that they raise their fish differently than we do.

While talking to "the people at Kispiox" on the phone, Shelby suggests that he received one answer to a question on one day and a very different, even logically incompatible answer to the same question on another day.

> So we went up there ... and visited the Kispiox hatchery. It was really useful, I now know what's going on. They had all my fish in one tank, they're covered up with the dark cover. They get natural daylight but no lights. The water was always six degrees Celsius. These things that I now understand about the system much better than I would before, that really helped as far as kind of understanding why my data was what it was makes quite a difference and it was really good I think from their point of view.

In this quotation, Shelby states that it his becoming familiar with the Kispiox hatchery and the conditions under which the coho fishes are kept allow him to explain his data. That is, the data did not speak for itself. To say anything about these, the familiarity with the original lifeworld of the fishes was required.

At a general level, the possibility for an alternative hypothesis already exists in this episode. The most rudimentary form of generalization exists in the transition from observation sentences ("The temperature is different," "the porphyropsin levels are different") to an observation categorical ("When the porphyropsin levels are different, the temperatures are different"). Such observation categoricals exist in the linkage of two observation sentences. In themselves, however, these do not yet constitute the fundamental condition for science, which is based on *focal observation categoricals* of the sort "Whenever ... it ..." (Quine 1995, p. 27). In the present instance, we already have the materials that we need, if temperature similarities and differences were to be tied to the similarities and differences of the porphyropsin levels at the Robertson Creek and Kispiox hatcheries. A hypothesis in the form "Whenever there is a temperature difference, *it* leads to a (significant) difference in porphyropsin levels" might have been possible at the time. It is precisely the presence of the essential pronoun "it" in the connection between the two observation sentences relating to differences in temperatures and porphyropsin levels that the possibility for the discovery lies. But our team does not (yet) recognize it as such at that time, even though, with hindsight, this connection might be said to have been plain and clear.

In terms of the model used here researchers have to provide a description of those features of the stage preceding the first qualitative change (#1, Fig. 7.2) that are relevant to the subsequent bifurcation. In this section, I describe some inconsistencies that members of our team noted at the time and also some of the alternatives members articulated for conceptualizing the changes in porphyropsin levels over the course of the year (i.e., temperature). This, then constitutes a ground in which the emergence of a bifurcation point – recognizable by the concurrent present of different alternative conceptions, though one tending to be the dominant one – can occur.

Fitting the Data to the Dogma

"The annual cycle looks like ... I suppose we can look at George's data."

We met again during the following month (July 13, 2001) at which point there were further data available. During this meeting, our team talked about the possibility of using eye pigment data for determining the release date of coho salmon from hatcheries. We discussed the data collected so far, with porphyropsin levels lying around the 47–50 % mark. Our data could be used subsequently in support of further studies that correlated the level of porphyropsin at the day of release to return rates. For this top happen, we would need: (a) our data to lie at about the halfway point on a dogma-like graph (Fig. 7.1a) or resemble the Alexander et al. results (Fig. 7.4). The data would need to be correlated to the results of the saltwater challenge experiments – used to determine whether the fishes can survive in ocean water. If these checks were positive, then the kind of eye pigment data that we produced could be used for determining ideal release dates.

At one point during the meeting, Craig approaches the graph projection on the screen and indicates by means of gestures that we do not have measurements for the full annual cycle (offprint in turn 01, Fragment 7.3a). This is the first time he articulates for everyone else something of this kind, even though the present data under discussion, those collected by our team and those presented in the Alexander et al. publication, do not suggest an *annual* phenomenon or anything cyclical but physiology-driven changes in preparation for the seaward migration. In fact, while producing the utterance about the annual cycle, Craig gesturally moves to the left and back to the right in the shape of the logistic (sigmoid) curve suggested by the Alexander et al. data.[5]

[5] A sigmoid or logistic function is defined by $f(t) = \frac{a}{1+e^{-kt}}$ and looks like, with alternatively $k < 0$ and $k > 0$, the following graph that approximately models the Alexander et al. data:

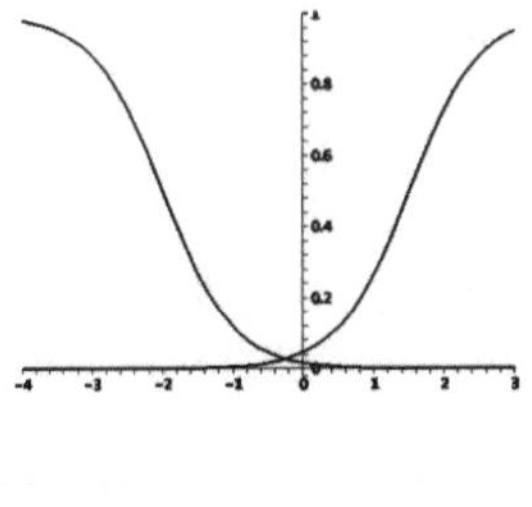

Episode 7.3a

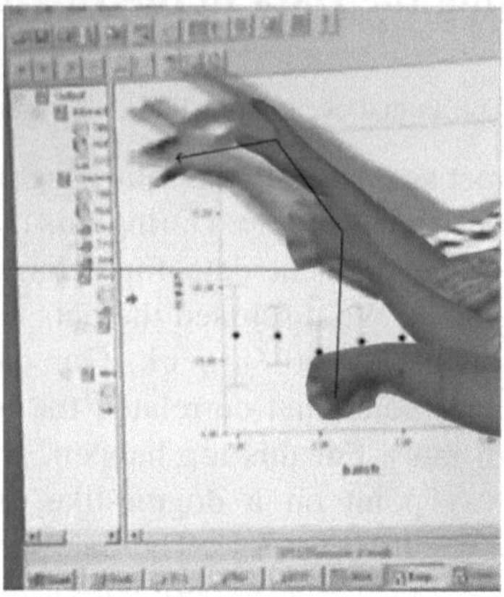

```
→ 01  C: unfortunately we are
         sort of trapped here
         mean that is a good
         thing that we started
         when we did, but we
         dont have the annual
         cycle * we dont know *
         what the annual cycle
         looks like i sppose we
         can look at georges
         data and
  02  S: yea, ive got it here
         and it it IT shows,
         you know, it shows
         what we are not seeing
         at all i=mean hes got
         something right up
         here at eighty percent and then just quickly
         dropping down in here ((follows curve on Alexander
         et al. graph)) and then its going across for five
         weeks and back up again we dont have anything like
         that, this is a period of thrEE mONThs. we were
         down here for thrEE mONThs. so that is what i was
         saying; something is different in the hatchery
         than this hatchery; thats for sure. thrEE mONThs.
         so that is what i was saying; something is
         different in the hatchery than this hatchery;
         thats for sure. that is this is comox hatchery or
         whatever is that river capilano
  03  C: um capilano
  04  S: capilano sorry. so I mean thIS hatcherys
         ((Alexander)) situation is completely different
         from this ((points to screen)) one um, and in this
         way you can look at this and say yea fifty percent
         you are dead on ((points to Alexander et al.
         graph)) fifty percent they are dropping down they
         are ready to go and way they go.
  05  C: oh well we
  06  S: thats cool say this starts in january um
```

In this fragment, Craig can be heard to belabor the fact that our team does not have data prior to the month of April when the current study actually started, even though he considers it to be a "good thing to have started when [our team actually] did" (turn 01). Immediately after stating that they "do not have the annual cycle" and while saying "we don't know" prior to completing "what the annual cycle looks like," his hand moves along a trajectory over and about the data that looks like the graph that Alexander et al. had published – like a sigmoid function where the current data constitute the lower arm (offprint in turn 01, lower part of left graph in note 3). He suggests that our team could look at "George's data," and, in so doing, invokes the paper by George Alexander for a second time and in a different, verbal modality. Shelby then says, gesturing towards his papers, "I've got it right here." He makes a statement that describes the graph and says "what [we] are not seeing at all," that is, the drop from 80 % and quickly dropping down, just as Craig has gestured earlier. Shelby suggests that there is something different in the hatchery that the team works with compared to the one in which Alexander et al. conducted their study (i.e., Capilano Hatchery, North Vancouver, British Columbia).

The team returns to the same issue a few minutes later (Episode 7.3b), and, even though Shelby has suggested that there might be something very different going on in our hatchery, Robertson Creek, Craig repeats the previous gesture to project where our team's data, if we had actually and already collected these, would lie.

Episode 7.3b

```
01   C: so I think probable
        the * swing is
        somewhere in * here
02   S: yea.
03      ()
04   C: on that time line
        ((points to abscissa))
```

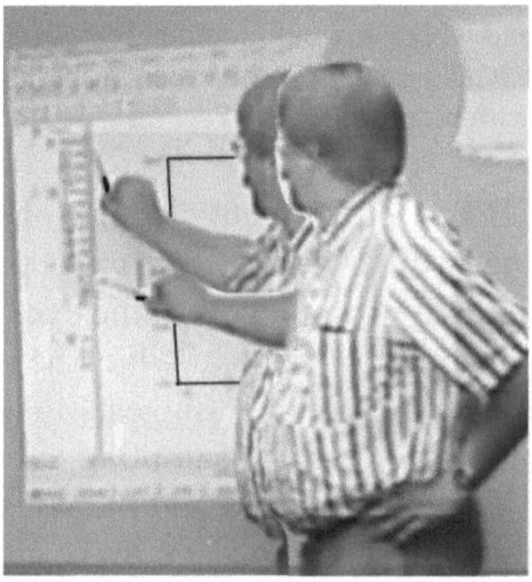

Craig locates the "swing" that Shelby earlier described as missing in their data just to the left boundary of the current graph given the timeline currently plotted. Shelby states what can be heard as agreement (turn 02). In this instance, the lead scientist on our team states the expectation that the data follow the pattern described in the Alexander onset al. paper. Craig gesturally shows that the steep incline up to the anticipated 80 % mark would occur just prior to the beginning of the time axis ("the swing is somewhere in here, on that timeline" [turns 01, 04]). This statement is logically consistent with the canon, according to which euryhaline fish rapidly change from the porphyropsin- to rhodopsin-based chromophores; and it is consistent with the graph Shelby has drawn for the Kispiox data (Fig. 7.6b) in a diagram that also contained a generalized graph representing the Wald/Alexander conception (Fig. 7.6a).

Later during the same year, during the early parts of the meeting on November 7, 2001, the team talks about the way in which their experiment will unfold, which data they will require, the frequency of data collection, and so forth. At this point, the discussion is entirely in terms of the dogma, which can be seen how the Wald/ Alexander conception appears in the graphs that Shelby draws for the team and that serves as a reference for denoting points in time when the data have to be collected. During this meeting, as part of planning the data collection for the new year, Shelby also summarizes what our team has done so far: he describes, pointing to the graph that he is concurrently drawing (Fig. 7.6a) how we have started collecting data around the 50 % point on the Alexander-like graph; and he projects that the most important issue for our team is to collect the data in the upper part, which he suggests to be from late January all the way to the point where they had started measuring the past year, that is, around 50 % porphyropsin content. At this point in the research, we see our data through the lens of the Alexander graph (Assertion

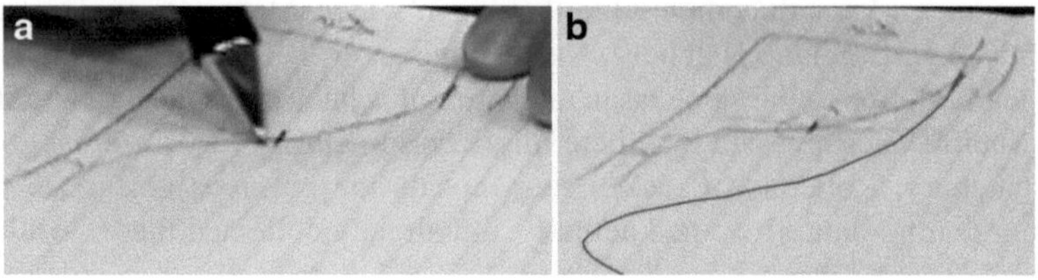

Fig. 7.6 (**a**) Graph drawn as part of explaining where the team started to collect data during the year and where it would anticipate collecting a lot of data. (**b**) Anticipated graph for coho parr, which would spend another year in the hatchery before being released (© Wolff-Michael Roth, used with permission)

1, p. 213). Our team anticipates collecting data only a few months hence from that day that would confirm the upper part of the curve, when the coho is expected to have porphyropsin levels in the eye of around 80 %. Moreover, instead of sampling once a month – a point at which he marks of one-month intervals on the lower part of his graph – Shelby suggests that we have to sample every two weeks. While uttering this intention, sampling every two weeks, he marks points along the curve (Fig. 7.6a). The dogma and the Alexander curves *are* shaping the lens that frames what we can see in its data: those already collected and those to be collected in the near future. The hand gestures mark, "on that timeline," where the sharp decline of the approximately logistic curve should be; and, these therefore, literally give the collected data a place in the existing theoretical paradigm. In this situation, the Alexander et al. curve *is* the reference. This is so even though in our final report, we will be noting that there were other studies – including George Alexander's dissertation – that reported seasonal, that is, circannual variations. Our team's decisions about data collection frequency are made *in terms of* the current paradigm and the existing experimental results rather than in terms of an emerging alternative. Moreover, the coho salmon currently in the hatcheries' ponds, which would be released during June of the coming year (2002), are predicted as per the diagram to begin with constant A_2 levels at around 80 % to drop only at a later stage. This, too, is made visible in the new line Shelby is drawing.

First Awareness of the Data as a Challenge to the Dogma

> "So does this have theoretical implications? Because it seems to contradict the paper with the reversion experiments."

A subsequent meeting took place on November 7, 2001 Episode 7.4). This will have been a turning point because it is in this meeting that we can locate the transition from the dominancy of the old paradigm to the new paradigm (3 → 4, Fig. 7.2; ④, ⑤, Fig. 7.3). It began with Shelby using a simile of the Alexander et al. graph as part of a statement that suggested our team's being on track to replicate the Alexander et al. work. But in the course of the meeting, doubts began to emerge about the suitability of the dogma to explain the data that we had collected thus far.

In fact, the data "seem to contradict the paper with the reversion experiments," that is, the Alexander et al. paper and the dogma it supported. During this meeting, the contradiction between the data and the dogma comes to be explicitly articulated.

Shelby presents "one of the better slides" that our team has of its results. The graph compares the results of the data collected during the summer on coho salmon specimens caught in the wild (Fig. 7.7). This fifth episode in the present microgenetic and historical account begins with Shelby presenting the data that we had collected since beginning the data collection ("date of release").

Episode 7.4a

```
01   S: this shows the date of release ((Fig. 7.7, #1))
        from robertson creek. and then it follows ((#2))
        the freshwater fish the green line and the marine
        fish ((#3)) the kind of burgundy line and it shows
        that these kind of point to the same place here
        ((#1)). this ocean point ((#4)); i think theres
        one more ((adds point #5)). yea this ocean point
        ((#4)) is, was collected off the coast of ucluelet.
        so and there were some fish in there that were
        also from the same hatchery; so it shows that over
        this time period ((moves cursor from #1 to #4));
        of two months; they ended up in the same location
        ((#4)). so that this is a natural variation; that
        we had it at the tanks here at the university
        represents that the natural transition ((moves
        cursor from point #1 to #4)) in photopigments,
        that they both end up at the same location.
        ((01:03:19)) um and then what i show up here
        ((#5)) is what was going on at the estuary for
        both the hatchery fish ((#5)) and wild fish
        ((#6)); in other words at both those times, both
        the hatchery and estuary were very similar in
        their pigment as were the freshwater fish that we
        have here ((moves cursor around #5 and #6)) that
        school. they all fell in the exact same location.
```

These data pertain to fishes from very different locations or conditions and were collected at different times: (a) coho at the date they were released from the fish hatchery in early June (Fig. 7.7, #1); (b) coho from the same population kept in fresh water (Fig. 7.7, #2); (c) coho kept in a saltwater tank (Fig. 7.7, #3); (d) coho from a marine environment ("ocean point," Fig. 7.7: #4); and (e) hatchery fish (Fig. 7.7, #5) and "wild fish" (Fig. 7.7, #6) caught in the estuary. Shelby states that all the fishes "end up in the same location" and that they therefore are observing "a natural variation and a natural transition." Shelby adds that the coho caught in the estuary – i.e., after having traveled 30 km from the hatchery – also "fell in the exact same location." This statement, therefore, challenged the dogma, which suggests that fishes held in fresh water should revert in the porphyropsin to rhodopsin ratio in the photoreceptors to a composition typical for the freshwater environment. In the slide Shelby projected, it appeared as if all fishes, independent of their origin, exhibited the same composition.

Scientists do not abandon a dogma unless the data contradict what would be expected (predicted, explained), and even then only with some difficulty (Kuhn 1970). The data that Shelby presents here might be seen as doing precisely that: the

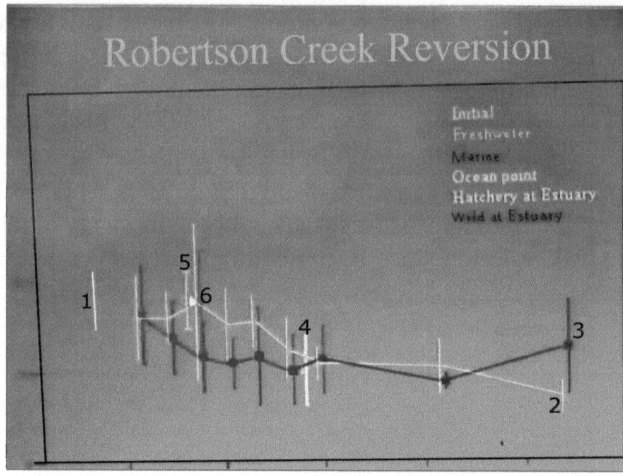

Fig. 7.7 PowerPoint slide projected during the meeting showing the results for all coho salmon (wild, hatchery, estuary, reversion experiment) from the Robertson Creek Hatchery. (Some lines slightly enhanced for reproducibility; numerical indices have been added for easy reference) (© Wolff-Michael Roth, used with permission)

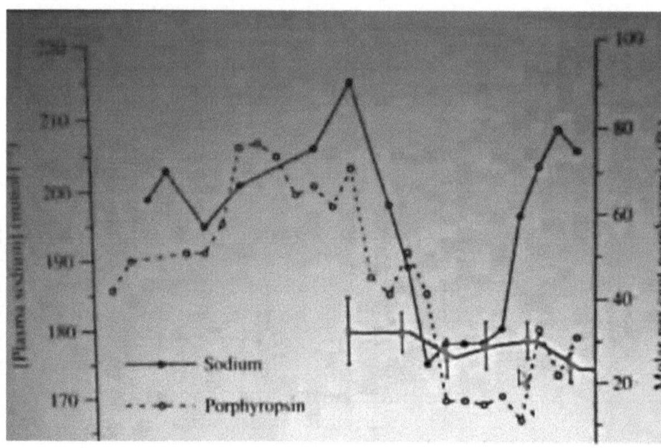

Fig. 7.8 This video offprint shows the instant at which Shelby projects a PowerPoint slide containing his data (*red, lower right*) superposed on previously published data and the associated graphs (© Wolff-Michael Roth, used with permission)

freshwater fishes are no different from the fishes collected in different marine environments, whereas the dogma suggests that there are significant differences between fishes in the marine environment (almost completely rhodopsin) and the freshwater environment (almost completely porphyropsin). But are scientists conscious that their data contradict the canon? The fact that these data contradict the dogma was made salient in a subsequent question that I address to Shelby (turn 01), and which leads him to project a graph that contains the Alexander et al. data as well as the team's data from the Kispiox samples (Fig. 7.8). The response shows that Shelby has heard me mention the reference paper, because he suggests that Alexander et al. also showed reversion for the freshwater fish (turn 05). I state the contrast with our data (turn 06), which leads Shelby to introduce a paper by "Allen"[6]

[6] The reference is to a study published by another group (Allen et al. 1973), which concerned a different fishes from the genera *Salvelinus* and *Salmo* (the same as Atlantic salmon) whereas our team researched coho salmon, that is, fishes from the genus *Oncorhynchus*. The life history strategies of two genera do not have to be the same, so that the Allen et al. results may not be pertinent to our study.

that showed seasonal variations rather than the life-cycle-dependent changes predicted by the canon (turn 07). One may therefore note a contradiction between explaining data collection in terms of the Alexander et al. study and the familiarity with at least one study that suggests a different pattern of variation (Assertion 6, p. 213). At this point in time, however, the salience of this other information is very low and does not have a significant effect on how the team sees its data (Assertion 4, p. 213). In terms of the present model, there are two possible conceptions, two branches in the graph of Fig. 7.2. Up to that point, our team still finds itself on the lower branch and the dominant classical theory of sigmoid changes in the retinal composition during migration between environments of different salinity.

Episode 7.4b

```
→ 02   M: so does this have theoretical implications?
          because it seems to contradict the paper that, uh
          with the reversion experiments
  03   S: which which paper? sorry i didnt hear that
→ 04   M: it was a paper ah on the reversion experiments;
          where we should have seen a reversion here
  05   S: yea so i mean alexander sweeny and mckeown showed
          a reversion as well. um in their stuff. so i just
          zip to that. ((selects graph from slide outline))
          to that graph, what they show here ((shows our
          graph and our data, Fig. 7.8)) is uh they suggest
          that the pigments after um june; if the fish are
          maintained in the freshwater start to revert back
          ((moves along graph)) um theyre showing them going
          back up again
→ 06   M: but ours didnt revert
→ 07   S: ours did not revert and in fact when i looked at a
          paper by donald allan done in the seventies he
          also, well although he wasnt trying to revert he
          did not see a reversion. and um what he saw was
          that this this whole data was a a change in
          pigments like this ((follows Alexander et al.
          curve with cursor)) that went down actually best
          fit the photoperiod.
  08   M: okay
  09   S: and thats exactly what were seeing in our data set
          in both tanks here and what i=m getting from the
          wild and everywhere else is that in fact the
          visual pigments follow the photoperiod not this
          physiological indi indicators that are here
          ((makes circular motion with cursor on sodium
          graph)) and when i deeply critiqued this alexander
          sweeney and mckeown paper, i discovered that by
          using thEIr vERy data and plotting it and doing a
          correlation coefficient between day length and
          visual pigments there was a better correlation
          between day length and visual pigments then there
          was between visual pigments and sodium content of
          the blood. um. so unfortunately that would suggest
          that the visual pigments ARent an indicator of the
          physioLOGical pROcesses but they ARe an indicator
          of where the fish is on its photoperiod change.
          which may be consistent.
```

Shelby subsequently describes that he has done a correlation between the Alexander et al. data as part of “a deep critique,” which brought out that there was a better correlation with photoperiod (“day length”) than with “sodium content of the blood” (plasma sodium levels are indicators for readiness to survive in saltwater), as those authors had reported. “Unfortunately,” Shelby states, “the visual pigments aren’t an indicator of the physiological processes but they are an indicator of where the fish is on its photoperiod change.” The results are unfortunate because the very purpose of our research had been to use the visual pigment information as a measure of readiness for release of the young coho into the river to begin their migration. For this to work, the graph would have had to have a sigmoid shape, at least during the time of the oceanward migration. But alas, as seen in the plot of our data (red) together with those of the Alexander et al. study (Fig. 7.8), there appeared to be no indication that we would see the same sigmoid relation between time and porphyropsin levels; instead, all of our data had values between 20 and 30 percent porphyropsin (Fig. 7.8).

My follow up question (turn 10) is concerned with the solidity of the data that our team has collected so far). The point clearly addresses the issue whether the data are strong enough to be mobilized in support of a claim that questions the classical paradigm that George Wald had established. In his replique, Craig states that the purpose of the research project and the paper to derive from it are changing. His statement notes that what we are doing is “extremely appropriate for publication,” but that we change the way “we are going to pitch [it] the way it was originally planned” (i.e. as a tool for testing readiness for release from hatcheries). This is equivalent to saying that the pitch will be different from the “dogma,” which describes precisely the initial “pitch” (turn 11), which Craig has made, in various settings, since I first met him in the fall of 1998.[7] Shelby’s subsequent statement summarizes the results of the discussion: the visual pigments are independent of the physiological changes of smoltification, even though he adds a modifier by saying that “we don’t have enough evidence” (turn 13), a statement itself modified by the addition of the adverb “yet” as if saying that it is only a matter of time that the evidence is forthcoming. This might indicate that the present results merely deviate from the established paradigm. But, because the number of data points collected is orders of magnitude larger than any other previous research project, it is unlikely that the difference is a mere deviation towards statistically more unlikely extremes. Rather, a revision of the theory might be in order, that is, a radical undermining of the dogma by the present results.

[7] Some (cognitively oriented) research appears to suggest that scientists are into hypothesis testing and reporting the findings of these tests (e.g. Dunbar 1999). However, scientists’ own accounts (e.g. Suzuki 1989) and reports on what happens in the natural sciences (e.g. Couzin-Frankel 2013) suggest otherwise: the “pitch” is made up with hindsight, when it is known what the data can actually be used to support.

Episode 7.4c

→ 10 M: what do you think about the, do you think that the argument holds in the literature? if you are trying to publish it.

11 C: well i think what were doing is extremely appropriate for publication. bu=we are not going to pitch the way it is was originally planned on being [pitch]ed [

12 M: [uh hu] [uh hu

13 S: its just a bit; i mean. the indication so far; and we dont have enough evidence yet; but the indication so far that the visual pigments um are independent of the physiological changes of smoltification that thats what were finding and well i need to get some more [data.]

14 C: [or un]coupled [or]

15 S: [yea] theyre jst theyre not as tightly

→ 16 T: but dont forget the temperature. you could get [prob]ably

17 S: [yea]

→ 18 T: a good correlation with [temperature]

19 S: [so if i] showed [yea so sorry]

20 T: [i see a good] correlation wiz ze sunspots he ha [he ha he]

21 S: [i always] neglect to show that

→ 22 T: one must be careful cause if you have a good agreement that it is statistical or like the birthrate of which increased with the birthrate of the incidence of (?) okay so, alright, so, i would like to warn you that the correlation wiz ze photoperiod could be because ze underlying mechanism has dis it could still be an indicator.

Up to this point, photoperiod (i.e., amount of day light) has been the main alternative variable discussed. Theo later reminds those present not to forget temperature (turn 16), which he had already proposed five months earlier as a possible candidate variable. He suggests that there might be a good correlation with temperature and he makes a statement that turns out to elaborate a simple statistical correlation with photoperiod, which could have its origin in some other underlying mechanism rather than with photoperiod strictly (turn 22). Shelby formulates remembering something he has read: there are circannual cycles that organisms keep track off even in completely controlled environments. He adds having read about experiments with parr and smolt from other genera and that even in an entirely controlled environment – with 12 h of light and 12 h of dark, in water from the tap as supplied by the city – the basic physiological measures remain on their natural cycles. He extends the argument to the Alexander et al. data, which, as an analysis he conducted showed, are better correlated with the lunar cycle than with plasma sodium content.

In this second part of the meeting in November 2001, the possibility that the current data contradict the dogma was articulated explicitly for the first time. But the sharp contrast – possibly a gloss of a scientific revolution – is not taken up in the subsequent turns, which merely consider the data as "extremely appropriate for publication" and the data as indicating a different trend for which they do not yet have sufficient evidence. The team cannot yet claim that the "visual pigments are independent of physiological changes of smoltification." Doubts are raised about the role of photoperiod and about the possibility of a correlation with temperature as the real, biologically relevant effect. It is only after becoming more familiar with temperature in the various specimen collection sites that our team begins to place greater emphasis on temperature as a true variable, which would be consistent with the slow take-up in the team's discussions of this variable that Theo had proposed already in the beginning.

In terms of the catastrophe theoretic model of conceptual change (Fig. 7.2), the dogma has been still the referent at the beginning of the meeting. That is, we still are on the lower branch following the bifurcation (②, Fig. 7.2) but with developments whereby the conception is under challenge. Consistent with the description of what the analyst has to do, I describe above the new conception that relates the data in a new way and, thereby, articulate an alternative branch that is still subordinate to the dominant one. Some alternative conceptions are salient now, which are modeled by the alternative (upper) branch in the graph (Fig. 7.2). The alternatives to the dogma include the independence of porphyropsin levels from the physiological changes and correlations with photoperiod and temperature. The tone was set for a major change in the way our team was conceptually talking about the porphyropsin levels in the course of the life cycle of coho salmon. At this point in the development, the new conceptual talk, though already existing as possibilities within the horizon of our work, was not yet dominant in any way.

From Dogma to Complete Speculation

> "This is complete speculation, but I am sure that's what's happening in the wild."

In the preceding section, the contradiction between the data and "dogma" came to be stated explicitly. Shelby also talked about having "deeply critiqued" the Alexander et al. paper. Later during this same meeting, he proposes an alternative, which he denoted to be a "complete speculation" about "what's happening in the wild." This "complete speculation" anticipates the ultimate claim that our team published in the scientific journal article that resulted from this work – even though substantial work lay before them before there was sufficient data to support the claim in public.

This episode in my microgenetic historical account exhibits the continued discussion about the differences between wild and hatchery-released fish, the former necessarily traveling through brackish water and the latter that, in some cases, are directly transported to and released into the ocean. The transcript fragment begins with Craig's summary of where our team is at the moment. He rearticulates the dogma and then states that the team is "starting to question that."

That is, it is during this meeting that the research explicitly states questioning the canon, with the possible implication that we would be changing our paradigm (④, Fig. 7.2). Craig also talks about the requirements, such as "captur[ing] fish in the open ocean" to "find out what sort of visual pigment composition they have" (turns 03, 05). He makes a statement that explains "what they have to do" (turn 03), "because [we] don't know if fish have a porphyropsin" (turn 08), that is, whether wild fish have the freshwater pigment.

Episode 7.5

```
→ 01   C: the dogma is that they have a marine pigment in
          the marine environment and a freshwater pigment in
          the frESHwater environment and were starting to
          question that.
  02   M: uh hm
  03   C: i mean we dont knOW what we have to do i guess is
          capture fish in the open OCean
  04   S: yea ((nods))
  05   C: and find out what sort of visual pigment
          composition they have [that would] that would be
          ideal
  06   S:                 [absolutely]
          [yea]
  07   M: [but] um
  08   C: because we dont knOW ifish have a porphyropsin
```

Shelby then states a prediction, using a sinusoidal line he is drawing on paper and showing to Craig (Fig. 7.9): "the fish are going through a cycle related to temperature/photoperiod ... regardless of whether they are fresh water or saltwater." This statement adds temperature dependence as a possibility. He continues by producing a full articulation of the dogma and, using a drawing of the team's data onto which he projects Wald's claims (Fig. 7.10), explains why the Nobel Prize winner might have been misled by his data. In essence, he suggests that Wald confounded points in thelife cycle of the fish with trends in the annual cycle of the fish, because the samples were caught in the spring (Fig. 7.10: #1, 5, 8), when the young fishes migrate out into the ocean, and in the fall, when the adults return (Fig. 7.10: #11, 12). These times also are the instants when porphyropsin levels are on the downward and upward trend in the annual cycle. In his diagram, Shelby hypothesizes the data collection Wald must have done to arrive at the dogmatic statement of porphyropsin as a characteristic property of the freshwater stages in the life history and rhodopsin as a characteristic property of the saltwater stages in the life cycle of euryhaline fishes (Fig. 7.10: #2, 3, 9). Shelby restates that Wald did not have "any of this" while gesturing along the sinusoidal line, which we can her as making reference to our data collected over the course of the year. Shelby's turn comes to an end in and with the statement that what he has presented is "complete speculation" but that he is "sure that that's what's happening in the wild" while gesturally following the sinusoidal curve twice (Fig. 7.10). Some time later in the meeting, Shelby reiterates that he is "speculating completely on what he has found so far and on what some of the literature says." Because of the existing literature, he is "very cautious about how [he] analyzes [his] data."

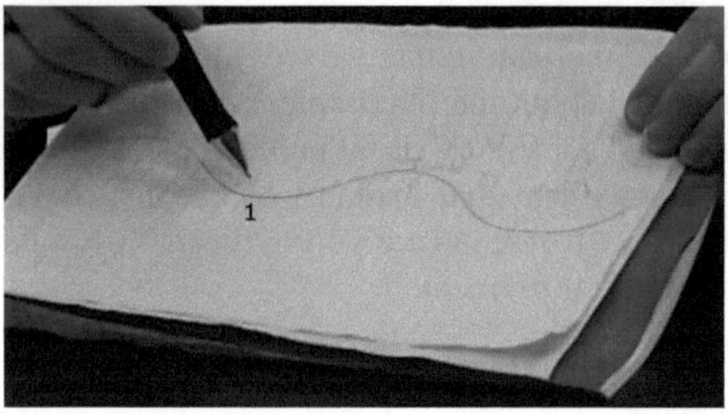

Fig. 7.9 This video offprint shows Shelby while accompanying his "complete speculation" with a graph displaying sinusoidal, circannual variation (© Wolff-Michael Roth, used with permission)

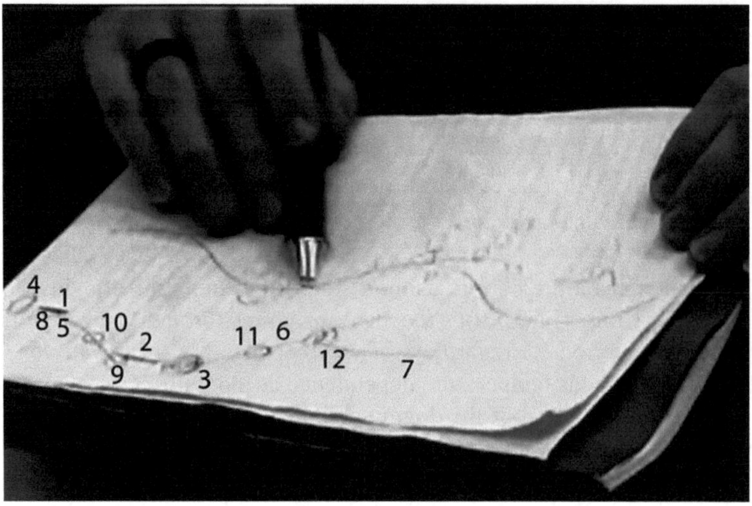

Fig. 7.10 This video offprint shows the important stages of coho salmon changes in porphyropsin levels as part of an explanation why the Nobel Prize winner George Wald erred (© Wolff-Michael Roth, used with permission)

Here, then, we are possibly in the middle of the transition between two conceptions (④, Fig. 7.2). But the demonstration of the full change in dominance from one form of conceptual talk to another form required as evidence for Stage 4 in the model is yet to occur. The materials for such a demonstration would become available during the weeks following this meeting.

Three months later, Shelby gave a presentation in the biology department, showing a slide containing more data and a graph that was already more suggestive of a circannual cycle of sinusoidal form (offprint Episode 7.6a, turn 01). Talking about the graph in which all his data points related to Robertson Creek fishes were plotted – including those kept in saltwater and freshwater tanks at the university, marine fish and parr (fish that have not "smolted") – a common trend appears: they all "follow the same trend."

Episode 7.6a

```
01   S: when i compared that data with data collected in the field
        so in this case wild fish collected from the estuary
        ((cursor at 1)) just after release it turns out that yes
        the laboratory and hatchery hatchery caught in the fish do
        reflect whats going on in the wild and then again here
        ((cursor at 2)) this point represents um marine fish both
        wild and ocean um just off the coast of tofino and again
        theyre representing whats going on in the wild same for
        the parr who follow the same trend.
```

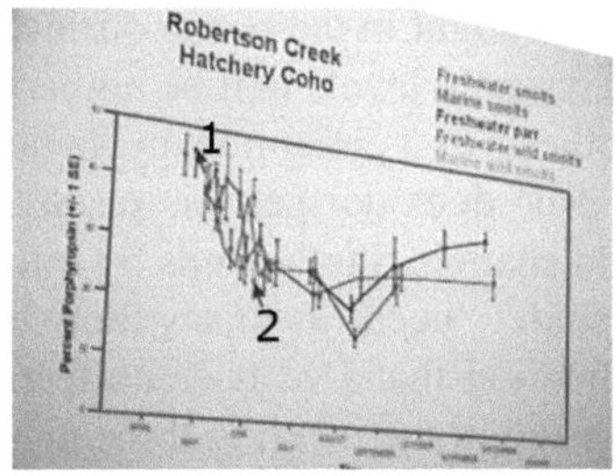

Fishes of very different environments exhibit the same trends in their porphyropsin levels, including fishes that were kept in a freshwater tank compared to those from the same capture but kept in saltwater. This strongly suggests that porphyropsin levels are independent of the preparation for the saltwater journey. Shelby then summarizes his results (Episode 7.6b): a circannual rhythm that is "more likely related to photoperiod or temperature" and that therefore "is changing with a seasonal effect." He raises several other possibilities that might be responsible for the changes in pigments and that he anticipates looking into (e.g., color of the light or hormones). At this point, therefore, our team, represented here by Shelby, no longer explains the data in reference to the dogma and the Alexander et al. study, but in terms of a circannual rhythm (offprint, turn 2, Episode 7.6b). That is, a new conception has become dominant, even though it is still associated with the hedge that coho salmon "appear" to follow this newly proposed function.

Episode 7.6b

```
02   S: so to summarize then it appears that the percent porphyropsin
        is changing on this circannual rhythm that its going through
        these trends up and down which means its more likely related
        to something like photoperiod or temperature which is
        changing with a seasonal effect. but the next step is to
        tease out what some of those effects are so to look at um
        whether or not things like color of the light or exactly
        whether or not temperature or photoperiod are effecting
        these things; also hormones play a role as well; so are
        hormones changing over the year throughout this system and
        so these will be some of the things i=ll be looking for in
        the future.
```

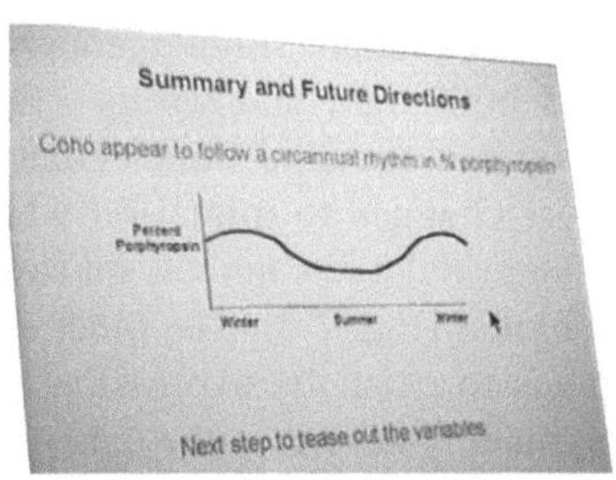

Shelby also states needing one more data point, which is a measure of adult coho salmon caught in the ocean during winter. It would suggest, he anticipates, that the porphyropsin levels will be higher in the winter than in the summer, supporting the hypothesis of a circannual cycle rather than the migration-related changes. He adds that if he does not get this result, then his talk anticipated for the following year would change. Shelby ends by referring to the controversial nature of the results of this study: "needless to say there may be some controversy now with the with this information that I've been collecting um it goes against some theories – as I had suggested in the meeting three months earlier – that were proposed by Nobel Prize winner George Wald um back about fifty years ago." At this instant, then, he expresses certainty about the status of our team's data: it can be used to undo the canon. But Shelby also states being aware that more data are needed to ascertain that the patterns shown are real. From that point on, our team spent much effort to consolidate its findings by searching for particular data points that would strengthen the claim about a sinusoidal, circannual pattern. That is, we are now in a new phase of the research where a new form of conceptual talk has become dominant and has led to a new trajectory in the evolution of the explanations that the laboratory used.

In this episode of the present microgenetic historical account, the "dogma" and the "complete speculation" come to be exhibited within the same turn at talk (presentation). But the "complete speculation" does not appear to have arisen simply from a set of anomalies (data) and then led, through a temporal hierarchy of processes, to a new hypothesis (as this might be suggested in the above-rendered description of a conceptual change in the sciences). Multiple issues become salient simultaneously. Shelby articulates our team's model of temperature as a driver of changes in the visual system: the noise to signal ratio increases in the porphyropsin-based receptor with temperature (summertime), which would explain the decrease of porphyropsin levels with a corresponding shift to rhodopsin. The salience of this fact co-appears with the salience of the seasonal variations in the porphyropsin levels. Such seasonal variations *appear* from the present data. Seeing the data differently is associated with the dominance of alternative explanations. It also becomes salient simultaneously with the possibility of explaining why Wald's study is both right and wrong simultaneously: Because it sampled at two time periods when seasonal trends coincide with migrations, the dogma attributed the chemical changes to migration when in fact a continuous data collection would have suggested a circannual, sinusoidal variation. We are now on the upper branch in the model (Fig. 7.2). The older conception, the dogma, has not disappeared and continues to exist. Its expression in data has become recontextualized so that is approximately describes the data correctly within a small range where the changes are largest (Fig. 7.11). That is, in this range, the sigmoid function consistent with the Alexander et al. study and with data that had led to the construction of the canon are not inconsistent with the present data better modeled by a sine function. What remains to be demonstrated in our analysis is that as the result of the restructuring process, that is, as a result of the conceptual change, there was a new evolutionary trajectory of the research team and its conception (i.e. ⑤, Fig. 7.2). In the next section I show how during the following 18-month period, our team consolidated its

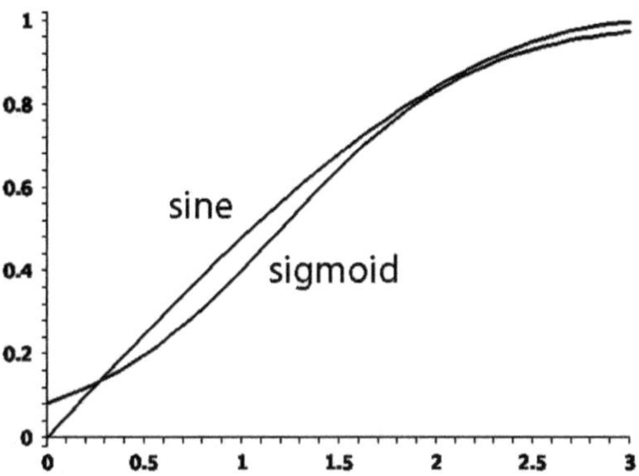

Fig. 7.11 A sigmoid function consistent with the canon and the Alexander et al., within the small range shown, does not differ significantly from a sine curve. In this way, the transitions G. Wald observed do not necessarily contradict those observed in the present study, though the theory would have to change substantially (© Wolff-Michael Roth, used with permission)

results, which it exhibited, together with the new conception, in posters at research conferences, local conferences, meetings with the hatcheries, and in the ultimate journal article that we published.[8]

Consolidation, Posters, and Publications

Following the November 7, 2001 meeting, there was a period of consolidation and reifications of findings. This is part of the new evolutionary trajectory (⑤, Fig. 7.2), where the data were no longer seen to support the old conception but in terms of the new one developed through this research project. The language used during repeated presentations exhibits increasing confidence with the data as supporting the claim of circannual, sinusoidal changes in the amount of porphyropsin (A_2) in coho salmon retina. At that meeting, Craig had articulated one of the important open questions: What variations, if any, might wild fish undergo when these live year round in a marine environment? The team had discussed, from early on, the "ideal situation" of getting marine fish, which was not an easy feat because it required access to a scientific marine vessel or to other sources of coho salmon that lived in a marine environment throughout the year.

On April 22, 2002, Shelby talked about the status of the research while he and I were in the wet laboratory conducting measurements on retinal cells. The videotape shows him drawing a graph to represent the data from the Robertson Creek

[8] It was also Chapter 2 in Shelby's dissertation, which, in what sometimes is referred to as the "natural sciences model," consisted of chapters that were also jointly published papers plus an introduction and a conclusion chapter written by the candidate.

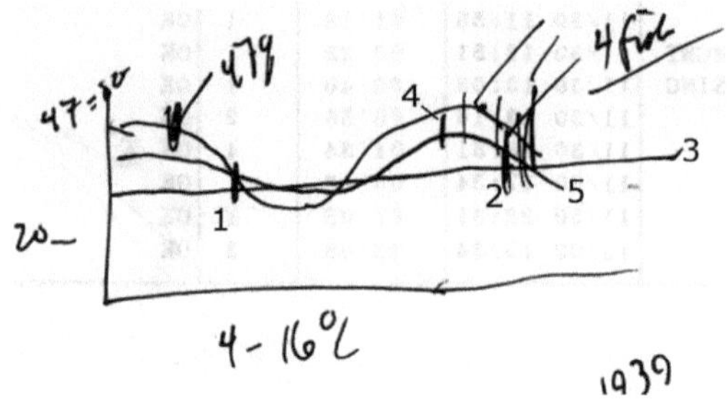

Fig. 7.12 On a piece of scrap paper, Shelby makes notes while explaining three possible scenarios that could arise from the data if coho salmon caught in the ocean were to be added to what he has at that point in time (© Wolff-Michael Roth, used with permission)

hatchery-related coho, with porphyropsin variations from about 20 to 47 % and temperature variations in the ponds from 4 to 16 °C (Fig. 7.10). He indicates the place in the graph where the marine fish from the previous summer would be found (Fig. 7.12: #1), that is, right on the curve that those fishes could be found that were held in the tanks and those fishes that had been captured in the estuary. He suggests – thereby reifying the a posteriori reconstruction of events that historians of science have described (e.g. Kuhn 1970) – having had the suspicion all along that the marine fish are following the same curve. This led him to measure the marine fish again, and the measurements were a little below the main curve (Fig. 7.10: #2). He states that it could mean that the fish do not change at all, a possibility he expresses by drawing a straight line (Fig. 7.12: #3). It could also be that the fishes caught in the marine environment increase their porphyropsin levels so that in the wintertime, Shelby would be measuring a value below the curve (Fig. 7.12: #4). As a result, our research team would have some indications that the marine fishes are following a circannual pattern with lower amplitude (Fig. 7.12: #5). A third possibility would be that the data leading to the measurement at point #2 might not be reliable – because there were only four specimens from which to generate the data – and that the point really should have been on the main curve. We see here, therefore, the new forms of explaining and describing investigations that have become possible with the transition from the canon to the new conceptual language.

Shelby states the hypothesis that there are multiple, interacting factors including temperature, photoperiod, photonic environment, and hormones. To tease these factors apart, Shelby suggests an experimental approach where each of these factors would be manipulated individually to see how it affects the changes in the porphyropsin levels. Together with another graduate student, Sam (who did his Masters degree on the project), he had already conducted relevant experiments to test the effect of hormones on changes in the visual pigments. A year later (May 23, 2003), during a conversation in the wet laboratory, he talks about how he found out that there may indeed be a "depressed cycle" (i.e. a cycle with lower amplitudes of A_2). And while working on this problem, a paper by a Japanese research group came to stand out. It provided our team with another "little piece that says okay there's an annual cycle that's going on whether they're in."

We presented the results of our research so far at the annual meeting of the *Association for Research in Vision and Ophthalmology* (ARVO), which took place

May 4–10, 2002. The poster findings included two graphs with associated descriptions and claims (Fig. 7.13). Related to the Robertson Creek fish, the graph does not include all data that the team had collected but contains the data from smolts at the point of release, smolts from the estuary, smolts from freshwater tanks at the university, and to temperature, then the Kispiox hatchery graph should have been higher than the Robertson Creek graph and closer to the Kispiox wild coho.[9] He concludes: "So there may be something else at play here." However the team does not pursue parr in thehatchery (Fig. 7.13). Although our team had already conceptualized its work as refuting the "dogma," it introduces its results by referencing three other studies that had found annual cycles (claim 5a, Fig. 7.13). The trend line in the graph is a sine curve and claim 5b states that the shift in chromophores is significantly correlated with temperature but not day length. Claim 5c (Fig. 7.13) includes additional information not contained in the graph about smolt from saltwater tanks and wild fish. These wild fishes are described to have "similar rhodopsin-porphyropsin ratios" as all the other fishes, suggesting that the hatchery-sourced fishes "closely matched the natural environment" (point 5d on the ARVO poster, Fig. 7.13). In contrast, the wild fishes from the Kispiox River environment differed from those raised within the hatchery, which exhibited an annual cycle with a "decreased amplitude."

The new conceptual language is so strong by now that some evidence concerning differences in the temperature between hatcheries and wild fishes are noted in the Kispiox case and overlooked (omitted) in the Robertson Creek situation. While talking about a side-by-side presentation of the same two graphs that also appeared on the ARVO poster, Shelby describes his confusion on June 19: If porphyropsin levels are related to temperature, then we should porphyropsin levels to be higher when the temperature is lower. In fact, the Kispiox wild fish disappeared from the final representations; and so did the data from the fish that had been kept in the laboratory. A graph similar to the one that appeared in the published paper was shown during a presentation to the hatchery personnel (Fig. 7.14). In this presentation, the temperature dependence is maintained even though the data from the Kispiox suggest that there are cyclical variations when the temperature is kept constant (Assertion 6, p. 213). The comparison between the two hatcheries may be used to hypothesize a possible dependence of the porphyropsin variation amplitude on the temperature (range). This would be consistent with the very high porphyropsin levels in wild coho from the Kispiox region, which live in water that is near 0 °C during five months of the year.

The story seemed to come to its end, as there appeared to be consensus about the variations of porphyropsin levels in coho salmon over the course of a year. However, six months later, Shelby still expressed uncertainty about the general nature of the pattern that nevertheless had already been reported during conferences

[9] There is empty space in the legend, which, during other presentations at the time, are filled. It is not clear whether the data, in the PowerPoint slide copied, had been forgotten in the display or had been left out on purpose.

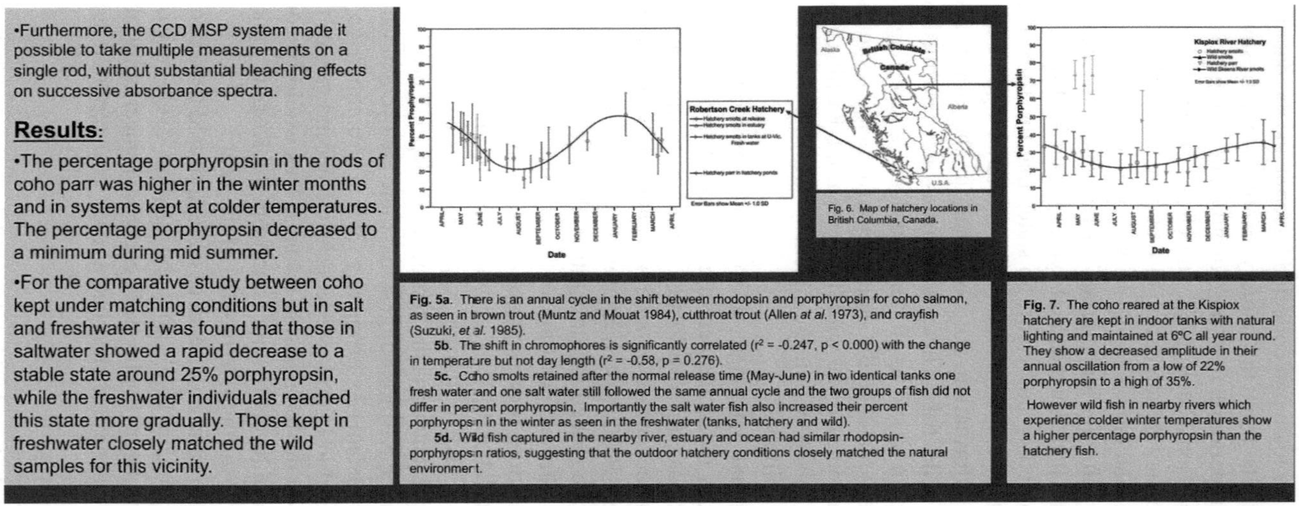

Fig. 7.13 Excerpt of the poster presented at the Association for Research in Vision and Ophthalmology (2002) (© Wolff-Michael Roth, used with permission)

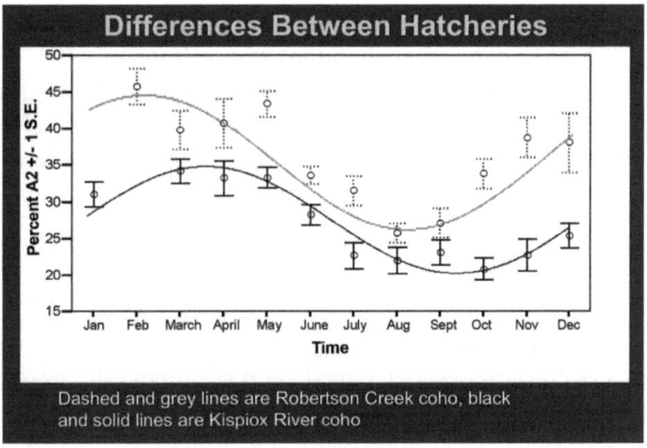

Fig. 7.14 This PowerPoint slide is part of the presentations that Shelby gives during 2004 to several audiences. It is similar to the graph ultimately published in a scientific journal (© Wolff-Michael Roth, used with permission)

and talks to various audiences. During an interview on November 8, 2002, Shelby says:

> So the other the other major thing that we have kind of what I've kind of stumbled on is the interpretation of my data to this point um until now I've been kind of going with, "Ah well *and I'm still I'm still not sure which one is which,*" but um with the idea that there's an annual cycle and that that annual cycle occurs in both the parr and the smolts. Up till now my conclusion was that if both parr and smolt go through this annual cycle, so, in other words, this would be a parr it hatches here ((points to a curve)) it decreases in it's first summer, increases again in the winter, and then decreases as it prepares to be released in the spring. And as I've already shown, if they're kept in the freshwater, they continue to decrease and then increase again in the next winter. Um, and this has been seen by numerous people.

Although the evidence is strong in favor of seasonal variation at the time, the one kind of data our research team has not had access to was from coho salmon during their life in the ocean. If the annual cycle interpretation were to be correct, then season-related cycles should be found in these salmon as well, and, thereby, contradict the Nobel Prize winning theory that Wald had established. On May 3, 2003, we find Shelby talking about how some of the data that he had found in the literature confirmed the pattern that was now accepted within the research team:

> These Japanese authors had collected salmon at sea, were measuring the absorption spectra of these salmon and they happened to find, like I looked at it and they said they found porphyropsin in the salmon at sea in chum, I think it was chum [salmon] and I was like, "Oh cool but they won't tell me the dates, that's the important, I need to know the dates." And sure enough, I went through and there were the dates, October 31 they found porphyropsin in their chum and right away I was ((*snaps fingers*)) "Click," the light went on. I said, "that's awesome because that falls exactly in the little window that I've predicted." So at this point without even collecting anymore data I'm gonna step out on a limb and say "I believe that ocean coho, ocean salmon go through an annual cycle but that it's depressed, that they still increase their porphyropsin probably between September and January."

During the same interview, Shelby also notes that the Japanese scientists had investigated salmon of approximately the same age as the coho salmon our team was researching. But the Japanese studied chum salmon, which went out in the ocean one year earlier than the coho salmon.

At that time, our research team also had some contradictory data that did not fit the temporal pattern. However, because we had an alternative explanation of why our data from coho salmon did not fall into the pattern Shelby was in a position to report:

> When I went out on the Richer [research vessel] on February of two years ago and went looking for fish they were five O three they were at where I didn't think they would be in the wintertime that's because I missed it I needed to go out in the fall.

Here, the coho salmon had λ_{max} levels of 503 nm, which is equivalent to saying that the porphyropsin levels were 0 % and the rhodopsin levels were 100 %. He did however suggest that our team would try to get further data should the paper that he was working on was not yet finished, which was to feature in his dissertation and become a journal article, then he would attempt to get again onto the research vessel to capture coho salmon in the open ocean during the October to December period.

There appears to be a contradiction in the published paper (Assertion 6, p. 213): our team presents an interpretation according to which the variations with temperature is privileged, even though the hatchery where the salmon are kept in water with constant temperature also shows variations and even though the amount of variance explained (R^2) is higher for day length than for temperature (see also Chap. 6). That is, in this journal article we report that there is a strong correlation between temperature and porphyropsin levels for the Robertson Creek hatchery even though the temperatures do not vary at the Kispiox hatchery. In the discussion section of the publication, our team admits that there is too little known for a full model of the phenomenon. But the article suggests that the smaller activation energies in porphyropsin would decrease the signal to noise ratio in warmer water temperatures. The shift to the thermally more stable rhodopsin chromophore would therefore constitute an advantage in the summer when the ambient water temperatures are higher.

Toward a Biologically Relevant Explanation

In this chapter, I provide a microgenetic historical account of a "radical" change as it unfolded in one scientific laboratory concerning the model of one part of the life history of coho salmon and the changes in the chemical composition of their photoreceptor pigments. The change is between two forms of conceptual language used to describe and explain the levels of porphyropsin in the photoreceptors of coho salmon. Although we may recognize elements of the classical description of conceptual change, the process observed here was not as linear as these traditional descriptions might lead some to think. Most importantly, the data that our scientific team collected were not discrepant in themselves; it is only once the (initial) data were seen differently than through the lens of the present canon that they become the drivers of the search toward alternate explanations. In this and the subsequent sections, I discuss three issues that complicate the classical conceptual change description for the conceptual change in the sciences. This complication may well be an outcome of the fact that Chi (1992) studies conceptual change from an after-the-fact perspective more typical to historians, whereas my study followed scientists in their day-to-day operations where the salient categorizations are not so clear. Although there are other learning sciences oriented studies that apparently followed scientists (e.g. Dunbar 1999), it turns out that these (a) did not follow scientists over time and microgenetic-historically as I have done here and (b) begin their work with presuppositions about the scientists rather than looking at the ways in which they are actually dealing with data and the inscriptions that are produced with and from them.

Scientists do not just "interpret" data, or describe the trend, for what the trend is also is a function of the underlying biology (Assertion 1, p. 213). Thus, there is a strong correlation between photoperiod and the data but shifted by a period of time. It is not satisfactory in itself, however, to be able to report the existence of a high correlation. This is reflected in the draft version of the published paper dated May 18, 2004: "With a lag of 7 months the correlation coefficient reaches a positive peak of 0.966, however, this second peak *is not biologically relevant* as the response time for chromophore shifting relative to temperature is on the order of days-weeks not months (Allen and McFarland 1973)" (emphasis added). What our team needed to know were possible *biological* factors that would produce such a shift. The issue was not just to model the data as a mathematical relation: it is modeling the *biology* of the fish *in terms of* the mathematical relations that can be demonstrated. In terms of the chain of translations, the claims and biology are described and articulated in language (far right of Fig. 1.1), whereas the supportive graphs are further towards the left and in the direction of the natural phenomenon. In a causal account of conceptual change that is possible only after the change process is completed (e.g. Romano 1998), this might be referred to as the reason why our team took so much effort in reconstructing the context from which the fish has been taken. This is so because the mathematical relation is supposed to be a reflection of the biology of the fish and its interaction with the environment. As shown in Chap. 6, a purely mathematical relation between porphyropsin levels and temperature or time of year is useless to biologists. Because biologists are interested in *biological* and *biologically relevant* causes they need to be familiar with and bring into the explanation the relevant ecological systems. In the interpretive section of its published report, our team writes about "seasonal effects" rather than unambiguously attributing the changes in porphyropsin to temperature. The data that the scientists published in the end could have been modeled, as shown in Chap. 6, by a mathematical function of the type

$$A(T,t) = A_1(T)\cos(\omega t + l(T)) + A_0(T) \tag{7.1}$$

where T = temperature, t = time of year in months, $A_1(T)$ = amplitude of cycle, $\omega = 2\pi/12$, $l(T)$ = lag time, and $A_0(T)$ = vertical offset. But a lot of our team's effort consisted in re-contextualizing the data so that we could produce biologically plausible explanations. This contextualization of the data and familiarity with the material dimensions of the experiment and the ecological setting of the data does not appear as such in existing models-of-data theories (e.g. Chinn and Brewer 2010).

The mathematical function (7.1) is not really of interest to our team, for biologically, the effects of temperature are on the order of days or a few of weeks rather than months. But without a keen background familiarity with the biological conditions, biologists therefore can say very little. Not surprisingly, our team initially spent a lot of time rebuilding and reconstituting the context that we had lost in the process of generating the data points. Rather than the construction of a "deep-featured" situation at an intermediate level of abstraction" (Roschelle 1992, p. 237), our team had to find out the concrete particulars of each individual situation from which the fish derived. Without a deep familiarity with the environmental context of the locations where the fishes were sourced, we had little to go by and found themselves confronted with difficulties in evaluating whether there was or was not an effect of temperature. Thus, the hatcheries modified the temperature of the water that they drew from nearby sources, some for the entire year (Kispiox), others for part of the year (Robertson Creek); the wild fishes underwent changes whatever the temperatures, which differed in range and absolute value for the river systems and ocean involved. In the end, our team reported an effect of temperature *even though the annual cycle was observed when the temperature was constant!* There existed evidence within our team that the effect was not temperature at all but something other related to seasonal changes, such as the photoperiod (length or amount of day light). Thus, Shelby had tested the temperature theory on another species of fish. In his case, he did an experiment with zebra fishes, which is in the same class Actinopterygii but from a different biological order (i.e., Cypriniforms) of coho salmon (i.e., Salmoniformes). Shelby stated on January 22, 2003:

> And then the next part of that is applying that to fish beyond salmon, you know I have a gut feeling that it might apply to other fish. Well recently I tested that idea and I thought that maybe temperature was maybe the be-all-end-all. *And so I tested it on zebra fish and its not!* It was a major slap in the face. It didn't work. Um. But what's interesting is that I have another gut feeling and that is well maybe it's not temperature per se but it might be correlated with temperature and maybe it's a seasonal thing for them as well. And so in fact when I look back at the data set that we collected umm back in June 25 of last year, we collected some information on zebrafish, and we got a higher percent porphyropsin than we got in January which right away made me go, "Well, hang on a second one is winter one is summer. Maybe I should try again this June and see what I get."

This hypothesis that "temperature was maybe the be-all-end-all" did not bear out. But this hypothesis was not pursued as an alternative in this research, which might have otherwise turned the team towards using a mathematical model as presented (Eq. 7.1).

The point therefore was not to merely to "interpret the data." Our team could have done a (mathematical) description of the pattern long before we actually came to the conclusion that the results were publishable. Our team had to come to grips with the biological implications of the study; and this required being familiar with and providing descriptions of the biological context of the specimen that we had analyzed or subjected to various experimental conditions. This was particularly the case because the canon had related the changes in porphyropsin to physiological changes in the euryhaline fishes. That is, in the canonical conception there was also an explanation for why there ought to be changes in the retinal composition: physiological changes and the life history strategies, whereby changes in porphyropsin levels and associated changes in the color spectra may have provided evolutionary advantages. Initially, our team did not have such an explanation. The emerging photoperiod and temperature explanation constituted a radical shift in the explanation, one of which was based on migration the other on seasonal variations. As a result, the change in our conceptual language took a considerable amount of time, involving a slow and incremental reorientation toward the biology at hand.

Confirmation Bias, Trends and Extrapolation

Extrapolation from existing data is one of those competencies that tend to be attributed to scientists. In the present instance, we see quite clearly that the issue about extrapolation is more complicated. Precisely when a scientist knows the theory or the biological context, the extrapolation may be inconsistent with the trends that a posteriori are clearly visible in the data. In an earlier meeting, Craig gestured what he expected the curve to look like such that it also fits with the present data (Episodes 7.3a and 7.3b). Later, our team exhibited the relationship within the data in a very different manner, by means of a sinusoidal relationship. That is, the very same data came to be modeled very differently; and Shelby exhibited these differences during one presentation when he exhibited the team's data in the context of the previous study (Alexander et al. 1994) and in terms of what the "dogma" suggested (Fig. 7.15). Craig's gestures (Episodes 7.3a and 7.3b) are consistent with the dogma, which provides a much poorer fit with the first 8 data points than the nearly linear portion of the sine function.

Interesting here is the extrapolation presented in the gesture versus the reasonable extrapolation that the data themselves would have suggested, which, at best, is a linear function with a much-reduced slope. The change in conceptual language conceptual involved, among others, precisely exhibits this changeover from seeing the data through the existing dogma rather than through the new relationship suggested by the new data themselves. This conceptual change is associated with and requires a fundamental change in the way in which biologists model the physiological changes associated with migration and the causes underlying the changes in the retina. These are no longer considered to be due to the *organism-centered*, physiological changes in anticipation of changes in salinity – as suggested by the correlation in the graph with plasma sodium levels – but with very different, *environmentally driven* factors, such as responses to temperature, daylight, and other factors in the life history strategies of the fish.

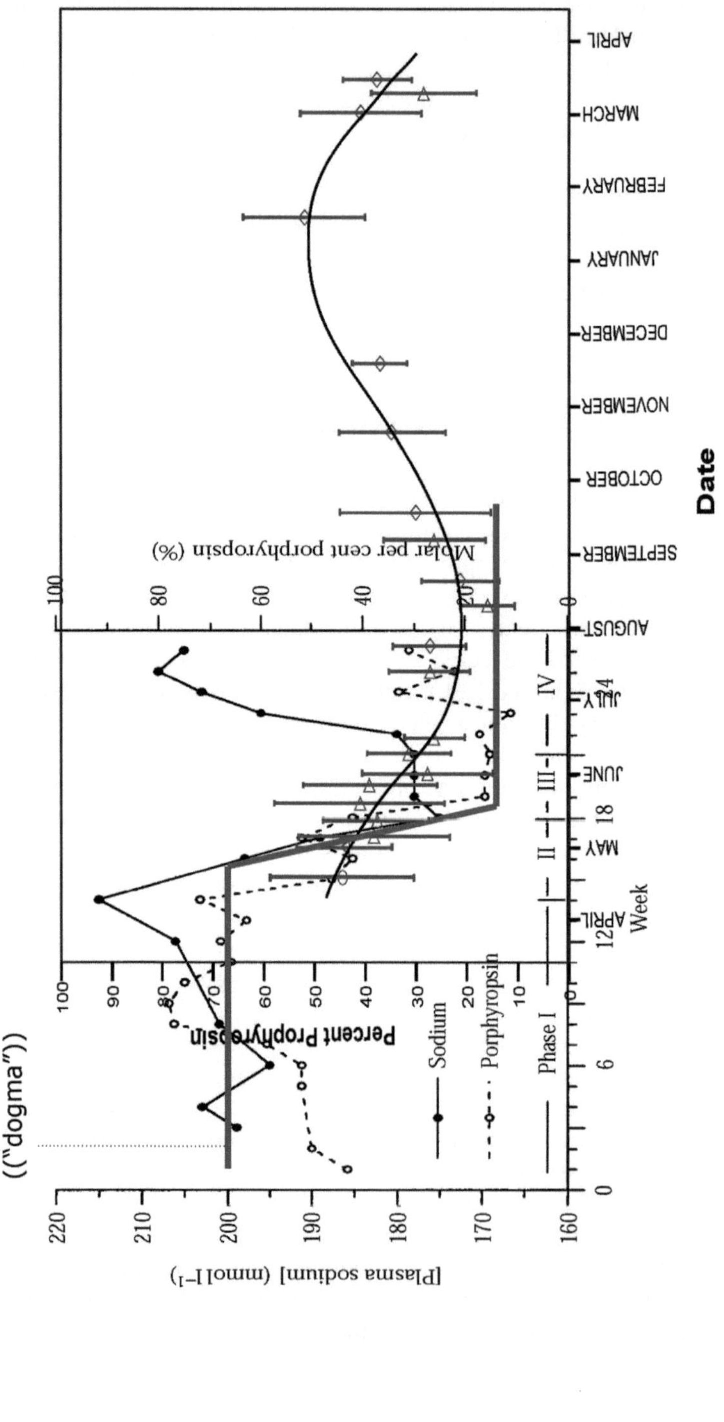

Fig. 7.15 In this PowerPoint slide, Shelby presents the "dogma" (step function, *red*) the new conception (sinusoidal curve), and the data from this and the Alexander et al. (1994) studies (© Wolff-Michael Roth, used with permission) (color figure online)

In these observations from my microgenetic historical study of scientific discovery work, we see that trends and extrapolations do not occur in some way that will be consistent with whatever is subsequently known. Rather, just as with perception, what scientists project from their data is a function of their current (disciplinary) view of data and how these fit in the context of the bigger picture, which is what the extrapolation aims at showing. In the psychological literature, the term *confirmation bias* is used when people seek or are oriented towards confirming their current theories and beliefs rather than seeking information to disconfirm those. A classical example of confirmation bias is the use of horoscopes and clairvoyant statements, where subjects see in whatever happens a confirmation of some previous articulation of the future. Research shows, however, confirmation bias also is a ubiquitous phenomenon in the sciences (Nickerson 1998). This should not come as a surprise to close readers of classical historical analyses and theories of scientific revolutions (Kuhn 1970). Interestingly enough, Shelby, soon after the team's interpretation of the data had changed, expressed a position consistent with the disconfirmation approach to the sciences. On January 22, 2003 he stated:

> I start everything with a feeling of skepticism (laughter) unfortunate to say! Um, I've always been a real critic, um about things and I always question them. Um and it, so and maybe that predetermined how, what results I was going to find? I've always often thought that? I always feel like I always go into the negative? Not believing what people say? I'm always trying to prove them wrong? But what's interesting is that that's how science works: the only proof is disproof. You cannot prove that something is, you can only prove that it isn't the way it is, right?

The concept of confirmation bias, however, may be much too strong and misleading if we look at the microevolution of events in a scientific laboratory. At first, our scientific research team did not have any reason to doubt that the data they were collecting would not be consistent with the scientific canon. Thus, whatever existed in early data not only could be seen as confirming existing theory and research but also allowed an extrapolation, which is actually a derivation from Wald's theory. It is true that there was some research that had shown circannual behavior in the amount of porphyropsin levels. But this work was not salient in and to our team, even though some team members were citing the work in other studies they were publishing at the time. As Shelby was saying, he did have a copy of the Alexander thesis, from which the Alexander et al. (1994) was produced, and he also had read other studies in which the paradigmatic step function was not confirmed. However, at the time of our research, the contents of those other works did not stand out to constitute a different basis for looking at the new data our team was collecting. The same is the case for Craig, who had spent much of his scholarly career on investigating visual pigments. But what had stood out for him, ever since we started working together, were the Alexander et al. publication and the work of U.S. scholars intending to use the stepwise change in porphyropsin levels to find out the optimal time for the release of salmon species from fish hatcheries. Our paper draft from May 18, 2004 did in fact exhibit awareness that there were two main categories of "hypotheses or models" — "those that correlate the shift in A_1/A_2

chromophore ratio with some type of migration or metamorphosis event, and those that find correlations with variables that are associated with seasonal changes."

In his own reflections, Shelby certainly was aware – at least after the fact and with hindsight – of the way in which existing ways of relating to an experiencing the familiar world shape what scientists see and how they see it. Thus, for example, during an interview on January 22, 2003, he noted:

> And it really goes that way! You you know you set out with an idea that you think might be right, and you have some forecasts as to how it might be applicable in the long run, and you start off to have a path. And as you go down the path, things don't work, some things do work, and some things just come out of the blue, and you don't know why or how. And it's usually things that are unpredictable, that didn't work, that ends up being the most interesting, so whereas you know, you start off with one idea, within a very short term you have a whole bunch of other ideas and those become more interesting than the original one and maybe even have more interesting applications.

Here, Shelby highlights setting out with an idea possibly or likely going to be "right," which shapes the path along which the research unfolded. He points out not only the contradictions that emerge but also the fact that "some things do work." From the shop-floor perspective of the actual research-in-progress, it is not apparent whether what does not work is due to the data or due to some artifact. These contradictions are unexpected within the long-range forecasts, and, therefore, "come out of the blue." Shelby also notes that "a whole bunch of other ideas" arise "within a short term," though the starting point for such an emergence was not predicted within our team, and would not be predicted within the catastrophe theoretic model introduced at the beginning of the chapter (Fig. 7.2). In part, Shelby attributes the emergence of new ideas to the evolution within a research team, especially with the arrival of new members who had their previous training in other laboratories. During the same interview on January 22, 2003 he stated:

> I would say that there is no question that it evolves and that this lab has certainly seen evolution, partly from these new things that you find, and partly from new grad students coming in. As new grad students come into the lab they have come from different places where they did their masters and undergrad and they have interesting ideas about how they think the world works and they try to apply that and we all have biases about the things in the work that we do. And so you're undoubtedly going to apply those biases to your everyday studies. They maybe right, they maybe wrong it doesn't matter it, pulls the research in different directions.

In this quotation, Shelby not only recognizes that the research team has looked at the data through the lens of things that were salient to the team at the time, its "biases," but also that the results that were consistent with a different paradigm could in turn constitute a bias in a part of the study that he was to add to make it substantial enough to warrant publication. Thus, given that there already existed studies that had reported cyclical patterns in porphyropsin levels, Shelby intended to extend our planned research and show that the new pattern was the same for coho salmon of a different age class. Thus, one of the two things he wanted the team to analyze over a period of time were the porphyropsin levels in retina from coho parr, a environmentally determined variable stage that lasted at Robertson Creek about 14 months.

> I already know what I'm going to find for one part of it, so I need to show that it's true. Because people have looked at this before and seen similar trends and I want to show that, yes it can work . . .. And I suspect that these other parts [of the system] are doing the same. So in other words [other studies] knew that smolts shifted pigments, I want to know do parr do that? And do they continue to do that once they go to the saltwater? Now, let me say it, you know when I say I know they're going to do that, well I'm making an assumption, based on my own personal bias: I know they are going to do that. I really hope that they do, ha ha, and it looks like they do so far, so that's good news.

Just as he outlines this research, which was to complement research with more-difficult-to-get-at coho salmon caught out in the open ocean, he says, with substantial self-awareness and not without irony, that he makes assumptions about the new data to fit with the new paradigm "based on [his] personal bias." Without adding a qualifier, he then states that at that point, the parr appear to be exhibiting porphyropsin levels consistent with the trend and extrapolation from the data that the research team has collected up to that date.

References

Alexander, G., Sweeting, R., & McKeown, B. (1994). The shift in visual pigment dominance in the retinae of juvenile coho salmon (*Oncorhynchus kisutch*): An indicator of smolt status. *Journal of Experimental Biology, 195*, 185–197.

Allen, D. M., & McFarland, W. N. (1973). The effect of temperature on rhodopsin-porphyropsin ratios in a fish. *Vision Research, 13*, 1303–1309.

Chi, M. T. H. (1992). Conceptual change within and across ontological categories: Examples from learning and discovery in science. In R. Giere (Ed.), *Cognitive models of science: Minnesota studies in the philosophy of science* (pp. 129–186). Minneapolis: University of Minnesota Press.

Chin, C. A., & Brewer, W. F. (2010). Models of data: A theory of how people evaluate data. *Cognition and Instruction, 19*, 323–393.

Couzin-Frankel, J. (2013). The power of negative thinking. *Science, 342*, 68–69.

Dunbar, K. (1999). How scientists build models: In vivo science as a window on the scientific mind. In L. Magnani, N. Nercessian, & P. Thagard (Eds.), *Model-based reasoning in scientific inquiry* (pp. 85–99). New York: Kluwer Academic/Plenum Publishers.

Holzkamp, K. (1983a). *Grundlegung der Psychologie* [Foundations of psychology]. Frankfurt/M: Campus.

Holzkamp, K. (1983b). *Lernen: Subjektwissenschaftliche Grundlegung* [Learning: Foundation in subject-centered science]. Frankfurt/M: Campus.

Kuhn, T. S. (1970). *The structure of scientific revolutions* (2nd ed.). Chicago: University of Chicago Press.

Nickerson, R. S. (1998). Confirmation bias: A ubiquitous phenomenon in many guises. *Review of General Psychology, 2*, 175–220.

Quine, W. V. (1995). *From stimulus to science*. Cambridge, MA: Harvard University Press.

Romano, C. (1998). *L'événement et le monde* [Event and world]. Paris: Presses Universitaires de France.

Rorty, R. (1979). *Philosophy and the mirror of nature*. Princeton: Princeton University Press.

Roschelle, J. (1992). Learning by collaborating: Convergent conceptual change. *Journal of the Learning Sciences, 2*, 235–276.

Roth, W.-M. (2008). The nature of scientific conceptions: A discursive psychological perspective. *Educational Research Review, 3*, 30–50.

Roth, W.-M. (2009). Cultural-historical activity theory: Toward a social psychology from first principles. *History and Philosophy of Psychology Bulletin, 21*(1), 8–22.

Roth, W.-M., & Duit, R. (2003). Emergence, flexibility, and stabilization of language in a physics classroom. *Journal for Research in Science Teaching, 40*, 869–897.

Sève, L. (2005). *Émergence, complexité et dialectique* [Emergence, complexity and dialectics]. Paris: Odile Jacob.

Suzuki, D. (1989). *Inventing the future: Reflections on science, technology, and nature*. Toronto: Stoddart.

Temple, S. E., Plate, E. M., Ramsden, S., Haimberger, T. J., Roth, W.-M., & Hawryshyn, C. W. (2006). Seasonal cycle in vitamin A1/A2-based visual pigment composition during the life history of coho salmon (*Oncorhynchus kisutch*). *Journal of Comparative Physiology A: Sensory, Neural, and Behavioral Physiology, 192*, 301–313.

Wald, G. (1957). The metamorphosis of visual systems in sea lamprey. *General Physiology, 40*, 901–914.

Chapter 8
Some Lessons from Discovery Science

> A *nonteleological view of intellectual history*, including the history of science, does for the theory of culture what the Mendelian, mechanistic account of natural selection did for evolutionary theory. Mendel let us see mind as something which just happened rather than as something which was the point of the whole process . . . "our language" – that is, of the science and culture of twentieth-century Europe – . . . and our culture are as much a contingency, as much a result of thousands of small mutations finding niches (and millions of others finding no niches), as are the orchids and the anthropoids. (Rorty 1989, p. 16, emphasis added)

In this Part II of the book, I investigate how scientists deal with uncertainties graphs and graphing in scientific discovery work, where ongoing graph readings cannot be checked against a normative one. I present a *nonteleological* view of the kind that Rorty, in the introductory quotation likens to the Mendelian reframing of evolutionary theory. Moreover, I describe the entire scientific process as something in flow, a dynamics, rather than as rational mental activity checked by another rational activity that runs as a control instance (i.e. "metacognition"). Whether the data in the scientists' hands exhibit a phenomenon or whether what is seen constitutes noise is an outcome of the members' embodied, irreducibly *joint*-actional work. Joint here is used to mark actions that cannot be reduced to the intention of the individual but are collective such that individuals find themselves contributing to these but the outcomes are always in excess of what they want, intend, or plan to do. Their living work, on the one hand, and the objects and accounts produced, on the other hand, are taken in my description to be an irreducible pair. That is, graphs do not (always) *stand for* something else but are objects of inquiry in their own right precisely because the persons working with and using them do not see what these graphs relate to. In the introductory quotation, Rorty explicates that we need to approach science, a specifically and culture generally in the way we approach nature and evolution. Science, as its language and product, is in a constant evolution[1] so that we do not know in any determinate sense what science *is* generally and what science *is* right now. This is why science, as a form of experience, is internally

[1] Saying that a process is evolutionary is equivalent to saying that it is nonteleological.

W.-M. Roth, *Uncertainty and Graphing in Discovery Work*,
DOI 10.1007/978-94-007-7009-6_8,

antinomic when taken as an irreducible whole rather than as an assembly of disjointed artifacts and actions (Il'enkov 1977). Science, not only as product but also as process, is a contingent becoming that can be rationalized and explicated in terms of its outcomes only after the fact and with hindsight, after everything has been said and done. This is so, as we see from the preceding chapters, because scientists themselves do not know with absolute certainty that what they intend to do is what they actually do; and when they do not know what they are doing, scientists are as much in the dark groping about as people do in other walks of life. Moreover, they start a project to test one hypothesis and, when their data do not bear out the feasibility of this project, they still get their work published but by "pitching it" differently. Even though scientists' own accounts, those that they provide in the journal articles that they produce, are teleological, written such that the ultimate findings are said to be intended. The story comes to be turned on its head, when scientists re-write history by choosing causes – with hindsight that is to say – to explain why they engaged in *this* study. But it has been suggested a century and a half ago that cause (intentions and agency) and effect can be determined only once a period and section of flow is stopped and, therefore, no longer flows (Nietzsche 1954). To paraphrase Nietzsche in the present context, rather than saying that scientists produce scientific facts, we ought to make statements such as "The sciences science." But I treat this question of modeling science as flow in the next part of the book generally and in Chap. 9 specifically. In the present chapter, I reflect on (some of) the lessons we may take from observing scientists when we do not allow received preconceptions to operate in our stories about what scientists do – which tend to be grand narratives of scientific heroes. Because discovery happens to an individual, those who become the heroes of science do so in the course of events that they are never in complete control of.

The dominant portrayals of science are written and told from the perspective of the Monday morning quarterback – always with hindsight. Scientists and those interested in what they do produce narratives where intentions lead to specific effects so that discoveries become nothing but effluents of the individual scientists' minds. But in Part II of this book, we see a lot of evidence of scientists in situations of uncertainty and without knowing what really is going on. They find themselves with malfunctioning equipment one morning even though the night before everything worked; and we find them without any clue about the underlying reason. To get themselves out of the situation, they tinker on in the dark until something makes their equipment work again. Then I show scientists at work collecting data with the conviction that these are consistent with some scientific canon only to find out later that their findings contradict the canonical description and, with it, the causal explanations that traditionally have been provided for it. Concerning descriptions of the course of history in the sciences, social constructivism, often mobilized in alternative accounts of science, has precisely the same problem, which is a problem of all forms of constructivism: the ontologizing of causes and intentions that are merely reified in actions and effects. In this section, I provide brief reflections on (a) the nature of scientific research process in terms of induction versus abduction, (b) what we need to consider in working towards a more appropriate description and theory of conceptual change, (c) on the salience of "facts" (that lead scientists to reconsider their conceptual talk), (d) the role and production of uncertainty in the course of scientific research, (e) the role and appearance of inner and logical contradictions in the sciences, and (f) the levels of control over the becoming or not becoming of data. I then move to reflect upon the use of graphs in discovery sciences and on communication at work.

Induction Versus Abduction

Science is concerned with producing concepts and theories that are valid independent of particular (geographical) locations and situations (specificity of the laboratory or other location where the phenomenon occurs).[2] This inherently requires active disattention to *particulars* and, as I quote Kant to have state in Chap. 5, calls for negative attention. As a result, scientists end up with data points, graphs, and means. The foregoing analyses show that scientists do not move from their specific observations (measurement) to the coho population in general and to a universal law – e.g. in terms of inherently mathematical properties. This constructive process would be referred to as *induction* (Fig. 8.1a). If scientific interpretation were to move in this manner, then they could derive a general law from their data without further ado. No additional information would be needed. The law would "jump out of" the page and "into their eyes." In Part II of this book, the evidence provided shows that science is not a purely inductive endeavor. Nevertheless, school science and mathematics courses are conducted as if it were. Students are provided with hands-on tasks designed for them to "construct" knowledge together with others and individually, as if in these "constructions" they could induce the current scientific canon on the topic under investigation. There is a recognition in the STEM field that students do "construct" "knowledge" based on what they know all the while there are expectations that students construct the scientifically correct conceptions. The present study shows that scientists appear to require a deep familiarity with both the generalization and the concrete situation to which it pertains. The process of the discovery or constitution of facts, therefore, is a different one. There is a slow, painstaking process at work in which scientists become familiar with their phenomenon and all the steps in translating from the phenomenon into the data and graphs, which they use to support their claims.

[2] Laboratories are places of abstraction, where the contingencies of the world are excluded and controlled. Should scientists actually venture into the field to conduct research equivalent to the one they conduct in the laboratory, they may transform the field setting into something like a laboratory (Latour 1988). The effect of the control over the conditions became quite evident in a study of mathematics where persons were observed in the supermarket while shopping, then were presented with items from the supermarket but on a table outside and apart from the normal shopping for groceries, and a third condition where "problems" were presented in pencil-and-paper format (Lave 1988).

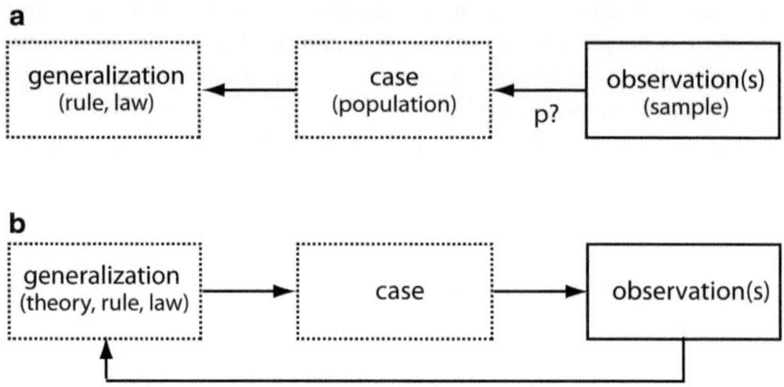

Fig. 8.1 (**a**) The logic of inductive reasoning and research. (**b**) The logic of abductive reasoning and research (© Wolff-Michael Roth, used with permission)

Students, on the other hand, generally are provided with little time to deeply delve into an investigation and phenomenon, preventing them from ever becoming familiar in the same way as the present scientists became familiar with the research process, environmental origin of the fishes, and the specimens themselves.[3]

In the process of data interpretation that I describe in Chap. 5, for example, the relationship between explanans, "sentences describing the phenomenon to be explained," and explanandum, the "class of those sentences which are adduced to account for the phenomenon" (Hempel and Oppenheim 1948, p. 137) comes to be reversed. Scientists use what is to be explained – the description of the difference between the porphyropsin in coho – to look for differences that would explain differential porphyropsin levels. Thus, scientists seek to produce descriptions of the differences between situations and the coho that derive from them. It is only in the ultimate write-up that the logic of induction comes to be observed, whereby statements of antecedent conditions and general laws derived from the observations are used to explain the phenomenon discovered. Thus, although the study began as an investigation of porphyropsin levels during the time of smoltification (physiological metamorphosis) and onset of oceanward migration, it was subsequently pitched as an investigation of porphyropsin levels "at three life history stages throughout the year to differentiate between migration/metamorphosis and seasonal variability hypotheses" (Temple et al. 2006, p. 303). As I show in this Part II, the seasonal variability – expressed in terms of day length and temperature – only arose as a consequence of the work being conducted. Moreover, the investigation at different stages of the life history itself was an emergent aspect of the work.

[3] There are exceptions, such as in Queensland, Australia, where the curriculum provides students with opportunities to research some phenomenon for an entire school term (Roth 2013a) or in my own curricular practice, where students conducted independent investigations for a great fraction of their time in class and throughout the entire year (Roth 1995).

What I show the scientists to actually do differs from the descriptions of others – e.g. cognitive psychologists – even when they claim doing "*in vivo*" studies of scientific reasoning (e.g. Dunbar 1995).[4]

The present study of scientific discovery work shows how an existing graph serves as a lens that frames the new data in terms of their contribution to values and relations already exhibited in the existing graph. The data are literally seen as following a curve even though the steep slope that should have been observed was not. In the course of their efforts, our team mobilized the familiarity of its members with the laboratory and their deep, practical familiarity with the situations from which the fish specimens have been taken. Thus, my personal detailed knowledge of the ways in which the coho salmon were raised – after all, I had personally participated in fertilizing eggs, transferring hatched fry and parr between the different tanks, watched the marking process, and personally spent hours feeding them – contributed to the data discussion sessions because it provided materials for describing and explaining the life history condition of salmon in rivers and hatcheries. That is, in our effort to explain some natural phenomenon, our team mobilized its familiarity with the natural environments rather than inducing knowledge based on the absorption spectra alone. In this study, therefore, becoming familiar with the data was taking more of the form of abduction, defined as the "tentative and hazardous tracing of a system of signification rules which will allow the sign to acquire its meaning" (Eco 1984, p. 40). The process of abduction – where results are explained through the hypothesizing of rules applied to specific cases leading to tentative results that are compared with actual results – has the structure depicted in Fig. 8.1b. This makes it reasonable to hypothesize that knowledgeability related to specific cases and their context supports rather than hinders abduction. In this ethnographic study, therefore, given the decontextualized nature of the data scientists were looking at, team members are shown in the attempt to arrive at an unknown rule by positing possible rules and seeing whether the concrete results are explained by the application of the rule to the specific context. That is, highly competent and successful scientists draw on their familiarity and personal knowledge of cases to learn (about) general laws (rules).

The logic of abductive reasoning (Fig. 8.1b) suggests that coming to know graphs involves a double ascension that operates simultaneously: from abstract to concrete and concrete to abstract. I use the adjective abstract to denote nevertheless inscriptions when these are used to make statements about large numbers of cases, all of which are treated the same, that is, independent of any contingency and specifics. A fish is concrete in the same way as a graph on paper, but it differs from the graph in that it is not used to stand for anything else than itself. The resulting double movement is depicted in Fig. 8.1b as going from rule to result via the case

[4] In Dunbar's case, he interviewed scientists before and after a laboratory meeting and videotaped the meeting. From the difference between the before and after interviews, he produced inferences about how the meeting changed their thinking. As the present data show, scientists' after-the-fact accounts often change objectively given historically documented facts and recorded statements.

and from result to rule. The two movements are manifestations of one and the same process, abduction, much as thinking and speaking are processes that manifest the process of signification. Learning to read a graph is equivalent to becoming familiar with the corresponding context and knowing a context is equivalent to coming to know the corresponding graph. The two mutually implicate each other. As I explain and develop elsewhere, graph interpretation involves a double movement from abstract (generalization) to concrete (observation) and from concrete to abstract (Roth and Hwang 2006a, b). That is, the movements indicated in the inductive process (Fig. 8.1b) are not separate and sequential but are part of one overall movement that co-implicates its parts, like the water in a river that may flow forward and backward (in eddies) in the overall movement from the source to the ocean. But, like in the case of the salmon investigated by our research team, the life history of fishes is only part of an even more encompassing movement of life on earth, the evolution of species, and the continual reproduction and transformation that occurs in and through the specific lives of individual salmon. This description is consistent with a philosophical analysis of the relation of the concrete and abstract in scientific research: "If we single out the phenomenon … and consider it in the abstract, that is, leaving aside all the circumstances that do not flow from its immanent laws, we shall *understand nothing in its motion*" (Il'enkov 1982, p. 104, emphasis added). "Concrete understanding," on the other hand, "coincides with taking into account all those influences exerted upon [the phenomenon] by all the developed and increasingly complicated forms" (p. 105).

The double movement from abstract to concrete and from concrete to abstract constitutes a contradiction. Il'enkov states the conclusion that contradiction is not a subjective phenomenon that "regrettably recurs in thought due to the imperfections of the latter" (p. 234). Instead, from the view of dialectics, contradiction is "the *necessary logical* form of the development of thought, of the transition from ignorance to knowledge, from an abstract reflection of the object in thought to an ever more concrete reflection of it" (p. 234, original emphasis, underline added). That is, contradiction and development and transition are but alternative manifestations of the same phenomenon. Scientists, in doing research, are also changing and developing, for example, in regards to the conceptual language they use to explain the world (see Chap. 7). When there is transition and development, then there inherently is a contradiction; and when there is a contradiction, there inherently is transition or development.

It may appear strange to see a team of highly competent scientists that did not inductively "construct" knowledge beginning with and based on their data. But similar observations have been made some time ago in a sociology research project where team members were tasked with coding outpatient clinic records for the purpose of producing knowledge about treatment criteria. It was noted that in the same way as in the present study, "coders were assuming knowledge of the very organized ways of the clinic that their coding procedures were intended to produce descriptions of" (Garfinkel 1967, p. 20). Moreover, familiarity with the clinic, as with the natural and social settings (hatcheries) in the present study, "*seemed necessary* and *was most deliberately consulted* whenever, for whatever reasons, the coders needed to be satisfied that they had coded 'what had really happened'" (p. 20, emphasis added).

Conceptual Change

In Chap. 7 particularly, I provide a description and an explanatory model of the change in conceptual language that our research team underwent while investigating the phenomenon of changes in the porphyropsin levels in the rod-shaped photoreceptors of coho salmon. The initially dominant conceptual talk was aligned with the description and explanation of the phenomenon that the Nobel Prize winner G. Wald had developed in the course of the 1930s and 1940s. Our team ended with a form of conceptual talk according to which the changes in porphyropsin were not shaped like logistic (sigmoid) curves during the two migration periods, but instead were sine-shaped curves with an annual cyclical pattern independent of life history stage. Traditional conceptual change accounts share a lot with an approach to historiography referred to as *Whig history*. In this approach, the history science is seen through an after-the-fact lens, with 20/20 hindsight, such that progress is the inevitable result of logic and reason towards the same endpoint that Whig history has chosen as its starting point of analysis. In conceptual change, the first marker for such an attitude to the analysis of development comes from the fact that the researchers already know what the endpoint of development should be – either the ultimately reported and accepted scientific fact or theory or, when students are the subject, the correct (i.e., scientific) conception concerning some phenomenon. The new conception already is presupposed to be had prior to choosing it over another conception. In such accounts, facts are taken as seen by scientists, and "correct" reason and knowledge is that currently accepted in the sciences. The Swiss psychologist J. Piaget was typical in this respect, as he assumed, for example, that every child interacting with a balance beam would or should – if it was not deficient in some ways – induce the ratio and proportion scheme to solve problems of equilibration. But children prior to and after some developmental changes do not *perceive the balance beam in the same way* – their perceptual processes change (Roth 1998).[5] Consistent with such an approach are scientists' own descriptions of the events when they look back – such as some of our team members provided them once it was clear that the dogma did not well describe

[5] Merleau-Ponty (1996) – whose descriptions and explanations of perception have been shown to be consistent with modern neuroscience – critiques Piaget for writing that "the passage to a superior perception and conduct can be explained by a more complete and more exact registering of experience, whereas this precisely supposes a *reorganization* of the perceptive field and the *arrival* of clearly articulated forms" (p. 104, emphasis added). That is, the new perceptual forms *arrive* rather than being constructed; and this arrival coincides with a reorganization of the perceptual field. That is, the forms are not the causes for the reorganization.

our measurements. In Piaget's approach to development, all human subjects develop to a stage denoted as *formal reasoning*, that is, the scientific forms of reasoning that Piaget saw as the pinnacle of human development. It is a teleological account of development. But the scientists here do not have a telos. They do not know where they will end up with their conceptual talk. Teleology is not a good basis for telling a story about the changes in the conceptual language of scientists and science.

In contrast to the Piagetian constructivist approach, I follow others (e.g. Garfinkel et al. 1981) who study the discovery sciences from a shop floor perspective, that is, from the historical perspective of the experiencing subjects themselves. This means that analysts place themselves in the shoes of the scientists and follow them along even though they may actually know the ultimate outcome of the discovery work. In this case, analysts of scientists' activities refuse themselves to use as a referent for earlier actions whatever was available and known to the scientists only subsequently. In other word, all knowledge other than that available to and considered by the scientists the time is bracketed for the purpose of the analysis. In this form of analysis, therefore, all actions are seen through the eyes of the scientists themselves, at the moment of the action. This is why this approach is characterized as the shop-floor perspective, where one sees the production in its *incourseness* and through the lens of the participating witnesses rather than from the bird's eye perspective of the Monday morning quarterback. Such an approach to analysis yields results that are better described by the catastrophe theoretical model provided in Chap. 7. This is so because neither the point of bifurcation nor the change over in the dominance from one conception to another can be predicted from the preceding history of the system, here, the participant scientists and our laboratory as a whole.

Consistent with the catastrophe theoretical model, my ethnographic study shows that the change in the conceptual language of the scientists, even when it appears to imply a radical shift in conceptual, theorizing talk about biological phenomena, does not have to be linear or to be experienced as a sudden insight – as might be taken from the discourses about shifts in conceptual hierarchies or restructuring of conceptual frameworks. Rather, the change in conceptual talk is the result of multiple transacting aspects involved – e.g. salience of different pieces of pertinent literature, salience of trends in data, or familiarity with the instrumentation and the setting from which the specimens were culled. My study shows that although each aspect may change in a linear, quantitative fashion, together these lead to qualitative transitions. Such changes where quantitative variations in several variables bring about a qualitative change from one conception to another can be modeled easily drawing on the mathematics of catastrophe theory that is well suited to model morphogenetic phenomena of all kinds and disciplinary fields (Thom 1981).[6]

[6] Thom provides examples from biology, the symbolic realm of the imaginary, philosophy, analogies, and magic in the use of natural peoples. Others use catastrophe theory to explain the transition from peace to war even when the participants involved do not desire the latter, the emergence of new biological species, or cognitive development.

In such models, quantitative changes (e.g., salience of data or piece of information) lead to the emergence of multiple qualitatively different states in a first form of an elementary catastrophe with subsequent transitions between the states as the result of infinitesimally small influences that constitute a second type of catastrophe. My study therefore bears great similarity with the descriptions in which there are continuous linguistic changes that are not entirely apparent to the subjects themselves until after these changes have occurred. Thus, during the Copernican revolution mean that scientists "gradually lost the habit of using certain words and gradually acquired the habit of using others" (Rorty 1989, p. 6). It was only some 100 years of "inconclusive muddle" that scientists "found themselves speaking in a way which took these interlocked theses for granted" (p. 6). Just as my ethnographic study suggests, Rorty, on philosophical grounds, describes the qualitative changes of the Copernican revolution as the result of quantitative changes in the ways certain aspects of discourse changed. This view is more consistent with conceptual change as arising from complex webs of dilemmas and beliefs that act together until they cross a critical threshold. The change in the conceptual language of science, therefore, is not the result of incompatible and radically changing "meanings" or of conceptual ontologies but rather, consistent with Wittgenstein's position of how language works, it is the result of a continuous transformation of the ways in which (concept) words are employed. The conceptual change described here therefore is consistent with a view in the philosophy of science according to which scientific revolutions, especially taxonomic changes, may in fact arise out of continuous changes. In this model, a network of attributes and values links a superordinate concept to the subordinate concept. In such models, continuous change at the level of empirical observations may lead to categorical changes at the level of taxonomy (Roth 2005).

The present study suggests the need for a different theoretical approach to those educators interested in the general phenomenon of learning where there are transitions between forms of conceptual talk. Rather than expecting changes to occur with exposure to one or the other "discrepant event," educators and learning scientists may want to take a long-term approach where the various contributing moments are studied in their emergence, unseen and unforeseeable from the perspective of those who *undergo* change. It may turn out that the qualitative changes in students' conceptual talk can be better modeled and theorized as arising from a variety of quantitative changes – much as this would be suggested by both catastrophe-theoretic approach to modeling the morphogenesis of new symbolic structures more generally (Thom 1981).

On the Salience of "Facts"

Learning scientists of all brands tend to look at the data to which some learner – here the scientists of our research group – is exposed as if these had factual status. That is, for example, learning scientists present "discrepant events" to their subjects as if the event in and of itself is discrepant. This fails to recognize the above-noted transformation of the perceptual field together with the recognition of facts that could not be perceived prior to the reorganization. That is, in the received account the world is taken to stand on its own and independent of the human agent; if anything is recognized then it is something like an individual "construction" or "interpretation" that makes the thing look different without actually considering it *as* different. A different approach consists in taking a unit of analysis that includes the material setting and how it is reflected in the consciousness and affects of the subject (Roth and Jornet 2013). In this approach – which implements J. Dewey's and L. S. Vygotsky's calls for inclusive units of analysis – only the transaction subject-experiencing-in-setting matters. Experiencing here not only equivalent to acting and bringing about change but also to being affected by acting and by the responses from the environment in which the subject (i.e., individual researchers and research team) is embedded and from which it is inseparable. Thus, what matters in accounting for human actions and decision-making is how the world is for the subject rather than how some aspect of the world would appear in a theoretician's description. Thus, when our research team saw the data as fitting with and into the dogma, this was not just the result of a force fitting, whereby human subjects consciously look at trends in raw facts in different ways. Rather perception has a passive component, where (an aspect of) the world appears to the subject who does not have to interpret raw facts at all. This is most evident when a person comes to experience some phenomenon for the very first time, even though it belongs to a world that we thought to have been very familiar with (Roth 2012). With the coming of the phenomenon our perceptual field, too, has been restructured.

The first stage in a scientific conceptual change is said to derive from the presence of a set of anomalies, unexplained phenomena, or impasse. This requires, however, that the anomaly is seen as anomaly. However, I am not aware of any conceptual change theory that explains, from the perspective of the learner, the emergence of new ways of seeing and modeling the data given the general view that perception is theory dependent. A change in perception and a change in conception go hand in hand. This would require the emergence of new structures that are not grounded in, and therefore cannot be anticipated based on, old structure – morphogenetic processes that tend to be captured in chaos and catastrophe theoretic models of cognitive change (Fig. 7.2). Minute variation within the subject or the environment can lead to the qualitative changes at points ② and ④. In Chap. 7 I show that data that have the potential to be seen as anomalous are "force fit" to the existing paradigm. Thus, as seen in the extrapolations produced during the early stages of the research, our chief scientist did not model the data from the trends that could be observed but extended them such that a very different curve was made compatible with the data at hand. It is only with hindsight and with additional data from the same fishes and fishes at very different stages in the life cycle that our team began to accept the possibility of a different pattern of porphyropsin levels along the life cycle.

Initially, however, our team saw the data precisely through the lens of what we were expecting to see. The same data were seen very differently together with the conceptual gestalt switch. Clearly, the process at work is not inductive but looks more as the reverse process of deduction with a certain level of aspect blindness. This is not unlike what I had observed in an Australian 12th-grade physics class, where those students expecting to see movement in a demonstration did see movement ($N = 18$) and those who did not anticipate seeing movement did not see it ($N = 5$) (Roth et al. 1997). Whereas the classical conceptual change formulation of the issue does allow for the fact that scientists themselves have to perceive something as anomalous to get the change process started, educators appear to be operate under the assumption that *the event* can be discrepant in and of itself and prior to the gestalt switch of a change in conceptual language[7]; the question about whether and what subject sees as discrepant is rarely asked in the conceptual change literature. In the mentioned Australian study, the movement was small so that it could be seen as insignificant or significant simultaneously. A similar effect may be at work during the initial data collection period of the present study, where the number of data points was small so that it was possible to view the trends in the data as consistent with the former conception. But as soon as some threshold was crossed, the data no longer lent themselves to the traditional way of seeing the data and associated interpretations. A rupture occurred in the lens and the way in which the measurement points gave themselves to see.

A shift in salience also was noted concerning the role of certain texts (articles, dissertations) and their content. Shelby talked to an interviewer about the Alexander et al. (1994) paper and noted that it in fact constituted the first chapter of the author's doctoral dissertation. It showed a "quick reversion back after, you know, when they're no longer ready to go to sea." However, another chapter in the same thesis shows a different result: "a clear annual cycle in his data." As a result, there were "two chapters, so one chapter shows this conversion, and the other chapter shows this annual cycle." Shelby stated that our research project had started on the premises of the first chapter of Alexander's thesis, when in fact the second chapter would have contradicted the assumptions on which our study were based. He added: "if you dig a bit deeper you go back to you know Beatty 1966 and Allen et al. they clearly show these annual cycles over time." Asked about the point in time when he read the Alexander thesis, Shelby responded to a query on November 8, 2002:

> I had a copy of his thesis for about a year and I kinda I read it a long time ago. And again you know, because my mind was in the same place, you know those people that thought about the project was, "that's it." So you overlook it 'cause you just assume that those other pieces, the first chapter is the exciting one, it is the right one. So you follow that right and then it's ony after you've done a year that you kinda go, "Well, hang on, I seem to recall that

---

[7] Heidegger (1927/1977) describes the relationship between language (logos) and perception as a dialectical one, where the former lets things be seen in the way they show themselves. This is not another version of the Sapir-Whorf hypothesis, according to which language drives perception. Rather, there is a mutually constitutive and implicative relation, where language allows us to see *what shows itself* and which is given to us without our intention.

when I looked at his thesis before, this might have happened." And I go back: sure enough there's his annual cycle in there.

Shelby also stated possible reasons for a decreased attention to and salience of the older studies from the 1960s and 1970s. Because the equipment used in those studies was not developed very well, he supposed that the data were "probably less accurate" and that, because their own "machine is more accurate, you are going to pick something up." On April 22, 2002, a conversation was recorded in the wet laboratory during which Shelby talked about our team's data in relation to the Alexander et al. paper, the dogma, and the studies conducted around 1970 (Beatty, Allen). He pointed to the cyclical variations reported in Chapter 2 of Alexander's dissertation. Thus, Shelby was indeed familiar with this work showing annual cycles. But at the time these other results were not sufficiently salient to drive the conceptual talk our team deployed. He also suggested that "those people that thought about the project [i.e. Craig and I] thought 'that's it'" and that led to the continuation of the dogma-related hypothesis. With the benefit of hindsight, Shelby did have an explanation for the fact that the earlier research reporting cyclical patterns had not been salient in our team and, therefore, did not constitute a lens through which we would model the data that we were collecting. Thus, on January 22, 2003, he stated:

I mean as far back as 1966, the same annual cycle trends had been published. Now, what I had, when I saw them, when I read them, and this was one of the things with science, I mean there's hundreds of papers to read, thousands of papers to read on every topic and so often when you go in and read a paper you're looking for a particular point. You want a point of view that's going to support some hypothesis that you have. And so when you're reading it, you can read all the stuff, but you'll miss certain chunk. It'll go through your head because it's not what you're looking for you. Uh it's like "Where's Waldo?" pictures, you know. You're searching for Waldo and you find him. But if someone asks you later on. "Name 5 other people that were in that picture." You probably couldn't name any of them, you're only looking for one.

It might be precisely for this reason that our study, though contradicting the dogma, was not considered worthy enough to attempt publication in one of the leading journals (*Science, Nature*): After the fact, we can say that there were already bits and pieces of research available consistent with an alternative conceptual language. Thus, what might have looked like a major discovery within our team, the reversal of a Nobel-Prize-winning dogma, was eventually described in a very simple, almost humble statement: "Our evidence shows that coho salmon do not shift their visual pigments strictly to accommodate migrations between fresh and salt water (Wald 1960; Hoar 1976)" (Temple et al. 2006, p. 311). Our paper continues by refuting the claims of another study that had supported the dogma: "We therefore suggest that the A_1/A_2 ratio is not a suitable indicator for the parr–smolt transition in salmonids as was recently proposed by Alexander et al. (1994)" (p. 311).

This suggests that a new hypothesis is not necessarily subjected immediately to experimental testing. It may exist among other statements, just as modeled in the co-presence of alternative conceptions in the catastrophe theoretic model (③,

Fig. 7.2) and only become salient simultaneously with other relevant aspects immediately preceding or together with the gestalt switch that occurs during the second qualitative change (④, Fig. 7.2). There is therefore a co-emergence of the salience of different pieces of information. In the present study, the hypothesis of a temperature effect existed as a statement long before it was actively pursued. It is not even certain that it was considered as a serious hypothesis, as there were no explicit discussions in those instances when Theo articulated this factor as a possible variable for explaining the trends in our measurements.

The data presented in Chap. 7 specifically and in this Part II of the book more generally therefore suggests that we may have to model the change in conceptual language in a new way. The change occurs when the *same* data come to give themselves differently under the same paradigm. These data, which initially were seen to support the dogma (older conception), now are seen as opposing it. This shift occurred before a new theory existed and even before another hypothesis could replace the existing one. The data are no longer seen *through* the lens but next to what looking through the lens would suggest; or, to extend this metaphor, scientists have two different lenses making the same data look different. The data belong to and support the dogma and already no longer belong to and support the dogma. It is precisely such a tension that allows and describes change in the unit of analysis approach provided in Chap. 6. And this is precisely the point where a rupture occurs in the perceptual field and in the old ways of looking at the world. In catastrophe theory, this is an initial catastrophe, that is, the emergence of new structural forms that cannot be deduced from an existing structural form. That is, in the conceptual development from left to right in Fig. 7.2, the first qualitative change in ② is not predictable from earlier events that would have *caused* it. Similarly, the transition from one conception to the other is not attributable to causes but emerges from minor variations that lead to a gestalt shift (e.g., similar to the "butterfly effect" in chaos theory).

Any change in conceptual language used for describing and explaining scientific phenomena arises out of the contradiction that the prior conception also constitutes the material from which its undoing is made. As shown in Chap. 7, the data initially were seen through the lens of the dogma so that it cannot be argue that the discrepancy just appears. To see the problematic, a sort of double vision had to occur: the data had to appear twice for a contradiction to appear: once through and once next to the dogmatic view. This is so because a deconstruction of an old discourse (theory) cannot be conducted in terms of that very discourse (theory), because it cannot ever be shown that a discourse, vocabulary, or theory deconstructs itself (Rorty 1989). It is only with hindsight that scientists have the language that allows them to provide a reason for the development that they have undergone: The "new vocabulary makes possible, for the first time, a formulation of its own purpose" (p. 13). A "half-formed" new vision, vocabulary, or discourse promising "great things" has to co-exist with the old to serve as a springboard that allows the old discourse, theory, or vocabulary to be dislodged. It is out of this contradiction that a new hypothesis, which, subjected to experimental tests, can evolve and lead into a new theory. This double vision is captured in Fig. 7.15, which exhibits the initially collected data in the light of the dogma (step function), the study confirming the dogma (Alexander et al.), and the new conception emerging in the researchers' present work (circannual sinusoidal pattern). Because of the same reason, there cannot be criteria for deciding between the difference conceptions. These can be available only as the result of the conceptual change, grounded in the prior conception, rather than as its antecedent.

Uncertainty

The perhaps most outstanding feature of my ethnographic study is the documentation of uncertainty that marked the studied research team from the beginning to the point in time where it was relatively certain about just what it had investigated and found. The study of graph interpretation among scientists specifically draws on fragments from team meetings as illustrative material to exhibit the processes by means of which scientists come to grips with relations in their data that they cannot immediately talk about and explain in the context of everything else that they are knowledgeable about. Our team had engaged in this study to document the precise nature of the changes in the visual system of salmon, which was to be used as a predictor of smoltification – transformation into the stage at which the coho salmon go to sea. To see what scientists do with data and graphs when they do not yet know the outcome of their experiment, the presented materials focus among others on the early stages of scientific discovery work rather than at the later stages, when scientists might be tempted to say that they have had a hunch (or knowledge) all along – in the way the study was subsequently written up as a test of alternative hypotheses, the second one of which I showed emerged as the result of the research activity. The fragments I mobilize in my account exhibit that what eventually would become evidence concerning the age- and weight-independent nature of porphyropsin levels was not initially considered. What might have been reason to pursue other routes to modeling the data was not considered during the earlier meetings and therefore were nothing at all. Our team did not attend to the similarity *as* similarity between the porphyropsin levels of different age classes. Having to act in the face of uncertainty, our team had no criteria for judging our own thinking and progress. That is, in the discovery sciences, precisely because of their discovery aspect, metacognition does not appear to be a promising concept in the explanation of the unfolding of the history of a scientific research group. The perhaps most outstanding case in my ethnographic account is the instant when Craig, Theo, and I returned to the laboratory one morning and the equipment that worked the night before no longer functioned. Our attempts to get it to work again was not marked by a particularly reflective process, not even as a situated *inquiry*, though it was a completely *situated* and *contingent* process of pushing here and pushing there, reading this and saying that until, all of a sudden, the equipment was working again.

Uncertainty also was a dominant feature in the description and explanation of data and the graphs that were used to summarize them. In the course of producing the data, our team initially had little to go by to explain what we had in hand. We were in a quandary that I also point out in Chaps. 2 and 3. It is only when we became intimately familiar with all aspects of the sourcing of the specimens, the transformations that the retinas underwent, and the trajectory of the transformations along the chain of translations that we gained more stable (certain) ground and some foothold together with it. In the course of my analysis the scientists involved are shown to express trouble providing reasons for why the data between the Kispiox Hatchery and the Kispiox wild coho were so different, whereas in the simultaneously conducted study at the Robertson Creek Hatchery, the data between wild and hatchery fish were very similar. Our team was confronted with data that in the process of the research had been decontextualized and the early data analysis meetings brought about the emergence of relevant context that assists us in explaining what we had lost or given up earlier.

My ethnographic descriptions in this Part II of the book provide evidence, too, for, uncertainty as a transactional feature. That is, for example, in one meeting discussed above Shelby did not merely produce an alternative hypothesis but used the hedge "this may be complete speculation" to introduce uncertainty. This, then, would allow anyone using the hedge to back away from a statement he made without loss of face and back out of the situation should there be insufficient evidence materializing from the discussion that would support the statement marked as (potentially) contentious. The discursive production and reproduction of uncertainty not only was a feature within this study but also has been shown to be a dominant feature of research into expert thinking (Roth and Middleton 2006) or of social processes, particularly where there is something at stake (Lynch 1998). Throughout my ethnographic description in the foregoing chapters, I point to uncertainty and its production by means of acts of hedging. Hedging modifies a speakers' declared commitment to the truth of propositions. At least two kinds of hedges have been discussed in the literature: the first is vagueness with respect to the propositional content – introduced by drawing on forgetting (e.g. Neisser 1981) – the other vagueness in the speaker's commitment to the truth of the articulated proposition (e.g. Jucker et al. 2003). Hedges offer the possibility to create uncertainty in the sense that the recipient of the message is left in the dark as to who is responsible for the truth-value of what is being expressed. Aspects of uncertainty available in the meetings including (a) the deliberate production of uncertainty, which allows interviewers to withhold assessment and therefore provides interviewees with the opportunity to continue producing an unaided interpretation and (b) the display of uncertainty on the part of members to the setting, which makes available to the participants the sense that they may lack the expected expertise.

The challenge for the scientific research team studied in this book is to provide some explanation of just what the graphs might be useful for. This is the sort of challenge that might be faced by any eleventh grader who is asked to explain to a teacher what some textbook diagram is used to stand for. It is immediately apparent in all examples provided here that just what is required to do this task, never mind what the graph represents, is far from straightforward. Thus, throughout the exchanges I feature in Chaps. 4, 5, 6, and 7, uncertainty is prevalent. Uncertainty is not a mere expression of some purported "misunderstanding." Uncertainty actually provides transactional resources to solve the "problem" of generating a discourse about the data and the graphs that were generated as best fit curves to explain the former. This does not simply involve mapping formal, domain-specific features of graphs drawn from the scientific canon onto some verbal explanation of what the data might be said to express. Rather, as described above, the ascending and descending movements between abstract and concrete themselves generate and are expressions of uncertainty. Much too little school science and mathematics currently requires students to act in the face of uncertainty and too much of their time involves matching their behaviors to canonical forms (Roth and van Eijck 2010). It is precisely the competent coping with uncertain situations that students come to be least exposed to and practice.

Contradictions in Scientific Research

In Chap. 6, I present a detailed description and analysis of the contradictions that exist in the modeling and interpretation of the data in scientific discovery work. The leading scientist specifically and the various groupings in his laboratory more generally have been highly successful in obtaining grants and publishing their work. Yet the study shows, with hindsight, contradictions in the modeling and explaining of the data in our study of porphyropsin levels in coho salmon. However, it would be mere speculation to suggest that the study was not published in a major research journal such as *Science* or *Nature* despite the fact that it overthrew a long-held dogma because of the contradictions it contained (see Chap. 7). In the study of the contradictions, I show how it was possible to have concurrent multiple descriptions contradicting each other without that our team of highly trained scientists or our representation- and context-savvy audience recognized the contradiction as such. In fact, contradiction is implicit in the catastrophe theoretic model where, when the developing system finds itself on one of several possible branches (Fig. 7.2), this possibility itself is equivalent to and an expression of an inner contradiction of the system. The entire endeavor of change was rooted in a dominant existing language (conception). Transitional language inherently contains contradictions. The period of change therefore may be characterized as "muddle" as a positive concept to denote the fact that multiple different discourses co-exist simultaneously and the contradictions that come with this co-existence. Muddle is but another word that marks the possibility and presence of contradictions. Muddle marks the presence of multiple interlocked theses, each of which may develop independently. From this perspective, then qualitatively new forms (e.g. of talking) emerge from quantitative changes in the old form. Both dialectical materialist logic and catastrophe theory have evolved discourses that handle qualitative changes as the result of quantitative change of a number of different dimensions simultaneously. Thus, although change is continuous, there are two forms of critical points, two forms of "crisis," that bring the new into the old. In the first, point a bifurcation occurs and another form becomes possible. Eventually, a more or less sudden change, the previously dominant form gives way to a less dominant form, and the relative dominance is inverted. Although these changes – emergence and transition – tend to be talked about as more or less sudden, they constitute periods in which the old and the new discourses more or less coherent, the contents and forms of speaking are in substantial flux so that, inherently, there is flow and, with it, there are inconsistencies.

I show in Chap. 6 how, after the fact, logical contradictions had been present but were not recognized as such by our team. This ought to engage educators in reconsidering the ways in which they think about *discrepant events*. Such events are used to raise cognitive conflict. But logical contradictions have to be recognized to lead to cognitive engagement. Even if a teacher were to denote two statements as logically contradictory, there is no guarantee that in students' perception and use these would actually be contradictory. In our situation, because the a posteriori noted logical contradictions were not apparent us, it is not surprising, therefore, that they did not influence the evolution of our scientific research project. In fact, I do make mention of a study I conducted in Australia, where demonstrations were presented to a whole class of physics students, who were asked to note on a sheet of paper what they had seen. Alas, the students did not see the same phenomena: where some saw movement others saw rest. The explanations that they subsequently provided differed, because some explained movement and others explained rest. That is, an event in itself is not discrepant; and (logical) contradictions are not inherently apparent – even, as shown in the present study, when the individuals involved are highly educated and experienced scientists.

There is a different form of contradiction at work, however. It is internal to the phenomenon when considered as a whole precisely because phenomena are spread out in the flow of time. Contradiction, then, becomes an expression of the changes in time that are observed. In the opening part of Chap. 6, I articulate dialectics as a form of thought that was developed to describe the flux of life generally and, in the case of Il'enkov (1977) the flux in the sciences more specifically. In the dialectical approach, older and newer forms are part of the same unit of analysis, are part of the same category (see Fig. 6.1). When we think of language rather than of a square that turns into a parallelogram in the activity of shearing, then we get a moving phenomenon: language as a moving system. What makes it move? We know that there are dead languages, and these are dead precisely because nobody speaks them. A living language is one that is being spoken and written. That is, speaking itself changes a language. The literary critic M. M. Bakhtin and the members of his circle, including V. N. Vološinov (1930), had the insight that to model the changing nature of language in this way. They thereby were able to account for the changing nature of literary forms such as the novel over historical times. In other word, it is when every word is considered as changing when used, then the changes observed in a living language become intelligible; and, simultaneously, the stasis of a language not spoken also becomes intelligible. But we tend not to notice such change while we speak until some time after it has occurred, sometimes a long time after. Again, this view allows us to see that the research team was undergoing change, and it is not surprising that its language concerning the topic of light absorption in the retina of coho salmon was undergoing continual change. A scientific article reporting the research, like any novel in the course of literary history (Bakhtin 1975), is but a reflection of a particular point in time of this development. It reflects the contradictions inherent in the discourse at the time, though these contradictions are not apparent precisely because this requires hindsight that inherently is not yet available.

Control Over the (Not-) Becoming of Data

In this study I describe and theorize how scientists produce their data by including or excluding some but not other measurements. Learning about data generation is important because the very nature of a scientific phenomenon depends on it. For example, if the scientists in this study had included all measurements they had made then their very phenomenon might have been lost in the variance caused by the data actually excluded from analysis.[8] If STEM students are to become more savvy about the nature of science and mathematical cognition in the wild, and if students are to take a more critical stance towards the results of scientific research, then they need to learn both to describe and explain the data that are included and to make judgments about the quality of evidence that includes considerations of data not retained for interpretation. Thus, for example, it is only under specific condition that Galileo's inclined plane experiment yields the data that support his claims about the quadratic increase of distance traveled with time (or linear increase of velocity with time) (Garfinkel 2002). There are many ways and reasons for losing the phenomenon. Students do not generally have experiences in learning to differentiate the conditions under which a scientific phenomenon appears und under which conditions it will be lost. In traditional laboratory exercises, students are held to produce data such that these support the scientific theory. Even in extended experimental investigations, which are premised on the idea that students learn to conduct independent research, teachers may disallow an experiment so that students get data that confirm some existing theory (Roth 2013a). Knowledgeably, reasonably, and accountably making distinctions between conditions that produce versus those that lose a phenomenon should be an integral aspect of literacy in the STEM education.

In contrast to much of STEM education, where students do what they are told to do, the scientists are in control over what to do and which measurements to retain

[8] We had made measurements but literally not recorded them, even though these were recorded and existed at least momentarily in computer memory. Our decision not to record them permanently excluded their effect on the explanations we could produce.

for the analyses that they ultimately report.[9] Scientists are in the position to exclude data, where students may not be. As I show in Chap. 3 particularly, scientists' inclusion and exclusion criteria are grounded in their familiarity with all those instances that do not even qualify for entry into the data sources. Scientists literally constitute the frame that allows only some measurements to enter into subsequent considerations. This frame therefore reduces the original messiness and uncertainty, which then permits the phenomenon to appear more clearly against the ground than it would if every measurement were included. This, therefore, constitutes a form of bias that overstates the evidence against the null hypothesis because the data that would be consistent with the null has been included. This bias is quite evident in the Bayesian approach to the relation between data and hypotheses, which tests the probability of a hypothesis given the data $p(H|data)$.[10] If data are omitted that would provide evidence for the null hypothesis, then omitting data overstates evidence against it. But there is then no guarantee that other potential phenomena are not lost when data are dropped. Without an integral knowledge of where the data come from, how they are generated, possible problems in the production of data, and how data differ from non-data, even scientists would be hard pressed to make conclusions and support claims.

This study, therefore, points us to the importance of learning to make decision especially in situations of uncertainty when there are no or not presently known normative answers. Making such decisions in the face of uncertainty is important, for example, in democratic decision-making processes. This became evident to me when the mayor, council, and engineers in my hometown based a decision for not constructing a water main to supply people with running water on the report of a particular scientist who only collected data on a single day and in only one-sixth of the homes concerned (Roth 2008).[11] In the ensuing public debate, some savvy citizens, however, did point out both aspects of the data collection as problematic issues. However, the mayor, council, and engineers not only disregarded the critique of the data collection and quality but also failed to take into account, and even omitted from entry into the data sources, more than 30 years of anecdotal information that locals had collected about the water. A Bayesian approach would have indeed be able to take into account this anecdotal information alongside other

---

[9] Even many science teachers, in part as a result of deprofessionalization, have to cope "with a top-down, assessment-drive curriculum" (Levinson 2011, p. 113).

[10] This approach is diametrically opposite to traditional hypothesis testing, which establishes the probability of the data given a particular hypothesis $p(data|H)$.

[11] The most extensive presentation of all issues involved in this case can be found in an article for municipal engineers on the construction of community health and safety (Roth 2008). In that article, I describe the politics that made it possible for community engineers and politicians to stall on the establishment of a water main, supplying the residents of one part of my community with running water. In that process, decisions were made on the grounds of engineers who had collected data on only 1 day during the year; all other data, collected at multiple time points during the year and anecdotal data from nearly 40 years produced by the residents living in the area were disregarded.

more quantitative information in the decision-making process. That is, these municipal officials could perceive a phenomenon emerging from their data rather than a different phenomenon that would have emerged if *all* the information available at the time had been taken into consideration. The citizens displayed exactly the kind of literacy that STEM educators might want to foster: Rather than simply accepting the claims of scientists and engineers, educators should want to develop a STEM savvy citizenry that raises questions about the data collection, demands public articulation not only about how claims had been produced but also about how and which data were dropped, or engages in queries about how the framing of the nature of data collection is related to the sociopolitical agendas in play. It was just such forms of scientific literacy that AIDS activists displayed and that led to changes in the scientific protocols for collecting data on the efficacy of new drugs. It would lead us toward a citizenry engaged in "more and more diverse trajectories of [scientific] fact construction and closure in controversies" as much as learning about the myth of "science as clean and elegant" (Roth and Désautels 2004, p. 154).

My ethnographic study exhibits the control scientist have over the definition and implementation of their research as a whole and the data that will be interpreted specifically. In fact, during the research our team no only learned when the phenomenon exhibited itself – which, as I show, they might denote as "beauty" – but more importantly we became familiar with the conditions that made us lose the phenomenon, such as when a dissection was recognized as having gone awry or when the minimal essential (MEM) solution they have been working with will have been recognized as being out of date. This control that scientists have over the definition of solutions and problems is similar to observations of mathematical behavior in other everyday settings, where an important aspect of problem-solving practices in the everyday world where "[p]ersons-acting are free to transform, solve or resolve a problem, or abandon it in favor of other options. In the parlance of the [Adult Math Project], they 'own' their own problems" (Lave 1988, p. 156). In an equivalent manner, our research team owned its problems and its data. We made decisions whether we wanted to include or exclude measurements that we had conducted for supporting the claims that we intend to be reporting in a research article. Without finding the relation between graphing (a social practice) and the setting (of research) we have little to go by for modeling how "cognition is constituted in dialectical relations among people acting, the contexts of their activity, and the activity itself" (Lave 1988, p. 148). Dialectical relation here means that there is a unity to the activity as a whole, which is the minimal unit to model any of its parts, including data and their graphical representation.

The implication for my own teaching of STEM-related subjects would be to turn over control to students so that they can define the question to be research, the data to be produced, the ways in which results are reported (e.g., mathematical or statistical modeling), and what significant results of their studies have been. In fact, this was precisely studying what students would do when placed in such situation that had led to me to write *Authentic School Science* (Roth 1995) and *Rethinking Scientific Literacy* (Roth and Barton 2004). These extended studies show that middle and high school students are quite ready to engage in such endeavors; and, most importantly, they learn a tremendous amount of science beyond the official, institutionally defined and regulated STEM curriculum.

Graphs in Discovery Work

Graphs and graphing are not merely integral to the sciences but in fact are constitutive of the sciences specifically and the STEM fields generally. Doing science and producing graphs co-implicate each other. Graphs and graphing, in different forms, are seen throughout the chapters in which I describe what actually happens in a scientific research laboratory and in the meetings of its members. Even when they communicate among each other, team members drew graphs to articulate ways of talking and thinking about phenomena. Graphs not only were the end result of the work, but, as shown here, these were the very starting point of the research project designed to measure with great detail and resolution the changes in porphyropsin levels in the retinal photoreceptors of coho salmon during the time the fishes undergo the physiological changes that precede and accompany their ocean migration at around 18 month of age. In scientific research, therefore, graphs have very different functions and show up in very different parts of the knowledge-productive activity (dejatel'nost' / Tätigkeit). Thus, in addition to resulting from and being representations of the research, graphs also serve as lenses (tools) for looking at and explaining data. Graphs also are objects of inquiry, such as when scientists attempt to figure out which of several possible curves best describes a plotted set of data points. This is especially the case in the present ethnographic study, where some graphs were intermediary products, such as the absorption curves that were then interrogated to reveal the position of the maximum along the wavelength continuum (λ_{max}) and in terms of their width at maximum height (half-max bandwidth). In fact, the graphs did not easily reveal their secrets and sometimes the data had to be cleaned – using Fast-Fourier Transformation and its inverse or polynomial fits – before they rendered the sought-after information. That is, these graphs were cleaned up, transformed, polished, and otherwise worked before being useful for subsequent work. In the following, I offer brief reflections on (a) graphs as tools, objects, and products of scientific research and (b) the deep background familiarity required to make any useful statement other than providing some superficial statements as to particular values (as is the case in school and international testing).

An important aspect of graphs is their nature as resources in learning, which derives from their usability in diagrammatic reasoning, experimenting, and reflecting on the results (Bakker and Hoffmann 2005). We see that the use of graphs that these do not appear to pre-exist their appearance during the session. In Chap. 4, when Craig draws what he later calls "just as sort of preconceived notions," there is little evidence that the image on the board is preceded by some other image in the mind of the speaker. Rather, there is a lot of evidence that the graph emerged and thereby *was* a form of thought in movement. Craig repeatedly wrote and erased axis labels on the board. Then, before placing the actual line that is said to be representing a relationship between "%A_2" and "HMB," he looked repeatedly from left to right and vice versa before the hand moved to produce the line. There also was a lot of long pausing. All of these aspects contrast those of the manner in which Theo's graphs comes to be recorded, in a quick and deliberate way. Finally, when Craig returned to his seat, he actually copied what he had been drawing – which he would not have had to do if all of what he has done had already existed as a clear, stable, and repeatable mental image. This form of producing graphs, therefore, is better theorized by means of a cultural-historical, developmental perspective on thought according to which thinking and expressing co-evolve (Vygotskij 2005). Philosophical considerations, too, encourage us to look at the dynamic nature of graphs because it is in and through the diagram that an already available mobility, the gesture, comes to be immobilized: "the gesture is put to rest even before it plots itself in a sign" (Châtelet 1993, p. 33). Accordingly, graphs do not have to be the externalization of a pre-existing idea. Instead, the subject of the graphed expression may find his/her thought in the expression or by analyzing its expression. Thought follows rather than precedes external expression and the person finds his/her thought in the embodied expression as a whole. This role of artifacts serving as (external) representations in the development of thinking is explicit in the manner in which Watson and Crick discovered the structure of DNA. They were playing around with cardboard shapes of the four bases they knew were part of the DNA. All of a sudden, in a flash of sight and insight, they saw that two specific pairs of bases produced identical shapes when pushed together in particular ways: these would be the rungs of the double helix. Such forms of action, whereby the external representation and representation use precede inner thought have been called "epistemic actions" (Kirsh and Maglio 1994). There was nothing in the head but all in the shapes and their constellations on the table. The process of discovery followed more the principle of "I know it when I see it" rather than the other way around. In this manner, the graph also becomes some more-or-less specified tool or, perhaps better, some material to think with individually and collectively. What it *is* arises from what scientists see *after it* has been produced rather than preceding its own production. Thus, Shelby suggests that the graph is "more or less" like this [as it is], "may be higher at one end, but same idea" and "more on that end." This could be seen as invitation to a modification of the graph, even though it did not happen here despite the verbal affirmation on Craig's part.

One of the least attended-to aspects of graphs and graphing are the events surrounding inscriptions and related practices in the discovery sciences. In this situation, scientists themselves do not know whether they have something in hand, like an independent Galilean pulsar (Garfinkel et al. 1981) or the transition points in amorphous alloys during constant heating (Woolgar 1990). This changes the problem of knowledge and knowing because one has to pursue how members to the setting come to recognize a discovery as discovery – even and precisely when they do not know in advance what there is to be discovered – or without knowing in advance whether a graph has or has not certain properties. Again, the order appears to be that of "I know it when I see it" rather than one of knowledge preceding the seeing. In one of the situations analyzed, the lead scientist worked out some "preconceived notions" as to the relationship between two salient variables, $\%A_2$ and half-maximum bandwidth of the absorption spectrum. Even though he suggested producing a preconceived notion, the production, which involved a lot of erasing, provides proof to the contrary, that the notion arose in, as, and with the production of the graph. The actual relation between half-maximum bandwidth and porphyropsin levels was important, because our research team was hoping to use it for the determination of the relative amounts of rhodopsin (A_1) and porphyropsin (A_2) in the pigment of the fish eyes, which are the relevant variables for determining – so the Nobel Prize winning canon that the research team was adhering to at the time – the readiness for the saltwater journey. According to Theo, he had not yet "checked what the bandwidth gives [them]." This is what our team pursued in the episode analyzed.

My analyses exhibit the work that the scientists produce for finding something in the data or graphs, for example, for finding "what the bandwidth gives them" (see Chap. 4). Craig made an argument for eliminating some of the data points, but Shelby resisted, for, as he argues, there are about 3,000 data points. This means that there will be some large variations, on the one hand, and relative stability of the data, on the other hand. Theo does provide an explanation of why some individual HBW measurements might be much smaller than they ought to be: His algorithm cuts off one part of HBW when the "ringing" that all of them are familiar with crosses the HBW line before the best-fit curve does. Craig provided a first reading of the trend, which, in his gestural movement, took on a curvilinear shape that matched his "preconceived notion." The team here attempted to control the variability by looking at a subset of the data. But the variability that was the salient aspect in our work remained when we looked at the data from one of the seven sources only. The next move was to establish a better determination of the variability within each of data point by "trying to plot with error bars." Here it turned out that two of the "spikes" did not have error bars and, according to the scientists, were "unique points." Once Shelby had changed the scale of the plot, two readings of the data were provided.

In this situation, the graph became a material form to think with. Shelby, by means of the cursor, traces out a more-or-less linear relation, "a general increase from zero to a hundred" that is "very, very subtle." The very deployment of the graph made it possible to move the cursor over the data. In this way, the trend was precisely of the kind that a dominant researcher in the field (Hárosi) and others previously published. It had, as predicted, a lower bandwidth for rhodopsin (A_1) on the left end of the graph and a higher bandwidth of porphyropsin (A_2) on the right end of the graph. This was also the way one would expect it to occur if one knew that the HBW values plotted here were not measured in nanometers, that is, on the same scale that the absorption spectra are plotted. On the other hand, HBW, as plotted here according to the values on the abscissa, would vary linearly with $\%A_2$ if the widths of the spectra, measured in nanometers, vary as a negative square function.

Craig also provided a reading; in fact, he provided two different readings of the same data set (see Chap. 4). He used gestures to produce an ephemeral graph that might best describe the trend in the data. There was a linear increase on the right-hand side of the plot similar to what Shelby had shown. In the center, however, Craig had a parabolic relationship, as exhibited in his "preconceived notion" and as gestured in an earlier plot. Finally, Craig first gestured a linear relation first with a negative slope at the left end of the plot; he then changed to a linear relation with a positive slope. In the course of the data analysis, there were different trends proposed. There were also trends that had been published in the research for nearly two decades. However, in this meeting, the question was never posed or answered why the measured curve did not match the anticipated curve, as articulated at the beginning. This situation occurred even though this very issue was a stated goal for the laboratory. There was also no discussion concerning the different proposals concerning the trends.

In this particular session, the scientists saw the data/graph through the expectation that HBW was one of the possible indicators of smoltification, the changes in the physiology of the fish getting ready for its migration to a different habitat, and the associated changes in the physiology of fish vision. Graphs therefore appeared in different ways, as objects to be investigated and as lenses through which the investigation was conducted. This is similar to the results of a study of oceanographers at work, who came to coordinate in their mundane practices multiple perceptual frameworks that included a variety of devices that plot aspects of the natural world (Goodwin 1995). But even if there was only one graph, it still constituted the confluence of many different aspects and perceptions. Team members would mobilize relevant references, draw graphs as a means to connect to work in other places, and so on. Thus, when the issue concerning variability arose, Theo drew and modified a graph to explain just where in his work variability became an issue. He made this variability visibly accountable in the "oscillations" that he added to the left part of the otherwise picture-perfect graph. The graph on the chalkboard, therefore, literally became a worksite, affording changing graphing practices with changing graphs.

In various places of this book, I make reference to studies concerning scientists' reading and interpreting of graphs. In contrast to other researchers, I have repeatedly pointed out in the past that graphing competencies are not somehow or primarily due to special abilities and brain cell configurations but can be attributed to familiarity with phenomena and the entire chain of translation from phenomena to signs that serve to make them present again. In fact, my research shows that even eight-grade students working in pairs on data tasks may outperform university graduates with BSc and MSc degrees when the former are more familiar with phenomena and research process (Roth et al. 1998). In the present context, precisely because background familiarity was missing in the research team, it had to be reconstructed from bits and pieces known to individual members. In the course of our analytic work, the members of our team exhibited to each other a great deal of biological content knowledge next to the mathematics involved in modeling their data and the graphs that these gave rise to. This is indicative of the background familiarity scientists generally have to bring to the task of describing and explaining their data and which they, when asked to explain a graph from their own research, voluntarily articulate *prior to* talking about a graph from their work (Roth 2003). It is not only indicative but also symptomatic. When scientists do not have this background familiarity, they struggle in their attempts to explain graphs to the point of suggesting that these inscriptions are poor or represent bad practice. In Chap. 3 we see how the scientists draw on their (a) intimate familiarity with the natural phenomenon and (b) detailed familiarity with the transformation processes that produce the various inscriptions that are part of their everyday work. In Chap. 5, where familiarity with the phenomenon has gone lost in the chain of translations – which begins with the capture of the fishes in hatchery ponds, creeks, or estuary to the extraction of retinal pieces and the recording of absorption spectra – the scientists have to reproduce this familiarity in the ascent from the abstract (general, imprecise) forms to the concrete and detailed familiarity with the provenance of the fishes. It is the (regained) familiarity with the entire research context that comes to be denoted by the graph in synecdochical fashion. When asked, scientists talk however much necessary or required prior to providing the requested explanations concerning what a graph is used for ("means") or the reasons to which particular features in it can be attributed to. The question of how "authentic practices" become supportive of mathematical reasoning can be answered in this way: "Authentic practices" allow learners to become very familiar with a particular context so that they can draw on this familiarity to produce cases that support abductive reasoning. The adjective "authentic" does not denote some special practice – e.g. what scientists or mathematicians do – but means familiarity with specific settings that supports "positive attention" to context and, thereby, supports general, non-context-specific mathematical and scientific conceptual talk as well.

Across several chapters of Part II we see that the existing canon "biases" what the scientists see in their data and how they see these – which would not be surprising given that confirmation bias has been observed widely in the sciences. It is indeed integral to the methods of those scientists who use traditional frequentist statistics (e.g. Rouder et al. 2009). Formulated hyperbolically, scientists do not give their data a chance to speak for themselves. In the early parts of this study, each data point was seen in terms of its possible location on an already published graph. It would take the scientists nearly 2 years until they found that the porphyropsin levels correlated highly with temperature levels and day length. Sinusoidal functions with temperature or day length as independent variables best fitted the data collected over the period. That is, just as Shelby suggested in one of the initial data interpretation meetings, there was no correlation between body weight and porphyropsin levels; and just as Theo hypothesized during one of the meetings, temperature would become a significant factor. It may therefore not surprise that Shelby, in a move that reconstructed the history of the discovery, would say in an interview after the paper was published that he has had a sense all along that the data did not fit the dogma.

It has been suggested that the study of a science and engineering (and we might legitimately include technology and mathematics) in action requires us to follow practitioners "either before the facts and machines are black-boxed or we follow the controversies that reopen them" (Latour 1987, p. 258). But studying such a practice during the controversy is not so easy because the existing paradigm (canon) has such a stronghold on the scientists that they have a difficult time seeing their data in a way that it would re-open a controversy. In fact, those individuals on our research team who did not have histories in the field as long as the lead scientist also were those who first raised doubt: the doctoral student (Shelby) and the research associate (Theo) with a background in physics. A second aspect is the canon, which oriented our team to see the data through a particular lens. A canon is integral to a discipline, with its characteristic visions and *di*visions, which discipline the (normal) scientist to view data in a disciplined way. The canon constituted a form of disciplined disciplinary vision that allowed us to see what was in the data – inappropriately so as we realized in the course of the project. Our team had anticipated getting something like the Alexander et al. data, and, therefore, saw the early data that we had in our hands in terms of that other study. Thus, the scientists' intimate familiarity with theory and previous research simultaneously was an affordance and a constraint simultaneously.

In a detailed comparison of a forest engineer's activity with introductory ecology graphs and graphs from his own work, I suggested that without practical familiarity, there would be no interpretation possible (Roth 2004). This is so because, as the analysis of everyday cognition shows, the development of practical familiarity *provides what is a necessary condition for an explanation*. It is in attempts to explain something that practical familiarity comes to grips with itself in an appropriate way: "In interpretation, practical understanding does not become something different but becomes itself" (Heidegger 1927/1977, p. 148). In attempting to explain, the possibilities only implied in existing practical familiarity come to be developed. Explanation articulates significations, which are already prefigured as possibilities in the practical familiarity with the world. Here I show that the members of our team presented to and for each other what they had learned and had become familiar with: two different types of creek systems, the windows in the closed rearing facility, the constant water temperature in the Kispiox hatchery that is much higher than the near freezing temperatures of the creeks, and the fact that the fish culturists release coho 20 months after egg fertilization. This is why the possibilities embodied in practical familiarity actually *make* the sense that makes sense. It does make sense that the coho in the Kispiox hatchery are different from those in the wild; and it does make sense that the wild and hatchery-raised coho are similar at Robertson Creek because they are reared in ponds that directly receive their water from the neighboring creek. In and through our effort, we became familiar (a) with an aspect of the concrete world as it manifested itself to us together with the surrounding material world and (b) with the way it was presented again in an abstract and abstracted way in our data. It is this same co-implication that has led others to suggest that we comprehend the world because it comprehends us (Bourdieu 1997). In this approach, new ideas arise from our material praxis so that any knowledgeability first exists *as* a societal and material relation before it comes to exist as conscious thought for individual team members.

Communication at Work

Throughout Part II of this book, I mobilize extensive transcriptions of communication from the wet laboratory and (data discussion) meetings. Communication is integral to getting the day's work done, and, in so doing, frequently is not so much *about* something as it is *for* something or getting something done. Communication does work in establishing the relations characteristic of work: *in-order-to* [Um-zu], *for-the-purpose/sake-of* [Um-willen], *what-for* [Wozu], *wherefore* [Wofür], and *with-what* [Womit] (Heidegger 1927/1977). Communication itself is at work and does work when we consider everyday workplaces. There were many features to the laboratory talk that were not at all of the conceptual type that STEM educators often seek to stimulate in their students. The communication had particularly little to do with the written aspects of science that some philosophers and science educators pursue. As I suggest in several chapters of Part II, the written scientific report had very little to do in content and grammatical form with the talk I recorded in the course of the inquiry. The scientists and hatchery workers in my study routinely made perceptual distinctions without being able to draw on a set of verbal descriptions that would suffice to articulate the perceptible gestalts. Moreover, the evidence provided in the course of the different chapters supports the contention that perceptual gestalts are part of the turn-taking patterns in laboratory communication. That is, verbal communication is a form of action intended to move material actions ahead and to the object of the activity on some desired trajectory. It is true that there exists a high degree of common ground and alignment between the participants in each of these settings even without talk – in the way one might find this in a couple that has lived together for many years. The participants – scientists or hatchery workers – have worked many years in their respective setting, and worked together for months if not years. The work often proceeds in silence and the members begin to use gestures and utterances to talk about the work only when trouble is evident. Gestural and verbal modalities are then used in conjunction with the perceptual modality to re-establish the alignment – which subsequently continues to be monitored by perceptual means. Most importantly, in this book I am not drawing on the concepts of "meaning" or "understanding" to explain the talk or to make statements what the scientists' talk was about – I only articulate their ways of using language in getting their job done, whether it was collecting the data, modeling them, or writing the ultimate research article. In fact, as I show elsewhere (Roth 2013b), "meaning" and "mental representation," as well as the associated notion of "understanding," are quite unnecessary for explaining what makes humans tick so well when working in familiar settings and why they do not tick so well in unfamiliar surroundings.

When people work together, they continuously make available to each other the rational, coherent, consistent, or knowledgeable character of their respective verbal and material actions. In this, the scientists in my study are no different from other people, although the extent of their joint work makes it possible to monitor the unfolding events by perceptual means. Material action (moving a microscope stage, perceived as changing focus on the monitor; opening, pulling down and closing software windows) not only brings about changes in the environment but also "tell" the respective other(s) what is going on. In these situations, verbal and gestural means emerge when someone spots trouble. It is therefore not sufficient to model communicative action as involving verbal, gestural and perceptual signs. The ethnographic work on which this book draws was conducted in a knowledge-producing workplace,. There is a continual change in the setting, tools and objects, in pursuit of new ways of modeling nature and research method. That is, our scientific discovery work was continually in a state of change rather than stasis, forcing scientists and scientific staff to learn to cope with and become familiar with new configurations. There were therefore repeated instances when trouble was more evident, requiring a lot of talk-in-action, until a particular form of action had become routine, at which time it largely went without saying. More and more explicit communication was necessary precisely when familiarity did not yet exist.

The transcripts also show interesting features pertaining to communication. For example, there were relatively long pauses between two speakers – up to 30 s long in the laboratory during the collection of the absorption spectra and other, not quite as long pauses during the discussions. There actually is a considerable disciplinary history of analyzing conversations, particularly pertaining to turn-taking patterns, repair, function of pauses and overlaps, and so forth (e.g. Sacks et al. 1974). In the laboratory work documented in my videotapes, many of the reported patterns do not hold. Unlike other conversations often analyzed in the literature, communication in the course of doing science is not an activity in itself but is part of the day's work of collecting data. The temporality of language use and turn taking is subordinated to, or rather, an integral part of the total situation to which the participants are attuned. It is the total activity (in the sense of Tätigkeit) to which speech activity is subordinated.[12] The purpose of the events in the laboratory is to produce the data for an article, and the goal for any one present day was to produce such data. Talk was used in so far as it assisted our team members in moving us toward that goal. Being co-present to one another in the laboratory or the meetings allowed us equal perceptual access to the material situation (displays on computer monitors, chalkboards, and projection screens; we also have had a considerable history of collective activity together. Many things therefore went without saying. Words and statements contribute to moving us along, sometimes bringing something into the light of a clearing, even if our statements did not constitute well-formed sentences and even if they did not constitute complete representations of whatever they might be said to refer to. What was transforming was not the content, as moved ahead and transformed by the verbal statements, but the research generally and the physical laboratory and our team members specifically. Because members were attuned to the relevant aspects of the setting, those aspects that went without saying did not have to be and in fact were not made salient, for any "statement is communicatively

[12] The totality of practical aspects of activity and language also has been referred to as *language-game* (Wittgenstein 1953/1997).

determining demonstration" (Heidegger 1927/1977, p. 156). This articulation actually flies in the face of common practice to search for "meaning" behind the words that students or teachers are said to make, construct, or communicate. We do not in fact require the concept of meaning (Rorty 1979), and "once we give up the notion of meaning, we also give up the notion of reference as determined by meaning – of the 'defining attributes' of a term picking out the referent of the term" (p. 274). All that matters is the usability of words.

Much of the laboratory communication I recorded – verbal, gestural, perceptual, or otherwise – had at least two functions. One was to point out something may have not been noted by another person. Here, "talk 'lets us see,' . . . from itself, what is being talked about" (Heidegger 1927/1977, p. 32). Not only talk but also any gesture or movement that lets us see something, from itself, needs to be considered. Thus, when Shelby moved the cursor along the data points, it was not only the cursor others could see moving but also the almost linear alignment of the data in the plot. The data showed themselves in this way, and the cursor was a means of allowing us to see it as such. But the cursor moved that way because of the way in which the data offered themselves. There is a mutually implicative relation in the cursor movement and the arrangement of the data points. Similarly, the sigmoid-shaped graphs based on the dogma and the results of the Alexander et al. study were communicative means that showed what could be seen in the data, even though we later chose a different tool to reveal a different relationship in what we had emerged from our measurements. With the communicative act, whatever its kind, one laboratory member assisted others in seeing a trend *in* the data that had been salient to him.

The communicative acts also functioned to maintain our common ground and mutual attunement to the situation and therefore to the current state in the unfolding of the events. But these very acts presupposed a common ground and intelligibility, because it would not have been of any use to attempt communicating in the first place. The longer we worked together, the less talk there was necessary to coordinate our respective actions while collecting data together. That is, our communication on any particular day cannot be seen and explained as a momentary phenomenon but requires historical analysis. When team members working together at some did not talk, this did not mean that nothing is going on. In fact, not talking generally meant that the participants have the sense that they know what is going on and that they assume it is the same for the others. It was only when someone had a sense that the common ground no longer existed that communication was necessary. Thus, for example, three of us were watching images of stuff from the microscopic slide pass across the monitor as a result of Craig's search for suitable cells without anyone speaking. Then suddenly, either Theo or I would make a statement about something that had just disappeared, engaging with Craig in an exchange about why he had not considered it for conducting a measurement. That exchange, thereby, made public that Craig considered whatever was seen as not worthy of further investigation, whereas Theo or my comment questioned such a consideration. When nobody raised an issue, it was an implicit agreement that the things coming and disappearing from the monitor were not worthy of our time. This points to the need to consider perceptual gestalts take as semiotic resources in scientific communication. Perceptual gestalts are not merely ground against which communication takes place and to which interlocutors refer by means of gestural and verbal deixis. Rather, perceptual gestalts constitute a sign form in its own right and require to be modeled parallel to the gestural and verbal modalities that have currency in pragmatic studies of workplace interactions (e.g., Heath and Luff 2000). Perceptual gestalts function as "affordances," information specific to (groups of) individuals that provides real opportunities for action. Laboratory members used these gestalts not by passively absorbing them from their environment; rather, gestalts and the customary routines for using them were structured, giving shape to the communicative interactions by means of which they are brought about. In this way, situated communication becomes an aspect of a much broader concern for the situated nature of human cognition.

Attending to the perceptual environment is an ordinary unremarkable activity in which the hatchery workers or experimental biologists engage in a continuous manner. It is so unremarkable that I initially received "funny looks" when, while being a newcomer to the laboratory, I was asking someone to explain what he was saying. In the early stages of my participation what a person had said was not necessarily following from a previous statement. It was only when I began to consider perceptual gestalts, (deictic and iconic) gestures, and words as co-deployed semiotic resources that I was able to follow the ongoing laboratory communication. Among the hatchery workers and biologists in the laboratory, perceptual gestalts went without saying. At best, they were glossed in ways that still left to the newcomer-me the work of learning to perceptually distinguish instances from non-instances of perceptual gestalts relevant in the situation.[13] Only when I was able to make those distinctions could I become an insider, knowing what was going on without asking "stupid" questions about it. At this point, speech let me see from itself what there was to be seen – e.g., a rod, a cone, a double cone, or something else on the slide. It did not really require speech, for the perceptual gestalts always and already were there to be seen.

The idea that perceptual gestalts are an integral part of (laboratory) communication may appear to some readers as odd (especially those trained in linguistics who focus on language). However, to visual artists (e.g., Bob Rauschenberg) and composers (e.g., John Cage), white spaces and silences have long been used and theorized as figurative rather than background elements (Cage 1990). Thus, traditionally the absence of played notes was taken to be the ground against which music becomes the figure. Pauses were heard as the absence of music. It was in the piece entitled *4' 33"*, in which a piano player enters the stage, sits down for 4' 33", then

[13] When I asked in the hatchery about the difference between coho salmon going to one of the machines injecting a coded wire tag rather than into another, the women operators told me "Look!" When I said, "Yea, but what is the difference," the operators asked me again to "Look!" I found myself in the predicament that I was asked to see the difference in the way these coho salmon looked, and that I asked for some description of *what* to look for the difference between see those specimens going into one device rather than the other.

closes the keyboard cover and leaves that John Cage called attention to the role of the "pause" as a figurative entity in its own right:

> in the case of 4'33" i actually used the same method of working and i built up the silence of each movement and the three movements add up to 4'33" i built up each movement by means of short silences put together . . . (Cage 1990, pp. 20–21)

The (extended) pause is not ground against which notes are played but figure in the same way and of the same order as any other note. This, therefore, leads us to a much broader concept of communication. In the scientific laboratory and hatchery situations researched here, the absence of words or gestures does not mean an absence of communication. Rather, (familiar) workplaces provide opportunities not only for detecting information but also for monitoring (mutual) common ground. In the course of their work, those working together evolve a set of background familiarity that provide resources against which features of particular scenes become noticeable and accountable. Even the absence of words and gestures is a form of communication – things are going smoothly and others know that this is the case. This is why ethnographic researchers have to have tremendous local competencies and knowledgeability so that they can hear what is communicated even in the absence of words.

Traditionally, language has been treated as the primary modeling system of communication, often considering non-verbal means as secondary, derivative, or partial translations of the primary system. In science education, for example, there is a movement that focuses primarily on "writing science," as if this were the primary mode of communication in the sciences. My research shows that in the laboratory and hatcheries – as in other workplaces that I have researched more generally – signs from all three modalities have to be accounted for to model (describe and explain) what is going on. I moved to the consideration of the three modalities (verbal, gestural, and perceptual) motivated by semiotic treatments of written and verbal communication. As a result, the relationship between the different modalities is symmetrical: the perceptual gestalt of a "double cone," the utterance "double cone," or the iconic gesture outlining a double cone on the monitor each may stand for another during communication. We can therefore think of workplace communication as occurring in any one or more of the three modalities. There is the world before interlocutors and collaborators, affording known perceptual gestalts that are salient on their own or are made salient by means of gestures and language. What my videotapes reveal is a multi-level communicative process that has to be modeled concurrently at the three levels *in the context of the particular setting*.

Gestures play an integral role in the search for trends. They are a way of orienting visual perception through the maze of the data that Craig does "not know what is going on here." The trends exist in their relation with each other, where they take on the *public* visibility of their characteristics. Thus, even if there are multiple accounts of possible trends, these are in any case articulated and therefore rationally accountable trends. When Craig moved his hand/index finger across the display (Chap. 4), he actually directed and taught others and himself how to look and make visible a possible trend. Similarly, in moving the cursor across the data display, Shelby exhibited in his movement instructions for *how* to move our eyes so that we could "get a general increase from zero to a hundred, very, very subtle." The movement of the cursor movement produced and exhibited the material that justified the description of the general but subtle increase from the very left to the very right of the display. The movements constituted the natural accountability that grounded the observation of the trend.

Non-verbal aspects have to be considered in a theory of communication because of their prevalence in face-to-face conversation and particularly in those exchanges when the conversational topic is a feature within the setting itself. Certainly, language is the most powerful semiotic device but there are dimensions that it does not cover as effectively as other devices, such as the perceptual gestalts that people orient to in the different chapters of this part of the book. In other words, the perceptual gestalts are intimately tied to basic processes of human transactions and participant frameworks. Leaving out these gestalts would not allow us to make sense of the interactive nature of laboratory work even in the absence of gestures and words. Perceptual gestalts shape the interaction at the same time that they shape the indexical ground.

References

Alexander, G., Sweeting, R., & McKeown, B. (1994). The shift in visual pigment dominance in the retinae of juvenile coho salmon (*Oncorhynchus kisutch*): An indicator of smolt status. *Journal of Experimental Biology, 195*, 185–197.

Bakhtin, M. M. (1975). *Voprosy literatury i estetiki* [Problems in literature and aesthetics]. Moscow: Xudoš. Lit.

Bakker, A., & Hoffmann, M. H. G. (2005). Diagrammatic reasoning as the basis for developing concepts: A semiotic analysis of students' learning about statistical distribution. *Educational Studies in Mathematics, 60*, 333–358.

Bourdieu, P. (1997). *Méditations pascaliennes* [Pascalian meditations]. Paris: Éditions du Seuil.

Cage, J. (1990). *I–VI*. Cambridge, MA: Harvard University Press.

Châtelet, G. (1993). *Les enjeux du mobile: Mathématique, physique, philosophie*. Paris: Éditions du Seuil.

Dunbar, K. (1995). How scientists really reason: Scientific reasoning in real-world laboratories. In R. J. Sternberg & J. Davidson (Eds.), *Mechanisms of insight* (pp. 365–395). Cambridge, MA: MIT Press.

Eco, U. (1984). *Semiotics and the philosophy of language*. Bloomington: Indiana University Press.

Garfinkel, H. (1967). *Studies in ethnomethodology*. Englewood Cliffs: Prentice-Hall.

Garfinkel, H. (2002). *Ethnomethodology's program: Working out Durkheim's aphorism*. Lanham: Rowman & Littlefield.

Garfinkel, H., Lynch, M., & Livingston, E. (1981). The work of a discovering science construed with materials from the optically discovered pulsar. *Philosophy of the Social Sciences, 11*, 131–158.

Goodwin, C. (1995). Seeing in depth. *Social Studies of Science, 25*, 237–274.

Heath, C., & Luff, P. (2000). *Technology in action*. Cambridge: Cambridge University Press.

Heidegger, M. (1977). *Sein und Zeit* [Being and time]. Tübingen: Max Niemeyer. (First published in 1927)

Hempel, C. G., & Oppenheim, P. (1948). Studies in the logic of explanation. *Philosophy of Science, 15*, 135–175.
Hoar, W. S. (1976). Smolt transformation: Evolution, behaviour and physiology. *Journal of Fisheries Research Board of Canada, 32*, 1234–1252.
Il'enkov, E. (1977). *Dialectical logic: Essays in its history and theory*. Moscow: Progress Publishers.
Il'enkov, E. (1982). *Dialectics of the abstract and the concrete in Marx's Capital*. Moscow: Progress Publishers.
Jucker, A. H., Smith, S. W., & Lüdge, T. (2003). Interactive aspects of vagueness in conversation. *Journal of Pragmatics, 35*, 1737–1769.
Kirsh, D., & Maglio, P. (1994). On distinguishing epistemic from pragmatic action. *Cognitive Science, 18*, 513–549.
Latour, B. (1987). *Science in action: How to follow scientists and engineers through society*. Cambridge, MA: Harvard University Press.
Latour, B. (1988). *The pasteurization of France*. Cambridge, MA: Harvard University Press.
Lave, J. (1988). *Cognition in practice: Mind, mathematics and culture in everyday life*. Cambridge: Cambridge University Press.
Levinson, R. (2011). Science education from people for people: Taking a standpoint. *Studies in Science Education, 47*, 109–117.
Lynch, M. (1998). The discursive production of uncertainty: The O. J. Simpson "dream team" and the sociology of knowledge machine. *Social Studies of Science, 28*, 829–868.
Merleau-Ponty, M. (1996). *Sense et non-sense (Sense and non-sense)*. Paris: Gallimard.
Neisser, U. (1981). John Dean's memory: A case study. *Cognition, 9*, 1–22.
Nietzsche, F. (1954). *Werke in drei Bänden* [Works in 3 Vols]. Munich: Carl Hanser.
Rorty, R. (1979). *Philosophy and the mirror of nature*. Princeton: Princeton University Press.
Rorty, R. (1989). *Contingency, irony, and solidarity*. Cambridge: Cambridge University Press.
Roth, W.-M. (1995). *Authentic school science: Knowing and learning in open-inquiry science laboratories*. Dordrecht: Kluwer Academic.
Roth, W.-M. (1998). Starting small and with uncertainty: Toward a neurocomputational account of knowing and learning in science. *International Journal of Science Education, 20*, 1089–1105.
Roth, W.-M. (2003). *Toward an anthropology of graphing*. Dordrecht: Kluwer Academic.
Roth, W.-M. (2004). What is the meaning of meaning? A case study from graphing. *Journal of Mathematical Behavior, 23*, 75–92.
Roth, W.-M. (2005). Making classifications (at) work: Ordering practices in science. *Social Studies of Science, 35*, 581–621.
Roth, W.-M. (2008). Constructing community health and safety. *Municipal Engineer, 161*, 83–92.
Roth, W.-M. (2012). *First person methods: Towards an empirical phenomenology of experience*. Rotterdam: Sense Publishers.
Roth, W.-M. (2013a). *What more? in/for science education: An ethnomethodological perspective*. Rotterdam: Sense Publishers.
Roth, W.-M. (2013b). *Meaning and mental representation: A pragmatic approach*. Rotterdam: Sense Publishers.
Roth, W.-M., & Barton, A. C. (2004). *Rethinking scientific literacy*. New York: Routledge.
Roth, W.-M., & Désautels, J. (2004). Educating for citizenship: Reappraising the role of science education. *Canadian Journal for Science Mathematics, and Technology Education, 4*, 149–168.
Roth, W.-M., & Hwang, S.-W. (2006a). Does mathematical learning occur in going from concrete to abstract or in going from abstract to concrete? *Journal of Mathematical Behavior, 25*, 334–344.
Roth, W.-M., & Hwang, S.-W. (2006b). On the relation of abstract and concrete in scientists' graph interpretations: A case study. *Journal of Mathematical Behavior, 25*, 318–333.
Roth, W.-M., & Jornet, A. G. (2013). Situated cognition. *WIREs Cognitive Science, 4*, 463–478.

Roth, W.-M., & Middleton, D. (2006). The making of asymmetries of knowing, identity, and accountability in the sequential organization of graph interpretation. *Cultural Studies of Science Education, 1*, 11–81.

Roth, W.-M., & van Eijck, M. (2010). Fullness of life as minimal unit: STEM learning across the life span. *Science Education, 94*, 1027–1048.

Roth, W.-M., McRobbie, C., Lucas, K. B., & Boutonné, S. (1997). Why do students fail to learn from demonstrations? A social practice perspective on learning in physics. *Journal of Research in Science Teaching, 34*, 509–533.

Roth, W.-M., McGinn, M. K., & Bowen, G. M. (1998). How prepared are preservice teachers to teach scientific inquiry? Levels of performance in scientific representation practices. *Journal of Science Teacher Education, 9*, 25–48.

Rouder, J. N., Speckman, P. J., Sun, D., Morey, R. D., & Iverson, G. (2009). Bayesian *t* tests for accepting and rejecting the null hypothesis. *Psychonomic Bulletin & Review, 16*, 225–237.

Sacks, H., Schegloff, E., & Jefferson, G. (1974). A simplest systematics for the organization of turn-taking in conversation. *Language, 50*, 697–735.

Temple, S. E., Plate, E. M., Ramsden, S., Haimberger, T. J., Roth, W.-M., & Hawryshyn, C. W. (2006). Seasonal cycle in vitamin A1/A2-based visual pigment composition during the life history of coho salmon (*Oncorhynchus kisutch*). *Journal of Comparative Physiology A: Sensory, Neural, and Behavioral Physiology, 192*, 301–313.

Thom, R. (1981). Worüber soll man sich wundern [What we should wonder about]. In K. Maurin, K. Michalski, & E. Rudolph (Eds.), *Offene Systeme II: Logik und Zeit* (pp. 41–107). Stuttgart: Klett-Cotta.

Vološinov, V. N. (1930). *Marksizm i folosofija jazyka: osnovye problemy sociologičeskogo metoda b nauke o jazyke* [Marxism and the philosophy of language: Main problems of the sociological method in linguistics]. Leningrad: Priboj.

Vygotskij, L. S. (2005). *Psychologija razvitija cheloveka* [Pyschology of human development]. Moscow: Eksmo.

Wald, G. (1960). The distribution and evolution of visual systems. In M. Florkin & H. Mason (Eds.), *Comparative biochemistry* (Vol. 1, pp. 311–345). New York: Academic Press.

Wittgenstein, L. (1997). *Philosophische Untersuchungen Philosophical investigations* (2nd ed.). Oxford: Blackwell. (First published in 1953)

Woolgar, S. (1990). Time and documents in researcher interaction: Some ways of making out what is happening in experimental science. In M. Lynch & S. Woolgar (Eds.), *Representation in scientific practice* (pp. 123–152). Cambridge, MA: MIT Press.

Part III
Retheorizing Graphing

> The gradual trial-and-error creation of a new, third vocabulary – the sort of vocabulary developed by people like Galileo, Hegel, or the late Yeats – is not a discovery about how old vocabularies fit together. That is why *it cannot be reached by an inferential process* – by starting with premises formulated in the old vocabularies. Such creations are not the result of successfully fitting together of a puzzle.... Someone like Galileo, Yeats, or Hegel (a "poet" in my wide sense of the term – the sense of "one who makes things new") *is typically unable to make clear* exactly what it is that he wants to do *before developing the language in which he succeeds doing it. His new vocabulary makes possible, for the first time, a formulation of its own purpose.* It is a tool for doing something which could not have been envisaged prior to the development of a particular set of descriptions, those which it itself helps to provide. (Rorty 1989, pp. 12–13, emphasis added)

Descriptions of the nature of the STEM fields often depict what happens in these fields as rational pursuits that teleologically strive towards pregiven outcomes. Engineering design, for example, is conceptualized in terms of problem framing and goal setting and the solution and achievements thereof, which is not how engineering design in the real world unfolds (Bucciarelli 1994; Sørensen and Levold 1992). In the introductory quotation, Rorty provides a different, phenomenological and pragmatic description of discoveries as emergent phenomena, which is more consistent with the observations ethnographers make when they follow engineers around. What creative activity – and learning something new is the result of creative activity, captured not in the least by the constructivist notion of learning as (auto-) *poiesis* – also does is this: creative activity has the effect of bringing about the evolution of a language that makes possible an explanation for what creative activity has done. As Rorty suggests, the creator *first succeeds in doing* something and then gets into the position to make clear *what* she wanted to do and has done.[1] It is only when everything has been said and done that the evolving language provides a tool for explicating what has been said and done and why it has been said and

[1] This is also the Marxist insistence on the primacy of praxis over theory. We can only explain what we are already familiar with – we can only learn a formal theory of language after having learned a language that can be formalized, a process that itself requires language as a tool.

done, including the evolution of language itself. A co-product of the creative activity, the descriptive language, could not have been envisaged prior to its actual emergence. But if the nature of the happening is uncertain – other than that it is scientific or mathematical research generally and without any specification – then the nature of everything else that is talked about after the fact in specific terms, including the objects, tools, means, and even identities of the subjects involved, remain open and "in-the-making" just as the encompassing activity itself (i.e., science-in-the-making). In this Part III, I present a way of looking at graphing through a lens that is itself in the making, so that phenomena no longer exist as such but always are phenomena-in-the-making. It is a way of approaching the event-nature of science and mathematics *as* events, that is, it shows them in their "evental/evential"[2] nature (Romano 1998). In other words, I approach the flow of scientific activity *as* flow rather than approaching flow in terms of the difference between two states of something (e.g., system, phenomenon) caused by a force outside the something (e.g., system, phenomenon).

Some historians study science, technology, engineering, or mathematics from an after-the-fact perspective, becoming interested in and explaining it through the lens of their existing knowledge. Historians themselves name such an approach Whig history. In the social studies of sciences, on the other hand, it has become common to study science-in-the-making (e.g. Latour 1987). Such research suggests that scientific knowledge is constructed. However, "to construct" is a transitive verb, which implies that there has to be an object towards which the actions of the subject are oriented. The problem with the social constructivist approach is that it does not theorize the STEM fields and their ways of operating through the lens of events*-in-the-making.[3] Such a lens does not even allow us to speak of an "event" because scientists do not know what they are witnessing. For the scientific research team that I describe in Part II, the nature of their activity was inherently uncertain. Initially, they were measuring the transition between rhodopsin–porphyropsin composition of the retina typical for freshwater and saltwater environments. Then they were certain to be in the middle of a scientific revolution worth a journal article in one of the two leading scientific journals. Finally, the team was merely reifying what others have already shown. That is, in the course of their activity, scientists do not know what is going to come from their actions – other than that they want to publish whatever they get. Even more important, the scientist as subject of activity no longer *is*, but always in a state of *becoming* so that there is nobody upon whom to pin intentional agency. It is only when everything has been said and done that the

[2] Romano theorizes the event as event, as something unfinished and in the making, which implies that all its constitutive parts are in the making. He uses a pair of adjectives, *événementiel/évé nemential,* the first of which denotes relation to event (here, as in the English translation of the book, as *evental*), and the second being a neologism (here, as in the English translation of the book, as *evential*).

[3] I use the asterisk to denote the provisional character of the event, for in the unfolding happening, participants do not know which kind of event is in the making.

scientists can write causal narratives according to which they designed a specific study based on what they already knew.

In Part II of this book, the scientists did not know what they ultimately understood as having done. Thus, an appropriate theory of learning and development in the sciences needs to theorize what is happening through the lens of the unfolding event*, the nature of which is yet to be determined. Cause and effect relations (we did this *because*...) can therefore be established only after the fact. But at that stage, the scientists themselves no longer are who they have been at the outset of the project. They, too, have changed, together with their science, tools, materials, and so on. In Chap. 9, I retheorize scientific work through the lens of the event*-in-the-making, which forces us to theorize people, materials, representations, instrumentation, and everything else as events*-in-the-making (Bakhtin 1993; Romano 1998). I emphasize, with Luria (2003), that even though our focus is so close as to show only the labeling of the abscissa, what we see needs to be related to and described as integral to the whole of *society-specific* human activity, with all its cultural-historical dimensions. Between the whole activity and its microconstitution at the second or millisecond level, we may be interested in focusing on movements that have their characteristic temporal dimensions on other scales, including minutes, hours, days, weeks, or months. From this perspective, even subjects, objects, or tools themselves have to be theorized as phenomena*-in-the-making, which obtain their specific characteristics only after everything has been said and done. Before that we have to consider these in provisional terms, that is, as subjects*-in-the-making, objects*-in-the-making, or tools*-in-the-making.

Graphing (production, interpretation, reading) tends to be approached as a skill or a set of skills that is somewhere located and represented in the mind (e.g. as procedural skill). However, from a cultural-historical perspective, the specificity of human practices is their *society-specific cultural* nature. The origin of individual practices and skills, therefore, is not the individual but culture. Where in culture would the individual find the skill or acquire them? Vygotskij (2005) suggests that *all higher psychological functions* are *societal relations*. That is, by participating in societal relations, human beings come to engage new forms of practices (skills) that later appear to be owned by individuals. This also allows us to explain why new skills, practices, and forms of knowledgeability can emerge even though these exceed individual competencies. The practices and forms of knowledgeability exist *as* and *in* relations first. We may observe them again looking at individual persons when they have become familiar with the new skills, practices, and forms of knowledgeability. At that point they exhibit these skills, practices, and forms of knowledgeability independent of the relations where they first appeared. This is also the case in scientific discovery research. I show in Chap. 10 how new forms of knowledgeability concerning graphs and explanations of graphs exist *in* relation before they exist *for* the individual.

References

Bakhtin, M. (1993). *Towards a philosophy of the act*. Austin: University of Texas Press.
Bucciarelli, L. L. (1994). *Designing engineers*. Cambridge, MA: MIT Press.
Latour, B. (1987). *Science in action: How to follow scientists and engineers through society*. Cambridge, MA: Harvard University Press.
Luria, A. R. (2003). *Osnoby nejrolpsixologii* [Foundations of neuropsychology]. Moscow: Isdatel'skij Centr «Akademija».
Romano, C. (1998). *L'événement et le monde* [Event and world]. Paris: Presses Universitaires de France.
Rorty, R. (1989). *Contingency, irony, and solidarity*. Cambridge: Cambridge University Press.
Sørensen, K. H., & Levold, N. (1992). Tacit networks, heterogeneous engineers, and embodied technology. *Science, Technology & Human Values, 17*, 13–35.
Vygotskij, L. S. (2005). *Psychologija razvitija cheloveka* [Pyschology of human development]. Moscow: Eksmo.

Chapter 9
Graphing*-in-the-Making

> No longer the pleasure of certainty but of uncertainty; no longer "cause and effect," but the continuously creative. (Nietzsche 1954, vol. 3, p. 437)
>
> The effect never is "conscious": the discovered and imagined cause is projected, *follows* time. (p. 473)
>
> The living body [Leib]. The thing, the "whole" constructed by the eye evokes the distinction between a doing and a doer; the doer, the cause of the doing, grasped in finer details left the "subject" in the end. (p. 484)

Many historians working in the area of science education tend to study scientific research from an after-the-fact perspective, becoming interested in and explaining it through the lens of their present, existing knowledge. In the field of history, this approach is referred to as "Whig history." Nietzsche deconstructed this approach[1] because, as he suggested, (precise) effects never are present to consciousness – a recognition that over 100 years later were described as the distinction between (mental) plans and situated actions (Suchman 1987) and has led to ethnographic studies describing scientific research in terms of its radical uncertainties (Roth 2009). It is only after the fact, once effects are known, that "causes" really can be attributed to actions and effects. In the social studies of sciences, a different approach to the study of the science orients us to its always-unfinished nature, to *science-in-the-making* (e.g. Latour 1987). During the 1980s, it has become fashionable to say that scientific knowledge is "constructed," generally within communities of practice and, depending on the bent of the researcher, might be said "to be constructed" by individual members of the community. However, "to construct" is a transitive verb. Transitivity implies a process beginning in the subject of action that affects the object towards which its actions are oriented. The resulting pattern may be denoted by the formula $S \rightarrow O$, which is the same structure realized in Western languages with their subject (intent) $\rightarrow$ verb $\rightarrow$ object structure.

[1] I use the verb "to deconstruct" not as synonymous with "to destruct," but as synonymous with "to unpack" and "to critique." Nietzsche showed that cause–effect talk is fundamental to the human endeavor to anticipate and control future events. The cause–effect figure therefore is equivalent to the human "will to power."

W.-M. Roth, *Uncertainty and Graphing in Discovery Work*,
DOI 10.1007/978-94-007-7009-6_9,

The problem with the (social) constructivist approach is that it does not theorize the sciences and its ways of operating through the lens of science*-in-the-making, which does not even allow us to speak of a "science" as something fixed. Participating in changing activity, scientists themselves change – evidence for which is amply provided in the chapters of Part II. Most specifically, I show in this book how scientists themselves do not know what is going to be the effect of their actions – even though, especially in later stages of the research, they are more successful in acting such that their plans (goals) become appropriate a posteriori descriptions of what they have done. In the chapters of Part II, the scientists are in the process of collecting and analyzing measurements without knowing what the results will be, whether these confirm theory, in which case the results are rather uninteresting. What we need are descriptions and explanations that allow us to theorize actions in the situation of uncertainty in which scientists operate. At the beginning of our study, Craig and I did not know that we would overthrow what he called the canon in the field. Thus, an appropriate theory of learning and development in the sciences needs to theorize what is happening through the lens of the unfolding event*, the nature of which is yet to be determined. But when the ontological status of the event* is uncertain, so will be the equivalent status of the subject and object. More radically stated, the event *is not* and, as a consequence, we have to say that the subject (object, instrument) *is not* because everything is becoming. This opens up opportunities for integrating the role of (radical) uncertainty into theories of (the nature of) science, for becoming is becoming only when there is novelty (Prigogine 1980).[2] Cause and effect relations ("we did this *because* . . .") can be established only after the fact. In this chapter, I retheorize scientific practice generally and graphing specifically through the lens of the event*-in-the-making, which forces us to think about people, materials, representations, instrumentation, and everything else as events (Bakhtin 1993).

Toward a Dialectical Theory of Learning and Change

> We exactly cannot think anything insofar as it *is* . . . (Nietzsche 1954, vol. 3, p. 497)

In Part II, we observe scientists in the process of creating new knowledge as the result of conducting research. In the process, they undergo what can be characterized as a change in their conceptual language (Chap. 7). However, the change had none of the characteristics that scholars associate with a conceptual change, such as the switching from one cognitive framework to another (e.g., Chi 1992). Rather, there were continuous changes on a number of fronts – best-fit curves, equations,

[2] In a simple physical system, where velocity v may be given as a function of time – such as in the equation $v(t) = v_0 + a \cdot t$ – there is change but no novelty (Müller 1972). The system is not changing but pregiven. There is not even history, for t may be replaced by $-t$ without any change in the physics involved.

explanatory talk, delimitations of data included – that together led to qualitative changes: first the emergence of alternative hypotheses and then the transition in dominance from one to another hypothesis concerning the shape of the relation between porphyropsin levels and time. The quantitative changes were associated with changes of many different aspects in and of the overall activity,[3] increasing familiarity with the instrumentation, method, data, original context of the data sources (coho salmon), and so on. The continuous change is unanticipated and internal to the activity itself. That is, when we observe research activity, then we cannot think of it *as it is*, following Nietzsche's quotation above, but only observe it in its continuous *becoming*. Thus, we need to talk about *graphing*-in-the-making* rather than as a constant "skill" that scientists somehow "have." In this section, I weave together some of the strands started in Chaps. 1, 6, and 7 to articulate a different way of talking about change that has arisen with dialectical logic. This form of logic underlies several theoretical frames for explaining human activities generally and language specifically, frames that have emerged in the early to mid 1900s in the Soviet Union. These theories have come to inform Western scholarship only in the 1970s and 1980s. But in the translation and adaptation of the theoretical literature, the dialectical logic at the heart of these theories has been lost or extirpated (e.g. Roth 2014). Therefore, prior to articulating a dynamic theory to scientific research activity generally and graphs and graphing more specifically, I begin by providing an overview of the dialectical approach. This description is in continuation of ideas initially articulated in Chap. 1 concerning unit analysis and in continuation of the role of inner contradictions (Chap. 6).

Characterizing Flow and Change

In the classic approach,[4] change has been characterized in terms of the difference between two self-identical states. Thus, for example, learning, tends to be characterized in terms of the difference between knowledge before and knowledge after some (curriculum) event. Some agent or force *outside* of the phenomenon – e.g., knowledge – is said to be the force of the change. Thus, when it is said that a "'learner' 'constructs' 'knowledge'," we have some subject *S* that acts upon an object *O* so that we get, as noted above, a subject (intent) → verb → object structure that also is the structure of Western languages. Thus, the logic of language and the logic of phenomena in the material world are identical (Heidegger 1927/1977;

[3] I define the category *activity* below in a way consistent with cultural-historical activity theory, as a societally motivated and organized form, rather according to the everyday and scholarly use of the term in the Anglo-Saxon literature, where the term might be employed to talk about children doing a school mathematics task involving Cuisinaire rods.

[4] The approach typical of philosophy termed "metaphysics," dealing in ideas and "meanings," which is inherent to the social and natural sciences.

Nietzsche 1954). Sometimes such learning – i.e., when it is not simple accretion – is said to exist in the form of a change in cognitive structure (e.g., Chi 1992). During instruction or as a result of the curriculum, learners (S) change this structure (Fig. 9.1). Each structure, the one prior the curriculum unit and the one following it, can be assessed and determined. The unit of analysis, therefore, is the element, the object of activity (O), here knowledge. The elements – in the figure the circles and relations – are put together by relating them hierarchically, yielding the overall structure. The difference between these structures is said to be a cognitive reconstruction or conceptual change. The figure shows that the actions or force that produce the change are *outside of the conceptual structure* that is made to change. The *cause* of the change is a construction on the part of the subject; the effect is a restructuring. The problematic nature of cause–effect thinking is apparent below.

The problem with this approach is that change itself is not theorized *within* the changing system (conceptual structure). That is, the conceptual structure itself is not involved in the change. Some agent outside the two structures is dissatisfied with the first and then exchanges it for the second structure. On the other hand, to make change part of the system requires a fundamental unit that is change itself. This is equivalent to saying that the category needs to embody change – it needs to be a unit of change. We might ask, for example, about flow what the minimum unit is (Fig. 9.2a). The minimum unit of flow, our category of flow, will have to be a unit that embodies *flow* itself (Fig. 9.2b). We can immediately see that change is internal to such a unit, for the left and right of this unit are not the same. A physicist, for example, might characterize flow – e.g., of water in the Heraclitean river, which never was the same when someone placed a foot twice – in terms of pressure. Within the unit, then, the pressure is not the same, here characterized by Δp, that is, any two values p_1 and p_2 although there will be infinitely many different pressures in that same unit (Fig. 9.2c). This is also consistent with the description of space as filled with (quantum) fluctuations rather than constituting something empty.

The reader unfamiliar with dialectical reasoning might think that the situation is the same as that above depicting conceptual change. But this is not so because the two pressures are *within* the minimum unit. That is, depending on when and where we look, the minimum unit *manifests* itself differently, sometimes as p_1, sometimes as p_2, and generally in any other pressure as well. This is where dialectical logic radically differs from classical logic: the same unit may *manifest* itself differently and, therefore, in contradictory ways. It manifests itself differently because it is not identical with itself: it *becomes* rather than *is*. Classical philosophers tend to say that these differences are due to differences between observers, or differences in the position of the participants – e.g., such as the same commodity that manifests itself differently to seller and buyer, namely as exchange-value and use-value, respectively (Il'enkov 1977). The point K. Marx made is that in dialectical logic, one and the same thing can manifest itself differently only because it is already *different within itself*. That is, the thing *is not*, that is, it does not have an essence but is becoming. We see the relevance of this statement in our consideration of flow (Fig. 9.2b, c), where the difference is *internal* to our unit.

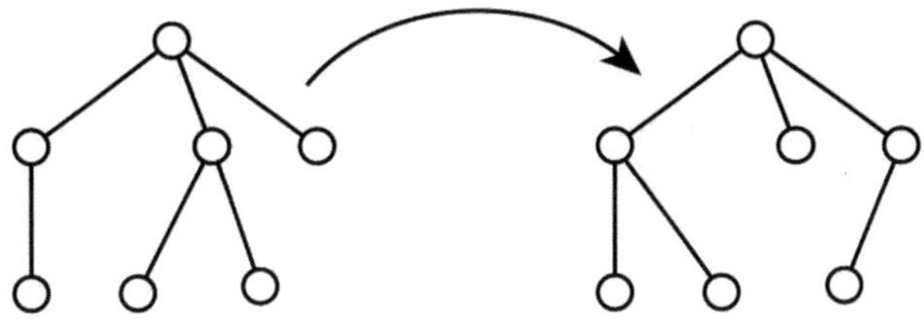

Fig. 9.1 Learning by means of conceptual change involves actions (*arrow*) that reconstruct the structure. These actions, the force of change, are outside of the structure, composed of the elements (*circles and relations*) from which the structure is built

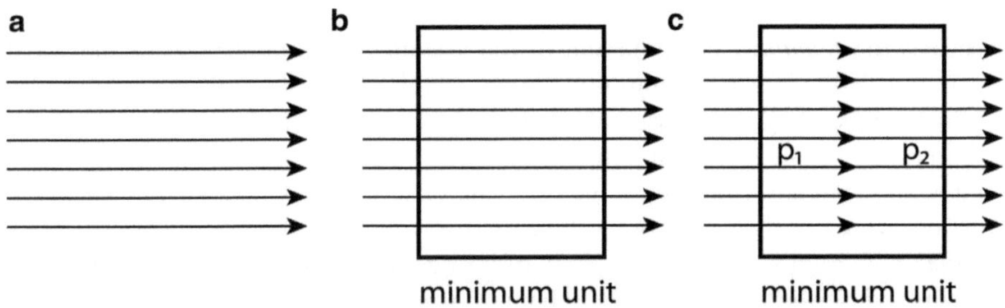

Fig. 9.2 (a) A depiction of flow, such as in a river. (b) A minimum unit of *flow* must itself be flow. (c) A physicist might characterize flow in terms of the difference of hydrostatic pressure in different parts of our unit. In the dialectical approach, different pressures are manifestations of *the same* unit

For scientists this phenomenon of one thing manifesting itself in different ways is not so unfamiliar. For one, depending on the way we look at it (in our experiment), light *manifests* itself as a wave. Thus, to explain how light propagates through the apparatus that we used in the research described in Part II to measure light absorption, we have to use a wave model. For example, the light was split into the different wavelengths, and this phenomenon of refraction requires the wave picture. However, to explain how the light meter within the two CCD cameras that we employed – one for recording the absorption the other for viewing the objects on the microscopic slides – we use a particle picture and the photo effect. This is so because explaining the operation of a CCD requires the modeling of light in terms of photons that knock electrons from the walls of the tubes that the CCD is made off[5]; and these electrons are accelerated and amplified into a measurable electronic signal. That is, light manifests itself as a wave or as a particle. Following N. Bohr, scientists refer to this as complementarity. We get a similar phenomenon – one even closer to our phenomenon of flow – in the story of Schrödinger's cat, a story that is now widespread in popular culture. As I describe in Chap. 6, the story is about a cat

[5] This is so because, as Max Planck and Albert Einstein have shown, waves never could transfer in a sufficiently short time the amount of energy necessary to knock an electron out of material.

in a box, where there is a vial with a deadly poison. The vial is broken by the decay of an atom also within the box. When we put the cat into the box it is alive. When we look into the box a sufficiently long time later, it is (more or less likely) dead. But how do we model – i.e., think about – the cat in the meantime? Saying that it is both dead and alive is not intelligible: A cat cannot be both dead and alive simultaneously. As shown in Chap. 6, physicists model the wave function ψ of the cat as a superposition of the two unobservable states *alive* and *dead.*

$$|\psi(t)\rangle = \frac{1}{\sqrt{2}}(|dead\rangle + |alive\rangle) \tag{9.1}$$

In physics, the wave function is not the same as an observation. It merely describes the *temporal* development of the system. We cannot observe the wave function. It is like our minimum unit above, which manifests itself differently – as p_1 or p_2 – depending on where and when we look. That is, when we actually make an observation some time following the closure of the box, we observe the cat either dead or alive but not a mixture of the two. The system is well described by the wave function, which itself is not observable. But it gives us an appropriate observable.[6] We may now return to another analogy of change involving a structure, such as the shearing I describe in Chap. 6 that changes a rectangle into a parallelogram. We may then describe it in dialectical terms such that the minimum unit of analysis is one of change (Fig. 9.3). The figure makes perceptually salient that the minimum unit can manifest itself differently, as rectangle, parallelogram, or as parallelogram in between. In this way, we think about shearing *as* shearing, that is, about flow *as* flow rather than as a difference between two states. In the present way of thinking, the different states are but manifestations of the same unit.

Describing change in this way is consistent with a statement J. Lave (1993) made about knowledge and learning, which has not been used as much in the learning sciences as I am pushing it here. She describes the difference between two conceptions of knowing and learning as an epistemological one,

> between a view of knowledge as a collection of real entities, located in heads and of learning as a process of internalizing them, versus a view of *knowing and learning as engagement in changing processes of human activity*. In the latter case "knowledge" becomes a complex and problematic concept, whereas in the former it is "learning" that is problematic. (p. 12, emphasis added)

When we consider this quotation in terms of the dialectical approach that I am working out here, we can see how these two statements are commensurable with each other. We can use Fig. 9.3 as an analogy of changing human activity and the geometrical figures as *manifestations* of knowledge. It is immediately evident that

[6] Physicists call the change from wave function to observation as the collapse of the wave function. It is calculated analogically to the length of a vector via the multi-dimensional equivalent of the Pythagorean theorem $c^2 = a^2 + b^2$. In the case of the wave function and the bracket notation, the observation a is obtained as $a = \langle\psi|\mathbf{A}|\psi\rangle$, where **A** is a matrix-like *operator* characteristic for the observations that can be made of the object.

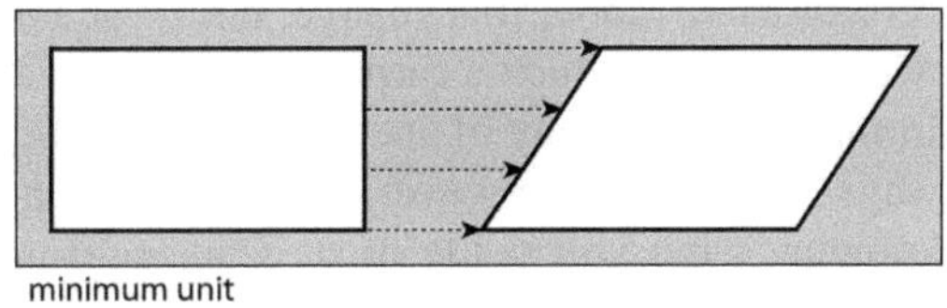

Fig. 9.3 The process of shearing thought in dialectical terms, that is, in terms of a minimum unit of analysis that *embodies* change

in this analogy, learning, as a form of change, is unproblematic. Learning, as change, is *internal* to our unit of analysis and the category that we use to denote it. It occurs by the mere fact that our fundamental unit of analysis takes a flow perspective on the river of human life. On the other hand, *knowledge* becomes problematic in this approach because one and the same unit expresses itself in different knowledge forms – depending on when and where we look. Lave therefore also uses the term *knowledgeability*, which emphasizes the capacity to mobilize what is required at hand rather than something stable stacked away in a storehouse underneath the skull and between the two ears. This contrasts the earlier analogy of the conceptual change as a change between different structures (Fig. 9.1), where the identification of the knowledge structures is taken to be unproblematic and the conceptual change becomes the problematic issue to be dealt with. The dialectical logic articulated here is useful, as is the foundation of two theoretical frameworks that are often cited in the scholarly literature: cultural-historical activity theory (Leont'ev 1983) and the (Marxist) sociological theory of language that was developed by the circle around M. M. Bakhtin (e.g., Vološinov 1930).

Evolution of Theory and Language: A Dialectical Approach

> As Kuhn argues in *The Copernican Revolution*, we did not decide on the basis of some telescopic observations, or on the basis of anything else, that the earth was not the center of the universe, that macroscopic behavior could be explained on the basis of microstructural motion, and that prediction and control should be the principle aim of scientific theorizing. Rather, after a hundred years of inconclusive muddle, the Europeans found themselves speaking in a way which took these interlocked theses for granted. Cultural change of this magnitude does not result from applying criteria (or from "arbitrary decision") any more than individuals become theists or atheists, or shift from one spouse or circle of friends to another, as a result either of applying criteria or of *actes gratuits*. We should not look within ourselves for criteria of decision in such matters any more than we should look to the world. (Rorty 1989, p. 6, original emphasis, underline added)

Language, including scientific language, evolves continuously, and without our explicit intentions. Commenting on Kuhn's historical analysis, Rorty (in the introductory quotation) suggests that the Europeans *found* themselves speaking a language that they had not intended. This language change, having occurred over roughly 100 years, constituted a scientific revolution. There is nothing "constructed" about this change, not if the term "construction" has anything to do with the way in which a carpenter constructs a house following the architect's plan. In the scientists' way of speaking, certain theses were taken for granted. The new ways of speaking, however, have *evolved* in the course of speaking and without knowing what will come of it. That is, at the beginning of this roughly defined period of 100 years, the other end was not in sight; it could not be anticipated because the language-in-use did not foresee what ultimately will be taken for granted. As the philosopher points out, this is "just a way of saying that our present views about nature are our only guide in talking about the relation between nature and our words" (Rorty 1979, p. 276). Rorty describes scientific language in terms of a continuous evolution that does not follow intention. In fact, the future states – e.g., whatever follows the revolution – are not implied in the preceding states and therefore cannot be predicted on the basis of what currently exists. Otherwise the consequences could be stated. Instead, the very concept of revolution suggests that something radically changes. But this radical change does not occur like the conceptual change in the conceptual change literature (Fig. 9.1). Instead, Rorty's description is more consistent with the image of flow that is continuous and where later states cannot be anticipated, especially if the flow is not linear and laminar, as in the simple analogy I provide above (Fig. 9.2b, c). The change between the theories – when the two theories (i.e., before and after) are held against each other – appears more or less radical, and perhaps even unintelligible. In Chap. 7, I describe a catastrophe theoretic model that does much better work in describing the kinds of changes that we actually observe. Thus, when Rorty looks at Kuhn's account, he finds neither intentional constructions nor reasons that would have driven the change in conceptual language.

The problem with the traditional conceptual change account is that the change process is not described to a sufficient degree. Why would or how could a scientist or learner intentionally orient herself toward and construct the new conception that is outside his or her horizon of familiarity? Who is the agent that compares alternative hypotheses and how is this agent related to the hypotheses? This new conception cannot be anticipated on the basis of what currently exists ("prior knowledge"), much like Christopher Columbus could not have set out to discover the Americas precisely because he did not know that these existed; and had he known the existence thereof, he would not have had to discover the Americas. How could one look for a needle in a haystack when one does not orient towards finding a needle or when one does not know (from hindsight) that one is searching for a needle? Rorty describes the change in evolutionary terms; and evolutions are contingent. The outcomes are not predictable or derivable by analytic means – which is the precisely the key point of the catastrophe theoretic description and the unforeseeable appearance of the two qualitative transitions. At any instant in time, there might be unanticipated qualitative changes, bifurcations in the catastrophe (chaos) theoretic framing of system development, or transitions between different states that are brought about by infinitesimally small influences ("the butterfly effect"). Although scientists know about contingencies, they do not conceive of their work in such terms and instead tie it to what they state to be a structured nature out there. Thus, in the present study, Shelby acknowledged contingencies in our team's and his own work:

> And as you go down the path, things don't work, some things do work, *and some things just come out of the blue, and you don't know why or how*. And it's usually things that are unpredictable, that didn't work, that ends up being the most interesting. So whereas, you know, you start off with one idea, within a very short term you have a whole bunch of other ideas and those become more interesting than the original one and maybe even have more interesting applications. I would say that there is no question that it *evolves*. And that this lab has certainly seen *evolution*, partly from these new things that you find, and partly from new grad students coming in.

Despite such contingencies scientific research will arrive at the truth – a realization of the "truth-will-out" figure of speech ("discursive device") (Gilbert and Mulkay 1984) that brings everything back into the order that appears in the depiction of the research in a scientific journal.

Rorty actually is not the first scholar to describe macroscopic change in terms of microscopic evolution. The literary theorist M. M. Bakhtin (1981) suggests that the change of the novel genre from its prehistory to the modern day *cannot be explained* when we line up all the novels that have been written and by going from one to the next in the way series of photographs are lined up on a movie reel, which are animated using a movie projector (or software program in the case of digital images). Bakhtin points out that going from one novel to the next, as going from one photograph to the next – or, in our context, from one theory or form of conceptual talk to the next theory or form of conceptual talk – does not explain *why* or *how* the novel form changed in the way it did. It is not by analyzing the structure of the novel during one period – equivalent to analyzing one of the photos in Fig. 9.4 – and then analyzing the structure of the novel of another period – equivalent to analyzing the structure of another photo in Fig. 9.4 – that will yield the *inner* dynamic of the change. Instead, Bakhtin proposes taking each novel as a reflection of its time, making use of the languages presently available refracted through the personal styles of author and protagonists (a farmer in one of Dostoevsky's novels will speak differently than a prince or a monk). Thus, the languages we find in a single novel

> are, in the main, period bound, generic and common everyday varieties of the epoch's literary language, a language that is in itself ever evolving and in process of renewal. All these languages, with all the direct expressive means at their disposal, themselves become the object of representation. . . . The author participates in the novel (he is omnipresent in it) with *almost no direct language of his own*. The language of the novel is a *system* of languages that mutually and ideologically interanimate each other. It is impossible to describe and analyze it as a single unitary language. (Bakhtin 1981, p. 47)

Here Bakhtin describes the language in a novel as a multiplicity, which is not a unitary phenomenon. Moreover, this language is undergoing continuous change, ever evolving, ever in the process of renewal, and ever *becoming*. In a way, therefore, language never *is*; it *is* not, because always changing and, therefore, always different from what is has been just fractions of a second before. Why, one might ask, is language changing? This is so because as an expression of life, as a

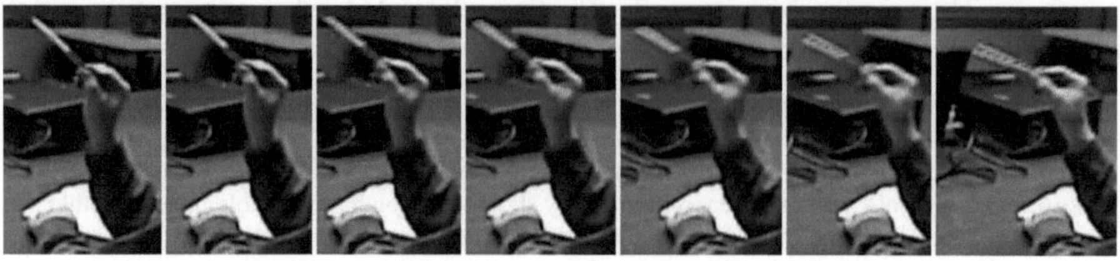

Fig. 9.4 Lining up and making a motor (software) go from one photo to the next does not explain the movement *from within* itself. The movement is produced by an outside force

living phenomenon, it is equivalent to change; only dead languages do not change. Life, and with it language, is flowing (living) and therefore changing. Dead are those languages that are not spoken; and when they are used, such as Latin in the Vatican, then they are no longer generative. Authors writing for recipients overcome "the superficial 'literariness' of moribund, outmoded styles and fashionable period-bound languages" (p. 49). The language of the author "strives to renew itself by drawing on the fundamental elements of folk language" (p. 49). Folk language, however, as the literary scholar shows elsewhere, undergoes continuous change (e.g., Bakhtin 1984).

To summarize, then, Bakhtin writes about language, its continuous change, in the way I describe flow. Language changes because it is spoken. But language is only one of the many different manifestations of a culture – paintings, sculpture, technology, music, or science being other manifestations (e.g. Mannheim 2004). Thus, scientific language – i.e., theories and conceptual talk – changes because it is used; and it is used in the course of doing science. As a consequence, *scientific activity changes when done*, resulting, when we only look at sufficiently macroscopic scales, in clearly noticeable differences, some of which are denoted by the term scientific (technological) revolution.[7] One theory that has been increasingly used to explain cultural change, as I point out in Chap. 1, is cultural-historical activity theory. In the following, I articulate this theory with a particular focus on its dialectical underpinnings, which allow us much better to appreciate what the theory was designed to capture than the static, structural ways in which this theory tends to be used in Western scholarship (Roth 2014).

[7] This raises questions about "the" "nature" of "science," for if anything at all, the nature is changing. The nature of science is better described in terms of a changing nature, which is equivalent to the always temporary and therefore provisional quality of *nature** and *science**.

Cultural-Historical Activity Theory: A Dialectical Perspective

Cultural-historical activity theory was developed to describe and explain development and change. Its fundamental underpinning is that of the societal nature of those characteristics that make us specifically human. Thus, what distinguishes humans from animals are not so much certain abilities – that we walk, breath, or perceive – as it is the fact that humans live in society and its culture. Societies continue to live on even though their individual members die. Thus, when analyzing characteristically *human* behavior and competencies, we need to work with categories that are characteristic of human society. The categories also have to reflect the changing nature of culture so that the evident historical changes we observe around us come to be built into the theory. Most well known among the psychologists with a dialectical bent are L. S. Vygotsky and his student and colleague A. N. Leont'ev. Their fundamental starting point is captured in the diction that "*the psychological* nature of man is *the totality of societal relations* shifted to the inner sphere and having become functions of the personality and forms of its structure" (Vygotskij 2005, p. 1023).[8]

Humans change their conditions in and through work. Labor and activity, therefore, are characteristically human. More specifically, the (increasing) division of labor already existing within animal forms – e.g. chimpanzees engage in the collective hunt of colobus monkeys and, in so doing, practice a division of labors where one is the chaser and the others are the killers – has become the epitome of human society. Society therefore can be explained in terms of the activities by means of which humans fulfill their basic and extended needs. As pointed out in Chap. 1, cultural-historical activity theory is often represented as a structure (Fig. 9.5), which has led to many misconceptions and inappropriate applications of the theory that Vygotsky and Leont'ev brought into being and developed in its earlier forms. The first and major problem is that the different *aspects* of an activity – read and heard as dejatel'nost/Tätigkeit – are treated as elements that could be explained in and of themselves. In fact, these are just manifestations of the minimum analytic unit in the same way as p_1 or p_2 were manifestations of the flow unit, or in the way that the live and dead cat was a manifestation of the temporal description of the cat function. To properly explain the function of the theory describing change – learning and development in schools or science – we have to view it in a way analogous to flow. That is, viewing activity in terms of the

[8] The English version of the chapter to which I am referring here translates the Russian adjective *obščestvennix* as "social," whereas it ought to be "societal," because Vygotskij directly refers to K. Marx, who used the German adjective *gesellschaftlich* [societal] rather than the adjective *sozial* [social]. The difference between social and societal is an important one, with many political ramifications concerning the source of psychological or sociological problems we observe. Thus, for example, in Vygotskij's view, the psychology of a criminal is a reflection of the totality of his societal relations reflected on the inner sphere. The source of the problem is society rather than the individual.

Fig. 9.5 Cultural-historical activity commonly is represented in the form of a triangle that depicts the main aspects of activity and its constitutive relations

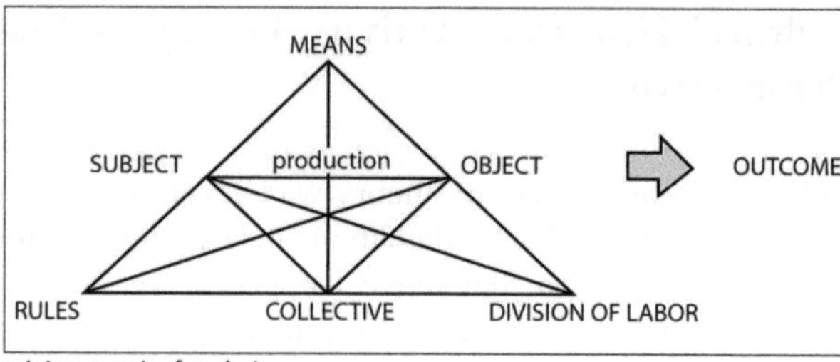

triangle is deceiving, as it easily gives the impression of a structure that remains constant while the product is manufactured.[9]

To move toward a dialectical way of using activity theory, let us return to the process of shearing (Fig. 9.3) and pretend for the time that it is one of those societal activities that provide for fundamental, generalized human needs. The square then represents the object, that is, the material condition at the beginning of the activity as a whole; the parallelogram represents the final state, after all transformations have been accomplished in and through human labor. Because activity is characteristically human, the anticipation of possible future states is also part of the object (square) – activity theorists speak of the object/motive.[10] The special characteristic of cultural-historical activity is that it takes the entire process, from the beginning to the end, as *one* unit: activity (dejatel'nost'/Tätigkeit). But just as the rectangle-transforming-into-a-parallelogram is continuously changing – *one and the same unit* thereby manifests itself differently depending on when and where we look – so the activity system manifests itself differently depending on when and where we look. In any event, wherever and whenever we look, there is only flux – no object, no subject, but object*-in-the-making and subject*-in-the-making, and, therefore, also community*-in-the-making, rules*-in-the-making, tools*-in-the-making, and division-of-labor*-in-the-making. All of these are continuously *becoming*.

First, as we look at an activity, its objects, means of production, community, or subjects may stand out. That is, in each of these gazes the minimum unit, the activity as a whole, will manifest itself in different ways. Second, the minimum unit will manifest itself differently as a function of *when* we look. For example, the object in its transformation toward the product will look differently in the early stages from later stages. But all these different looks belong to the *same* minimal unit. These are different manifestations of this same unit. Because of the nature of the unit, when the object changes, so does everything else in the minimum unit. This is so because there is a whole–part relation and every manifestation is a function of all other manifestations. An easy way of portraying this is by

[9] One of the reason for the incorrect uptake is the website of the *Center for Activity Theory* in Helsinki, which denotes the nodes as "elements," which flies in the face of Vygotskij's (2005) call for *unit* analysis that is to replace analysis in terms of *elements*.

[10] Marx says that the difference between the worst human builder and the best bee lies in the fact that the human builder builds his cell beforehand in his head (Marx and Engels 1958).

considering the energy that a human subject consumes while doing something, and this reconfigures her body; and because the object is becoming, so is the reflection of object-related reality in her mind. That is, in the course of the productive process, not only the material world is becoming but so is the way in which the world is reflected in consciousness. All these different manifestations are part of this minimal unit of analysis – activity. Theorizing activity means theorizing change. That is, time is built into the minimal unit (category) rather than constituting an a priori to experience (as Kant wanted it) or being constructed in and by the experience individual (as Piaget wanted it).[11] When we look at activity, we see change occurring all over. This is precisely the same thing as saying that language changes in speaking: activity changes in doing. It is precisely this aspect that is highlighted in the above-quoted statement that portrays "knowing and learning as engagement in changing processes of human activity" (Lave 1988, p. 12). That is, engagement is synonymous with change, which, when it pertains to human consciousness, normally is denoted by the concept *learning* (or development). Knowing, too, is a manifestation of the changing process of human activity, but it never is the same. At the most basic level, someone doing some manual task (slowly) gets better at doing it in the course of doing it (e.g. Lee and Roth 2005) – just as someone gets better at speaking a language by speaking the language.

The immediate consequence of this description is that none of the manifestations of our basic unit of analysis of productive human activity – subject, object, means of production, rules, community, division of labor, and product – ever *is*; rather these are manifestations of a process of life continuously *becoming*. Constancy and stasis are problematic ideas; change, evolution, learning, and development are inherent and immanent in the dialectical way of portraying human activity. In the following, I take such a perspective to the contents of Part II of this book, exemplifying this form of theorizing to phenomena at different temporal scales.

Even before starting the next section, we can already anticipate from the aforesaid that neither graphs nor graphing are to be thought in terms of constant entities or phenomena – as fixed things and processes. At best, we might denote these by terms such as graphs* and graphing*, where the asterisk underscores the always provisional nature of our description. In the same way, everything else will be described in terms of change, because the entire research project underwent continuous change, from the point in 1998 when Craig and I first met and talked about relating porphyropsin (A_2) levels at the point of release of young coho salmon from hatcheries to the publication of our study in 2006. Readers find in Part II descriptions evidencing change in the subjects – our laboratory competencies, theories, physical bodies – objects, means of production (the instrumentation), practices (community-of-practice-specific actions), and division of labor changed and nothing was the some in 2006 as it had been in 1998. The change is described, here, in terms of an evolution, part of which manifests itself in the changing

[11] It has been shown that becoming aware and time co-implicate each other, because awareness of something as something always is delayed (e.g. Romano 1998).

language-in-use when talking about the changes coho salmon undergo in the course of their life history (see Chap. 7).

A Dynamic, Cultural-historical (Dialectical) Perspective

Graphs that result from scientific research activity cannot be explained in and of themselves but only in their relation to everything else that was happening in the research laboratory between 1998 and 2006. The graphs were becoming; and so were our research team's descriptions and explanations thereof. Graphs were not stable entities but graphs*-continuously-in-the-making: on material and ideal planes. Explanation, projection from available data, and other aspects of graphing also changed in the course of and in concert with all the other parts of the scientific research activity-realizing project that we were undertaking. We too, the subjects of the activity, were changing, talking differently about the life history of coho salmon, differently about what we were in the process of doing, and we related differently to each other. Saying that our knowledgeability was changing is equivalent to saying that we were changing: subjects*-in-the-making. We were changing also in terms of our material practices, which became more competent – a phenomenon especially visible in the newcomers, including Shelby and Sam, who became familiar with the project by participating in it. The practices were practices*-in-the-making, and those related to graphing were graphing*-in-the-making. The laboratory practices also changed in their content, so that Shelby and Sam, for example, some time into the project, began to sample photoreceptor cells only from the dorsal part of the retina rather than taking just any photoreceptor as we had done in the beginning. Eventually becoming part of the methods section in the ultimately published article, these instructions for particular sampling practices also became available for others to follow. The scientific equipment changed over time, and, with it, the tool-related actions (practices), the quality of the data, and, ultimately, the entire research project. The division of labor also changed, as the work that initially Craig, Theo, and I were doing was shifted first to Shelby and, once Sam had joined, from Shelby to Sam. Theo, who had been part of the data collection aspect increasingly shifted to analysis, writing software, and modeling phenomena. I started focusing more on the hatchery partially in the attempt to better explain the life history of salmon there and the practices* of raising salmon from the day of the egg and milt take to the day of release.

From Initial (Ephemeral) Idea to Scientific Research Outcome

In cultural-historical activity theory, the entire process of production from the initial conception (motive) and the materials at hand (object) to the final product lies *within* the minimum unit of analysis. In our case, this spans the entire period from when Craig and I first met in 1998 to the publication of the article in 2006, where we report the changes we observed in the composition of photoreceptors of coho retina (percent A_2). This period is captured by means of images taken at two points: The images show, side by side, the ephemeral gestures Craig used when I met him to make present in our talk the canonical view of retinal changes and the final journal article (Fig. 9.6). This entire span, everything that has happened between these two points in time, spanning nearly 8 years, *constitutes our fundamental unit of analysis*. This unit cannot be broken down further into smaller elements that could be understood independently (Vygotskij 2005). Everything within this unit is thought as part of a whole. This has as a consequence that if the whole changes, all the parts change and the relations between them; and when any single part changes, the whole and all other parts change. This is quite a challenge to think activity in the way Leont'ev (1983) asks us to do, for what we have come to theorize and explain as relatively stable *elements*, such as the subjects of activity, now have to be thought *in the context of everything else* and *across time*. That is, if we assumed for the moment that the subject of activity – Craig, Shelby, and the entire research team – could be considered independently, they still have to be thought of *as one* from the beginning in 1998 to the final paper in 2006 (Fig. 9.6).

In the preceding paragraph I write about parts that relate to the whole. This may actually lead to a wrong image if a reader were to depict the whole in the same way that we often picture a crowd of people: It is still the same crowd if we take away or add a person. First, to appropriately employ the theory, we need to include the entire activity as one whole. Then everything that we might want to consider – such as the subject, object, division of labor, or means of production (tools, instruments) – is but a *manifestation* of the whole. The phenomenon we consider is to the entire activity what a raindrop is to the entire world that is reflected in it. In other words, the phenomenon we consider functions like a metonymy: It is part of the whole, which it reflects and stands for (temporarily). It is a document of that whole in the way it has concretized itself in one of its parts (Mannheim 2004). But we cannot describe this part in and by itself unless we take into account the whole. The figure of metonymy is actually quite instructive, as the following example shows. When a waitress says to her colleague "the ham sandwich over there is quite cute," where the colleague hears her talking about the man eating a ham sandwich, a metonymy is employed. The ham sandwich is part of a whole – man-eating-a-ham-sandwich – and is used to refer to the whole, stands in for the whole. It is the man currently eating the ham sandwich that she finds cute, not the ham sandwich itself. Here, neither the man nor the ham sandwich is independent of the other. Without the ham sandwich, the reference to the man makes no sense; and the ham sandwich is not what is the point of the commentary. It is only the ensemble that is intelligible. In a similar way, the subject* or object* of activity is intelligible as a concrete phenomenon only in the metonymical relation that we must think between the activity as a whole – here from the inception of the scientific investigation in 1998 to the publication in 2006 – that we may intelligibly consider Shelby and "his"

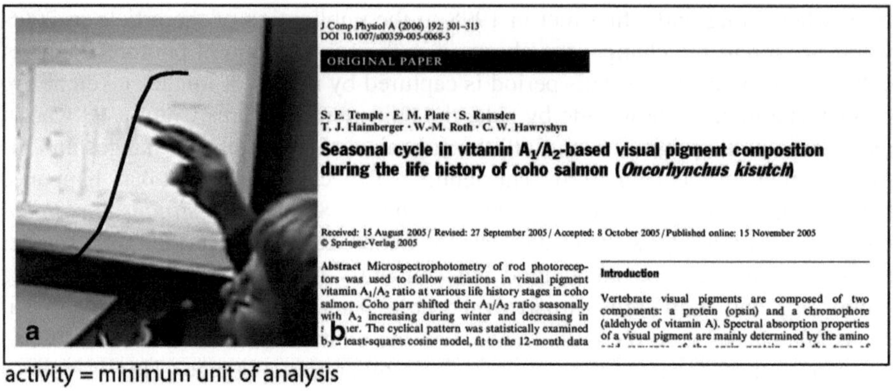
J Comp Physiol A (2006) 192: 301–313
DOI 10.1007/s00359-005-0068-3

ORIGINAL PAPER

S. E. Temple · E. M. Plate · S. Ramsden
T. J. Haimberger · W.-M. Roth · C. W. Hawryshyn

Seasonal cycle in vitamin A_1/A_2-based visual pigment composition during the life history of coho salmon (*Oncorhynchus kisutch*)

Received: 15 August 2005 / Revised: 27 September 2005 / Accepted: 8 October 2005 / Published online: 15 November 2005
© Springer-Verlag 2005

Abstract Microspectrophotometry of rod photoreceptors was used to follow variations in visual pigment vitamin A_1/A_2 ratio at various life history stages in coho salmon. Coho parr shifted their A_1/A_2 ratio seasonally with A_2 increasing during winter and decreasing in [illegible]ier. The cyclical pattern was statistically examined b[illegible] least-squares cosine model, fit to the 12-month data

Introduction

Vertebrate visual pigments are composed of two components: a protein (opsin) and a chromophore (aldehyde of vitamin A). Spectral absorption properties of a visual pigment are mainly determined by the amino

Fig. 9.6 The entire movement from an initial gesture Craig produced for me during a meeting in 1998 (**a**) to the final product of the scientific research activity – the peer-reviewed article (**b**) – has to be theorized *within* one and the same (minimum) unit of analysis. It is this one unit (category) that captures the changing processes of human activity

development.[12] If we do not take this approach, considering the development of the whole activity and Shelby thereof as an integral part, then we do not need cultural-historical activity theory. We can take any other theory of human development that looks at change independently of activity – of course, without any hope of really being able to explain or theorize our everyday becoming in the world.

A useful way of thinking about cultural-historical activity theory has been for me the idea of a constraint satisfaction network (Fig. 9.7). Such constraint satisfaction networks can be described in terms of an analog (as opposed to digital) system of elastics connected at nodes. The system can be modeled rather easily and quickly in a digital computer system (Roth 2001). In such a constraint satisfaction network, the value of any single node depends on all those other nodes to which it is connected and on the nature of the connection. Thus, for example, in Fig. 9.7, node 6 affects node 5; and node 5 affects node 6. The two types of forces do not have to be symmetrical. For example, the relation between two individuals may be hierarchical, and even though they are both required to model the relation, it will be such that one individual comes to be in a power-over –the-other situation.[13] In cultural-historical activity theory, connections tend to be made between any one node and all the others that make an activity system (Fig. 9.7). If we give each node

[12] Idealism abstracts because it does not consider the real, concrete individual or phenomenon, which, inherently, only exists in this material world, interconnected with everything else in this world. It is only when we no longer consider the real material person or the material phenomenon that we can pretend to explain a person without reference to his or her living conditions.

[13] Mathematically, the constraint satisfaction network can be implemented in vector form (Roth 2001). The activity "triangle" (Fig. 9.7) might then be a vector $\mathbf{x}$ with 6 dimensions. Two consecutive states are then related by means of the equation $\mathbf{x}_{n+1} = \mathbf{A} \cdot \mathbf{x}_n$, where $\mathbf{A}$ is a matrix encodes the weights $a_{i,j}$ that mediate between dimension i and j. In the case of a symmetrical relation $a_{i,j} = a_{j,i}$; generally, however, $a_{i,j} \neq a_{j,i}$.

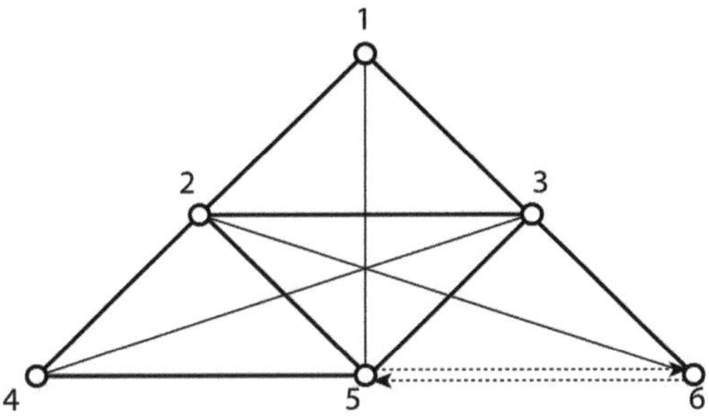

Fig. 9.7 Cultural-historical activity theory may be modeled by means of a constraint satisfaction network, where the value of each node is a function of the effect of all other nodes connected to it and, therefore, of the network as a whole. As long as there are constraints, the network will change. The entire process from initial setting to stable state is *one* unit. The effects between any pair of nodes do not have to be symmetrical but in most instances will differ, such as shown for the case between node 5 and node 6

some initial value and then let it adjust itself, the values of the individual nodes will change as a result of the effect other nodes have on it. These changes will continue until the network as a whole has found a minimum such that the values of the nodes no longer change. We can use this phenomenon, the network trajectory from the beginning until it has settled into a stable state as an analogy for an activity (dejatel'nost'/Tätigkeit). The whole process, from the beginning to the end, is *one* unit. This unit is not stable but continuously changes. We cannot ever say what it is, for it continuously becomes.

When using the constraint satisfaction network as an analogy, we immediately are constrained: we cannot talk about an individual node in isolation from all other nodes, as its value is a function of all the other nodes including its own and, therefore, is a function of the network as a whole. Moreover, the value of a node is not constant in the unit but continuously changes. The connections in the network mediate between nodes, which is the equivalent to the mediation that we find in cultural-historical activity theory. In constraint satisfaction networks, the effect of one node on another frequently is held constant. However, to be a better analogy of an activity system, the relations between the nodes have to be made a function of the unfolding activity. This then reflects the observable fact that a change in the means* of production changes the practices* we observe in the laboratory. Thus, when Craig purchased a CCD-based[14] microscope for doing the dissection of the, it became possible to select the tissue from different parts of the retina: The team then took photoreceptors only from the dorsal part whereas before that photoreceptors were obtained from whatever part of the retina happened to be on the operating table. When Craig purchased another CCD camera to be connected to the microscope where the samples were mounted for obtaining the absorption spectra, the

[14] CCD stands for charge-coupled device. It is the basis of digital cameras. Fundamentally it consists of an array of pipes that amplify the light – actually, individual photons – falling into it.

training of new team members became easier as the team members present in the dark laboratory now could orient to the image as a means for establishing common ground whereas before one had to look through the ocular and find the relevant entity on one's own. That is, in the course of this particular instantiation of research activity – from inception to outcome – the actions that link subject (team member) and object (photoreceptor) changed.

Once the network is let go from an initial state so that it relaxes, it is in continuous movement until it finally settles. This is entire process from beginning to end thought as *one*. If we were to look at the value of a node between beginning and end, that is, *within* the unit then we do not get it "as it really is." All we get is an instantaneous image. But this image does not tell us anything. It is not even a state. Moreover, we do not get at the dynamic of the network if we take consecutive values of a particular node by imagining some kind of animation. To explain the node, we have to consider it in the context of all the other nodes, that is, in the context of the network as a whole and from the beginning to the end of the relaxation process. Once we commit to this view, it becomes evident that *nothing is the same, everything changes continuously*. We now have a model for describing activity in the way that Leont'ev proposed. We explain the change in the subject* – the research team as a whole and its individual members in particular – in terms of the entire activity.

We can better see what is happening in the model of an actual situation. For example, what happens in and to a constraint satisfaction network during a classification, as I observed while collecting coho specimens in the river together with Shelby, Elmar, and four staff members from the Robertson Creek Hatchery (Roth 2005). In this situation, six salmon specialists were trying to classify a fish that we had caught in our net. The specialists were generating and weighing the evidence for two hypotheses, one that the specimen was a young chum salmon, the other one that it was a sockeye salmon. A constraint satisfaction network model shows different trajectories of the unit – the decision whether the specimen is a chum or a sockeye salmon. Because the model is sensitive to initial conditions, the same end results may be achieved by different trajectories (Fig. 9.8a.i, ii); or, from the same or nearly the same starting point, the opposite decision (classification) might be reached (Fig. 9.8a.iii). What I actually observed was a "hung jury," where the group left the nature of the specimen up in the air because they could not reach agreement even though many different environmental factors had been considered. Figure 9.8b shows how from very different starting points of the network as a whole, corresponding to the different initial predilections of participating specialists, the same "hung decision" is reached. The important point is that each of these trajectories needs to be considered as part of *one-and-the-same* unit rather than as different stages thereof. Only if the classification (decision-making) process is considered as *one* are we able to describe and explain the *internal dynamic* that drives the system and allows us to model change appropriately.

In this last example of a decision-making process, I have already moved from considering the activity as a whole – conducting research into the physiological changes of coho salmon – to considering a manifestation thereof: deciding on the

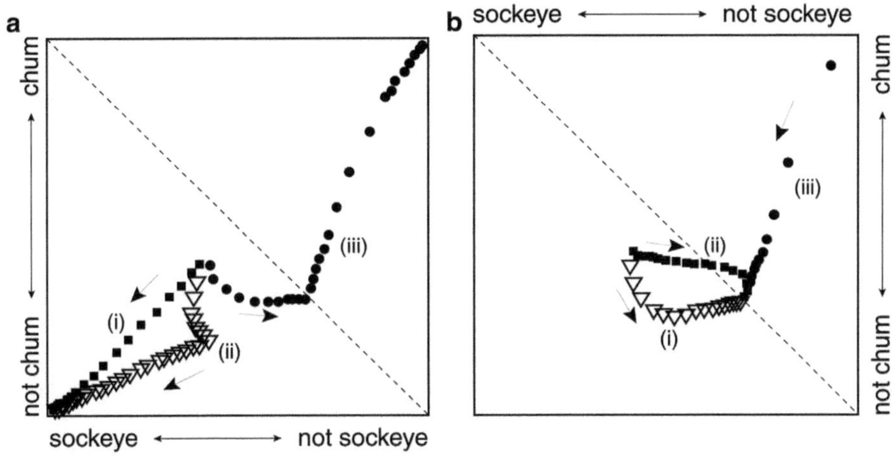

Fig. 9.8 A constraint satisfaction network is used to model the decision-making process in a classification involving Shelby, Elmar, and four staff members from Robertson Creek. (**a**) The model is sensitive to minor variations such that from nearly identical initial conditions, the decision processes differ, sometimes reaching the same result (*[i] & [ii]*, it is not a chum but a sockeye) at other times reaching the opposite conclusion (*[iii]*, it is a chum and not a sockeye. (**b**) Very different initial conditions may end up with the same result (product)

nature of a specimen caught and excluding it as belonging to the coho species. Of course, once we consider the movement from the inception of the research to its final product as the minimum unit, there will be movement when zooming in on part of the trajectory. This part of the trajectory is not intelligible in itself, however, but only as part of the whole trajectory. At the same time, the whole trajectory consists of parts. Thus, we think of the whole as being constituted by parts (i.e., manifestations of the whole). In cultural-historical activity theory, this is reflected in a hierarchy of productive movements: activity > actions > operations > physiological (neural) change. Activity is the minimum unit, but within it we can identify actions that constitute societally motivated activity; but goal-directed actions are executed only because of the activity. These two levels therefore mutually constitute and presuppose each other. We observe certain actions, such as sacrificing a fish, extracting the eye, and dissecting the eye to retrieve the retina only because of the overall motive of producing scientific knowledge about the changes in the retina of salmon in the course of their life cycle.

Goal-directed actions can be thought of as being constituted by conditioned operations. Thus, when Craig, Shelby, or Sam macerate the retina to obtain small pieces with photoreceptors to be placed on the microscopic slide, the action "macerating" involves hand-finger movements coordinated via observing the pieces on the monitor with the microscopic image of the operating table contents. These micro-movements are not present in consciousness but unfold on their own and in response to the present condition, like a (kinetic) melody initially triggered and then unfolding on its own (Luria 2003). The image of the melody is appropriate here, as there is no melody without individual notes, and function of each note depends on their place in the melody. There is no equivalent to the individual image in the world of sound, which consists only in melody: one note is a note only in melody. Thus, if we so desire, we can look even more closely to see physiological and neurological changes in any one of the research team member (subject). The point of cultural-historical activity theory is that even at this level, we can explain the changes only if we view them as irreducible moments (manifestations of) the whole activity. In the following subsection, I consider one such micro-movement and show how thinking changes in the course of doing a graph. However much we are used to break out of activity such a micro-event, considering it in a decontextualized manner, cultural-historical psychologists such as Luria (2003) insist that these cannot be considered independently. A melody does not consist of a seriation of individual notes. A melody involves notes, themselves phenomena in and producing time, related not only to their immediate members but to the melody as a whole.

Microconstitution of Graphing

One can often hear, on the part of certain colleagues, the complaint about doing microanalyses – such as those conducted in Part II – as if the microevents were standing on their own. But the microanalyses are not conducted for their own sake. As pointed out in the preceding subsection, the microevent is a manifestation of the whole. It is not an independent part, the seriation of which with other parts would make the whole activity. Rather, this whole is the minimum unit; and a microevent is but a reflection of this whole. Microanalysis makes sense only when it is revealing something about the whole activity. It is like an individual cord played by a particular instrument in the course of a symphony. To hear *this* cord appropriately, one needs to consider it in *this* place played by *this* instrument in the symphony as a whole – which is the analogue of the scientific research activity. The "same" cord in a different symphony or in a different place of the same symphony or played by a different instrument (group) is heard and experienced very differently.

In the course of the entire research project, our minimum unit of analysis, change in the talk about the phenomenon has occurred. The contradiction between the two forms of talk, from a dialectical perspective, is merely the external expression of an inner difference, which is the movement of science (Il'enkov 1982). In Chap. 7, I describe this change as an evolution, which I model using a catastrophe theoretic approach suitable for modeling the emergence of new forms (morphogenesis); and I contrast this account with traditional conceptual change approaches. My explanation takes into account that "the genuine beginning of a new branch of development" – such as the one on branch in Fig. 7.2 – "cannot be understood as a product of a smooth evolution of the historically preceding forms" (p. 218). That is, there is movement in and of activity; activity is a *self-changing* phenomenon. The unit of analysis is sufficiently large so that all the forces are within rather than operating from without the unit. But aspects of this movement may be observed at very different scales as well; and, in fact, the overall movement of talking about data and graphs cannot be explained independently of the movement of talking about data and graphs at much shorter time scales, which constitute the overall movement. But the movement even at very short time scales can be explained only because of the overall movement. Consider the following fragment from a meeting analyzed in Part II.

In this meeting – which was in fact our first data analysis meeting (June 18, 2001) of the project that was ultimately published in print 5 years later – the issue at hand is the variation in the porphyropsin concentration data (%A_2). The relative amount of porphyropsin in a photoreceptor is obtained from one or both values that can be obtained from the light absorption spectrum: the location of its maximum (lambda-max or λ_{max}) or the width of the spectrum at half its height (half-maximum bandwidth, HMB). As the talk comes to half-maximum bandwidth, Craig states what the presupposition about the relation between it and the porphyropsin concentration is, makes an approximately parabolic hand gesture, and then gets up and walks to the board. He stands there staring at the empty board for a while (Fig. 9.9a), and then writes "½ m." In fact, the first mark below the fraction line is difficult to decipher and looks like a 3, though it is accompanied by the word "half"; Craig writes a "2" over it, as if realizing the ambiguous nature of the first thing written and without erasing what has been there so that the traces of the previous mark still are visible (Fig. 9.9b). He gazes at what he has written, then gets an eraser, wipes that part of the chalkboard clean (Fig. 9.9c), and next writes "HMB" (Fig. 9.9d) while saying "half-max bandwidth." All of those team members present hear it to be a variation of the "HBW" that we normally use. Although only 13 s have passed between the beginning and the end, we observe considerable movement and change. While reading the following paragraph, readers should keep in mind that when he sat down, Craig copied what he had just produced into his own notebook as if he were copying the notes a lecturer put onto the chalkboard.

First, we might theorize that Craig had something already configured in his mind, which he then made available for others in the form of writing "½ m" and by saying "half." However, we see that as soon as he has finishes writing, he orients downward toward the eraser, gets it with his writing hand, and then erases what he has just written. He then writes something else. Why would Craig go through the extra effort of writing "½ m" and saying "half" only to "backtrack" and get the chalkboard into the same state as it was before and then to write something else on it? Why, if he had a particular conception in mind, did he change what he was putting on the board? It may be more productive here to follow activity theory – and, incidentally, phenomenological philosophy (e.g. Merleau-Ponty 1945) – and consider this as an instant of consciousness in movement (Vygotskij 2005). From this perspective, it is *after* his doing (writing) and saying that Craig is finding out what he is thinking, considers it inappropriate, and changes it into something else. The final "HMB" is the trace left by a thinking in movement not by a stable

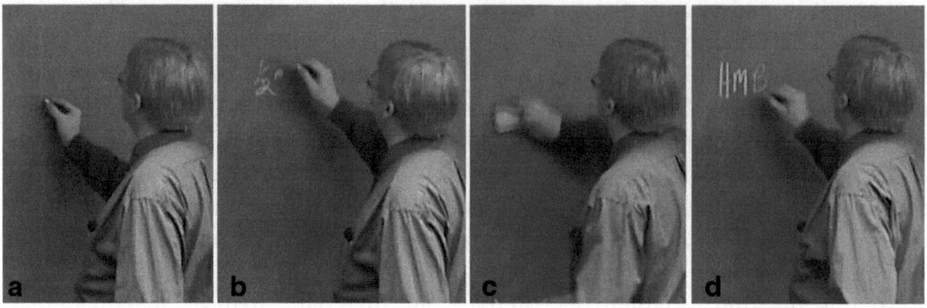

Fig. 9.9 Craig confronts the empty chalkboard, then notes ½ m, which he erases, and then writes "HMB." In the course of writing/erasing, thinking develops

prefinished thought already in mind, which, somehow made it first to the outside in imperfect form only to be revised subsequently. It is in the *result* of the writing that thinking discovers itself rather than being the direct cause of an explicit intention that led to this writing. As Nietzsche (1954, vol. 3) suggests, a force (e.g., intention) is not constituted in and by itself; instead "cause" is "the translation of *effect* into a totally different language" (p. 485). The cause–effect figure that tends to be employed to relate thinking and acting (speaking) is in fact a matter of interpretation rather than a reflection of the real relations between the two forms of movement. The observed sequencing of certain observations does not imply a law but a relation between two or more forces. "We are not dealing with a succession – but with a one-in-another, a process, in which the individual sequential moments do *not* constitute each other as cause and effect" (p. 489). Saying and writing, thereby, are explained in terms of an overall process of one-in-the-other. Thinking and communicating are but manifestations of the internal dynamic of activity (Vygotskij 2005). Thinking and expressing (saying, writing) are two unfolding, mutually implicating movements. These are not independent, *inter*acting movements. Rather, these are manifestations of a higher-order, more integrative *signifying* movement that reflects the movement of the material activity (see also Leont'ev 1983).[15] The movement on the ideal level (what is conscious to the person) and the movement at the material level (what the person is doing in the world and what is therefore available to everyone else and to the acting person) are but manifestations of the activity as a whole and cannot be explained independently.

We may in fact use *writing* as an analogy and metonymy for the movement of this scientific research activity specifically and for the movement of life more

[15] English translations often use "word meaning" to render the Russian expression "slovo značenie" that Vygotskij (2005) uses. Signification, however, is a better and more consistent rendering of "značenie." The scholar uses the term "slovo značenie" in the same way as others in his culture and the same time use "tema," theme (Vološinov 1930), as shown in the mutual discussion of precisely the same episode from *A Writer's Diary* (Dostoevsky 1994). In the little story, drunken peasants use the same obscene word six times in a row, never expressing the same but contributing to an unfolding scene, followed by a seventh use on the part of the writer himself.

generally. Writing is a unit in the same way as activity is. It is always spaced out and spacing, as shown in the "½ m" or "HMB," globally moving from left to right, though there are retrograde movements from right to left, such as in writing the "2" or "B." This situation is actually interesting in terms of our overall analogy, because if we were conducting a microanalysis reduced to the left-ward movements in the writing of "2" or "B" disconnected from everything else, we would have gained very little; but it helps us to explain the overall movement and the content of the writing once everything has been said, done, and written. The analysis of this fragment also has to be seen in relation to the overall activity, because it might stand to the latter in the way as the retrograde movement in the writing of the "B" might stand to the writing of "HMB" specifically and to the graph that will have been on the board when Craig returns to his seat. Writing – in which I am including the drawing of lines – also always occurs in time and makes time. Writing *is* movement itself, and includes erasing (un-writing). *Writing* embodies the difference between writing and erasing because it always goes hand in hand with the latter, just as life does not go by itself but always goes hand in hand with death. In writing "½ m" (Fig. 9.9b), Craig is not just putting something there on the chalkboard but he is erasing what has been there before: what has been an empty chalkboard disappears with writing (Fig. 9.9a); and in erasing (Fig. 9.9c), Craig is "un/writing" what he has written before, thereby "writing" an empty chalkboard.

Of interest here is not this fragment in and of itself but the fragment in its relation to the activity of making a discovery as a whole, that is, the entire period from 1998 when Craig and I first met to 2006 when our research article was published in print. The fragment is a manifestation of the whole, reflects the whole as a raindrop reflects the world of which it is part. We see the movement in graphing*, which, in fact, is graphing* in movement. This movement here is considered at a small scale, which reflects the whole activity as an individual note or cord reflects a whole symphony. Just as we hear any *this* note in a symphony only in relation to the symphony as a whole, we hear this fragment involving the writing and erasing of "½ m," the writing of "HMB," and the erasure of the empty chalkboard in its relation to the overall activity. Even if we were to look at what happened at the time in Craig's brain, the particular neuronal activity, it would be there at that instant in time only because of the overall activity (Luria 2003). It contributes to realizing, and reflects, the productive activity (dejatel'nost'/Tätigkeit) as a whole. Thus, the study of "the true cerebral mechanisms of the highest forms of mental activity" required "the study of the elementary forms of physiological processes" that are "adequate for the study of human conscious activity, *mediated by its social-historical* [sozial'no istoričeskoj] *origin, and its complex hierarchical structure*" (p. 73, emphasis added). That is, Luria suggests that we cannot explain fundamental physiological and cerebral mechanisms unless we study them in terms of conscious human activity, which has a cultural history and is structured hierarchically (i.e. activity > action > operation > physiological processes).

A Day's Work in the Laboratory

In the two preceding subsections, we see the activity as a whole unit, which is indivisible by analysis into elements – lest we resign to losing our phenomenon. This is so because anything characteristically human is the result of living in human society. The smallest unit of analysis, our fundamental category, is that of activity. Thus, the entire process of the production, from initial conception in 1998 to the final product in 2006, lies *within* the unit. We may identify a manifestation of this unit, but this manifestation relates to the whole as a part relates to a whole in a metonymy. Here I focus on the movement of activity at a courser-grained level than I do in the preceding subsection. It makes sense to select temporal periods that form a whole. For example, we may think of everything as belonging to a particular melody – within the whole symphony – that makes one data point in the graph that ultimately appears in Shelby's dissertation and in our joint publication. It turns out that as our team became more efficient, individuals more skilled, and as we automated some aspects of the data collection, the number of cells required for a single data point (20 receptors from each of 10 fish) could be taken in the course of one day's work. We may therefore consider a day's work as an identifiable and repeated segment within the whole activity. The different aspects of one day's work include holding the fish in darkness in a bucket (Fig. 9.10a), sacrificing it and excising the eye (Fig. 9.10b), removing the retina and macerating it (Fig. 9.10c), adjusting the equipment (Fig. 9.10d), locating a photoreceptor on the microscopic slide (Fig. 9.10e), recording an absorption spectrum from the difference between the intensities of a beam going through the cell and next to the cell (Fig. 9.10f), extracting the porphyropsin content (%A_2) and plotting the levels in a histogram (Fig. 9.10g), and ending with another point in the graph that will be representing A_2 levels for a 1-year period (Fig. 9.10h). To describe this day's work, we need to picture it as we would hear a particular melody or development, played by one instrument, within a whole symphony played by an orchestra.

Whatever happens in the course of the day, whatever intentions scientists may articulate, what they are *actually* doing is available only after the doing is finished. Then the doing is described in terms of the traces that it has left. That is, what happens in the course of a whole day's work is movement resulting from forces internal to the activity taken as a unit and in that moment of it relating to the day. The need to approach practical competency in this way is very clear from the description in Chap. 3, where I show that at the end of a day, Craig came to the conclusion that he had done his dissection of the fish eye differently than he had thought he had done. Thus, although he might have had (and expressed) the intention (plan) of doing a particular dissection, *what* he had actually done was different. This difference was available to him only in and through the results of the day's work. It was not his thinking (intending) that determined his actions. The nature of his actions was determined in and by the effect that he had produced – which turned out to be no data or unusable data. Thus, the appropriateness of the actions depicted in Fig. 9.10a–f is established on the basis of the graph in Fig. 9.10g.

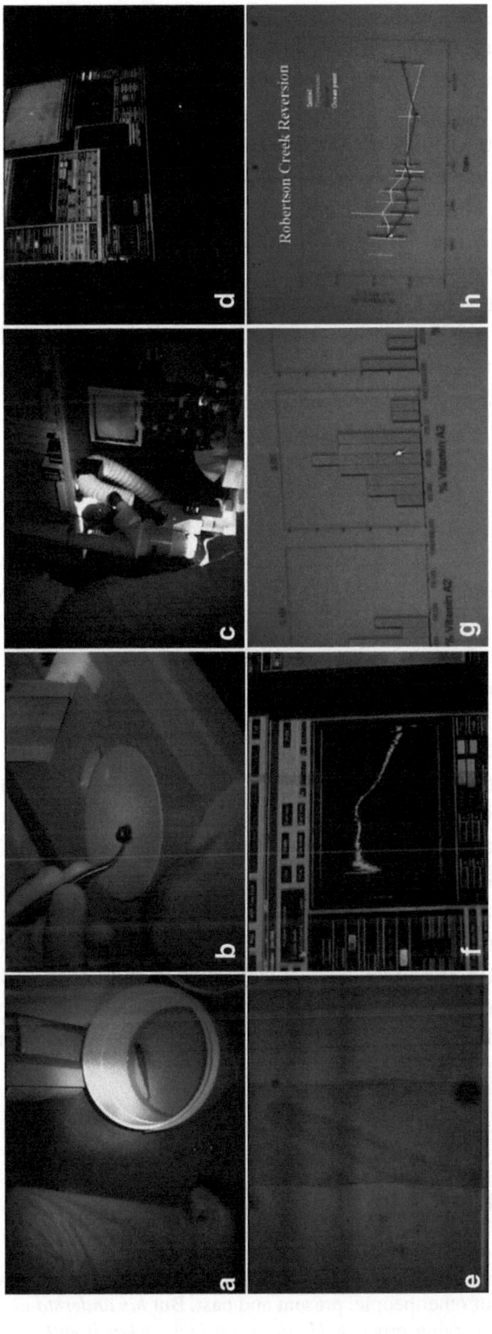

Fig. 9.10 Moments in a day's work in the laboratory. (**a**) A living coho salmon. (**b**) An extracted eye. (**c**) In the process of macerating the retina to extract an appropriate piece from the dorsal part. (**d**) Checking the measurement instrumentation, with graph information on the *left* and the image of the photoreceptor under the microscope to the *right*. (**e**) A *rod-shaped* photoreceptor slightly can be seen slightly on a diagonal. (**f**) An absorption curve has been recorded from the difference between reference and sample and the associate data are stored. (**g**) The totality of one day's work of measurement is represented in a histogram exhibiting variation of porphyropsin levels. (**h**) A day's worth of work ends up as one data point on a graph exhibiting the changes in porphyropsin in the course of several months

That something has gone wrong may not be available beforehand. In fact, in some instances, it took months to find out that some actions were inappropriate, leading to the inappropriateness of several days' work and the unavailability of 3 months worth of data. This was the case when Shelby discovered that Sam and he had taken retinal tissue that fell outside the desired band in the dorsal area of the retina. This, as he found out months later, led to the data points having consistently higher values than all the other data points collected over the course of the year. Thus, although their intention was to extract retinal tissue and photoreceptors in a consistent way, that they were not doing so became apparent only much later when Shelby had the opportunity to see the results (i.e., effects) of their actions.

In this case, we easily describe and explain the whole of the activity, which might be resumed and denoted by the graph ultimately published, as a series of "one day's work," many of which followed in more or less the way described here. We must not use causal reasoning, however, to make relations within a day's work concerning intentions and situated actions, and we must not make causal relations between any individual day's work and the activity as a whole. Just as modeled by the constraint satisfaction network, there is a continuous, mutual constitution of any individual node by all other nodes of the network, and a continuous constitution of the network whole by all of its parts (Fig. 9.7). We must investigate, if using cultural-historical activity theory is to be useful at all, the activity as a whole and describe its *internal* movements in the way we describe the changes within the constraint satisfaction network described above. Because of their mutually constitutive nature, we cannot get independent causes and effects but only mutually determining forces and relations.

The quality of the data at the end of our study, including on the one day's of work considered here, has evolved from discussions of the variance in the data, such as the one from which Craig's articulation of the preconception concerning half-maximum bandwidth was culled. That is, the nature of the data (quality) of the whole project (the activity as a whole) cannot be explained independently of what happened in microevents such as the one depicted above or what happened on intermediate-level events such as during one-day's-work described in this subsection. The microlevel events and the many one-day's-work considered can be appropriately explained only, in their specificity, as a function of the activity-realizing project as a whole. These scientists did *this* on any *this* day because they were in the process of realizing *this* project that Craig and I had conceived some time in 1998.

Through the Lens of the Unfinished

> The blacksmith's practices as he creates a skimming spoon draw on rich resources of experience, his own and that of other people, present and past. But *his understanding* of the skimmer also *emerges* in the forging process. *He does not know what it will be until it is finished.* (Lave 1993, p. 13, emphasis added)

Most accounts of practical action – including those describing scientists at work and those describing teaching and learning – are determinate in the sense that attributions about intentions of specific actions are made to the acting subject. The structure of such accounts is based on the very structure of those languages based on the subject-with-intentions–verb–object grammar: “Craig macerates the retina to get a photoreceptor.” There is a problem, however, because, as Craig himself says 8 h later, he apparently had not done what he had wanted to do (i.e. intended). His plan (intentions) differed from his actions. Similarly, every reader will have many examples of situations in everyday life where what we intend to saying is heard very differently – which leads in many situation to conflict, such as when an intended question is heard as an insult – or where what we intend to do is different from what we actually achieve (as in cooking). That is, even if a speaker intended to ask a question, if it is heard as an insult, the participants in the event now are faced with a new situation and have to deal with it; and this dealing essentially changes what might have occurred otherwise. What Lave says about the blacksmith, who does not know what he is forging until the object is finished, is true for every profession specifically and for life more generally.

In the preceding section, several leads are developed all of which are consistent with the fundamental idea that from within an event (activity), the world has to be considered as unfinished and therefore needs to be described and modeled in terms of becoming. Scientists do not know in the morning whether or not the dissection they intend to do and are doing will have been in that way at the end of the day. Sometimes they find out only months later that something they have done was other than intended. If what they have done turns out to be as intended, if there is no counter-evidence or reason to believe otherwise, then the intent (plan) turns into an after-the-fact-attributed appropriate description of what has happened. As I quote Nietzsche above – whose insight had been restated subsequently by Suchman (1987) in her findings about the abyss between plans and situated actions – the attribution of causes always follows in time. A cause cannot be known beforehand, not if we have to consider the activity (event) as a whole. As long as this whole is not given, which it is not as long as it is not finished, the relations between parts and whole are not given: they cannot yet be given inherently and without remedy. If this is the case, then the psychological and sociological grammars of actors and events have to be re-written, including that of the social and individual construction of knowledge. Because subjects are not in control over activities*-in-the-making and events*-in-the-making, they also become subject and subjected to these open and yet-to-be-namable happenings. This introduces an essentially passive aspect into all consideration of activity, and, therefore, into development and learning.

In Part II of this book, I provide an account of the emergent nature of science generally and of the emergent nature of graphs and graphing specifically. Initially, our research team models the data in the classical way, following “the dogma,” that is, the theory established by G. Wald who subsequently received a Nobel Prize for his work on physiology. As becomes salient to our team *after the fact*, this is not even the only pattern that has been reported in the scientific literature concerning the changes of porphyropsin levels in the life history of animals migrating between freshwater and saltwater environments. Other researchers and research teams have reported cyclical patterns before, including the very author on whose work our research on coho salmon was based. There were instants during the unfolding research where one or the other member of our team had a strong sense that the results would be overturning the existing dogma. To describe what was happening in this research group, why our team did what it did, one therefore needs to take the perspective on what was happening *through the lens of the unfinished*. After the fact, in 2006, but already beginning around 2004 when it became clear that there had been other work suggesting cyclical patterns, our team began to situate what it was doing in a different way. There was therefore also a change in our team's self-explanation of the research – one that considered it to be a conceptual revolution to one were it added just a bit more to the existing knowledge. That is, from *within* the event, to us participants as witnesses*-in-the-making, what was happening looked very differently than it looked once everything was said and done. At the end, when the paper was published – or rather, accepted for publication in October 2005 – the activity had come to a close. It could now be looked upon in its totality.

Once available in its totality, once the activity (event) does no longer change, the end result (effect) is known and, because the activity now can be *grasped* as a whole (com*prehended*), causes (intentions) may be attributed. This comprehension actually presupposes that everything is comprehended, that is, included – which presupposes the end of the activity. The cause–effect relations will remain in effect unless there are reasons for re-opening the case, re-opening the black box that has been closed with the acceptance and publication of the scientific article that resulted from the work. Such a re-opening might occur, for example, if there were suspicions that the data in a published paper were "cooked up," so that these come to be questioned and that a researcher or research team has doctored the results.

The present considerations, therefore, question all forms of description that rest on a cause–effect figure, grammatically embodied in the subject–verb–object structure of theoretical discourses. Although such talk is useful in gaining control over conditions, it does not correspond to empirical evidence. Our research team as a whole, and its individual members, did not just *construct* the graphs and the knowledgeability that comes with them. Graphs*, as graphing* – which we might consider as knowledgeability related to the use and deployment of graphs – were continuously in the making. Not only is the verb "to construct" transitive, it also implies knowledge of what will come out in the end. It also presupposes the self-identity of the intentional subject, who can verify whether its intentions map onto the achieved outcomes. This was not so for scientific research other than that the researchers intend *some* results without precisely knowing what they will get (e.g. Suzuki 1989). Although Craig and I had been out to correlate the porphyropsin levels in coho salmon smolts at the date of their release from the hatchery to the return rates, we ended up doing something very different: The demise of the theory that had framed the design of the study. In the process, our doing itself changed and needs to be viewed as doing*. Despite all of this change, as Craig was saying to Shelby and to me when asked about whether the data were making our study invalid, "We just publish another paper."

Scientists are not just subjects of research actions and research activity: they are also subject and subjected to the research activity. They come to be confronted with phenomena and situations that they had not anticipated. This is precisely because from within an event, it is unfinished and cannot be grasped as such. This shift is known in the literature and also is observable in the present study. The experimental accounts in the journal articles that result from scientific work tend to be causal, linear stories that leave out the contingencies and make it appear as if the entire study as reported originally had been designed as such. Take the following two excerpts from the final paper.

> Coho salmon were chosen for this study *because* their range is restricted to temperate, seasonally variable regions and they possess a migratory life history strategy, which enables a comparison between the seasonal versus migration/metamorphosis hypotheses. (Temple et al. 2006, p. 302, emphasis added)
>
> In the present study, we tracked the change in λ_{max} as a correlate of A_1/A_2 ratio in rod photoreceptors of coho salmon at three life history stages throughout the year *to differentiate between the migration/metamorphosis and seasonal variability hypotheses.* (p. 303, emphasis added)

In the first quotation, a causal attribution is made to the selection criterion of coho salmon over other available salmon species. When Craig and I designed the study, there was no talk about a particular species. The problem of varying return rates is a general one pertaining to all salmon species raised for release in fish hatcheries. It turned out, however, that the commonality between the two hatcheries willing to participate in the study was the raising of coho salmon. It was defensible to investigate the same species to be able to make comparisons – which was underlying our decision-making at the time to focus on coho salmon. Because our team was depending on consent to participate, the nature of the salmon species to be investigated was a function of the hatcheries that agreed to participate rather than the result of a deliberate focus on coho salmon because of its migratory life history strategy and the possibility to compare between two hypotheses, the one postulating seasonal changes, the other being about migration-related physiological changes according to Wald's theory ("the dogma").

In both quotations, the differentiation between the two hypotheses is said to have been the reasons for the study ("salmon were chosen *because* . . . which enables comparison" and "[in order] to differentiate"). That is, although the nature of the species studied was contingent, depending on the hatcheries that would be agreeing to participate, and although the research design was for a very different purpose, when it was actually reported the rationale was set up such as to appear logical and aimed at the actual outcome achieved. The account rewrites the history of our research in a teleological fashion. That is, intentions for doing the research were written such that they came to be the causal antecedents of the actual outcomes (effects). The account fits the cause–effect pattern – which has its use in anticipating future events and therefore controlling the human condition (Nietzsche's "will to power") – it does not lead to good theory in the study of science*-in-the-making. It does not lead to good theory in the study of learning and development of human beings generally (e.g., Piaget's developmental theory, according to which "formal reasoning," as he described it, is the necessary end point of development) because these accounts are just as flawed as the accounts scientists provide concerning the work that has led to the results reported in their research article. Evolutionary accounts, which inherently work with an open, undetermined, and indeterminate future, oriented towards continuous *becoming*, and which are characterized by contingency, do a much better job as in the following quotation:

> The craftsman typically knows what job he needs to do before picking or inventing tools with which to do it. By contrast, someone like Galileo . . . is typically unable to make clear exactly what it is that he wants to do before developing the language in which he succeeds in doing it. (Rorty 1989, p. 12–13)

Rorty strongly argues for a nonteleological view of intellectual history generally and for a nonteleological view of intellectual history of the sciences specifically. Thus, culture, language, theories, or concepts all are the results of contingencies, as is any particular plant or animal; and none of these phenomena ever *is* but continuously *is becoming*. Rorty strongly encourages us to

> resist the temptation to think that the redescriptions of reality offered by contemporary physical and biological science are somehow closer to "the things themselves," less "mind-dependent," than the redescriptions of history offered by contemporary culture criticism. We need to see the constellations of causal forces which produced talk of DNA or of the Big Bang as of a piece with the causal forces which produced talk of "secularization" or of "late capitalism." *Those various constellations are the* random factors *which have made some things subjects of conversation for us and others not, have made some projects and not others possible and important.* (pp. 16–17, emphasis added)

The constellations that Rorty describes are those that after the fact can be found in the activity when it has come to an end, such as particular constellations in the various manifestations of cultural-historical activity theory (Fig. 9.5) that drive some accounts of learning and development in the learning sciences literature. There are emerging and emergent constellations to which cause–effect attributions can be made only a posteriori, but such attributions do not describe intentions (goals) that emerge and pass when a happening – such as a research project in a discovery science – is described *from within*. An emergentist account of activity generally and graphing* specifically would pursue the recommendation "that we try to get to the point where we no longer worship *anything*, where we treat *nothing* as a quasi divinity, where we treat *everything* – our language, our conscience, our community – as a product of time and change" (p. 22). Through the lens of graphs*-*in-the-making* and graphing*-*in-the-making*, what the ultimate result will be is open while *there is happening* so that it is incorrect to merely write graphs-in-the-making and graphing-in-the-making.[16] The latter assumes that the end product, graphs and graphing are somehow achieved in a determinate manner. However, as

[16] Our language is not well suited to articulate this openness. *There is happening* rather than something is happening is perhaps the best way I have found to articulate what I am after. If we were to write some*thing* is happening, there is already definiteness implied with respect to the *what* that *thing* is that is happening.

pointed out throughout this chapter, the end product is unavailable while the event* (activity) unfolds until after everything has been said and done and at least a temporary closure has been achieved. It is for this reason that I add an asterisk – graph* and graphing* – which marks the unfinished and indeterminate nature of that which will have been in the making. Uncertainty in the nature of both reigns until, when everything has been said and done, is settled to the point that a more definitive *a-posteriori* account can be provided. If the experienced scientist in this study – just as the experienced blacksmith that Lave is referring to in the above quotation – does not know what they will have done until that instant when they will have come to an end, whether this is the end of the day or the end of the research project, then the results of the work has to be marked as open, unfinished, and inherently uncertain in our analyses until the very end. There are many measurements and data that get never published. This could have also been the case here. Thus, in the happening and while there is happening, we cannot know whether what we are doing is actually going to be published, which is the motive of scientific research activity. The corollary to the unfinished event* (activity*) and unfinished object* is that of the unfinished subject* (of activity*), which is just as much in-the-making subject to contingencies as everything else. Thus, "the word 'I' is as hollow as the word 'death'" (Rorty 1989, p. 23).

In this chapter, we move from considering flow to cultural-historical activity theory, which was designed to describe and explain human learning and development as something specifically human, which it is only in the context of society. The smallest unit that retains all characteristics of a society is *activity* [dejatel'nost'/ Tätigkeit], from the beginning to the end of a production cycle and realized in the concrete production of some outcome. Both aspects, the irreducible unit of everything partaking in an activity and the integration over time leads to an emergent account of production, learning, and development. This unit needs to be considered as *one*, which, analytically, implies that we cannot attribute what happens to this or that cause. As Nietzsche (1954, vol. 3) points out, "all unity is *only* unity as *organization and interplay*: not different than when human sociality is a unit: that is *opposition* to atomistic *anarchy*" (p. 499). Organization and interplay that arises from the forces tugging the nodes in different directions (Fig. 9.7) lead to models of development that are subject to contingencies, influenced by minor variations that may take the system into very different directions, as any chaos or catastrophe theoretic account would predict.

References

Bakhtin, M. (1981). *The dialogic imagination*. Austin: University of Texas.
Bakhtin, M. (1984). *Rabelais and his world*. Bloomington: Indiana University Press.
Bakhtin, M. (1993). *Towards a philosophy of the act*. Austin: University of Texas Press.
Chi, M. T. H. (1992). Conceptual change within and across ontological categories: Examples from learning and discovery in science. In R. Giere (Ed.), *Cognitive models of science: Minnesota*

studies in the philosophy of science (pp. 129–186). Minneapolis: University of Minnesota Press.
Dostoevsky, F. (1994). *A writer's diary*. Evanston: Northwestern University Press.
Gilbert, G. N., & Mulkay, M. (1984). *Opening Pandora's box: A sociological analysis of scientists' discourse*. Cambridge: Cambridge University Press.
Heidegger, M. (1977). *Sein und Zeit* [Being and time]. Tübingen: Max Niemeyer (First published in 1927).
Il'enkov, E. (1977). *Dialectical logic: Essays in its history and theory*. Moscow: Progress Publishers.
Il'enkov, E. (1982). *Dialectics of the abstract and the concrete in Marx's Capital*. Moscow: Progress Publishers.
Latour, B. (1987). *Science in action: How to follow scientists and engineers through society*. Cambridge, MA: Harvard University Press.
Lave, J. (1988). *Cognition in practice: Mind, mathematics and culture in everyday life*. Cambridge: Cambridge University Press.
Lave, J. (1993). The practice of learning. In S. Chaiklin & J. Lave (Eds.), *Understanding practice: Perspectives on activity and context* (pp. 3–32). Cambridge: Cambridge University Press.
Lee, Y. J., & Roth, W.-M. (2005). The (unlikely) trajectory of learning in a salmon hatchery. *Journal of Workplace Learning, 17*, 243–254.
Leont'ev, A. N. (1983). *Dejatel'nost'. Soznanie. Ličnost'*. [Activity, consciousness, personality]. In *Izbrannye psixhologičeskie proizvedenija* (Vol. 2, pp. 94–231). Moscow: Pedagogika.
Luria, A. R. (2003). *Osnoby nejrolpsixologii* [Foundations of neuropsychology]. Moscow: Isdatel'skij Centr "Akademija".
Mannheim, K. (2004). Beiträge zur Theorie der Weltanschauungs-Interpretation [Contributions to the theory of worldview interpretation]. In J. Strübing & B. Schnettler (Eds.), *Methodologie interpretativer Sozialforschung: Klassische Grundlagentexte* (pp. 103–153). Konstanz: UVK.
Marx, K./Engels, F. (1958). *Werke Band 3* [Works vol. 3]. Berlin: Karl Dietz.
Merleau-Ponty, M. (1945). *Phénoménologie de la perception* [Phenomenology of perception]. Paris: Gallimard.
Müller, A. M. K. (1972). *Die präparierte Zeit. Der Mensch in der Krise seiner eigenen Zielsetzungen* [Prepared time: Man in the crisis of his goals]. Stuttgart: Radius.
Nietzsche, F. (1954). *Werke in drei Bänden* [Works in 3 volumes]. Munich: Carl Hanser.
Prigogine, I. (1980). *From being to becoming*. New York: W. H. Freeman.
Romano, C. (1998). *L'événement et le monde* [Event and world]. Paris: Presses Universitaires de France.
Rorty, R. (1979). *Philosophy and the mirror of nature*. Princeton: Princeton University Press.
Rorty, R. (1989). *Contingency, irony, and solidarity*. Cambridge: Cambridge University Press.
Roth, W.-M. (2001). Designing as distributed process. *Learning and Instruction, 11*, 211–239.
Roth, W.-M. (2005). Making classifications (at) work: Ordering practices in science. *Social Studies of Science, 35*, 581–621.
Roth, W.-M. (2009). Radical uncertainty in scientific discovery work. *Science, Technology & Human Values, 34*, 313–336.
Roth, W.-M. (2014). Reading activity, consciousness, personality dialectically: Cultural-historical activity theory and the centrality of society. *Mind, Culture, and Activity, 21*, 4–20. doi:10.1080/10749039.2013.771368.
Suchman, L. A. (1987). *Plans and situated actions: The problem of human-machine communication*. Cambridge: Cambridge University Press.
Suzuki, D. (1989). *Inventing the future: Reflections on science, technology, and nature*. Toronto: Stoddart.
Temple, S. E., Plate, E. M., Ramsden, S., Haimberger, T. J., Roth, W.-M., & Hawryshyn, C. W. (2006). Seasonal cycle in vitamin A1/A2-based visual pigment composition during the life history of coho salmon (*Oncorhynchus kisutch*). *Journal of Comparative Physiology A: Sensory, Neural, and Behavioral Physiology, 192*, 301–313.

Vološinov, V. N. (1930). *Marksizm i folosofija jazyka: osnovye problemy sociologičeskogo metoda b nauke o jazyke* [Marxism and the philosophy of language: Main problems of the sociological method in linguistics]. Leningrad: Priboj.

Vygotskij, L. S. (2005). *Psychologija razvitija cheloveka* [Pyschology of human development]. Moscow: Eksmo.

Chapter 10
Graphing In, For, and As Societal Relation

> To paraphrase Marx: the *psychological* nature of man is the ensemble of societal relations shifted to the inner and having become functions of the personality and forms of its structure. (Vygotskij 2005, p. 1023, original emphasis, underline added)
>
> For us – the social person = *the ensemble of societal relations, embodied in the individual* (psychological functions, built according to social structure). (p. 1028, original emphasis)

Specifically human capacities, such as "scientific process skills," tend to be theorized from the perspective of the individual, who *has* or *does not have* the skills and associated competencies. Psychology and the learning sciences use the individual as the unit of analysis. What happens in dyads and larger groups, which may be so large as to constitute society, is thought in terms of the *inter*action of individuals as elements that compose the dyads and larger groups (Roth and Jornet 2013). The notion of *inter*action implies definable entities – e.g., human individuals – that engage in relations with each other. It is an atomistic approach, much as it exists in the natural sciences where the molecule of water is thought of as the result of a chemical reaction between hydrogen and oxygen (Vygotskij 2005). What such an approach does not consider is the fact that the properties of water cannot be derived from the properties of oxygen and hydrogen at the same temperature. Moreover, the behavior and nature of the "hydrogen" or "oxygen" atoms in the molecule are very different than those in their elemental form. The nature of the constituents has changed. This is even more so the case in the animal kingdom generally and the human species specifically where we do not just observe random and fixed behaviors but where individuals-in-relation react back toward another agent and this agent reacts back toward the recipient. The relations are of *trans*actional nature, which means that the structures of actions involving different persons are *interdependent*. The nature of a person's actions, therefore, is a function of the relation within a more global context of the situation as a whole. In Chap. 9, I present a constraint satisfaction network as an analogy for an activity system, which is considered to be the smallest unit of analysis that retains the characteristics of human society. In it we can see that the value of a single node – e.g. the one used in analogy for the subject – is a function of all the other nodes such that we cannot say

W.-M. Roth, *Uncertainty and Graphing in Discovery Work*,
DOI 10.1007/978-94-007-7009-6_10, © Springer Science+Business Media Dordrecht 2014

what the node is independent of all other nodes and independent of the historical trajectory of the network as a whole. In this chapter, I suggest the usefulness of a relational way of thinking about activity as a whole and about the subject – individuals and groups and their relation to communities – more specifically. Graphing*-in-the-making is an integral aspect of societal relations and changes in the praxis of doing science. Graphs*-in-the-making are changing objects with use-value and exchange-value *in* and *for* reproducing and transforming the societal relation and, therefore, living, immortal society as a whole. Much of what humans do exists in the form of practices that are specific to their societal (cultural) and historical contexts. This way of articulating human practices and structured actions goes against the common grain of theorizing development in terms the "construction" of knowledge and competencies, sometimes independent of social relations at other times preceded by a "social construction" within the group. Whatever sort of constructivism is used, the individual construction ultimately is considered to be the bearer of knowledge and competencies.[1] A radically different perspective emerges when we follow Vygotskij (2005), who, in the introductory quotations, references Marx as the source for his inspiration to a *concrete human psychology*.

Societal Nature of Psychological Functions

> In the course of this century, an *objective psychology* was founded the fundamental rule of which is *to study mental facts from the outside, that is, as things*. This should be even more for social facts; for consciousness cannot be more competent to know these than of knowing its own existence. (Durkheim 1919, p. xii, emphasis added)

In the introductory quotation to this chapter, from a chapter entitled "Concrete Human Psychology," Vygotskij (2005) states that the *psychological* nature of human beings has its origin in the ensemble of the *societal* relations that an individual has been part of and which, unstated here, have left a trace in the body. The founder of sociology, E. Durkheim, also refers to concrete human psychology in his argument that mental facts are external, social facts and therefore should be studied in the way we study material things (objects). Taken together, the recommendations of Vygotskij and Durkheim lead us to study psychological facts not just *in* or *for* relations but *as* societal relations; and these lead us to study societal relations in the way we study material phenomena.

Saying that a person is social is the same as saying that she is the aggregate of *societal* relations. Here, I emphasize that Vygotsky uses the adjective *societal* rather than social, which is further put into perspective by his statement that

[1] This is observable, for example, in a common way of talking about Vygotsky's *zone of proximal development*, where a developmentally less advanced student learns while "interacting" with a developmentally more advanced teacher or peer – as if the latter made available her knowledge for the former to take up and internalize for himself.

individual humans are the result of social *structure*, including the relations of power, rather than mere, arbitrary social relations that do not imply relations of power (e.g., in a class society or in a laboratory). Moreover, I emphasize that Vygotsky does not say the individual somehow constructs for itself what it previously has constructed with (or under the guidance of) others. Instead, he says that whatever we find as psychological phenomenon *is* a societal relation subsequently shifted to the inner [sphere], where it becomes a psychological function and a form of its structure. Some operation, whether it is a tool to operate on an object or a sign to work on another human being (e.g., to give an instruction, to ask for advice, to declare a fact) "*is always a social effect on oneself, using the means of social communication, and is disclosed in its full form as a social relation between two people*" (Vygotskij 2005, p. 1026, original emphasis). Thus, even if Shelby or another laboratory member has a "crazy idea," Vygotsky considers this having of an idea as a *social* effect of the individual on the individual himself, because the means of social communication have been employed. An idea is an idea only if it can be expressed, and being expressible is equivalent to communication, which always requires material form. This makes any thinking (being conscious) a societal phenomenon. This would become the main point in the work of a student and co-worker of Vygotsky, A. N. Leont'ev, who stresses the use of the adjective societal. The word, as Vygotsky points out in the same book, is a thing in consciousness impossible for one but possible for two. All consciousness is, as per the etymology of the word, a form of knowing (Lat. *sciēre*) together (Lat. *con-*).

One can indeed frequently read references to these writings of Vygotsky, but which appear in denatured form, in statements that are contrary to what the psychologist actually wrote. Thus, in the way the process of learning tends to be described there is a temporal order: first a construction in the social sphere, then an internal construction. But, we might ask, how can someone participate in enacting a practice if the body as a whole and the mind/brain specifically does not already *participate* in the relation from the beginning? Etymologically, the word participate comes from the Latin *participāre*, to share in; the second part of the word derives from *capere*, to take, to seize. Thus, the verb *participate* immediately suggests that the thing or practice is seized, shared in, at the very moment that the individual encounters it for the first time in and as a relation with another person. That is, an alternative account focuses on the fact that from the perspective of the individual, some psychological phenomenon first appears *in* and *as* relation. It appears and exists in relation and as relation. Everything is already there both outside and *within* the developing person. The talk about a transfer from the outside to the inside – i.e. *internalization* – refers to the fact that at some point the individual participates in the practice even when the other human being or beings is (are) no longer is present so that it *appears as if* something has been transferred from the outside to the inside. In fact, however, it has been there all along, appearing for the first time *as* concrete *societal* relations with others and in this relation, before apparently appearing in the absence of such relations. However, Vygotsky points out that even when someone is writing for herself, she is relating to herself as to another. That is, the relation still exists but now in the absence of a specific or generalized other.

Psychological characteristics are to be described as "interiorized relations of *social order*." Thus, not only individual psychological functions but the person as a whole, her personality, has *societal* characteristic through and through. This is the essence of the position Vygotsky takes on human development. He even ridicules the search for psychological functions in the individual:

> (1) it is funny to look for specific centers of higher psychological functions or supreme functions in the cortex (frontal lobes – Pavlov); (2) they must not be explained on the basis of internal organic links (regulation), but in external terms, on the basis of the fact that man controls the activity of his brain from without through stimuli; (3) they are not natural structures, but constructs; (4) the basic principle of the functioning of higher mental functions (personality) – social, the *interaction* of functions, having taken the place of human interactions. (Vygotskij 2005, p. 1024, original emphasis, underline added)

In contrast to other theoretical approaches, Vygotsky is explicit about deriving individual psychological functions from forms of *collective* life, which is not merely social but a life structured by *societal* relations. Anything a person does in the absence of another – e.g. during an exam – is to be explained in terms of a societal relation: Self relating to Self by societal means, graphs and words. Ordinarily, in the context of the present study, these relations are with others in the laboratory or with others in another configuration of people typical for human society, such as a scientific conference or meeting people during a visit to a fish hatchery. In conversations with others, scientists get to talk to others, and, in the process, new ideas emerge. This is precisely the experience Shelby has been making during his dissertation:

> What I find most interesting about conferences in particular, or going say to the hatcheries and meeting people there is that people will often ask you questions that you wouldn't ask yourself, at a much broader level that make you question what you are doing, "Why are you doing it? Have you thought about this other species altogether?" Or looking at a different perspective. And I find those things add a new dimension to your research and sometimes make you go back and redirect, rethink you approach and you learn new techniques, that's the other interesting thing. So I was just at a conference in January in Toronto, and learned a whole other way to measure how a fish, how, what stage a fish is in, smolting stage.

These societal relations are most fully developed, for Vygotsky, in the literary form of the drama. The dramatic nature of exchanges with others is especially salient when these appear in accounts where someone engages others directly to see whether his own thinking "has a gaping hole." In such a situation, there is an explicit engagement in a societal relation rather than a form of cogitation, where the person thinks and writes for his own in totally individual and subjective forms. Shelby talks about engaging with others when he has some "crazy" idea:

> But one of the things that I find myself doing when I have an idea even if it's not really well developed is I go out and tell somebody because right away I want to know what they think. I want to get their– is there a gaping big hole in my idea that they can see just like that? (*Snaps fingers.*) Then tell me, I want to know right away. And so I often search for an argument I go out looking for an argument, I go out looking for a critique. Um and I take that usually to Ted or to Craig, or Theodore, some of the things to Jim Plant, other people in the lab and I go to them and say, "I got this crazy idea, 'dah dah dah.'" I get really excited about it you know and they look at that, you know, often maybe the way I present it doesn't

give them an opportunity to really criticize it the way they should? (*Laughter*) So I have to work on how I present my ideas but generally what happens is, you know, they'll say, "Well, I think, I read somewhere about something like that, try this paper," and they send me off and I find this paper and I read about it, get more information and more details. And I think that's part of the process you know. It's something you have to do, you have to share your ideas and get feedback right away and see whether or not they have any worth?

Already prior to seeking out others to make them listen to his latest crazy idea, Shelby is oriented towards these others. The statements constituting the "crazy idea" are inherently assumed to be intelligible, despite the craziness involved, that is, these have use-value for the other. Vygotsky does in fact say that the *I* is a social relation of the "me" to itself. Thus, if Shelby's idea is comprehensible, enabled by the form of discourse used, it is not quite his own but already part of the general possibilities of speaking and thinking. In the relations with others, those forms of talk are further worked out, tested for the logical consistency with other forms of talk and statements. It is important here to retain the adjective *societal*, because it highlights the fact that there are institutional relations between the members towards which they orient themselves. That is, Craig is Shelby's supervisor; and Shelby, as Elmar or Sam, is funded through the research project that Craig and I have jointly received. Theo is an employee, whereas the relations between Craig and Shelby or Elmar are of the mentor–mentee type and their incomes are considered to be stipends rather than salaries.

There is a close relationship between cultural development and societal development and, therefore, between macrodevelopment and the development of societal relations. Such relations always already are relations*-in-the-making and, therefore, undergoing evolution and development. The development of any relation also means development of culture. Thus, for example, the advance of cultural knowledge is an advance in the societal relations of the research team. In the present study, we find ample evidence that aspects of graphing – i.e. the knowledgeable use thereof[2] – first exist *as* relation among people before existing for individuals. Moreover, following Vygotsky, even when someone is writing or thinking for himself, such as we might imagine Shelby to have been doing while reflecting about the study that would make an integral part of his dissertation, then the movement in thinking is irreducible to himself or the generalized other. The Other is both the origin of any higher function and towards which the products of Shelby's thinking are directed. In other words, the language Shelby uses and that is in relation with his thinking has come from the Other and, when he offers a "crazy idea" to someone else in the laboratory, returns to the Other. The thinking happens in the form of societal relations and leads to a societal relation concretized within

[2] Consistent with a pragmatic approach, neither *understanding* nor *meaning* are required for understanding everyday communication practice (Wittgenstein 2000). This is consistent with the more recent insistence on examining "first the many ways through which inscriptions are gathered, combined, tied together and sent back" (Latour 1987, p. 258) before attributing anything to cognitive factors. Finally, the approach is consistent with the contention that knowing a language is indistinguishable from knowing one's way around the world more generally (Rorty 1989).

the laboratory. The particular form of thinking about nature presented in the ultimate research paper, the scientific theories and facts it supports and discounts, have their origin in societal relations, within the laboratory and between the laboratory and the wider scientific community, which in a first instance acts in and through peer review.

In Chaps. 1 and 9, I present and work out some aspects of cultural-historical activity theory, which takes societally motivated activity as its minimal analytic unit. Everything within the activity, which covers the entire production process from inception to completion, is marked by the fact that an activity is oriented toward a generalized societal need. In the context of scientific research, this need concerns explaining, predicting, and therefore controlling natural phenomena.[3] In our particular case, the inexplicably varying return rates of salmon – as counted in the hatcheries or as appeared in the accounts of the capture rates from commercial, indigenous, and sports fishermen – was at the heart of our scientific research project. The knowledge that should have come from correlating specific physiological parameters to return rates was intended to assist in increasing these rates for the purpose of increasing economic returns and catch quota (control). Because the unit reflects society and societal order, everything within activity is marked by society and societal structure, including consciousness, practices, or knowledgeability. It should therefore not surprise the reader when I state that graphs* and graphing* emerge in, for, and as societal relations. Throughout Chap. 4, for example, I highlight those points where the societal nature of graphing stands out more than it might in other places – e.g. when Shelby draws one in the solitude of his office.

Societal Relation in the Laboratory

In Chap. 4, while presenting scientists' struggle to get a grip on the variability in their data, I repeatedly emphasize the societal relation not only as the place where this grasp emerges and comes to exist but also as the very form of graphing*-related, always developing practical competencies. Thinking about knowing and learning in terms of societal relations comes with advantages to research, for if higher psychological functions exist *as* societal relation first, we can study their emergence in carefully chosen settings. These psychological functions, as both Durkheim and Vygotsky suggest, are concrete and objectively available for the taking in relations with others. This often tends to be done with children and students, who are in the process of individualizing societal functions (as opposed to the socialization of the individual that Piaget described). However, learning may also occur among adults, and again, we can study the emergence of psychological

[3] Marx suggested that humans are not merely subject to conditions but in fact capable of changing the conditions to which they are subject and subjected. This capacity to anticipate and change conditions becomes a "will to power" in the work of Nietzsche.

functions by focusing on societal relations *as, for,* and *in which* these functions first appear. We are not considering learning *in* the individual but learning as synonymous with changing participation in changing practice. This was the case in Chap. 4. However, even when people are competent are we able to study these psychological functions when they find themselves in relations with specific other where they appear *again* between people. This is especially the case in situations that might be characterized as breakdowns, where what normally worked somehow did no longer so that members to the setting had to make available to each other whatever was required to concretely exhibit the orderly properties of the world or lack thereof for the purpose of getting things back on track.

Notable in the protocols provided in Chap. 4 are the different roles that the participants play, especially with respect to articulating the trends within the half-maximum bandwidth (HMB) data. Shelby presents the data that the team is working on, which he has collected with another graduate student in the project (Sam). Theo has already processed the raw spectra and then extracted the λ_{max} and HMB information. In fact, Theo has established the relative fractions of A_1 and A_2 on the basis of λ_{max}. He returned the results of his analyses, purely concerned with the mathematical aspects, to Shelby, responsible for ascertaining the biological relevance of the data. With respect to establishing the trends in the HMB data, we see the two biologists present possible alternatives, whereas the research associate – though regularly contributing to the publications – is not providing or withholding an assessment. One might have anticipated similarities and differences because of the disciplinary differences between the members of the group (Hall et al. 2002). Graphs, as other representations, function as "infrastructural resources at the boundary between disciplinary communities" (p. 206). One might have anticipated such work especially because of the different levels of mathematical and interpretative "competencies" the participants displayed when working independently and together to "coordinate layers of technical devices ... graphical and textual conventions ... and normative ways of making/reading representations of phenomena of interest to research specialists" (Hall et al. 2007, p. 104). It is precisely the collaborative work-organization that allows the concrete distribution and mobilization of conceptual and laboratory practices across groups.

Throughout Part II, but especially in Chap. 4, the knowledgeability we observe *is* a societal relation produced in and through the sequentially ordered turn taking, as the different members of our team bring to the table very different background familiarity and knowledgeability. This is so because the sequentially ordered talk not only is *about* something – i.e. has a topic – but also is for and produces the relation. New forms of seeing and talking about the graphs*-in-the-making exist in and as relation and prior to any individual member talking about the issue in front of some audience. Theo is the person responsible for data processing and extraction of relevant information. Shelby and Craig are the biologists familiar with the extraction of the retina and the physiology of the fish. Craig has had over 30 years of experience and publication record on fish vision and was a successful scientist as per publication record or funding record. Knowledgeability apparent in the relations allows for complex problem solving across disciplinary boundaries. Whether

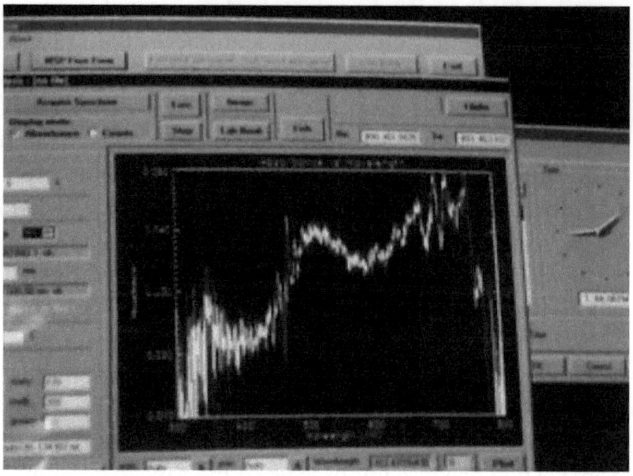

Fig. 10.1 A typical screen display of an absorption curve as it appears to the scientists while collecting data in the wet laboratory. The absorption curve appears to "sit" on an incline. When Craig calls "baseline," Theo's algorithm removes the incline to the extent possible so that the curve comes to sit on a horizontal baseline

the relation shows up as individual psychological function in the actions of one or the other laboratory member is itself a question of the practical requirements and contingencies. In some instances, it may be – e.g. when Theo notes that his algorithm under-estimates some bandwidth measures. In other instances, it *is* not – e.g. knowing about how HMB is calculated and the difference with the band-widths when thought of in terms of the width of (nearly) Gaussian distributions.

This has consequences for theorizing knowing related to graphs and graphing. When something *is* a relation first – rather than existing *in* – then it cannot be ascertained what individuals know because the relation inherently is a heterogeneous phenomenon and category. When individuals are tested for or manifest knowledgeability, this may differ from what the team as a whole expresses. For example, in the wet laboratory, Craig tended to describe the removal of the background (Fig. 10.1) as a rotation of the graph, which he made visible by means of a hand gesture. In the figure, this would be a rotation in clockwise direction with an associated shift of λ_{max} to the right (shorter wavelengths). However, Theo removed the background by subtracting a linear function fitted to the "foot" of the visible absorption. Such a subtraction moves λ_{max} to the left and shorter wavelengths. This can be seen in a simple model of the situation (Fig. 10.2) that includes two functions,

$$y = 0.5e^{-\frac{x^2}{s^2}} \quad \text{and} \quad y = 0.5e^{-\frac{x^2}{s^2}} + 0.3x, \tag{10.1}$$

where the second one is identical to the first but for the addition of a linear function $y_2 = 0.3x$. We clearly see that the addition of a linear function with positive slope moves the maximum of the peak to the right. As this function is analogous to the data we observe in the laboratory (Fig. 10.1), we begin our consideration with it. A clockwise rotation around the origin (0,0) would move the maximum further to the right and with a lower y value. On the other hand, subtraction of the linear function would give us the original Gaussian around $x = 0$ with a maximum at (0,0.5), that is,

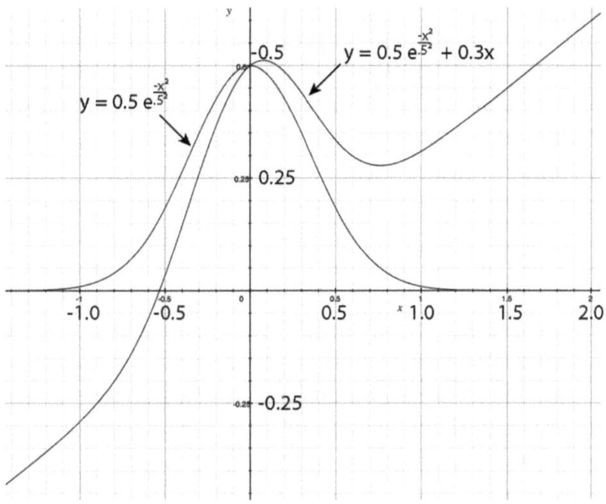

Fig. 10.2 This model shows that when a linear function is added to a Gaussian, then the maximum of the peak moves to the right. That is, if we were to make a "baseline" call in this situation, removal would mean a movement of the peak to the left

to the left. That is, as a team, the removal of the background that appeared in the form of an approximately linear function was completed correctly with respect to the mathematical canon. Individually, however, the practice of describing the removal of the background – concretely made this available by means of a hand gesture indicating a rotation – differed.

One of Durkheim's key concerns, which led him to found the new discipline of sociology, was the description of mental facts from the outside; in other words, he strived towards the abandonment of a mentalist discourse in favor of a discourse about concrete available facts accessible by all participants or witnesses to a situation. In the data analysis session that I present in Chap. 4 specifically and throughout Part II generally, we observe phenomena of order. In fact, the working session are ordered phenomena and the (orderly) natural phenomena that the collective sessions established were irreducibly related. This order did not exist outside the corporeal practices of the scientists who somehow made the orderliness of the phenomena available to each other. Spikes, trends, and even disorder as a phenomenon of order (lack of order against the background of order) were demonstrably (objectively, materially) available in the discussion as its topic and as resource in the maintenance of the topic. The orderly phenomena were produced *in*, *through*, and *for* the sequentially organized and ordered turn-taking routine. This "includes such practices as the assignation of alternative versions, the invocation of relevant mediating circumstances, and so on" (Woolgar 1990, p. 137).

The trends in the data were the upshot of the sequential turn-taking routines, that is, the relational, *joint-actional* work that the participants produced. In this study, this involved varying descriptions of what there was to see. The fact that the team members did not treat alternative versions of seeing (trends in) the data as something to be confronted suggests that at least at this stage of the work, possible alternatives even if these were contradictory, could co-exist. This co-existence of alternative models (hypotheses) is theorized in the catastrophe theoretic model that I offer in Chap. 7. What we do not observe here, before the gestalt switch between the dominant forms of the pattern seen in the data, was the kind of *transactional closure* that have been observed in other situations. When it matters whether an absorption spectrum was to be kept and saved or dumped, the team members come to a relational closure concerning the adequacy of the data and then took the relevant action. This therefore shows that "ending[s] without closure" was not only part of the "dark matter of lab work" in educational settings (Lindwal and Lymer 2008) – where both students and instructors may struggle to make sense of the relationship between a moving cart and a line graph – but also in the very "authentic" settings that science education takes as the norm to be attained. We may in fact see in my case examples of abandonment – when not pursuing the *reasons* for the malfunction of equipment or the transactional openness during data analyses – that are typical of everyday problem solving, where it constitutes one possible way of going about problem as coming up with some definite answer (Lave 1988). It stands, of course, in striking contrast to school situations, where students normally are required not only to have but one single answer in contrast to multiple answers, but where they generally are not allowed to abandon what the teacher has designed as the problem at hand.

The Multiple Functions of Graphs

In the analysis of empirical data, social scientists of all brands tend to take what someone says as a property of that individual, that is, as an expression of an interior state of "mind," "knowledge [structure]," "understanding," "conceptions," "feeling," and the likes. It is as if what appears in and as talk were pressed out of the individual – like toothpaste is squeezed out of the tube – with some kind of mechanism that serves to exteriorize what already exists on the interior. We might also say that communication, in this approach, is taken as an *ex*scription – a writing (saying) on the outside – of what has its origin on the inside irrespective of the outside. Clearly, this is a form of analysis that takes the individual as the unit of analysis, a form that has been widely criticized by language philosophers of very different brands (Derrida 1967; Wittgenstein 1953/1997). But this is not the only way of looking at scientific communication, especially the kind in which we are interested here where graphs play an integral part. Conversations, as meetings, are forms of societal relations. Following Durkheim, I take such relations – the concrete order of which participants display to each other – as *social* facts that cannot be reduced to individuals. Instead, although there are individual persons in a dyad, small group, or society, what happens in this resulting collective is not the simple sum of individual characteristics or the *inter*action of such characteristics. If it were, a computer could calculate the outcome of a meeting based on the inputs from each member. The fact that there are *new* forms of everyday practice emerging from team meetings– practice always and already is practice*-in-the-making – implies that we need theorize in which new forms emerge as results because of participation. Another term for the emergence of *new forms* is morphogenesis, which, as I show in Chap. 7, is well modeled by a catastrophe theoretic approach.[4] The dynamics of our research team, viewed from the perspective of individual participants, has emergent properties – i.e. is dynamics*-in-the-making. One therefore studies the group *as* group, the societal *as* societal – in the same way as we study water by using the water molecule, its electromagnetic properties (e.g., polarity), rather than by somehow adding up the properties of oxygen and hydrogen (gas) at the same temperature.[5] A relation between two people does not just exist, somehow, formed by ephemeral forces that glue participating individuals into a relational whole. Instead, a societal relation requires concrete work that makes available structure in concrete ways; and such work is performed all the while participants in a data analysis meeting communicate about graphs*, with graphs* (as a sign form), and with graphs* as background. In this last statement, readers already find three functions that graphs may have in relation. First, we can talk *about* graphs*-in-the-making, which makes graphs* a topic in their own right. Second, we can communicate with graphs*, where graphs* are signs among other signs (e.g. words, [iconic] gestures).[6] Third, graphs may serve as background so that something else becomes salient as a figure and obtains sign function. All three functions have figured in my account provided in Part II of this book. Thus, for example, a particular hand or pencil movement obtains sign function when it follows some line on a graph.

All the while two or more individuals communicate, with, over, and about graphs, they do relational work. Already, although we may hear only one individual talking, the analysis has to take into account that there is also one or more individuals *concurrently* attending to and receiving the statement. We therefore have to analyze a unit that includes articulating and receiving simultaneously, a unit that takes articulation and receiving as two manifestations of an overarching unit. To emphasize the unitary nature without creating a new term, I use a bar to form an ensemble that is based on the way in which the unit manifests itself: communicating = articulating | receiving. Communicating is an exchange relation where words and graphs are taken to have use-value and exchange-value, a view that completely eliminates the need for *meaning* or *understanding*. Let us take a look these issues by considering a fragment from a meeting that I had with Shelby, where he described

[4] Bakhtin's dialogism also models the emergence of new forms, new ideas, from the relation of two or more voices (Bakhtin 1984). My own research shows that the dialogical approach is fruitful for explaining the emergence of new discourse among science students (e.g. Roth 2009).

[5] Vygotskij (2005) uses this analogy when encouraging the use of *unit analysis* to replace traditional analysis in terms of elements.

[6] As seen in Chap. 4, when Theo modified an existing graph on the chalkboard, the graphs are graphs*-in-the-making whatever the particular function they currently have or serve.

to me an emergent sense of where our study was heading in terms of its findings. The meeting occurred prior to our entering the wet laboratory, in the joint office that led to the two wet laboratories of the team. While talking, Shelby also produced a graph of the pattern of the porphyropsin levels (%A_2) in juvenile coho salmon from Robertson Creek over the course of a year (turn 01). These levels changed between a maximum of 47 % A_2 down to 20 % in this hatchery where the annual water temperatures range from 4 to 16 °C (turns 03 & 06). These two turns, and the simultaneously drawn graph, constituted the background against which what is to come becomes figure: the two measurements of A_2 levels that Shelby had completed with older coho salmon caught from a research vessel in the open ocean (turns 08–12).

Fragment 10.1

01 S: but the robertson creek fish that went from up to about fortyseven *
02 M: uh hm
03 S: down to twenty percent

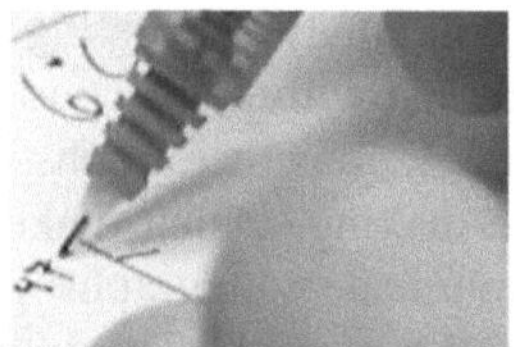

04 M: uh hm
05 (0.76)
06 S: and they range in temperature from four, this is degrees celsius
07 M: yea yea yea *

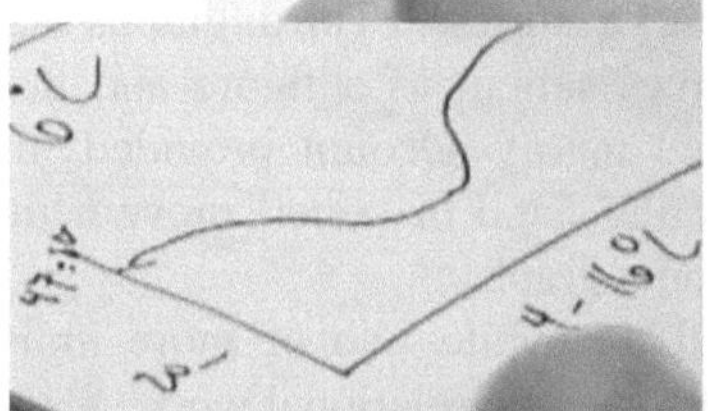

08 S: nOW the marine fish that i went=t=get from the ocean; the ones i got last summer wore sitting * almost deadon that curve

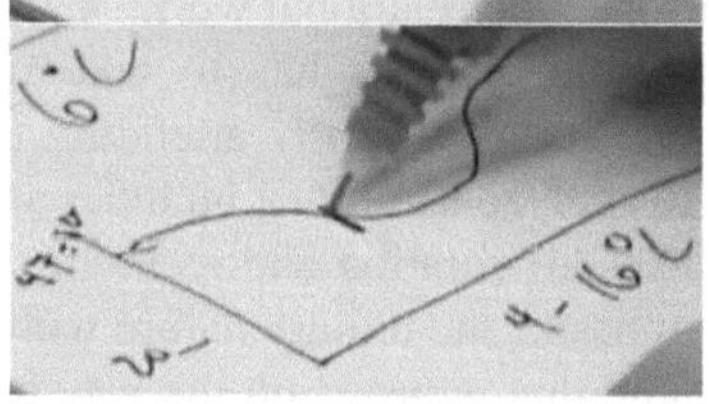

09 M: yea
10 S: so my suspicion was so hang on if they are dead on that curve may be in fact they follow that trend all year round ((*follows curve*)). and so i was measuring them this year and this section went down * to (r?)

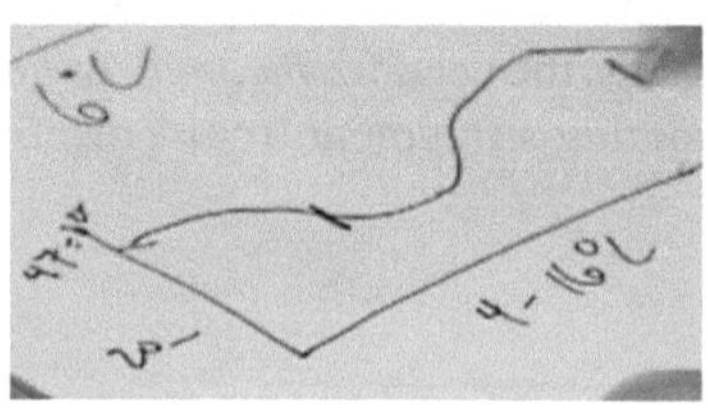

11 M: uh hm
12 S: i was measuring them
this year and they are
sitting right * HEre
(1.09) an:d sO::::
(1.59) i=m not exactly
sure how to interpret
that. (0.54) it could be
that they dont change. it could be that the marine
fish once they get there they stay there

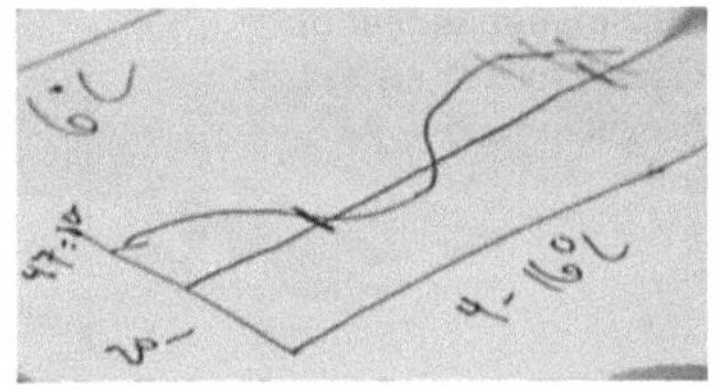

13 M: uh hn
14 S: thats one possibility.
<<all>the=other
possibility is that they
are shifting> but that i
missed their peak. their
peak may have been back
here * a little while

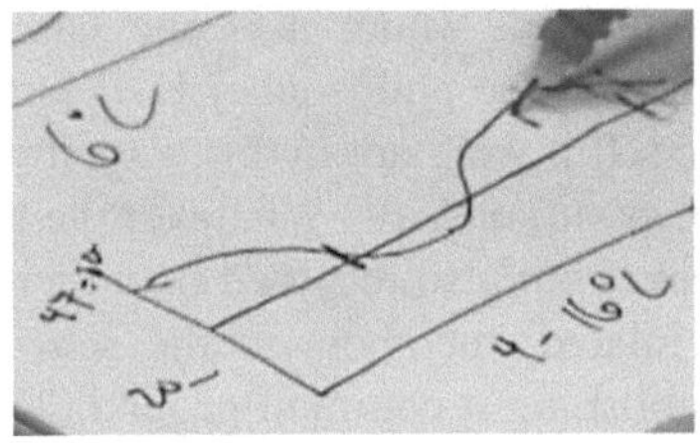

15 M: oh, okay

His first measurement of A_2 levels in saltwater coho – recorded around April or May, as we can see from the graph, but which remains unstated here – were almost the same as those that he had obtained from juvenile coho salmon from the Robertson Creek hatchery (turn 08). Shelby then says that he had a suspicion: the marine coho salmon were following the same curve. To test this hypothesis, he measured the porphyropsin levels again, but this time, as his mark shows (turn 10), it fell below the curve describing the A_2 levels in Robertson Creek coho salmon (turn 12). He then talks about not knowing how to interpret this observation and provides two possible explanations: (a) marine-based coho salmon do not change their porphyropsin levels (as the theory established by Wald suggests) or (b) that the porphyropsin levels do change but that the cycle is shifted with respect to that of the freshwater-based coho salmon, so that their peak in $\%A_2$ would have been at an earlier point (turn 14).

Transcription fragments such as this one often appear in the literature. However, these are used to make attributions to what *individuals* do rather than the *joint* action that makes this societal relation witnessable (from the inside, as participate) and observable (from the outside, as analyst). *Joint action* is a phenomenon of and exhibiting *societal* order, inherently transcending any individual participant. It is not just Shelby and I who, almost independently, add something that makes this *con*versation. We are doing something together, which, like dancing, cannot be explained by using the individual as the unit of analysis. First, the words and graphical signs – lines, Cartesian grid, numbers, degrees, and vertical markings – are *inherently* presupposed as exchangeable with the other person. Most of these signs come from the generalized other – Shelby did not invent the words he is using, nor did I – and, in communicating, return to someone other than the Self. Even when a sign is made up on the spur of the moment, such as when Shelby makes the

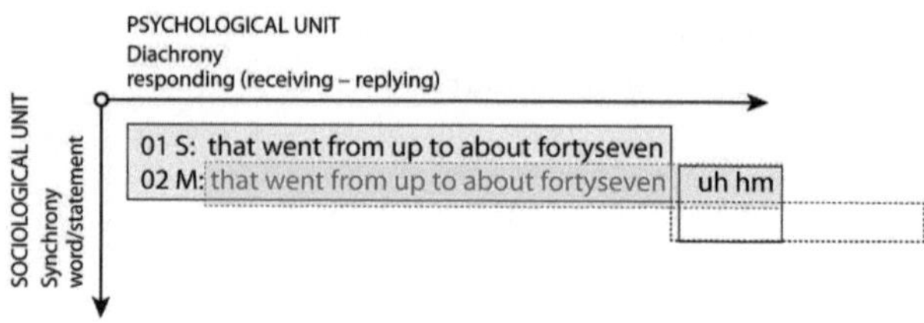

Fig. 10.3 This model shows the minimum unit of an exchange, which always includes author and recipient

marks to indicate the A_2 measures taken from saltwater (adult) coho salmon (turns 08 & 10), the articulation itself implies its intelligibility (i.e., its use-value for the recipient with whom it is and has been exchanged). The word, Vygotsky says drawing on the philosopher L. Feuerbach, is a reality for two: it never only exists for one. If it were not in this way, it would make no sense whatsoever for Shelby to speak, point, or draw. Whatever the transcription shows was not only articulated by a speaker but also received by a recipient (me). In general, if such presuppositions are violated, members to the setting act to make this known and to work on reestablishing a common ground. That is, whatever happens here is based on the implicit assumption of its use-value to the other, both on the part of the person articulating something as on the part of the person receiving. This is so even if there are differences in whatever individual knowledgeability and practice there might be. This relationship of the communicative sign taken as a unit spread over different members to the relation is expressed in the vertical dimension of the theoretical unit describing a conversation (Fig. 10.3).

In this model, each word, each statement belongs to speaker (black print) and recipient (grey print). The word – or any other sign that appears in Fragment 10.1, including those appearing on paper (movements of the pen that leave traces), and ephemeral movements of hand or pen over the existing traces – *always* and *inherently* is a *socio*-logical phenomenon. That is, its *logic* is social! If it did not belong to both participants, then sound, hand/arm movements, or traces would not be signs. If it is a word or sign, then it is a societal-historical phenomenon. Even if something is produced for a very first time, it is a sign only when it already can be assumed to have use-value for the other, that it is and has been a possibility for two. This implies that even if the two markers for the ocean data had never been produced before, these would be signs only when they existed for both Shelby and me. Moreover, even if Shelby were to use these markers for himself, these would be signs only if the were external to be used in communicating with others or with himself.

The second, horizontal dimension in Fig. 10.3 reflects a psychological aspect, the temporal unfolding of the saying (communicating) in time, and the temporal unfolding of the response in the transition from receiving to replying. We therefore observe a double dehiscence.[7] One part of the dehiscence derives from the diachrony of the response, which spreads active reception and reply across time. The

[7] A gaping opening between parts of a whole.

diachrony also expresses itself in the difference between saying (signing) and the said (sign). The other aspect of the dehiscence derives from the diastasis[8] of the word, its *dis*location that moves it across different people, with different positions and dispositions. The effect of this situation is that any word (sign) is different for author and recipient – not because "meaning" depends on the perspective, as those trained in classical forms of thought will want to claim, but because of the inner difference of a word with respect to itself. The word (sign) always stands in for something else, makes something present other than itself together with its own present. The word, in fact, shares a lot with a commodity, being, as it is, part of an exchange (communion) with others (Roth 2006). Thus, for Marx value is "the *relation of a commodity to itself*, rather than to another commodity" (Il'enkov 1982, p. 266), a relation that is such that commodity "emerges as a living, unresolved and insoluble inner contradiction" (p. 266). In a similar way, the signification of a word[9] is the relation of the word to itself rather than to another word or thing.

Initially, Shelby is talking *about* the graph while drawing it. He says that it denotes the variations from 47 to 20 % [A_2]. At the same time, he is using the graph as a sign to stand for the measures he had collected in the course of the year. Finally, when he moves his pen along the curve while saying "may be they follow that trend all year round," the graph is the ground against which the pen movement becomes salient (which, in turn, makes salient the curve). So we see the three functions of a graph at work, and, here as generally, these functions are in operation simultaneously and inseparably (as in the case of the pen movement following approximately the line on the paper). But, as suggested, there is more to the graph (with marks, letters, numbers), spoken words, and hand/arm/pen movements. This relation between two laboratory members exists because of the communication; and the communication exists because of the relation. Communicative activity and activity more generally are mutually constitutive (Leont'ev 1971; Roth 2013). This is why Wittgenstein (1953/1997) writes about the interweaving of language and practical activity, neither of which can be explained on its own. This interwoven something he calls language game (*Sprachspiel*). Communicating with, over, and about the graph*-in-the-making also establishes and maintains the relation that includes Shelby and me. We therefore do not need to get into the individual minds of the participants, but see both relation and psychological functions at work. As there is no objection or sign of lack of alignment, any statement, whatever its kind, belongs to both participants. The graph*-in-the-making is part of the material and glue of the relation, where it has in addition certain functions. It is precisely for this fact that it is also a graph*-*in-the-making*, continuously and without remedy evolving in terms of its exchange-value and use-value.

---

8 Technically, the separation of bones without fracture, slight dislocation.

9 Readers may want to use "meaning" instead – provisionally.

Here, the communication has a special purpose: to fill me in, one of the two supervisors of the project, on the latest development in the data. The special nature of this situation is clear when we compare it to others that occur in society, often denoted as "small talk," the purpose of which is to establish or maintain a particular relation. In (single's) bar talk – as talk with the same purpose in other places and at other times – the communication is designed precisely for establishing a relation that has not existed before. When the talk is aborted, the exchange is the beginning and end of a relation simultaneously. In the present instance, the talk contributes to constituting us as members of the same research team, where a lot of the required communicative ground is already shared enabling *this* relation, and where further communication is used to increase what can be taken as shared. That this is so can easily be gauged from the differences that exist in my hearing (reading) of the transcription and that of an unsuspecting reader. The latter would not be in a position to follow our talking, what the 47 and 20 refer to, what Robertson Creek is, what the communicative function of "marine fish" is, what species the fish is from, and so on. Similarly, Shelby does not talk – and does not have to talk – about the units and extent of the abscissa, for, as a member of the laboratory, I can be assumed to see and hear that the graph makes present the model of an entire year's dataset.

In this excerpt, there is no communicative trouble apparent. This is why we do not observe a special effort for making the relational work more explicit. There are signs, however, that the work is not merely individual work. The relational work, precisely because it is relational, requires irreducibly joint action. To take up the analogy of the dance once more, it takes two to dance (e.g. a tango). It is not that one dances and another one does not. Tango emerges from the joint, social action that cannot be explained by means of individual actions even though each participant has to move this foot or that or bend forward as the other bends backward. In this fragment, the second part of this joint work is made visible (hearable) in the form of interjections "uh," "yea," "hn," and "hm," which tend to have the function of acknowledging receipt and comprehension. There also is the interjection "oh" (turn 15), often acknowledging something communicatively novel, and the interjection "okay," which tends to be used as acknowledgment of hearing (both in the traditional sense of receiving and comprehending). Consistent with the model (Fig. 10.3), each communicative sign constitutes a juncture of two turn pairs, completing the preceding one and starting off the subsequent one. Each communicative sign belongs to both participants, the one producing it with his lips and the other one receiving it with his ears. Thus, if "okay" acknowledges receipt and marks comprehension (perhaps mixed with a little surprise), then it also constitutes the beginning of the next turn, in fact inviting the other to continue to communicate. In these turn sequences we therefore produce the work that maintains the relation.

The dialectic of the societal relation, realized in and through the communication, also allows us to explain the phenomenon of innovation and scientific discovery. When a research team publishes its results, these are inherently assumed to be intelligible within the community. Even if there is a scientific revolution, the communicative effort both assumes the communicability and the recognition of novelty on the part of the recipients. Even though a graph may appear for a first time in the literature, its intelligibility, its sign nature, has to exist as a possibility. But this possibility exists within and for a collective body (e.g., a scientific community) not merely for an individual scientist. If some finding expressed in a graph exists as a general possibility, it does not belong to the individual. Intelligibility refers us to society and cultural possibility. This implicit assumption that statements have use-value and exchange-value and that someone else can develop them is apparent in the following interview excerpt, where Shelby talks about ownership over and sharing of ideas.

> I firmly believe that science is meant to be done [by sharing ideas or getting ideas out there] and that if you have an idea, and you're not going to do it yourself, to get it to somebody so they can do it. It doesn't matter who does the idea it matters that it gets done especially if it's a worthwhile idea. Um and so I'm not, I don't have the same ownership that some people have about my ideas. Some of them might do, some of them might don't.

We get a further hint of the special function of talk in societal relations when Shelby distinguishes talk that serves to "schmooze" others and talk in and through which one learns and develops.

> I used to think that conferences were just a chance for people to go off and schmooze and you know, have a couple, a little bit of a vacation and you know, show off their work and I don't think I really understood the impact that they can have for sharing and like that and the little inspirations that come from seeing other peoples' work on different species or different systems? And kinda go, "Oh, like it sparks things (in you)." I think a lot of that stuff you don't even know you learn at the time? You walk away from there and months later you go, "Oh, that's a neat idea, I'm going to take this and mix it with that and . . ." So it's a different level of learning, it's what I mean, much more subtle than that.

The relation as origin of data analytic and graphical knowledgeability is apparent from the following account that Shelby provided during the same interview with another graduate student. He was in the process of talking about his learning since he had joined the laboratory, where he was introduced to the kind of work Craig and I wanted him to do. He talked about developing competencies in doing the dissections and conducting the measurements. There was then a sudden jump when he wanted to get into the data analysis, which had generally be the domain of Theo's participation. It is in the exchanges with Theo that Shelby first came to integrate certain spikes with everything else in his laboratory practice, the functioning of the instruments, and relating the graphical features to certain parts of the instrumentation.

> The next jump came when I started to learn how to do the analysis of the data because up to a certain point Theodore was doing all the analysis. And that was fine except that I was missing out on a big part of data collection, which is knowing what to collect and how it should look when you get it in the analysis. And then I learned the analysis part and then I added a whole new level because now I'm bringing it all together and learning the finer details of the mechanism and why it gets certain spikes where I do. And I started asking Theo a lot of questions about how the device works and that expanded my knowledge once again. And I would say more recently through collaboration with other people in the lab,

> I'm starting to add a whole other level because now I'm looking at going next door and doing some, umm, more, umm, well they call it intercellular recording where actually measure the electrical output of an individual cell. And so starting to keep that in my mind that's where I'm going I need to start building up my literature database and my knowledge a bit better in that area as well, and so, there's no question that, its kinda that whoosh, skyrocketing.

We can see from the account that Shelby experienced another level of practice in relation with team members working on other projects, the members of which shared a common front room that led into the different wet laboratories ("next doors"). This new form of practice is not only adding "a whole new level" but also de facto turns out to be a "skyrocketing" experience. This would not have been the case had he not communicatively established and maintained the societal relations, relations that were the basis of the communication and that was produced and maintained by the communication. Whatever Shelby learned and ascribed to himself, it existed in and more importantly *as* the relation first. We can say *in* and *as* relation because of the dialectic of communication and relation, which are merely manifestations of the minimum unit of analysis: scientific research activity.

In the end, therefore, the finally produced graphs – notice, "graphs" without asterisks – are the result of this multitude of societal relations with others in the same laboratory, across laboratories, and with researchers at conferences. The graphs that appeared in Shelby's dissertation and in the published research article are but traces that the activity as a whole left behind. They are marked by the activity as a whole such that unpacking any one of these graphs requires opening up the black box and unpacking what the activity has been. The finished graphs bear a metonymic relation to the activity, but only after everything has been said and done.[10] They are the communicative analogue to the Said after the Saying has ended. In analogy to the work that a laborer exchanges for an income, the author divests himself in and with the work. *Divest* here translates Hegel's (1807/1979) German *Entäusserung*, which implies at least three different dimensions in the philosophical tradition that developed dialectical reasoning: (a) creating of something new, (b) giving away or getting rid of something one has owned, and (c) opening the inside to the outside. It also connotes estrangement such that although the result of activity, as the said that results from the saying, relates to its creation as to a stranger. With respect to graphs, therefore, these also are the products of work that in a way abandons its product. The products bear the mark of the work and also denotes the movement of graphing*-in-the-making in an estranged way. Abstracted from the work of production, it lacks reference to the contingencies and *societal relations* that brought it to life.

[10] The graphs that our research team produced are both finished, as end products of our scientific activity, and unfinished, as signs that readers can use in the various functions that I describe in this chapter. Other scholars citing the works read and use these graphs in support of statements that advance their own agendas.

Exchange and Circulation

Graphs*-in-the-making, conceived as means and products of societal activity (dejatel'nost' / Tätigkeit), enable us to use Marx's philosophy, which underpins cultural-historical activity theory. As I was able to show, exchanging Marx's *commodity* for the linguist's *sign*, and examples of commodities with examples of signs, leads to texts that exhibit considerable resemblance with recent philosophical texts in the interpretive philosophical and post-structural domains (Roth 2006). One can actually describe the (always material) sign as another form of commodity in human (communicative) exchange relations, sometimes paralleling or reflecting the exchange of other material goods.

> Essential to *circulation* is that exchange appear as a process, a fluid unit of purchases and sales. Its first presupposition is the circulation of commodities themselves, as their circulation that emerges from many parts. The precondition of commodity circulation is their production as *exchange-values*, not as *immediate use-values*, but as mediated through exchange-value. Appropriation through and by means of divestiture [Entäusserung] and alienation [Veräusserung] is a fundamental condition. Circulation as the realization of exchange-values implies: (1) that my product is a product only in so far as it is for others; hence suspended particularity, generality; (2) that it is a product for me only in so far as it has been alienated [entäußert], has become for others; (3) that it is for the other only in so far as he himself alienates [entäußert] his product; which already implies (4) that production is not an end in itself for me, but a means. Circulation is the movement in which the general alienation [Entäußerung] appears as general appropriation and general appropriation as general alienation. As much, then, as the whole of this movement appears as a societal process, and as much as the individual moments of this movement arise from the conscious will and particular purposes of individuals, so much does the totality of the process appear as an objective connection, which arises naturally; arising, it is true, from the mutual influence of conscious individuals on one another, but neither located in their consciousness, nor subsumed under them as a whole. Their own confrontation produces an *alien* societal power standing above them; their interplay as a process and power independent of them. Circulation, because a totality of the societal process, is also the first form in which the societal relation appears as something independent of the individuals, but not only as, say, in a coin or in exchange-value, but the whole of the societal movement itself. (Marx/Engels 1983, p. 127, original emphasis, underline added)

In this quotation, we find several of the central ideas that I mobilize in the present work generally and in this chapter specifically. Exchange is a movement, a *societal* process that requires theorists to model it as such. Just as purchases and sales constitute a fluid unit, so do the articulation and reception of signs (words, graphs, symbolic gestures). Even though there appears to be a conscious will of individuals, in their confrontation, a movement occurs that *transcends* the individual, who therefore finds herself subject and subjected to the movement. Not only is there movement but also the appearance of a societal power that exceeds the individual will; and the collective work constitutes processes and powers that are independent of individuals. This also applies to the societal relation, which appears to individuals as independent of them even though they contribute to bringing it about, maintaining it over time, and concluding it to bring about its (temporary) end.[11] Graphs*-in-the-making are part of the "currency" or "commodity" *on* and *with* which the societal relations are built. Just as Marx's product of labor is a product only when it is for others, so the graph Shelby will have produced is a graph only in so far as it is a graph for others – her me but in the publication for the peer community at large. Marx points out that the product, here the graph specifically and the sign more generally, suspends particularity and constitutes generality. Any concrete sign, any concrete graph, is a realization of a general possibility, and, as such, it is the product of societal history; it is marked by this origin through and through. However, as Marx notes elsewhere, the mediating movement of the exchange disappears in its own product and does not leave a trace behind (Marx/ Engels 1962).[12] Paraphrasing Marx by changing his commodity into our sign, one might say that without their own doing, the signs find their value form – i.e. that these are standing for something other than themselves – as something that exists outside and independent of their signifying body. Thus, in use, the graph in Fragment 10.1 is not meant to draw attention to the ink trace as such but to the fact that it allows us to make momentarily present all the measurements that the research team has collected over an entire year and to mark off with respect to it the new data from saltwater-based adult coho salmon.

References

Bakhtin, M. (1984). *Rabelais and his world*. Bloomington: Indiana University Press.

Derrida, J. (1967). *La voix et le phénomène* [Voice and phenomenon]. Paris: Presses Universitaires de France.

Durkheim, E. (1919). *Les règles de la méthode sociologique septième édition* [Rules of sociological method 7th ed.]. Paris: Felix Alcan.

Hall, R., Stevens, R., & Torralba, T. (2002). Disrupting representational infrastructure in conversations across disciplines. *Mind, Culture, and Activity, 9*, 179–210.

[11] Contrasting the individualist and *inter*actional approaches to understanding speaking or graphing in the research literature, everyday common experience has its ways of describing meetings as socio-logical phenomena, such as when, for example, meetings are said to have "their own dynamic" or when groups or companies are said to have something like a personality. This personality of a group, as Durkheim (1919) suggests, has to be understood as social fact and modeled by *socio*-logical rather than *psycho*-logical categories.

[12] This is why an analysis of unfolding relations alone does not completely get at ruling relations, for the signs used in and as communication embody histories structured by ruling relations that are not apparent in the concrete exchange (Smith 1990). By using and accepting use of "single parent family" to describe her own situation, Smith contributed to the reproduction of inequities. Only through the historical analysis of the provenance of the term was she able to change her relation with the school, which had categorized her child in this way with consequences for particular treatment.

Hall, R., Wright, K., & Wieckert, K. (2007). Interactive and historical processes of distributing statistical concepts through work organizations. *Mind, Culture, and Activity, 14*, 103–127.
Hegel, G. W. F. (1979). *Phänomenologie des Geistes* [Phenomenology of spirit]. Frankfurt/M: Suhrkamp. (First published in 1807)
Il'enkov, E. (1982). *Dialectics of the abstract and the concrete in Marx's Capital*. Moscow: Progress Publishers.
Latour, B. (1987). *Science in action: How to follow scientists and engineers through society*. Cambridge, MA: Harvard University Press.
Lave, J. (1988). *Cognition in practice: Mind, mathematics and culture in everyday life*. Cambridge: Cambridge University Press.
Leont'ev, A. A. (1971). *Sprache, Sprechen, Sprechtätigkeit* [Speech, speaking, speech activity]. Stuttgart: Kohlhammer.
Lindwal, O., & Lymer, G. (2008). The dark matter of lab work: Illuminating the negotiation of disciplined perception in mechanics. *Journal of the Learning Sciences, 17*, 180–224.
Marx, K./Engels, F. (1962). *Werke Band 20* [Works vol. 20]. Berlin: Karl Dietz.
Marx, K./Engels, F. (1983). *Werke Band 42* [Works vol. 42]. Berlin: Karl Dietz.
Rorty, R. (1989). *Contingency, irony, and solidarity*. Cambridge: Cambridge University Press.
Roth, W.-M. (2006). A dialectical materialist reading of the sign. *Semiotica, 160*, 141–171.
Roth, W.-M. (2009). *Dialogism: A Bakhtinian perspective on science and learning*. Rotterdam: Sense Publishers.
Roth, W.-M. (2013). An integrated theory of thinking and speaking that draws on Vygotsky and Bakhtin/Vološinov. *Dialogical Pedagogy, 1*, 32–53.
Roth, W.-M., & Jornet, A. G. (2013). Situated cognition. *WIREs Cognitive Science, 4*, 463–478.
Smith, D. E. (1990). *Conceptual practices of power: A feminist sociology of knowledge*. Toronto: University of Toronto Press.
Vygotskij, L. S. (2005). *Psychologija razvitija cheloveka* [Psychology of human development]. Moscow: Eksmo.
Wittgenstein, L. (1997). *Philosophische Untersuchungen / Philosophical investigations* (2nd ed.). Oxford: Blackwell. (First published in 1953)
Wittgenstein, L. (2000). *Bergen text edition: Big typescript*. Accessed November 30, 2013 at: http://www.wittgensteinsource.org/texts/BTEn/Ts-213
Woolgar, S. (1990). Time and documents in researcher interaction: Some ways of making out what is happening in experimental science. In M. Lynch & S. Woolgar (Eds.), *Representation in scientific practice* (pp. 123–152). Cambridge, MA: MIT Press.

Part IV
Uncertainty and Graphing in STEM Education

Philosophers have hitherto only interpreted the world in various ways; the point is to change it. (Karl Marx, 11th thesis on Feuerbach)

In Part II, I provide the results of an ethnographic study of one scientific laboratory over the course of many years in the process of creating a new explanation of salmon life history for itself and others concerning the relative amount of porphyropsin in photoreceptors of coho salmon in the course of their life history. Central in this account are the pervasiveness of uncertainty and graphing. Both of these themes also are central to Chaps. 8, 9, and 10. In this fourth part of the book, I take these features as starting points for reflecting on possible implications of my ethnographic study for science, technology, engineering, and mathematics (STEM) education. Before getting into the two chapters concerning the ways in which we may change STEM education as a consequence of the observations reported in Part II of this book and the retheorizing enacted in Part III, I provide some reflections concerning such a move.

Some scholars with interests in science education and working from ethnomethodological perspectives have critiqued simplistic translations of lessons learned by studying the discovery sciences to school STEM education. The brunt of the critique lies in the recognition that schools, as the discovery sciences, are arenas in which participants display to each other the orderly properties and properties of order for the purpose of making practical action (do its) work. All the while I recognize and emphasize the differences between the two different societal activities, I do permit myself to take what I learn in one situation for the purpose of reflecting on the other situation. It is for the purpose of breaking open new ground for STEM education rather than to replicate the structures of science in STEM education or to increase the throughput in the STEM "pipeline." Nothing would be further from my intentions in writing this Part IV of the book. Instead, I see my own endeavors as a scholar in terms of two movements. On the one hand, I consider myself a researcher of practical action without any commitment other than to describe its structures as these are made available by the members to the settings investigated for other members therein. On the other hand, I am (sometimes) a

STEM educator concerned with bringing changes about in the world (of schooling) and, thereby, take a normative stance in identifying goals to be achieved in and by STEM education settings. When working with colleagues and graduate students, I am the first to emphasize that as an analyst I work without letting my analytic work be affected by any stance taken as an educator; and when I am a STEM educator, I exhibit interests that are very different from those that I would as the analyst.

Ethnomethodology has contributed a lot to the way we describe and explain the natural sciences. Part II of this book is strongly influenced by the interest in the everyday practices by means of which members to a setting make available to each other the ordered and orderly properties of the world that they inhabit. But I am not a maniac of one method only. I draw on whatever method my research object requires. Thus, I actively recognize that ethnomethodology discipline and method has done much less well in uncovering the political and ideological dimensions of human practices though there have been, as referenced below, exceptions to this (e.g. McHoul 1994). Such work has generally drawn on critical (neo-) Marxian theories, which also underlie cultural-historical activity theory, the theoretical framework that has shown fruitful in my own work. In my reading, the ethnomethodological critique does not make a different argument than the literature on situated cognition, cognitive apprenticeship, and authentic science by highlighting how constitutive features of societal life bring these about in moment-by-moment fashion. This, however, happens at the interface of action and operation. Such analysis is generally blind to the political dimensions of societal life, which require analyses at the interface of activity and action and historical analyses of activity (see Chap. 9). However, the descriptions of the above-mentioned literatures and the ethnomethodological critique provide are in stark contrast to much of science education today – both as education and research practice – which assumes that real science knowledge is in the head and generalizable across situations and independent of the particular object of activity.[1] The situated cognition movement simply pointed out that scientific knowledge, too, is embodied, situated, and contextual; and it pointed out that what students master is not knowledge of a cultural-historically relevant object. Instead it focuses on changing participation in changing societal practices as synonymous with learning wherever this participation might occur. The literature on legitimate peripheral participation further suggested that this needs to be taken into account when teaching and researching science in schools.

In its critique of the "authentic school science" movement, the ethnomethodological critique provides support to those who argue that schools need to be changed without providing the tools for constructing new visions, or how to

[1] The philosopher Georg W. F. Hegel was very critical about claims to *abstract* thinking, suggesting that only the uneducated but not the educated person think abstractly. In the mind of the experts, concepts always are *concrete* universals embodied in their actions rather than abstract thought forms. The latter pertain to general forms of things independent of the world we inhabit, they are merely theoretical, and therefore characteristic of the gap between theory and practice, school and the everyday world outside.

deconstruct its (bourgeois) ideological dimensions that lead to the reproduction of inequality along the lines of gender, race, culture, or socio-economic factors. For the very reason that the ethnomethodological critique suggests the school is a context that has to be taken in and by itself, school knowledge cannot be taken as a privileged non-context where generalizable knowledge is taught but has to be viewed as a special situation that is highly situated but in a different way. Schools can no longer claim to impart generalizable knowledge that can be used in various everyday situations outside of school. This is a serious question that has yet to be addressed, and a question to which the ethnomethodological critique does not provide and in fact avoid an answer.

The ethnomethodological critique of translations of research in the social studies of science into STEM education is legitimate and correct, though, if it targets any simplistic application of the situated cognition, authentic science, or apprenticeship models. What really is required is a conception of the learning problematic as arising from the discrepancy between the participative possibilities provided for in the structures of schooling activity and the individual's own participation possibilities with respect to some learning object. From my subjective learner perspective, increasing my own participative possibilities, and thereby coming closer to the (cultural-historically mediated) possibilities for participating comes with an emotional valence for me: it is in my own interest to engage and participate to decrease and diminish the discrepancy between what I can do at the moment and what can possible be done. Education is not democratic when the learning tasks are imposed from the outside and when an outside agent measures the degree to which I succeed, keeping me out of the decisions about why and what I should learn. In the final analysis, a true learning problematic can never be determined from the outside, because although the targeted learning objectives may be common to two or more individuals, the relation individual-object is always personally relevant (Holzkamp 1993). It is partially for this reason that I advocate for project-based learning and student control.

The ethnomethodological critique makes a legitimate point in highlighting that the analysis of school and out-of-school learning needs to be approached more symmetrically. The situated cognition movement was right in problematizing school learning, but failed in equally analyzing out-of-school situations (often viewed as mere field of the application of learning). One must not neglect considering the possibility that out-of-school situations, too, can interfere with and work against expansively motivated participation (e.g. Lave and Wenger 1991). In the apprenticeship model associated with the situated cognition and authentic science movement, this was not yet realized. There were unresolved discrepancies between different forms of participation (Roth and Lee 2006). The learning problematic thereby no longer arises in the constitutive tension between generalized possibilities of participation and those currently realizable within the specific relations of ruling that school produces. This leaves unquestioned the characteristic power–knowledge-related practices of schooling that contribute to inequities along the lines of gender, social class, ethnicity, or culture.

Ultimately, the learning situation that is most consistent with democratic ideals is tied to the description of learning as an individual's expansively motivated changing participation in changing societal relations and practices. Here, one does not need to abandon asymmetries in participation but need to allow asymmetries to be articulated and negotiated continuously in interpersonal relations.[2] The question about who knows what may change from instant to instant even when any two individuals seemingly are asymmetrically located with respect to their knowledge (e.g. Roth and Middleton 2006). From my position as an individual person, only I (who else?) can ultimately decide whether I can subsume my own learning problematic to a collective, collaboratively pursued learning problematic or whether I have to insist on its difference and otherness.

Ethnomethodology tends to be a passive social science in the sense that it takes social practices in their own terms without realizing the human capacity to change conditions in addition to being subject and subjected to them. In the limit, as ethnomethodologists themselves note, the discipline leads to the opposite of intelligibility: "The radical stress on observable details risks becoming an unprincipled, descriptive recapitulation devoid of significance ... minute descriptive detail is assembled in a hyper-realist profusion, until the reader loses any sense of meaning" (Atkinson 1988, p. 446). I take human existence in a different way. Unlike animals who respond to their life situation, I understand human beings as standing in a dialectical relation with their life situations both determining it and being determined by it. As a human being I am empowered with agency, which allows me to change my environment and myself. This Marxist stance is therefore decidedly optimistic because it allows us to conceive of change and a brighter future, where inequality and social injustice have given way to a more democratic society in which the pursuit of partial (capitalist) interests have been replaced by generalized interests that serve common interests. In which direction does this line of argument lead us? A possible avenue to take has been framed, in my view, by groups outside formal education.

A former Club of Rome president Ricardo Dietz-Hochleitner (2001) points out that we need forms of education that will make an effective contribution to democratic coexistence, tolerance, dialogue, empathy, solidarity and cooperation, in a context of rapid and far-reaching change and growing interdependence among different countries. The Club of Rome does not build on the existing structures of formal education but on the collective wisdom of informal community-based groups of learners (citizen movements, environmentalists, anti-globalization). In this respect, the task of making STEM education more appropriate and relevant is complicated by the fact that although educational systems have broken partially with the past these have not freed themselves from their own past, not even in those cases where broad reforms have been implemented. Education still is living in the past because it prepares for success in situations that are totally different from those

---

[2] In a philosophy of difference approach, asymmetries will always exist, for the only thing common to any two individuals is the fact that s/he is different from every other individual (Roth 2008).

for which education supposedly is designed – the everyday world outside school.[3] This seems to call for a more explicit linkage of learning for adult life before we fully participate in it as adults, a direction in which we might want to rethink science education. In various places I have therefore argued for the deinstitutionalization of school science (Roth and McGinn 1997) and for STEM contexts that allow students to develop *démerdise* and *débrouillardise*[4] (Roth and van Eijck 2010).

Recommendations from the discovery sciences for STEM education might be discussed in terms of "authentic science." So what kinds of future might the concept of "authentic science" have in science education? My early work was concerned with providing science students with authentic science experiences – the heart of the argument of *Authentic School Science* (Roth 1995). My interest in this topic actually predated by a decade the article that subsequently constituted a trigger for research on the situated nature of learning (Brown et al. 1989). When, in 1980, I began teaching science in a 45-student middle school of a small fishing village in southern Labrador, I wanted students to experience science in the way I had experienced it during my graduate work rather than in the boring, lecture-oriented ways that characterized my own school years. At the time, I had no educational background. I was a research physicist with a Masters of Science degree who, because of a lack of opportunities during an economic down, took a teaching job. My idea was to allow students to investigate phenomena in the way scientists do.

Over the years, my enthusiasm for turning students into scientists waned, and I actually became critical when I realized that the *enculturation* into scientific practice also meant enculturation into the blind spots of science (Roth 2001). Moreover, my environmental interests made me aware of the fact that many scientists were involved in the development of unethical technologies for the sake of financial profits to their companies – including "Round-up-ready seeds," sterile seeds, genetically modified organisms that are released disregarding precautionary principles. I also have come to realize that there is something fundamentally askew in schooling, which is, as I showed, less into the transmission of cultural knowledge and values and more into the reproduction of societal inequities (Roth and McGinn 1998). School science, in the same way as school mathematics, is a filter that sorts the student population providing those who come out on top with symbolic capital (grades and grade reports) that they can convert into accessing coveted spots in universities of their choice. After years of teaching science and after more years of conducting research on science learning I had come to the conclusion that school science needed to undergo deinstitutionalization (Roth and McGinn 1997). It had to serve some real purpose in the context of society – such as children's and students' participation in environmental activism, the running of community gardens, or the

[3] Already the Romans knew, *Non scholae sed vitae discimusx* (We learn not for school but for life).

[4] Literally, the French colloquial terms *démerdise* and *débrouillardise* may be translated as "ability to get oneself out of shit" and "ability to get oneself out of the fog." We used them to envision a very different kind of science education than most students experience today.

creation of salmon habitat and hatching of salmon. Here, graphs*-in-the-making and graphing*-in-the-making do play an integral role because of the evidentiary power that is associated with them when presenting research results in public arenas. All of this work led us to revisit and rewrite the function of "authentic" in "authentic school science" (Roth et al. 2008).

Over the years, I also have come to recognize school for what they are in themselves. I had been a fervent advocate of using research results from social studies of science – e.g. the ethnographic reports from laboratories and about laboratory life – as guides for designing ("authentic") school science. But in part because of my research I have shifted to considering school science as a particular form of practice, especially as one that reproduces societal inequities. Moreover, I no longer view students as individuals with knowledge to be fixed – in the way this is presented, e.g., in the conceptual change paradigm – but as whole beings, with their personal lifeworlds and needs. Even if their forms of participation and discourse differ significantly from science, the associated practical familiarity with how the world works is the ground – one is tempted to say the nourishing soil, as the nutrient agar on which bacteria are cultured in a Petri dish – for anything scientific to emerge subsequently. Even if the science overturns everyday familiarity with the world, this very overturning is possible only after everyday familiarity has developed (Husserl 1976).

And yet: I think there are lessons to be learned when science education looks to the discovery sciences – even for those students who are not in what scientists often refer to as the "science pipeline," which may or may not be leaking (e.g. Mervis 2012). This is so especially with respect to the development of a more critical stance towards the sciences, a stance that is recognizant of the fact that the sciences are just another human endeavor, just as fraught with uncertainty and foul-ups as any other dimension of human life. Learning about the nature of science means developing insights about why and how to take scientific research results with a grain of salt, and learning how one might go about taking more critically what comes out of scientific laboratories and investigations, including the data and the ways in which they are translated into series of graphs and other inscriptions. Ultimately then, and returning to the opening quote, my own life is more purposeful if there is the possibility not only for describing and explaining the world and thinking about change but also for actually bringing it about. Merely describing and explaining school science in its own terms is not what I want to do. The possibility that I can contribute to making science education and the world more equitable – characterized by the realization of collective common rather than particular partial interests – is a more hopeful gestalt for orienting my life than the despair that comes with accepting the world as is. I therefore say, let us change science education for the better and let us do it together. The following two chapters, as well as the concluding chapter, are offered up in this spirit.

References

Atkinson, P. (1988). Ethnomethodology: A critical review. *Annual Review of Sociology, 14*, 441–465.

Brown, J. S., Collins, A., & Duguid, P. (1989). Situated cognition and the culture of learning. *Educational Researcher, 18*(1), 32–42.

Dietz-Hochleitner, R. (2001). Education for the 21st century: A lifetime to learn. In D. B. Rao (Ed.), *Education for the 21st century* (pp. 13–18). New Delhi: Discovery Publishing House.

Holzkamp, K. (1993). *Lernen: Subjektwissenschaftliche Grundlegung (Learning: Subject-scientific foundation)*. Frankfurt/M.: Campus.

Husserl, E. (1976). *Husserliana Band VI: Die Krisis der europäischen Wissenschaften und die transzendentale Phänomenologie. Eine Einleitung in die phänomenologische Philosophie* (Husserliana vol. 6: The crisis of the European sciences and the transcendental philosophy. An introduction to phenomenological philosophy). The Hague: Martinus Nijhoff.

Lave, J., & Wenger, E. (1991). *Situated learning: Legitimate peripheral participation*. Cambridge: Cambridge University Press.

McHoul, A. (1994). Toward a critical ethnomethodology. *Theory, Culture & Society, 11*, 105–126.

Mervis, J. (2012). What if the science pipeline isn't really leaking? *Science, 337*, 280.

Roth, W.-M. (1995). *Authentic school science: Knowing and learning in open-inquiry science laboratories*. Dordrecht: Kluwer Academic.

Roth, W.-M. (2001). "Authentic science": Enculturation into the conceptual blind spots of a discipline. *British Educational Research Journal, 27*, 5–27.

Roth, W.-M. (2008). Bricolage, métissage, hybridity, heterogeneity, diaspora: Concepts for thinking science education in the 21st century. *Cultural Studies in Science Education, 3*, 891–916.

Roth, W.-M., & Lee, Y. J. (2006). Contradictions in theorizing and implementing "communities". *Educational Research Review, 1*, 27–40.

Roth, W.-M., & McGinn, M. K. (1997). Deinstitutionalizing school science: Implications of a strong view of situated cognition. *Research in Science Education, 27*, 497–513.

Roth, W.-M., & McGinn, M. K. (1998). >unDELETE science education: /lives/work/ voices. *Journal of Research in Science Teaching, 35*, 399–421.

Roth, W.-M., & Middleton, D. (2006). The making of asymmetries of knowing, identity, and accountability in the sequential organization of graph interpretation. *Cultural Studies of Science Education, 1*, 11–81.

Roth, W.-M., & van Eijck, M. (2010). Fullness of life as minimal unit: STEM learning across the life span. *Science Education, 94*, 1027–1048.

Roth, W.-M., van Eijck, M., Reis, G., & Hsu, P.-L. (2008). *Authentic science revisited: In praise of diversity, heterogeneity, hybridity*. Rotterdam: Sense Publishers.

References

Atkinson, P. (1988). Ethnomethodology: A critical review. *Annual Review of Sociology, 14*, 441–465.

Brown, J.S., Collins, A., & Duguid, P. (1989). Situated cognition and the culture of learning. *Educational Researcher, 18*(1), 32–42.

[illegible]

Chapter 11
Uncertainty, Inquiry, Bricolage

> The real issue is that scientists themselves are in a situation of radical uncertainty concerning their own actions – expressed hyperbolically, they do not know what they do all the while the artifacts and tools (technology) they use in the process have quite durable certain character to them . . . Interpretive flexibility of technology is the *result* of the fact that practical action is radically uncertain in principle, so that its outcomes never can be anticipated with certainty. (Roth 2009b, p. 335)

Much of Part II of this book deals with the ways in which scientists approach data, interrogate them, grabble with variability in the data, transform data into series of inscriptions following a chain of increasing abstraction, and, ultimately use the final graphs they produce to support claims about factual matters of nature. Throughout Part II, we see that variability and uncertainty are integral to the effort of producing knowledge claims supported by graphically presented evidence; and logical contradictions may be present but hidden from the scientists' awareness. My ethnographic study also shows how scientists do not just use data to induce patterns and theory but, instead, tend to be subject to bias vis-à-vis their measurements. At various stages in the philosophy of the natural sciences, it has been recognized that natural phenomena are bounded by theory and, particularly relevant in the present context, by human activity. Thus, asserting that the ontology of natural phenomena is circumscribed by human activity means denying "the hallowed independence of the world of representations from the world of embodied practices" (Gooding 1992, p. 66).[1] The study presented here underscores those findings emphasizing human activity as a phenomenon bounded by the material nature of the human body, which thereby belongs to the world of natural objects (facts) and the inscriptions it creates (artifacts). Actions, though seemingly ephemeral, have the same material nature as objects and inscriptions, as these are constituted by the materiality of the human bodies that bring them forth. The nature of the action is of the same type as that of

[1] *Ontology* – from the Greek ὄν [on], present particle of εἶναι [einai] to be, + λόγος [logos], word, science – is the science that constitutes *what* things are. In the information sciences, an ontology is a framework for organizing the available knowledge and things in a domain.

W.-M. Roth, *Uncertainty and Graphing in Discovery Work*,
DOI 10.1007/978-94-007-7009-6_11,

the objects, and therefore both are involved in their mutual implication, stabilization, and constitution.

Actions can be talked about and explained based on the same procedures used with texts, without, however, having the capacity to fully provide for their own explanation (by others) in any one situation. The present study shows that this is not only the case for recipients and observers of actions, but also for human actors (e.g. Chap. 3). Thus, the laboratory members could know that they had done what they intended to do when the visual image of a photoreceptor and its graphical representation stabilized as successful data. However, this event was rather unremarkable and unnoticed by the scientists. The ontological fragility and uncertainty of their own actions – just what is it that they have done? – become evident only when the sought-after objects and correspondences were absent. The scientists ventured on, however, and did not elaborate a self-critical epistemology of action. They simply ascribed subsequently recognized failures of having done something different than usual or than it was supposed to be.

In the everyday world, when we go out eating in a restaurant fancied for its chef or visit our dentists, we assume that the specialists know what they are doing. If my observations in the scientific laboratory turn out to generalize to such other situations, we might begin and take a more critical stance in and to our everyday lives as well. Increasing malpractice suits and reports that there is a high likelihood to get sicker when going to a hospital are consistent with the assumption that experts in areas other than the discovery sciences are caught in the same double bind that separate actions and plans (intentions), although, like their peers, scientists do not tend to make this widely known. From my study we may take that the mutual implication of finalized object and action is a more pervasive feature of the human condition and that we need to change our presuppositions and hopes when dealing with experts. Preparing students for a world full of ambiguities rather than a world where there is some right way that an authority can and does check of would be much more advantageous than what schools do today. What I envision schools could do is to enable students to engage others, such as experts, and to make decisions in the face of the inevitable uncertainty. This may be way better than teaching any specific STEM curriculum content in the manner that this has been the historical practice to the present day.

It is this state of affairs, the tinkering nature of the sciences – as those of medical doctors, dentists, and other specialists – that became salient to me in the course of the research reported in this study. Being considerably ill at the time, struck by a severe chronic fatigue, I realized that the doctors did not know and were dealing with the data they collected in and from my body in a similar manner in which the scientific research team reported on here did with its data.[2] This may have been the

[2] This statement is typical: "Try this [vitamin B preparation] and when it does not work come back in 2 weeks and we try another one." I realized that I was paying the doctor to *learn*, in a trial-and-error fashion: what is behind the kind of symptoms I displayed and how you can treat them (cf. Roth 2014).

most profound lesson for me: there is uncertainty everywhere, and most professionals hide it behind their apparent expertise. This may also be the main lesson I would want students to learn in our science classes, not only about science but about every human endeavor: *there is an unbridgeable abyss between our practical familiarity of the world as we daily encounter it, including our bodies and situated actions, and the world as it appears in our ideas in the form of facts, patterns, instructions (plans), and retrospective descriptions.*

Uncertainty – Literacy, Graphicacy, Contradictions

In this section, I reflect on the possible implications of radical uncertainty, as found in the sciences, on literacy, graphicacy (i.e. graphical literacy), and (logical) contradictions in STEM education.

Uncertainty and Literacy

> Among the essential phenomena of the modern times are the sciences. An equally ranked phenomenon is mechanical technology [*Maschinentechnik*] ... Mechanical technology remains the most visible consequence of the being/presence [*Wesen*] of modern technology, which is identical with the being/presence of modern metaphysics. (Heidegger 1977, p. 75)

The most popular image of science is that of a rational activity that proceeds as described in the methods sections of scientific articles, which – as I show here is consistent with what social studies and history of science and technology scholars have said for some time – are a posteriori accounts in which every action is described in terms of actual achievements rather than the initial goals and intentions. In the opening quotation, Heidegger suggests that science and mechanical technology embody – i.e. are identical to – metaphysics, the idealist Platonic world of talk and ideas. That is, the sciences, as technology, presuppose a perfect structure of the world. This structure is said to map onto and to be reflected in mathematical structure according to the couplet {fundamental structure $\leftrightarrow$ mathematical structure}. This structure often can be expressed in graphical form. Thus, for example, in the ultimate research article, our research team presented the combined measurements from the Robertson Creek Hatchery as a plot together with a cosine curve (Fig. 11.1). In this figure, I also inscribe the function, which, in the article, appeared only in the text in terms of its parameters rather than as a mathematical function: "Robertson Creek hatchery data fit a cosine function; with a period of 1 year, a mean of 36.6 % A2, amplitude of 8.1 %, and a peak in the month of February and a minimum in August" (Temple et al. 2006, p. 307). Such equations are taken to represent the real structure. Deviations from the perfect world – quite obvious in the data – are taken as error; and scientists, accordingly, have developed methods to clean up their noisy data so that these can reflect the real structure thought to

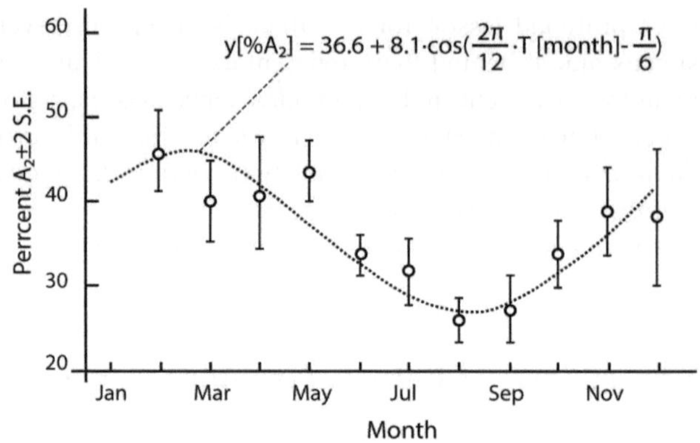

Fig. 11.1 The measurements from the Robertson Creek Hatchery were reported to be best modeled by a cosine function given in the research article only in textual form

underlie them. Human actions are thought (about) in the same way – in terms of ideal plans rather than as something practical that fit plans and descriptions only within considerable margins of variation. Thus, science textbooks often begin with a chapter on "the scientific method," usually specifying a sequence of actions including posing a question, hypothesizing, designing the experiment and so on. However, this image of science has very little to do with the way it is done as day-to-day praxis. In fact, it may be precisely this idea that there is a perfect world that scientists reproduce that may lead science students to "cook" their data if these do not correspond to the expected ideal (e.g. Roth and Bowen 2001a). A number of observational studies of science in the making articulated the contingent nature of decision-making, choice of ("do-able") research problems, making experiments work, politics of funding, or filiations of research paradigms (Knorr-Cetina 1981). Because this research focused on the social links between scientists, their peers, and the remainder of society, science and scientific knowledge have been described as "socially constructed." This notion, however, also is problematic in the sense that many science educators use the notion of construction to connote not only assembly but also assembly completely under the intentional control of the scientists.

Throughout Part II of this book, the scientists were confronted with and apparently talked about uncertainty, which was perhaps the most dominant feature of this scientific inquiry. Uncertainty also is of interest to some learning scientists, especially when children conduct inquiries of their own design (e.g. Metz 2004). Metz identified four major categories of open inquiry that also showed up in my study of scientists at work: uncertainty arising from data (determination of HBW), trend identified in data (multiple trends identified in scatter-gram with error bars), generalizability (all HBW data vs. those from Kispiox), and theory that best accounts for the data ("dogma" vs. alternative). In my wider study, however, additional uncertainty sprung up, for example, about the precise nature of the object seen under the microscope, the source of error variance in the absorption spectra, the precise source of the retinal cells studied (dorsal or ventral rods), the theory supported by the data ("dogma" vs. alternative), and so on. In fact, in this study there was uncertainty emerging from the particular transformations that the data underwent prior to being subjected to trends. The episodes analyzed throughout Part II exhibit the scientists' lived work of coping with this uncertainty. Coping with uncertainty may in fact be the least attended to attention in school STEM education (Roth and van Eijck 2010). Metz is clearly correct about the importance of exposing even young children to experiences where they are required to deal with uncertainty and, thereby, develop skills for doing so more broadly.

In the episodes featured throughout Part II of this book, the scientists first shifted their attention from collecting data to the instrument that did not work as it had done on the previous day. For example, in Chap. 2 I feature a 15-min period that shows how scientists literally groped in the dark, unaware of what might have caused the change in their instrumentation. Through their movement in and actions on software parts, the scientists came to create a clearing (in the dark), where they perceived and articulated features that began to recede as soon as the breakdown was cleared. In fact, in the face of one or more unknown reason that led to the malfunctioning equipment, the scientists' only hope is groping about in the dark, deal with the uncertainty, and through their actions bring some clearing to the situation. Without acting, there was little to no hope to ever get out.

Human agents generally know what they are doing only through the outcomes of their actions, that is, *after* the (f)act (Suchman 1987). We follow a new recipe to cook a dish only to find out *through the results* that what we must not have done what we intended to do; students follow instructions only to find out that they did not do what they were supposed to do through their findings (Roth et al. 1997a). This has serious consequences for social life. In the everyday world, when we visit, for example, our family physician or dentist, we assume that they know what they are doing. We assume that they are doing "the right thing" when they diagnose and treat our ailments. If my observations in the scientific laboratory turn out to generalize (transfer) to these other situations, we might begin and take a more critical stance toward actions in our everyday lives as well. When we go to our dentists, we hope that the gap between their plans and practical action is as small as possible. But this gap might in fact remain considerable: the media carry an increasing number of cases of medical litigation and other instances where doctors and hospitals admit having made mistakes. I take this to be an indication that doctors are in the same precarious situation not to know with 100 % certainty or precision what they are currently doing. Hospitals constitute a context where the relationship between actions and (deleterious) outcomes might be studied. In Canada alone, an estimated 8,000 individuals die from in-hospital acquired but supposedly preventable infections. A better explanation of the relation between actions and the degree to which we can know them while they unfold therefore is not only of theoretical interest but has considerable implications to society at large, which, in the case of hospital deaths in Canada, bears the costs (an estimated $100 million every year). This gap is particularly at issue in situations when people work at the limits of their knowledgeability. In the sciences, that something has gone awry in the chain of material transformation therefore is available only through the product of the actions rather than through the experience and evaluation of the actions while they are enacted. Once trouble has been experienced, specific past actions can be problematized and therefore articulated.

If physicians, dentists, and other practitioners know what they are doing only *through the effects* their actions bring about, then we have to develop different dispositions when seeking help, purchasing services, or watching courts that attribute causation and liability to human action a posteriori. Although *skilled practice* also denotes the closing of the gap between situated actions and anticipated effects, some research shows that even highly skilled engineers may not be able to reliably reproduce intended products (Sørensen and Levold 1992). The present study suggests that we can never be certain to have closed the gap. Between the *doing* and the gloss that is used to tell the doing (e.g. "measuring the absorption spectrum") in the irreducible {"*doing* [measuring the absorption spectrum]"} pair, there is a gap without remedy. Litigation is the predominant means by which the adequacy of action–gloss pairs is interrogated and blame is attributed to agents for not knowing what they were doing in the process of doing it. In this case, we can learn from the present study that the dialectic of object and action is a more pervasive feature of the human condition and that we need to change our presuppositions and hopes when dealing with experts.

We may therefore consider two possible forms of implications of this study for STEM education. On the one hand, we might want students to become more appreciative of the uncertainties of science and scientists to be more forthright in highlighting the uncertainties in the media reports. On the other hand, we might want to have science students have experiences with uncertainty and to learn that this uncertainty is due not to their own insufficiencies but rather to the in-principle uncertainty associated with human action. STEM education would in this way become a very different endeavor, developing a more critical stance, including a more critical stance towards the effort of the STEM fields rather than merely reproducing forms of thinking and representing through indoctrination.

Uncertainty and Graphicacy

The ethnographic evidence provided in Part II of this book exhibits the very different roles and functions graphs (as other inscriptions) may (have to) take during scientific discovery work. As students are to learn the use of graphs in different ways and at different stages of a STEM activity, offering opportunities in which they can mobilize a multiplicity of graphs as resources in their sense-making activities may allow them to develop the kinds of meta-representational competencies that some learning scientists have been describing. In Chap. 10, I propose theorizing graphs*-in-the-making and graphing*-in-the-making, because the developmental aspect of learning is thereby captured. For example, in the initial phase of finding an algorithm for the growth of an animal population, students use gestures that may have emerged somewhere in the production of the graphs, from enacted or perceived hand/arm movements.[3] As I have shown in a series of studies (Roth and Lawless 2002a, b, c), as soon as a body movement has sign properties – when the movement transcends itself and becomes sign used in communication – then the basis is present for subsequent transformations that increasingly shift towards linguistic representations. That is, in a second stage, the primary function of graphs* and gestures may be to support explanation and justification of their work. A secondary function of these inscription means may be to provide additional justification of descriptions by making visible aspects of previously shared images and experiences. Both functions serve to generate an algorithm. In the final stage, when students engage in generating predictions, graphs* and gestures serve as means to summon experiences common to the participants within the same classroom.

It has been observed that we "can expect *learners* to generate multiple interpretations of inscriptions, use multiple meanings of words, and focus on different aspects of inscriptions" (Moschkovich 2008, p. 578, emphasis added). As the present investigation shows, this state is not only characteristic of students – who are in the process of learning something that they do not know and therefore cannot evaluate – but also of highly experienced and successful research scientists confronted with some data* first time through when they are in the process of establishing possible "deep structures" therein. The present work also shows that there may be alternative and contradictory explanations for the data and visible patterns articulated in the same meeting *without* there being a contrasting evaluation that might lead to further learning. Yet research has shown that learning arises precisely from the opposition and confrontation of different discursive propositions or while providing reasons supporting alternative hypotheses. In STEM education, this might be the place where teachers – in the process of observing and overhearing students – may decide to enter their discussion and explicitly make any differences and contradictions the topic of discussion. It is precisely in the reflection that new descriptions and explanations may evolve in the course of inquiry. Do we have to conclude that scientists, too, may need such support? The very fact that scientists work at the cutting edge of their and their field's knowledgeability and the fact that these meetings focus on topics in real time that prevents a reflective approach may mitigate such attempts in the discovery sciences.

There are many presuppositions about what students of science *ought* to do when working with graphs. That they ask questions about aspects of a graph is a central

[3] While conducting an early study on graphs and graphing in physics lectures for elementary education majors, Ken Tobin and I analyzed how the professor, during his demonstration-based lecture, moved from showing Galileo's experiment of balls rolling on an inclined plane to lists of numbers and graphical representations: position-time, velocity-time, and acceleration-time graphs (Roth et al. 1997c). During a subsequent lecture, the professor referred back to the earlier one making a hand gesture. In our analyses, we first reproduced the gesture as a way of making it present again for our discussions in the absence of the video. We also had evidence that the students did not know what he was talking about or referring to. Thus, we soon used the hand/arm movement as a way to signify students' lack of understanding in lectures.

constituent of graphical literacy (Roth et al. 2005). For example, one study provides a list of questions students are to draw on in their argumentation about graphs (Chin and Osborne 2010). The questions that the authors provide are categorized according to some of the basic science process skills – e.g. observing, comparing, analyzing, predicting – and include: "'Is there anything unusual or unexpected?' 'How are A and B similar?' 'How are A and B different?' 'What is the significance of these similarities and differences?' . . . 'Is there anything I am puzzled about?' . . . 'Which is the better graph? Why?'" (p. 242). As the analysis of the transcripts shows, the research scientists I observed and worked with did not ask such questions, at least not in the sessions I recorded. They were able to live with apparent contradictions in the descriptions its various members provided in articulating possible trends in the data and with the apparent contradiction between the theory or a priori conceptions and the actual data. The kinds of processes observed and postulated in the Chin and Osborne and in other studies – including disagreement, challenge, cognitive conflict, and puzzlement – are not (as) explicit in the data analysis sessions that I report on in this book. But this does not mean that these processes *ought not* be fostered in schools, integral aspects of "immortal society" in their own right, where they might function as a means of developing changing forms of participation that are currently not generally observed in most working place contexts, including scientific discovery work.

In Chap. 7, we observe Craig projecting from the data to possible values that might have been measured had we started earlier in the year. The projection was not at all consistent with what we found out during the following years. This shows that STEM educators should not assume extrapolation and trends to be self-evidently arising from the data at hand. Even though extrapolation from existing data tends to be thought of as an "essential scientific skill," my research shows that scientists do project trends and extrapolate in ways that will turn out not to have been correct. In the present studies we see quite clearly that the issue about extrapolation is more complicated. Precisely when a scientist knows the theory or the biological context, the extrapolation may be inconsistent with the trends that are clearly visible in the data with hindsight. In an earlier meeting, Craig gestured what he expected the curve to look like such that it also fits with the measurements we had collected thus far. Later, the team exhibited the relationship within the data in a very different manner, by means of a sinusoidal relationship. That is, the very same data came to be modeled very differently; and Shelby exhibited these differences during one presentation when he exhibited the team's data in the context of the previous study (Alexander et al. 1994) and in terms of what the "dogma" suggested. Craig's gestures were consistent with the dogma, which provides a much poorer fit with the first eight data points than the nearly linear portion of the sine function.

To be noted here is the extrapolation presented in the gesture versus the reasonable extrapolation that the data themselves would have suggested, which, at best, is a linear function with a much-reduced slope (e.g. Fig. 7.8). The conceptual change involved, among others, precisely this changeover from seeing the data through, and describing the data in terms of, the existing dogma rather than through the new relationship suggested by the new data themselves. This change in how the data are perceived and conceived of is associated with and requires a fundamental change in the way in which biologists model the physiological changes associated with migration and the causes underlying the changes in the retina. These are no longer considered to be due to the *organism-centered*, physiological changes in anticipation of changes in salinity – as suggested by the correlation in the graph with plasma sodium levels – but due to very different, *environmentally driven* factors, such as responses to temperature, daylight, and other factors in the life history strategies of the fish.

Uncertainty and Logical Contradictions

In Chap. 6, I articulate some background on the ways in which contradictions are thought within (a) classical logic that underlies much of science and mathematics (education) as we know it today and (b) dialectical logic, which is the foundation of the soci(et)al psychological theories that Russian scholars including L. S. Vygotsky and A. N. Leont'ev began to develop. However, the dialogical logic, although Western educators have apparently appropriated the theories of these scholars, has not been taken up in the process.

In mathematics education – as in other areas of STEM education – contradictions are thought from the perspective of logical contradictions. Thus, either some statement is true or it is false. A third option is not given such that a statement is, in its very essence, both true and false simultaneously. For example, mathematics educators identify situations where a student's answer is inconsistent with a previously given answer or action (Ron et al. 2010). Typically, such inconsistencies or contradictions are attributed to some problem of the student, a mental deficiency or "partial and partially correct knowledge." For example, these authors report students' answers to a series of questions concerning a game with two dice. They note that in a second game, one of the students "considers another set of outcomes in each case he is not aware of *the* inconsistency in his answers, and does not see any reason for reconsidering the fairness of the games" (p. 80, emphasis added). My question would be, why should the student reconsider the fairness of the game of dice if he does not see any inconsistency?[4] From the perspective of the student, to view his actions and hear his discourse intelligibly, one has to assume that the contradiction does not exist so that a reconsideration of an earlier question is not necessitated. However, approaching such situations from the perspective that *the* inconsistency actually exists, mathematics educators then feel compelled to develop models to account for deficiencies, such as that of *partially correct constructs*. In such cases, whatever a student – or scientist, if I were to extend this approach – is doing is not taken as something in its own right but as something

[4] I use the indefinite "any" rather than the definite "the" to show that I do not assume that one actually exists.

lesser than an ideal where the contradiction is absent. To the researcher imports not what the student thinks, how she thinks, or how she does something and talk about it, but rather what she does not do with respect to some norm. But to think developmental possibilities – especially when we think these in terms of the zone of proximal development in the context of changing participation in changing practices – we need to be able to see how the world looks like to the individual and, therefore, what makes most sense to do as any next step. Thus, P. Freire's success with the peasants he taught did not come from the fact that *he* told them that they could not read but rather from the fact that he assisted them in becoming conscious of their situation. This allowed the peasants to realize by and for themselves that they needed to learn to read so that they could expand their action possibilities and thereby better help themselves.

The conclusions we can draw from the ethnographic work reported in Part II has implications for STEM education in the sense that it calls for much more caution of denoting a situation as containing a contradiction or of identifying contradictions between theory and data. It has been noted that being partially correct – which is the case here, too – are powerful tools in the analysis of inconsistencies in students' answers. This may be true for an outside perspective, but STEM educators need to ascertain that a contradiction appears in the consciousness of the learners, stated by *them* as such, before it can be exploited for educational purposes. As the present study shows, even successful scientists may state different descriptions and explanations but do not articulate the difference *as* a contradiction. One important reason for not noticing a contradiction as a contradiction until after the fact is the way in which an event*-in-the-making eludes its own grasp until it has come to an end when everything has been said and done. That is, what evidences itself within an unfolding event*-in-the-making as unease or dilemma may subsequently, once scientists grasp the finalized event as a whole, be recognized as a true contradiction. Without contradiction explicitly stated, however, a change in thinking may not be required. Thus, for example, students may talk about velocity–time line graph intersections as instances where two cars collide on the racetrack ("iconic interpretation"). However, this may not be a *mis*conception at all when viewed from within the perceptions and reasoning of the students, just as the differences in the description and explanation of the data in the present study was not a misconception when viewed from the perspective of the participants. What constructivist scholars such as Piaget have not sufficiently appreciated is that during human development not only does the (perceived) world change but so do the perceptual processes (Merleau-Ponty 1964). We see how the scientists' perception of their data changed. Such a change also was evident in the above-noted study in an Australian physics classroom (Roth et al. 1997b), where the predominant number of students saw motion when no motion should have been seen. *Within* their reasoning, these students were consistent, even though from without, where motion was not seen, observers noted a contradiction. The contradiction between predicted observation (motion) and actual fact (no motion) was not operative because the students made observations consistent with their (theory-laden) predictions. This also may require us to be more cautious when describing situations as harboring "built-in contradictions."

The work I report in Part II shows that there may be instances when a scientist is confused without such confusion leading to some change in conceptual language. But in many school STEM situations, teachers frequently know that two different descriptions or explanations constitute contradictions. To assist students in making these contradictions salient *as* contradiction, teachers may actually intervene while students talk about relevant phenomena without seeing (hearing) the contradictions that the teacher sees (hears). That is, this study implies that mathematics STEM, when evaluating and analyzing what students do during data acquisition and interpretation tasks, need to think more about learning from the perspective of those who do not know the final story and who do not have a god's eye view of the unfolding events in which they take part and therefore can explain their own learning only after the fact (e.g. Rorty 1989; Roth and Radford 2011). The studies of graphing reported in Part II also show that without deep familiarity with the settings in which data were collected – here, e.g., the temperature regimes at the different sites from which coho salmon were sourced – scientists also talk about the trends in their measurement in contradictory ways. This is very clear in the case of the temperature differences between the coho salmon raised in the Robertson Creek hatchery and those that were sourced in the creek and in the river into which Robertson Creek flows. This ought to have considerable implications for school mathematics, where students tend to be asked to interpret data without also being familiar with the context in which, or how, data were collected. It is precisely familiarity that allows the detection of contradictions, then this ought to encourage STEM educators more than ever to let students explore phenomena at great depths prior to asking them to describe and explain the trends or relations that might be observed in graphs.

What We Might Want to Foster in Student Inquiry

In this section, I reflect on some of the aspects of inquiry that we might observe and want to foster when students engage in inquiry of their own design. First, laboratory talk will differ from the texts that can be found in journals and textbooks. Second, the temporality of scientific practice is not at all like clock work. Third, learning is more like groping about in the dark. Finally, while engaging in open inquiry, students also learn *about* the nature of open inquiry.

Laboratory Communication

In Chap. 8, I comment on the nature of communication at work generally – including in the fish hatchery and in relations between scientists and others – and in the scientific laboratory more generally. I discuss the fact that this talk is very much unlike the texts that the same scientists subsequently produce for publication purposes. School science, however, tends to privilege fully articulated speech, almost as if students had to be speaking in the way scientific articles and science textbooks are written. Such an approach does not take into account the fact that societal activity and speech activity co-implicate each other and, therefore, develop together (Leont'ev 1971). Thus, if we theorize activity in terms of an emergent account, as activity*-in-the-making, then speech activity, too, needs to be taken as speech-activity*-in-the-making. With speech activity, individual speech and collectively available language also change (Roth 2013). The traditional accounts of classroom science talk actually had caused me quite some anguish while I was a teacher. I still remember having been baffled and disappointed when I transcribed in 1990 the first classroom videotapes while I was still a high school teacher. My students' talk was highly elliptical, often only one word at a time – much like what I reported from the science laboratory in the course of Part II. I still remember my first set of transcriptions, which were the verbal protocols of concept-mapping sessions.[5] Although the recommendations to teachers at the time had been to have students use these mapping tasks individually, I found the results frustrating. I decided to see whether working in groups would not allow students to achieve better results, both in the construction of these maps from given key words and the science they were learning. What I obtained at the time looked something like Fragment 11.1 (from Roth 2009a).

Fragment 10.1

```
024  R:  in pair production are all the different things here,
025  K:  they are different.
026  R:  photoelectrons.
027  K:  get all the;
028  M:  i am so sweaty.
029  K:  complementary?
030      (.)
031  R:  complementary?
032  M:  look around the guy
033  K:  complementarity.
034  M:  well what type? uh, jesus, we should put x=rays up
         here with the light, like that. should we put x=rays
         there?
```

It can easily be seen that in this exchange, the talk is highly abbreviated, often only one word. At the time, I was looking for "meaning" and "understanding" but could not find these unless I was making a lot of assumptions about what was in the minds of these students. I had read *Talking Science* (Lemke 1990), which presented ways of analyzing classroom talk in thematic patterns, where there were units of

[5] Having had a background in neo-Piagetian developmental theory, I was using at the time (radical) constructivist theory to understand learning. However, it was precisely out of this research that I evolved one of the first social constructivist accounts of learning in school science classrooms (e.g. Roth and Roychoudhury 1992, 1993).

actors/agents–process, process–goal, proves–means/materials, location–process, or classifier–things. We easily see that in Fragment 11.1, such units are absent. I had mixed feelings, both about my students and about myself. I was hesitant writing up my work for publication because the students did not speak in the ways that I found research articles to be reporting students to speak and write. Perhaps I was not a good teacher? Perhaps I was not helping my students enough to be able to speak science in fully articulated sentences?

At that same time, much of my teaching was oriented to open inquiry. This meant that students spent about 70 % of their time designing experiments within the domain of the prescribed curriculum – e.g. kinematics, electricity, or forces – but on topics of their interest. However, although I had the impression while teaching that there were really good things happening in my classroom, the transcriptions seemed to be lacking what the science education literature on exemplary science teaching was about, and the kinds of things conceptual change researchers were reporting. Thus, for example, I had recorded the following notes and transcriptions during an investigation where one group of four students looked into the relationship between potential energy (work) of an object and the velocity of this object at the bottom of the slope after descending an incline.[6]

October 29–30

Matt uses a protractor to measure the angle of incline. Martin assists with finding the appropriate points on the track and the protractor to find the angle. They use books to prop up the inclined air track. Tzen, who sits on the Apple [computer], familiarizes himself with the [data recording] program. Matt and Martin experience problems where exactly to measure the angle, the teacher comes by on his round trip to check the angle measurement with the protractor. [He] then suggests another way of finding the angle from a measurement of the length of the air track and the height of the incline by using an inverse sine function. Matt begins to calculate, Cam, the recorder, is emphasizing Tzen's function as a technician. The following is part of the dialogue that went with the last paragraph:

Teacher: There is another way to find out [the angle]. This one [side of air track] touches, and do you have a ruler, you measure how high it is at this point (pointing to the other end of the air track). You know that the length is 2 meters; then you can calculate the inverse sine.
Matt: We're trying to find the height (inaudible)
Do you have a meter?
Cam: Stan [Martin's last name], the technician should be doing that [to get the meter stick].
Matt: (*Has the calculator, calculates something.*)
Teacher: Do you have?
Cam: The technician should be doing that.
Tzen: You want me to measure it, we're only doing the height.
Martin gets the ruler.
Martin & Tzen measure the length of the base of the triangle protractor-table.
Matt comes in and asks to measure the length of the track.
Martin: Of this one, its right there, 2 meters (*impatiently*)
Tzen: We have to find the height.

---

[6] These data were collected as part of a study that I kept in a folder entitled "Laboratory Life," following Latour and Woolgar's (1979) description of scientists, and led me to publish one of the first qualitative studies of learning in open inquiry school science classes (e.g. Roth 1994).

Martin:	Ah, the height.
Matt:	The height and the length.
Martin:	The length at the table [one side of the triangle]?
Tzen:	The length of this [track].
Matt:	(inaudible)
Tzen:	He [teacher] said to measure the height and then you find the inverse sine (*proud gesture to have remembered the teacher's comments?*)

Matt calculates and measures height; Martin comes in to help him.
Martin measures height on his own
Cam and Tzen playing

As in the case of the concept-mapping recordings, the talk was very elliptic. For example, in the second part of the session protocol, many statements consist of nouns only, one following another such as in the sequence "the height," "the height and length," "the length at the table," and "the length of this." Although I was initially interested in doing the kinds of analyses that I found in *Talking Science*, it is evident that I could not use the scheme I had found in the book given the nature of my transcription. I had to pursue different routes to the analysis of laboratory talk. It was only over the course of time that I came to see and hear language as an integral part of the world, to be integrated with what we are doing such that what goes without saying does not have to be articulated. For one, in the concept mapping sessions, my students were focused on paper snippets with physics words printed on them, which were lying in front of them on the laboratory bench in an increasingly ordered way. My students also used gestures both to point and to indicate the nature of linkages between concept words to be made. Moreover, I learned that talk not only is about but also over, against, and integrated into the work: it gets things done rather than being *about* things. I learned that much of language use is not *about* something but is *for* work, to get on with business, to move business ahead. It is therefore not surprising that I was led to the description that knowing a language becomes indistinguishable from knowing one's way around the world (e.g. Davidson 1986). I came to take language in its function to be about the world only when scientists, or students in science courses, are writing up a report. This was precisely what I observed again during the years of ethnographic study of a discovery science reported here.

In the literary and hermeneutic sciences, there is a clear notion of the difference between orality and literacy, which derives from the fact that speaking occurs in the world of everyday being whereas texts constitute a world of their own (e.g. Ong 1982; Ricœur 1991). In everyday situations, the distinction between knowing a language and knowing one's way around the world more generally completely disappears. We do not have to interpret our neighbor's saying, "What a nice day today!" when the sun is shining and we are in the process of cleaning up the yard. We do not have to "construct" "meaning" or "understanding" at all, if something like these existed apart from the way in which we use words specifically and language generally. This integration of language and practical actions in the world is what Wittgenstein (1953/1997) refers to as *Sprachspiel* [language-game]. In this pragmatic context, as much as in the context of the cultural-historical explanation of higher psychological functions and personality to have their origin in societal relations (Leont'ev 1983; Vygotskij 2005), we are immediately led to the conclusion that students change participation in science talk by talking science while doing science. Rather than making students memorize words or engage in efforts to "construct [ephemeral] meanings," STEM educators ought to let them investigate *personally* interesting projects in the course of which they communicate with each other. When students subsequently report their studies and findings to peers, they again engage in communicative efforts in the course of which their communicative forms change. If teachers were to encourage presenting students to make the best and most convincing case while reporting, they might find – as my fellow teachers and I found out – that students develop high degrees of communicative and representational competencies (e.g. Roth and Bowen 1995). Our main concern had not been getting students to speak correctly, in the way the writing-in-science and conceptual change movements appear to recommend, but in getting them to talk science at all. Everything else, for example, correctness in expression and mastery of scientific representation actually emerge in the process – much like immigrants may learn a language and become very proficient without ever taking a formal course and merely by participating in using it. We knew this to be the case because, as noted elsewhere in this book, the eight-grade students of our school outperformed college graduates when it came to data analysis tasks merely on the basis of participating in the production of scientific claims that they were asked to present and defend in small-group and whole-class configurations. In the same way, I would anticipate students to develop coping practices when they are continuously confronted with uncertainty in situations where they have to make decisions and choices without being penalized because the had not done whatever the teacher might have wanted them to do.

Temporality of Scientific Praxis

Recent analyses of moment-to-moment engagement in science, grounded in dialectical theories of activity, highlight the emergent characteristics of skill, knowledge, and world (Gooding 1990). That is, at the outset of their research scientists frequently do not even know the phenomena they will ultimately discover. They do not have the skill to produce them, let alone describe and theorize them. In the face of the inherent uncertainty scientists face, this also shapes the temporality of the scientific research process. In the course of tinkering on, material practices, conceptual practice, and material world come to be intertwined in a "mangle of practice" (Pickering 1995). The three dimensions of activity form a dialectical unit, which is produced (changed) and reproduced with each practical action. Whereas existing studies already describe the phenomenon in more fruitful ways, they have relied on the analysis of historical records, for example, Faraday's notebooks. What they do not give us is an explanation of scientific discovery work as it unfolds in real time, which is in part made difficult because few analyzable records exist of scientists working in real time. When they do exist, interesting accounts of the way in which the analyzability of a phenomenon emerges from the day's work (Garfinkel et al. 1981). In Part II of this book, I provide descriptions of laboratory work unfolding in real time at a moment that scientists are (metaphorically) in the dark.

The temporality in the "discovery sciences" is very different from that observed in school science. Yet there have been calls for STEM educators to allow students to learn in project-centered investigations that integrate the different curriculum subjects (e.g. Wagenschein 1999). Learning in our laboratory did not occur in terms of clock cycles, interrupted every hour to move us – the subjects of activity – to a different task. Rather, scientific research work depends on the enchainment of actions that had as goal the transformation of the object of activity. In fact, one might say that the practical actions made time and determined the tempo and rhythm of praxis (Bourdieu 1980). For example, in school science, Craig would have been scolded for his late arrival and the team as a whole for its slow progress and for not being able to get back and continue where it had left off on the previous day.[7] The rate of our team's actions was also set by necessary adjustments to the special requirements regarding light and light intensity, which had to be in the far-red part of the spectrum and at very low intensities, requiring a rather long dark-adaptation period during which the team members could do little else but wait in the dark and perhaps talk. One might think that this temporality is an anomaly. But reading ethnographic studies will convince readers that this temporality is the norm in scientific work – which I have had the opportunity to observe and to live.[8] Deep familiarity, as the present study shows, takes long periods and deep engagement and participation with the subject matter to develop. There are long periods during which scientists engage in what they call "grunt work," deal with experiments that do not work, or engage in repetitive data collection (Roth and Bowen 2001b). Their learning, as expressed in the slow rate of producing new facts, is rather sluggish compared to the learning rates that STEM educators (prescribed curricula) demand of their students.

The work I present in Part II of this book– especially in the descriptions in Chaps. 2 and 3, but also in the chapter on the (logical) contradictions in and during scientific research work – shows that there is a lot of time involved when scientists operate in unfamiliar terrain, which in fact they know as a terrain only after their extended inquiries have come to a (temporary) end. Through their actions, scientists not only explore but also constitute a portion of the world, and in the process learn

---

[7] There are no reasons other than administrative ones that make the beginning of the school day the same for all students in a class or school. But this does not have to be. For example, I repeatedly described and theorized the practices of several French schools, at the elementary and secondary levels, where students themselves decide when it is time to come to school (e.g. Roth and Lee 2006).

[8] I would not be able to write books or research articles if I were jerked from my desk every 60 min to do something else.

to articulate, describe, and perhaps theorize segments of it.[9] Students, on the other hand, tend to be required to conduct particular experiments without being given the opportunities of finding out what the relevant segment of the world is like. Students are supposed to recognize when inherently ambiguous instructions describe what they have done so that one could say that they have followed these instructions and seen what they were supposed to see (e.g. Amerine and Bilmes 1990). In the light of the present study, it then appears as if schools made unrealistic assumptions about the temporality involved in learning activities, particularly learning by doing specific tasks in STEM related fields. In schools, the temporal unfolding of the day is highly and strictly regimented, changing subject matter in the regular intervals of airplanes or trains departing from the runway or station, and even within lessons having particular modes of temporality. In schools, students are expected to move through their STEM-related tasks quickly and accurately and to see precisely those structures that the teachers expect them to see, and then make inferences about the laws that are said to underlie the phenomena.

Some readers may want to suggest that schools cannot be organized differently. But, again, there are many schools where the days are organized differently, where students engage with topics for long periods of time and from the perspective of whatever discipline is most pertinent to the current topic. The German science and mathematics educator M. Wagenschein (1968) suggests long-term uninterrupted engagement that allows students to pursue a topic in depth and from many perspectives. He provides many examples of topics from the sciences and mathematics that put students into situations where they explore, beginning with wherever they currently are in terms of their practices*-in-the-making, a phenomenon in substantial breadth and depth. Other educators have asked Wagenschein asked about the amount of time required: “Do we have so much time?” (p. 47). The educator begins his response by saying:

> Everyone has the right to a foundational education in the natural sciences. This means: [the right to] to some supporting, exemplary experiences of rooting, *genetic* understanding. But not *every* scientific acknowledgment has to be educative (and cannot be, as there is not enough time). (p. 50)

Wagenschein continues by emphasizing that in communication with others and by means of experimenting, some basic pillars of education are built. These serve as the foundation to the remainder of the students’ education.[10] Whatever is made available as additional resource that students can employ ought to be such that it allows linkage with what they have learned in communicating with others over and about the fundamental and founding investigations that are completely framed in

---

[9] This is not unlike a world emerges for the person who finds herself in an unknown place and, by groping about, comes to be familiar with, and give shape to, the new lifeworld that she comes to inhabit.

[10] Wagenschein uses the term *Bildung* for which there is no exact equivalent in English, which often remains untranslated in the English scholarly literature (Elmose and Roth 2005). “Education” heard in opposition to “schooling” perhaps best translates the German word.

terms of their current knowledgeability. If we were to go faster, students would not follow what is unintelligible to them but rather would follow what they can really "see" (as in "I see!"). The observations in the science laboratory offered in Part II confirm that the scientists could only move from where they currently were, even if they subsequently realized that the canon that had framed their early work turned out to be inappropriate for describing and explaining the data that they themselves were collecting. There is, in other words, no supersonic or tachionic[11] route to a general education.

Learning Is Groping in the Dark

Educators have shown interest in self-regulated learning. In this approach, metacognition is of interest because it bears on learners' active selection of strategies for solving problems. Self-regulated students are said to be good learners because they solve the tasks, conceptualize their opinions, and adapt their strategies to task demands. In contrast, previous research on other aspects of the research process in the discovery sciences shows that even highly trained and successful research scientists grope about in the dark rather than make decisions about best strategies (Chaps. 2 and 3). In actual practice, even researchers of metacognition may not exhibit high levels of metacognitive behavior.[12] The process of the research scientists in Part II of this book more resembled a process of groping about rather than engaging in the kind of metacognitive behavior that are expected from mathematics students. Rather than attempting to come to grips with the totality of their data, this research team dropped those data from consideration that resisted integration into their model. The present results are more consistent with other research on mathematical behavior outside schools, where abandoning rather than persisting with problems and difficult issues is one of the legitimate choices people make (Lave 1988). Educators also noted that partially correct constructs are useful to explain the origins of students' inconsistent answers on mathematical tasks. However, the present studies suggest that discovery scientists who are the quintessential learners, precisely because they do not yet know the end result of the learning process, cannot use partial correctness or inconsistency to select learning strategies.

The ethnographic study in Part II shows how the scientists operate in the absence of knowing where what they do leads to and without any referent as to whether what they are doing is correct. Precisely because they do not yet know what they will

---

[11] Tachions are hypothetical particles traveling faster than the speed of light, which is the limiting speed for the movement and transmission of information.

[12] I once had the opportunity to review a book on metacognition – with examples from science and mathematics education – which had as much as 40 errors per page (Roth 2004). I concluded that these researchers of metacognition displayed very little evidence of metacognition.

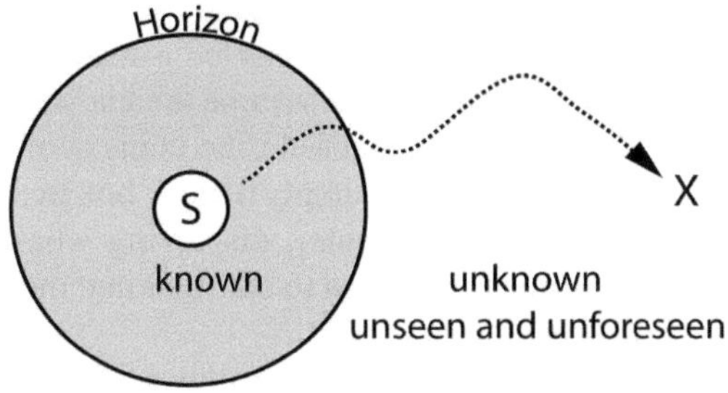

Fig. 11.2 An analogy of a subject on the path to learning something over its current horizon and, therefore, unknown, unseen and, therefore, unforeseen

have come to know once everything is said and done, they cannot *intentionally orient* to whatever it is that they will learn without knowing what it is until they have learned it (Roth and Radford 2011). We may use the following analogy and its visual depiction (Fig. 11.2). A learning *subject S* has a current horizon, which delimits its perception and intelligibility. What it will eventually know, *X*, is beyond the subject's horizon, therefore not yet known, therefore unseen and unforeseen, even unintelligible. In the ethnographic study, the circannual sinusoidal relationship between porphyropsin levels (percent A_2) and day length or temperature would have been something like *X*. Because the object that the subject will have come to know lies beyond the horizon of intelligibility, it lies in the region of the unknown. It is unseen and therefore unforeseen. Precisely because *X* lies beyond the horizon, *S* cannot actively orient towards it, aim at it, and make it the object of its intentions. More importantly, because *S* does not know where it will end up, it also cannot know whether it is "on the right track" or "doing the right thing." Precisely because the scientists did not know that they eventually would end up with a sinusoidal rather than a sigmoid curve, they could not orient and correct their research trajectory during its early part. Learners who do not yet know what they will eventually know are in the same boat as the scientists in the present study. Learners cannot orient toward the unknown or correct their path of getting there. They have to grope about in the hope of being on a fruitful path. This is the same reversal as saying that in and through their actions persons learn what instructions (recipes) were intended to make them do.

Once we conceptualize learning as creating clearings in a movement characterized as groping about – i.e. once we view changing participation from a learner and first-time-through perspective – our expectations of what a learner does or can do also changes. First, we come to recognize the necessarily contingent nature of knowledgeability*-in-the-making, for what comes into the clearing and what becomes clear is an emergent outcome that arises from the form of participation itself shaped as it has been in the continuity of experiences (Roth and Jornet 2014). Second, we come to see the limitations of certain theoretical constructs such as metacognition and self-directed learning in the face of the uncertainty linked to learning outcomes. For example, from a metacognition perspective, the members of our research team should have critically analyzed their actions. Between the 2 days (when the monitor was versus was not working), the only thing that they had changed was the monitor. From a normative perspective, the scientists perhaps should have begun to look how the exchange of monitors might have affected the display. But whereas this may be a reasonable thing to do in the ideal world of educational theorists who assume the knowledge already to be there as a normative fact, it was not reasonable in the context of everyday scientific activity. Accordingly, the monitor was simply a black box in the actions of the scientists, which was connected to the computer, displaying what the monitor card tells it to display. There was no good reason to consider that there might be an interaction between the software and the monitor.

The present study therefore allows STEM educators to learn that they often work with unrealistic metaphors for learning, such as the "information transfer" theories characteristic in university science courses – many science professors still talk about trying to *get* material and points *across* and about students who *don't get it* – or the "construction" of "knowledge." We need to reflect upon the fact that those in the process of learning, that is, in the process of creating for themselves an active language for getting around the world, also are in the process of creating a language that explains the learning process.

On Learning About the Nature of STEM

The cognitive apprenticeship movement was about allowing students to participate in the practices of specific disciplines (e.g. Brown et al. 1989; McGinn and Roth 1999) and, in so doing, also become familiar with the nature of the field of inquiry. For example, many science educators agree that one of the possible or desirable learning outcomes of school science courses ought to be familiarity with the nature of science – though there are widely varying conceptual discourses in the scholarly community as to what the nature of science is.[13] When I started teaching, as noted above, I wanted my science students to experience science in the way I had done while doing my MSc thesis research and while working in an advanced research laboratory on heat conduction in gum tissue and the mechanical brushing of teeth. I still remember long periods of wait until the equipment I had designed was completed in the machine shop, periods of data collection when I measured 36 h straight because starting up the equipment took such a long time and many times it did not work. It was a time of cogitating and (computer) modeling, and of search and frustration (when things did not work out). It also was a time of relating to others, including the postdoctoral fellow in the neighboring lab working under the same professor and my fellow graduate students with whom I already had attended high school physics. The interviews that my graduate students conducted with Shelby, the doctoral student on the research project that Craig and I had designed,

[13] It should be noted that many of those implicated in the oftentimes acrimonious science education debate over realist or constructivist conceptions of the nature of science never have had scientific training, done scientific research, or done ethnographic studies of science.

reveal similar learning to have occurred. In the following excerpt, part of which I also quote in Chap. 10, Shelby talks about ideas, sharing and testing them, critiquing ideas for the purpose of further developing them.

> I firmly believe that science is meant to be done; and that if you have an idea, and you're not going to do it yourself, to get it to somebody so they can do it. It doesn't matter who does the idea, it matters that it gets done especially if it's a worthwhile idea. Um and so I'm not, I don't have the same ownership that some people have about my ideas. Some of them might do, some of them might don't. But one of the things that I find myself doing when I have an idea, even if it's not really well developed, I go out and tell somebody because right away I want to know what they think. I want to get their [thoughts about] is there a gaping big hole in my idea that they can see just like that? (*Snaps fingers.*) Then tell me, I want to know right away. And so I often search for an argument I go out looking for an argument, I go out looking for a critique. Um and I take that usually to Ted or to Craig, or Theodore, some of the things to Jim Plant, other people in the lab and I go to them and say, "I got this crazy idea, 'dah dah dah.'" I get really excited about it you know and they look at that, you know, often maybe the way I present it doesn't give them an opportunity to really, criticize it the way they should? (*Laughter*) So I have to work on how I present my ideas but generally what happens is, you know, they'll say, "Well, I think I read somewhere about something like that, try this paper," and they send me off and I find this paper and I read about it, get more information and more details. And I think that's part of the process you know. It's something you have to do, you have to share your ideas and get feedback right away and see whether or not they have any worth?

Doing science is not just about having (good, correct) ideas: it is about generating ideas for the purpose of testing these and exposing these to others for the purpose of finding disconfirming evidence. These ideas may sound and, in the end, may turn out to have been crazy. But some of these ideas will have been worth having had, such as Shelby describes:

> But I think for me most of all is when you have that crazy idea and you put it together and you create a hypothesis that you can test it. You test it and find out whether it's true or not. And when it is, "Wow," it feels really cool. And even, even if it's something silly like, I mean, for instance, we had always kept our [photoreceptor] samples you know, for just a couple of hours on ice. And that was we figured that they just didn't last and I was kinda thinking about it. You know, we got them on this essential medium and they got everything they need, and if we cool them down, they should lower their metabolism so they shouldn't break down, I thought why can't we keep them longer? And so I tried. I had this idea, so I threw them in the freezer and put them down, like, its probably like for degrees C[elsius] or zero degrees C and came back two or three days later and tried it and it worked. And it was so exciting because it changed what we could do because before we had to have fresh, fish that were fresh and we had to do everything right here, and do it right away. But now I can go out and catch fish offshore, or in the hatchery if I need to, make them dark, take out their eyes, put it on the MEM [minimum essential solution] and bring it back frozen. And work with it on Monday if I got them on Friday, if I want to. Now I know there is some degradation in there but the basics is there I can still get what I need from it if I have to. So those kinds of little, little experiments that you do and they work is certainly exciting. And then beyond that you know the idea that I had say of the saltwater fish having a shift in pigment, now that's being a big one for me, because I had that idea for a long time. And I'm finally getting to test it. And so that first time when I went over to Target Marine [commercial supplier of young fish] and got fish and brought them back and I made sure they're alive, and I tested it, that afternoon I sat there and analyzed my data right away, because I wanted to know! And I was right, and it was so exciting to be right you know, to

> have had an idea. Not because you're right, but because you had an idea and it worked. Your prediction, you're able to predict something about nature, is just really neat. It means that you're starting to understand how the system works. And that's really rewarding, that's what it's all about I think. That's why I do it anyways.

This account Shelby provides about having had an idea about preserving retinal tissue without having to keep alive and transport the young fishes themselves turned out to have had benefits to his work. Other accounts about other ideas are similar. That is, this form of experience is not specific to the context in which it is told here. I therefore do not see any reason why children and students could not have the same kinds of experiences while being in their science classes. In this case, Shelby learned about how science is done by participating in doing it with others, in the laboratory, by talking to others during conferences, and by traveling with other team members to the hatcheries. He learned how science is done and about science in ways that he has not while attending high school, thereby confirming what others have written about the difference between all education preceding graduate programs and those that persons undergo during their work on graduate degrees in a science (Traweek 1988). But does science education have to operate in this way? Is it not possible to have students engage in inquiry where they pursue truly relevant questions that have grown out of their own interests and on the basis of what they currently know? My own work in high school science classrooms has shown that it is indeed possible to engage students in personally relevant inquiry without sacrificing deep familiarity, on the one hand, and a tremendous capacity for coping with uncertainty and the unknown, on the other hand (e.g. Roth 1995).

Bricoleur, Bricolage

> *Non scholae sed vitae discimus.*[14] (Roman saying)
>
> [The purpose of the science education curriculum is] to prepare students for further education and for their adult lives. (Ministry of Education British Columbia 2006, p. 11)

In Part II, my descriptions of how scientists work is more consistent with what we know to be bricolage than with the idealist talk about science as (hyper) rational inquiry. But do students learn to cope in the uncertain environments where scientists appear at their best? Do our science students learn to make do, getting themselves out of trouble in part by reframing situations so that these no longer are troublesome, for example, by discarding some of the data before even considering them? Probably not, as I argued in a recent article about lifelong education (Roth and van Eijck 2010). Yet just about every curriculum statement that I have read, whether in Canada or the US, contains purpose statements that include preparation for life. As the first opening quotation shows, even the Romans had a saying emphasizing that learning was to be for life rather than for school. However,

[14] We study not for school but for life.

a historical analysis of the emergence of schooling as we know it today shows that what really happens in schools is a hierarchical ordering within the student body – achieved through the disciplining of students' bodies – for the purpose of controlling access to coveted positions in the next stage of their lives (Foucault 1975). As a more recent analysis specific to science education argues and shows, science education, as practiced today, precisely has this effect of ordering the student body rather than constituting a real preparation for life (Roth and McGinn 1998). Science education, when it focuses on the preparation for real life, once it takes a whole-life perspective, pursues a different goal than trying to inculcate skills, knowledge, and attitudes according to Bloom's Taxonomy – in the way this is stated, for example, in the curriculum guidelines of British Columbia, which has been one of the top-scoring provinces in Canada on the 2006 PISA test with a mean score similar to Hong Kong-China, the second jurisdiction ranking after Finland. When science education focuses on real life, from a fullness-of-life perspective, then there is a primacy on forms of participation required for making do in complicated situations, defining problems and working out solutions, and getting oneself out of difficult situations with the means at hand (Roth and van Eijck 2010). As indicated in the introduction to Part IV, the French have concept words that create a field of phenomena that I have in mind when thinking about useful education: *se débrouiller* and *débrouillardise* (literally "to get oneself out of the fog" and "capacity to get oneself out of the fog") and *se démerder* and *démerdeur/euse* (literally "to get oneself out of shit" and "a person who gets himself/herself out of shit"). They also have concept words that have made it into English, such as *bricoleur* and *bricolage* from the verb *bricoler* (etymologically, from doing something in a back-and-forth manner, in little disconnected steps).

Throughout my ethnographic descriptions, we see the scientists at work, making do, looking around to find out why an instrumentation is not working, tentatively talking about variability that they cannot explain, grappling with trends in contradictory ways only to go on when things seem to work and without somehow assuring that all (logical) contradictions have been resolved. As long as things seem to work and work out, laboratory life goes on. Sometimes someone has a "crazy idea" and then they will try making it work out with the means at hand. At other times scientists go out and spend a lot of money in the attempt to solve a problem – like purchasing an especially expensive, very stable lamp to get rid of some ringing signal imposing itself on their data – only to find out that the expense has not resolved the issue. In the preceding section, Shelby talks about coming up with a different way of transporting specimen from the source to the laboratory. Rather than having the fish sent – which, in the case of the coho salmon from the Kispiox hatchery turned out to be very expensive – he sacrificed the fish right after catching them, extracted the retina, placed it in the minimum essential solution (MEM), and then deep froze it for the transport. This, too, was the result of making do in the context of having done a lot of the transports alone or with me traveling the

250-km distance from Robertson Creek to the university after my ethnographic stays.[15]

But we do not have to observe scientists to see people work in the same way on defining problems and finding solutions, or identifying solutions to problems that are not even known. Life is full of situations where what is thought to happen or is thought to be the case does not happen and is not the case. Real life is full of little accidents and mishaps; and it is full of situations that a little ingenuity could solve but where many individuals need to get a specialist or spend a considerable amount of money to solve the issue at hand. Take the following situation where a pocket door had become unusable because three screws had come lose that used to be holding the track, located inside the 4-cm-wide pocket. This pocket is too narrow to get the hand/arm into it with a screwdriver to tighten the screws. I consulted with my neighbors, who turned out to suggest that the wall itself needed to be opened to have access to the screws. This would have required cutting a hole into the drywall, tightening the screws, closing the hole, repairing the drywall, and then painting it again with the possible consequence to have to paint the entire hallway again to match the paint color. Proceeding in this way might require a drywaller and a painter. Handy persons can do it by themselves – with a considerable mess deriving from the removal and reinstalling of drywall and associated plasterwork.

There are other solutions to the problem of the loose screws – but these may not be self-evident and also require assembling tools and materials to do the job. The solution, which situationally emerged from the things (tools, materials) at hand, is illustrated in Fig. 11.3: A piece of ¾-in. plywood fashioned in such a way that it could hold a battery-driven screwdriver held to a backing by a bicycle inner tube and triggered by means of a string. The resulting tool is the product of a bricolage, a bringing together of discarded materials – a piece of painted plywood from a broken and dismantled deck, an already-cut-up inner tube unusable because it was irreparable, a piece of string used in garden works, and a borrowed screwdriver. The solution is not the result of a logical approach that in some rational way defines a problem space, decomposes it into its elements, and constructs a solution. I actually do not even remember how it emerged – but it did *come (appear) to me* rather than being intentionally constructed. I could not have constructed the solution because the problem space was not really defined other than in terms of the screws that no longer held the track and made the door unusable. In part, of course, the solution is a function of the particular type of screw driver, which I did not own, but which I had borrowed because I had become aware of the fact that it was impossible to turn a regular screw driver in the confined space I was working with. My arm being to bulky to fit into the pocket appeared to call for an artificial arm, with consequences for its "grasping the screwdriver" and pulling its trigger. In the end, the solution came together from a set of fortunate circumstances and the willingness to try fixing the pocket door without having to go through the messy removal of the wall and,

[15] For very large batches of fishes – such as we required for laboratory experiments – we rented a truck with large water tanks to transport about 1,000 coho from the hatchery to the university.

Fig. 11.3 The result of a bricolage: a battery-driven screwdriver is held against a notched ¾-in. piece of plywood by an old bicycle inner tube and triggered by a string. The tool is narrow enough to fit into the opening of a pocket door (© Wolff-Michael Roth, used with permission)

perhaps, the costly hiring of a craftsperson. These circumstances and the possible costs involved, more than anything else, oriented me towards seeking/finding other approaches than the ones my neighbors had suggested and the ones I initially had considered.[16]

Everyday there are little problems to be solved, and sometimes these problems have to be stated in ways that they are recognized as problems that might have solutions. In the summer, wasps are a problem for those who like to spend time in the garden or on a deck. These are even greater problems for those who keep honeybees in their backyard, because wasps attack the bees in the attempt to rob them of their honey. Now, it is possible to go out and buy a yellow jacket trap, which in my area sold for $13 during the summer when I was working on this chapter. The trap works for about a week and then a refill has to be purchased. The bricoleur can do better, and much less costly, using whatever is at hand to make a suitable and reusable trap. For example, an old 4-l vinegar container will yield a good result (Fig. 11.4a), but a discarded soft-drink bottle holding about 2 l of liquid, or any other container that can be cut also will do the job. Whatever can be used to create a darker environment on the inside of the container will do to prepare the lower part of the trap, together with anything working like an X-Acto or drywall knife to cut an upward directed, wide U-shape entry into the container (Fig. 11.4a). Put some sugar water, preferably with some fruit scraps into the container and lo and behold, there is a wasp trap: within 36 h, I had trapped over 50 wasps and none of my bees had ventured into it. Another way of dealing with wasps around the Sunday brunch table and around the honey bee hives consists of a swat made from anything suitable and available around the home, such as the weather and swat-resistant one I made from plasticized paper, a few little nails (because the ½-in. staples come out to easily, as experiences has shown), and a piece of scrap wood (Fig. 11.4b). Coping with the challenges of life ought to be an educational

[16] Ten years later, during a renovation project, the track was replaced after opening up the wall. The screws still were very tight, a situation that certainly would have outlasted the owners of the house given the length of the screws and the tightness of their fit at the instant of removal.

Fig. 11.4 (**a**) An old vinegar container, with other materials from around the home, has been converted into an efficient wasp trap sitting here on top of a beehive. (**b**) Scrap materials have served building this cheap swat – dealing so efficiently with wasps that it has become known, with a critical political eye across the border, as the "weapon of mass destruction" in the kingdom of wasps (© Wolff-Michael Roth, used with permission)

goal not only for getting students ready but already for solving personally interesting challenges during their school experience. We can learn quite a bit from the scientists how to do this, at least when we follow them around at work in their laboratories and meeting rooms.

In some instances, a little science comes in handy. For example, knowing that wasps attempt to fly towards the light will help and guide the bricoleur in trying the best she can in building the device. Knowing about characteristic sizes – such as the $^{3}/_{8}$-in. "bee space" typical for honeybees – helps create entries to the trap that are just large enough for wasps to squeeze themselves through but too small to be found from the inside to get back out again. Knowing about resistance and the bending of materials helps in searching for and identifying materials around the home suitable for making up a swat. Willingness and creativity in defining problems and seeking solutions thereof, more so than anything I learned in school science, help me cope with the demands of everyday life. Getting myself out of the troubles of everyday life, looking for things that solve issues that I might not have yet appropriately framed as problems, and engaging in searches to get myself out of the dark have helped me more in life than knowing how to factor polynomials or solving a quadratic equation, by whichever means I had been asked to do in school mathematics. What science, technology, mathematics, or engineering might come in handy never is quite clear, and oftentimes we do not need any one of these. What matters is our willingness and disposition in looking up what we might need even and precisely when we do not know what we need – just like the scientists when confronted with the problem where their apparatus did not work from one day to the next (Chap. 2).

Getting students into the mode of being creative posers of problems and solution finders has been a goal of mine when I was a high school teacher of physics, where I used open inquiry as a learning context. Provided with the opportunity to design their own experiments, pursuing personally interesting phenomena, the students got themselves into situations where it was not clear what needed to be done to continue their agenda. Oftentimes, students had "crazy" ideas to deal with a situation at hand – which were just as little or just as much crazy as those that Shelby had while working on the coho research project. For example, in one instance, a group of eleventh-grade physics students decided to investigate the effect of air friction on an oscillating object and, for getting a significant effect, needed to have their objects move in a controlled manner. Their equipment needed to be stable and free from shaking. Someone apparently suggested stacking two extremely heavy laboratory tables one on top of the other; and they implemented the solution even before I had realized what they were doing. They mounted the equipment to the table legs – and, as a result, there was no wobble at all. For a few weeks, while the group of students was doing its experiments, the two laboratory tables remained in this position.[17]

Another group wanted to measure the falling of differently shaped objects in different liquids. They needed to mount part of their measuring equipment up high so that it could follow the object falling in water and oil held in a tall cylinder (Fig. 11.5). The video offprint shows part of the equipment, an air supply hose, going up to the air track just above the image frame; and the offprint shows the cylinder and the wire hanging done, held by one of the three students visible. Another student stands on a chair, enabling him to work on the equipment. Again, whatever worked was used to get the investigation going, even if it required making changes to the experimental design. Students sometimes went to the school's workshop, where the craftsperson normally working around the school helped them in manufacturing the required item from wood, metal, or whatever they considered as helping them along in their project.

Over the course of their 2 years in the physics course, students really became good at dealing with any sort of situation on the spur of the moment and with the things at hand. They became real bricoleurs. I was able to see this repeatedly at the high school engineering competitions in which some of my students participated. Given their busy lives, they often did not spend as much time on preparing for a competition in the way students from other schools would do. However, once on the site where the competition took place, my students exhibited great competencies in dealing with any one of the contingencies that cropped up in the course, or between qualifying events and finals. In fact, they always brought a toolbox that contained,

[17] I do admit that the chemistry teacher in the school found my classroom "messy" and did not find it inspiring although the students really liked it and although the provincial inspector of private schools recognized exemplary teaching and learning going on and recommended visiting my classroom to science teachers of other private schools in my geographical area.

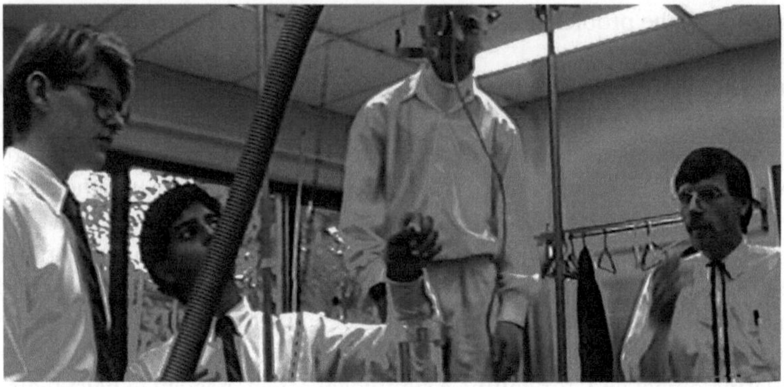

Fig. 11.5 In the midst of a bricolage, the students are in the process of getting themselves of a problem that would have prevented them from doing the measurements as intended (© Wolff-Michael Roth, used with permission)

in addition to the some tools, all sorts of material that could be used, abused, and deployed in a great variety of ways (e.g., duct tape is a good example of a material that has as many uses as the phantasy allows). Thus, even though my teaching was highly successful in common terms – as shown by the results of students, many of whom successfully completed advanced placement (AP) courses in physics – my proudest moments as a teacher were those where the students were exhibiting their resourcefulness and competencies in dealing with problems on the spur of the moment. It may be this experience, and my own inclination to solve problems as I go, that have led me to the present recommendations for science educations based on the observations of scientists at work.

Two types of narratives come to my mind when thinking about what kind of lessons to take from science*-in-the-making. One are the complaints of employers about university graduates who are ill prepared to cope with the demands of a real job, where problem definition and a search for a problem space is as much or more required than finding solutions. Students in STEM education, more often than not, are asked to solve well-behaved problems in highly constrained contexts, which does not allow them to develop the dispositions for coping with the uncertainties characteristic of everyday life. The second story was told round about the time I received my Masters of Science degree. It concerns the German industry often preferring to hire physicists rather than engineers as "general problem solvers," for the physics graduates were said to be more resilient in ill-defined settings and less prone to "think within the box." This story is consistent with the first employments of many of my fellow graduates, who found jobs in what might be though to be unlikely industry. For example, one was hired by a street car manufacturer to work on reducing the noise the wheels made in curves; another found a job in a toilet paper manufacturing companies for dealing with technological and engineering issues that were arising on a daily basis; and yet another was a member of the team responsible for decommissioning the first German nuclear submarine. One of the characteristics of the physics programs at the time were open-book examinations, an indication that something other than recall was to be assessed; such examinations are in fact closer to what we do in out-of-school situations, where we have access to means that tend to be barred in school testing situations.

References

Alexander, G., Sweeting, R., & McKeown, B. (1994). The shift in visual pigment dominance in the retinae of juvenile coho salmon (*Oncorhynchus kisutch*): An indicator of smolt status. *Journal of Experimental Biology, 195*, 185–197.

Amerine, R., & Bilmes, J. (1990). Following instructions. In M. Lynch & S. Woolgar (Eds.), *Representation in scientific practice* (pp. 323–335). Cambridge, MA: MIT Press.

Bourdieu, P. (1980). *Le sens pratique* [The logic of practice]. Paris: Les Éditions de Minuit.

Brown, J. S., Collins, A., & Duguid, P. (1989). Situated cognition and the culture of learning. *Educational Researcher, 18*(1), 32–42.

Chin, C., & Osborne, J. (2010). Supporting argumentation through students' questions: Case studies in science classrooms. *Journal of the Learning Sciences, 19*, 230–284.

Davidson, D. (1986). A nice derangement of epitaphs. In E. Lepore (Ed.), *Truth and interpretation* (pp. 433–446). Oxford: Blackwell.

Elmose, S., & Roth, W.-M. (2005). Allgemeinbildung–Readiness for living in risk society. *Journal of Curriculum Studies, 37*, 11–34.

Foucault, M. (1975). *Surveiller et punir: Naissance de la prison* [Discipline and punish: Birth of the prison]. Paris: Gallimard.

Garfinkel, H., Lynch, M., & Livingston, E. (1981). The work of a discovering science construed with materials from the optically discovered pulsar. *Philosophy of the Social Sciences, 11*, 131–158.

Gooding, D. (1990). *Experiment and the making of meaning: Human agency in scientific observation and experiment*. Dordrecht: Kluwer Academic Publishers.

Gooding, D. (1992). Putting agency back into experiment. In A. Pickering (Ed.), *Science as practice and culture* (pp. 65–112). Chicago: University of Chicago Press.

Heidegger, M. (1977). *Gesamtausgabe. I. Abteilung: Veröffentlichte Schriften 1914–1970 Band 5: Holzwege* [Complete works. First section: Published works 1914–1970 vol. 5: Off the beaten track]. Frankfurt/M: Vittorio Klostermann.

Knorr-Cetina, K. D. (1981). *The manufacture of knowledge: An essay on the constructivist and contextual nature of science*. Oxford: Pergamon Press.

Latour, B., & Woolgar, S. (1979). *Laboratory life: The social construction of scientific facts*. Beverly Hills: Sage.

Lave, J. (1988). *Cognition in practice: Mind, mathematics and culture in everyday life*. Cambridge: Cambridge University Press.

Lemke, J. L. (1990). *Talking science: Language, learning and values*. Norwood: Ablex.

Leont'ev, A. A. (1971). *Sprache, Sprechen, Sprechtätigkeit* [Speech, speaking, speech activity]. Stuttgart: Kohlhammer.

Leont'ev, A. N. (1983). Dejatel'nost'. Soznanie. Ličnost' [Activity, consciousness, personality]. In *Izbrannye psixhologičeskie proizvedenija* (Vol. 2, pp. 94–231). Moscow: Pedagogika.

McGinn, M. K., & Roth, W.-M. (1999). Towards a new science education: Implications of recent research in science and technology studies. *Educational Researcher, 28*(3), 14–24.

Merleau-Ponty, M. (1964). *Le visible et l'invisible* [The visible and the invisible]. Paris: Gallimard.

Metz, K. E. (2004). Children's understanding of scientific inquiry: Their conceptualization of uncertainty in investigations of their own design. *Cognition and Instruction, 22*, 219–290.

Ministry of Education British Columbia. (2006). *Science grade 9: Integrated resource package 2006*. Victoria: Author. ISBN: 978-0-7726-5687-2. Accessed August 15, 2013 at: www.bced.gov.bc.ca/irp/pdfs/sciences/2006sci_9.pdf

Moschkovich, J. (2008). "I went by twos, he went by one": Multiple interpretations of inscriptions as resources for mathematical discussions. *Journal of the Learning Sciences, 17*, 551–587.

Ong, W. J. (1982). *Orality and literacy: The technologizing of the word*. New York: Routledge.

Pickering, A. (1995). *The mangle of practice*. Chicago: University of Chicago Press.

Ricœur, P. (1991). *From text to action: Essays in hermeneutics, II*. Evanston: Northwestern University Press.

Ron, G., Dreyfus, T., & Hershkowitz, R. (2010). Partially correct constructs illuminate students' inconsistent answers. *Educational Studies in Mathematics, 75*, 65–87.

Rorty, R. (1989). *Contingency, irony, and solidarity*. Cambridge: Cambridge University Press.

Roth, W.-M. (1994). Experimenting in a constructivist high school physics laboratory. *Journal of Research in Science Teaching, 31*, 197–223.

Roth, W.-M. (1995). *Authentic school science: Knowing and learning in open-inquiry science laboratories*. Dordrecht: Kluwer Academic Publishers.

Roth, W.-M. (2004). Theory and praxis of metacognition. *Pragmatics & Cognition, 12*, 157–172.

Roth, W.-M. (2009a). *Dialogism: A Bakhtinian perspective on science and learning*. Rotterdam: Sense Publishers.

Roth, W.-M. (2009b). Limits to general expertise: A study of in- and out-of-field graph interpretation. In S. P. Weingarten & H. O. Penat (Eds.), *Cognitive psychology research developments* (pp. 1–38). Hauppauge: Nova Science.

Roth, W.-M. (2013). An integrated theory of thinking and speaking that draws on Vygotsky and Bakhtin/Vološinov. *Dialogical Pedagogy, 1*, 32–53.

Roth, W.-M. (2014). Personal health – personalized science: A new driver for science education? *International Journal of Science Education, 36*, 1434–1456.

Roth, W.-M., & Bowen, G. M. (1995). Knowing and interacting: A study of culture, practices, and resources in a grade 8 open-inquiry science classroom guided by a cognitive apprenticeship metaphor. *Cognition and Instruction, 13*, 73–128.

Roth, W.-M., & Bowen, G. M. (2001a). "Creative solutions" and "fibbing results": Enculturation in field ecology. *Social Studies of Science, 31*, 533–556.

Roth, W.-M., & Bowen, G. M. (2001b). Of disciplined minds and disciplined bodies. *Qualitative Sociology, 24*, 459–481.

Roth, W.-M., & Jornet, A. G. (2014). Toward a theory of *experience*. *Science Education, 98*, 106–126.

Roth, W.-M., & Lawless, D. (2002a). Scientific investigations, metaphorical gestures, and the emergence of abstract scientific concepts. *Learning and Instruction, 12*, 285–304.

Roth, W.-M., & Lawless, D. (2002b). Signs, deixis, and the emergence of scientific explanations. *Semiotica, 138*, 95–130.

Roth, W.-M., & Lawless, D. (2002c). How does the body get into the mind? *Human Studies, 25*, 333–358.

Roth, W.-M., & Lee, Y. J. (2006). Contradictions in theorizing and implementing "communities". *Educational Research Review, 1*, 27–40.

Roth, W.-M., & McGinn, M. K. (1998). >unDELETE science education:/lives/work/voices. *Journal of Research in Science Teaching, 35*, 399–421.

Roth, W.-M., & Radford, L. (2011). *A cultural-historical perspective on mathematics teaching and learning*. Rotterdam: Sense Publishers.

Roth, W.-M., & Roychoudhury, A. (1992). The social construction of scientific concepts or the concept map as conscription device and tool for social thinking in high school science. *Science Education, 76*, 531–557.

Roth, W.-M., & Roychoudhury, A. (1993). The concept map as a tool for the collaborative construction of knowledge: A microanalysis of high school physics students. *Journal of Research in Science Teaching, 30*, 503–534.

Roth, W.-M., & van Eijck, M. (2010). Fullness of life as minimal unit: STEM learning across the life span. *Science Education, 94*, 1027–1048.

Roth, W.-M., McRobbie, C., Lucas, K. B., & Boutonné, S. (1997a). The local production of order in traditional science laboratories: A phenomenological analysis. *Learning and Instruction, 7*, 107–136.

Roth, W.-M., McRobbie, C., Lucas, K. B., & Boutonné, S. (1997b). Why do students fail to learn from demonstrations? A social practice perspective on learning in physics. *Journal of Research in Science Teaching, 34*, 509–533.

Roth, W.-M., Tobin, K., & Shaw, K. (1997c). Cascades of inscriptions and the re-presentation of nature: How numbers, tables, graphs, and money come to re-present a rolling ball. *International Journal of Science Education, 19*, 1075–1091.

Roth, W.-M., Pozzer-Ardenghi, L., & Han, J. (2005). *Critical graphicacy: Understanding visual representation practices in school science*. Dordrecht: Springer-Kluwer.

Sørensen, K. H., & Levold, N. (1992). Tacit networks, heterogeneous engineers, and embodied technology. *Science, Technology & Human Values, 17*, 13–35.

Suchman, L. A. (1987). *Plans and situated actions: The problem of human-machine communication*. Cambridge: Cambridge University Press.

Temple, S. E., Plate, E. M., Ramsden, S., Haimberger, T. J., Roth, W.-M., & Hawryshyn, C. W. (2006). Seasonal cycle in vitamin A1/A2-based visual pigment composition during the life history of coho salmon (*Oncorhynchus kisutch*). *Journal of Comparative Physiology A: Sensory, Neural, and Behavioral Physiology, 192*, 301–313.

Traweek, S. (1988). *Beamtimes and lifetimes: The world of high energy physicists*. Cambridge, MA: MIT Press.

Vygotskij, L. S. (2005). *Psychologija razvitija cheloveka* [Psychology of human development]. Moscow: Eksmo.

Wagenschein, M. (1968). *Verstehen lehren: Genetisch – Sokratisch – Exemplarisch* [Teaching understanding: Genetically, Socraticly, exemplarily]. Weinheim: Beltz.

Wagenschein, M. (1999). *Verstehen lehren: Genetisch – Sokratisch – Exemplarisch* (5. Auflage) [Teaching understanding: Genetically, Socraticly, exemplarily, 5th ed.]. Weinheim: Beltz.

Wittgenstein, L. (1997). *Philosophische Untersuchungen/Philosophical investigations* (2nd ed.). Oxford: Blackwell (First published in 1953).

Chapter 12
Data and Graphing in STEM Education

The discussion of what STEM educators should do concerning data, graphs, and graphing in STEM classrooms cannot be detached from the questions of why we teach STEM, what we want students ultimately to know, and the epistemology that we bring to this project. Thus, for example, if we were to believe that knowledge exists in declarative and procedural form stored in long-term memory, it would make sense for us to teach basic STEM skills and facts. If, on the other hand, we believe (and have evidence for) knowledge only to exist in situated practices and that the whole concept of knowledge transfer from school to everyday life is a sham, then putting students in situations where their changing participation in changing STEM practices that best serve their and their "clients'" needs is a much more viable option. I tend to go with the latter approach, as I know both from personal experience, observations of people at work, and from the literature that much of what we learn in schools is neither used nor leads to other forms of competent STEM practice.[1] Thus, my own predilection tends to favor making it possible for students to place themselves in situations that have some value to them, allow them to articulate problems, and let them find solutions to these in the way students themselves see fit. I even want to see students involved in defining the criteria by means of which their work ought to be assessed, in the way I describe this in the preceding chapter. Much of the curriculum is still full of statements in the form of Bloom's taxonomy that require students to "explain and apply the exponent

[1] There are three episodes from my own mathematics education that I still remember 50 years later – and that have been traumatic. Perhaps the most important one – because it contributed to my abandoning plans of becoming a teacher – was in fifth grade, where I was asked to calculate the amount of wallpaper needed for a room renovation given a door and a number of windows of specified dimensions. I could not do it. Even when I did use wallpaper later in life to make an apartment more attractive, I never used the calculations that we were asked to use in fifth grade. Second, I used even less the processes of factoring polynomials that I had to do a few years later. Finally, I learned much of linear algebra that was so hard in school later, when I was interested in writing different pieces of computer software for analyzing data I was generating in research projects of my own interest.

W.-M. Roth, *Uncertainty and Graphing in Discovery Work*,

DOI 10.1007/978-94-007-7009-6_12,

laws for powers of numbers including $x^{m} \cdot x^{n} = x^{m+n}$," statements that we can find even in a province such as British Columbia that scores very well in international tests such as the *Programme for International Student Assessment* (PISA).[2] Yet as a society, we continue to graduate students who do not unpack the simplest graphs, numerical relations, or presentations of statistics that they encounter in one of their favorite medias. Similar statements can be made concerning just about every topic that students encounter in their STEM courses. Given what I see as making successful scientists, what I would like to see instead are individuals participating in the deconstruction of claims in the media based on statistical or graphical evidence or in the deconstruction of claims that are made in local politics (see example below). For this, they do not need to know every bit of mathematics that someone has invented at some point in time in the history of humankind. With Wagenschein (1999) I take it that a few really worked-through examples of mathematical thinking and transformation will teach most of the principles that are at the core of this discipline. Doing so in the context of real-world problems, where mathematics becomes a useful tool, may be an ideal entry point for subsequent inquiry in the mathematical properties of data, graphs, or functions.

Unpacking Black Boxes

In Chap. 1, I introduce the idea of graphs as black boxes that are outcomes of work and stand to this work in a metonymic relation. A metonymy is a figure of speech where some part of a bigger whole is used to refer to or stand in for this whole. As suggested, a classical example is that of the restaurant server who refers to a customer as "the ham sandwich," that is, to the customer who can be identified because he is eating a ham sandwich (whereas others surrounding the person may not). If graphs are theorized in terms of their metonymic relation to the research process, including the natural phenomenon, the data collection process, the instrumentation, the transformations – along the Latourian chain of reference – and so on, then there is more to interpretation than taking the graph as a sign that stands in for, and makes present, something else currently and otherwise absent. Graph interpretation then means recovering that instrumental whole, including the natural phenomenon that it is used to make present again. In essence, "graph interpretation" then constitutes the unpacking and opening up of a black box to recover the translation work that was required to prepare the package in the first place. Scientists and mathematicians are not unfamiliar with some aspects of this process, for example, when they attempt to recover the signal of a phenomenon when they know that what they actually measured is a convolution of that raw signal with the

[2] This focus on Bloom's taxonomy might in fact be the very reason why the province's students do so well on these examinations thereby differing little from other countries where reproduction rather than novel production is the prime.

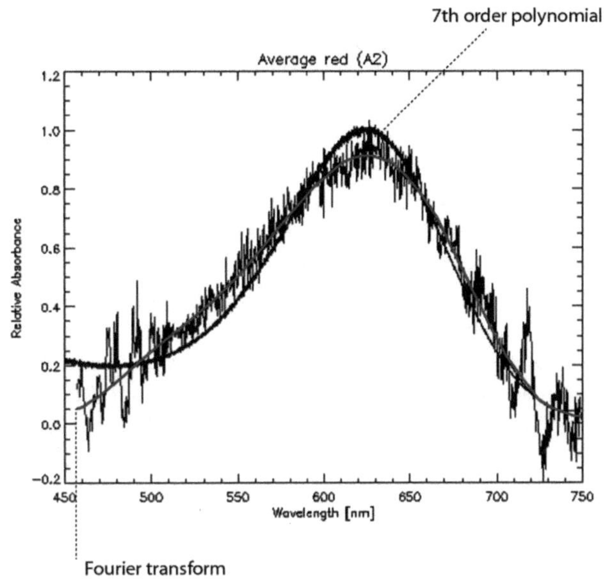

Fig. 12.1 An "average" curve for a photoreceptor absorbing in the red part of the spectrum, a 7th order polynomial used to fit the curve, and the curve obtained from the actual data after Fourier and inverse Fourier transformation with higher order frequencies lopped off (© Wolff-Michael Roth, used with permission)

mathematical function representing the instrumentation. In mathematical terms, the process is called a deconvolution. Thus, if the recorded signal is a function of the form h, thought to be the result of a convolution of the true signal f and the instrument function g – i.e. $h = f \circ g$ – then deconvolution involves finding f given g and h. In actual cases, some error variance ε will muddy the signal as well, so that the problem becomes recovering f from $(f * g) + \varepsilon = h$. In the case of graph interpretation, however, g, which encapsulates data collection, instrumentation, and transformation processes, is not available to the reader of a graph. Only what lies before them, the result (h) is available. To recover the phenomenon, assumptions about g and ε have to be made – at least implicitly. For example, the scientists featured in Part II recovered the absorption of light in the retina from a difference spectrum rather than from taking the light intensity prior to reaching the microscopic slide and after. In this way, they also controlled for all the light that their instrumentation absorbed, the slides, or the liquid surrounding the photoreceptor of interest. What they could not get rid of in this way was the error term ε. But they achieved this, in part, by using a widespread approach of Fourier transforms assuming that the error term is constituted by white noise, the equivalent of which in the Fourier transform are very high frequencies. By lopping off frequencies higher than the seventh term in the Fourier transform, they got rid of much of the noise while retaining the core of the signal (Fig. 12.1). That is, the red curve ("Fourier transform") represents the signal f, the true absorption curve, once the noise in the raw data (ε) has been mathematically removed.

We can use this example as an analogy for what individuals have to do when asked to interpret unfamiliar graphs. They have to recover some natural phenomenon that the graph is said to be representing. This requires not only some familiarity with the process that might have produced a graph but also some level of control over the *why* and *what-for* of the graph and its interpretation. In contrast to much of school mathematics, where students do what/as they are told to do, the scientists described in Part II are in control over what to do and which measurements to retain for the analyses that they ultimately report.[3] As I show in Chap. 3, scientists' inclusion and exclusion criteria are grounded in their familiarity with all those instances that they do not even want to qualify for entry into the data sources. Scientists literally constitute the frame that allows only some measurements to enter into the consideration of trends. This frame therefore reduces the original messiness and makes the phenomenon appear more clearly against the ground than it would if everything had been included.

Based on the studies presented in Part II, I strongly suggest allowing students to make decisions about which measurements to include or exclude from subsequent interpretations and claims. I also make this recommendation because an important dimension in and of changing participation (i.e. learning) appears to be the level of control that humans tend to have over framing the questions that are to be answered through the inquiry. In this form, changing participation is student-centered and satisfies the students' needs to seek and find answers to their own questions. Student question-based schooling environments "[afford] many possibilities for transforming classrooms into active learning environments where there is a dynamic interplay of questioning, explanation, argumentation, design of investigations, communication of ideas and findings, collaboration, and reflection" (Chin and Chia 2004, p. 725). In advocating data generation as an integral aspect of students' mathematics experience, I do not however abandon the idea that teachers are not essential, especially because it has been shown that the mere introduction of tools does not necessarily lead to inquiry. Rather, if Vygotskij (2005) is right in stating that *all* higher psychological functions are societal relations first (see Chap. 10), then arrangements in which STEM students relate to other individuals is essential to their changing participation in graphing practices.

Using graphs to produce verbal or written interpretations is one of the key competencies that students are to develop as part of their mathematics instruction. Many scholars approach the problem of graphing in a way that might be termed "semiotic," where the graph is taken as a sign that is related to some referent in a process called semiosis (Roth and Bowen 2001). During semiosis, other signs are produced – technically referred to as interpretants – such that the {new sign ⇔ original sign} relation elaborates the {original-sign ⇔ referent} relation. When approached in this way, the problem of graphing competencies is easily reduced to a cognitive problem, which has led many scholars to attribute difficulties to students' cognitive deficits. With others who take a pragmatic approach, I prefer looking at practices by following the actual deployments of graphs and other forms of inscriptions. Graphing, that is, using graphs in appropriate contexts, then requires

[3] Even many science teachers, in part as a result of deprofessionalization, have to cope "with a top-down, assessment-drive curriculum" (Levinson 2011, p. 113).

not just interpreting a sign but in fact opening up the black box, that is, recovering the particulars of the production process. But opening up a graph means revisiting all the steps that went into its production, being familiar with the natural phenomena represented, the equipment, the circumstances of the data collection, the data transformation, or the means by which it has been achieved. Without a deep familiarity of where the data come from, how they are generated, what kind of means were used in their production, possible problems in the production of data, and how data differ from non-data even scientists are hard pressed to make conclusions and support claims as my own studies show (e.g. Roth 2009; Roth and Bowen 2003). Making such choices about inclusion and exclusion of data is important, for example, in democratic decision-making processes where mathematical knowledgeability might come in handy.[4] This became evident to me when the mayor, town council, and town engineers in the municipality where I live based a decision on constructing a water main to supply people with running water on the report of a particular scientist who only collected data on a single day and in only one-sixth of the homes concerned. They did not take into account, and even omitted from entry into the data sources, more than 30 years of information that the local residents had collected about the water. That is, these municipal officials could perceive a phenomenon emerging from their data rather than a different phenomenon that would have emerged if *all* the available information had been considered. Some savvy citizens, however, engaged in the struggle with politicians and the scientists and engineers who worked with them by pointing out the problems in the method of the geologist. In fact, it was in a public meeting that some of the non-scientist citizens came to participate in unpacking the black box of the data that the geological consultant collected and interpreted. The following is a brief account of the events.

In that particular watershed of my municipality, the groundwater levels are continuously monitored and publicly available in graphed form (Fig. 12.2). This graph shows considerable variability between the winter months (November–February), when most of the rain falls in the area (90–143 mm/month, 1971–2000 average), and the sometimes exceedingly dry summer months (July–September), when there are few rain fall events (20–28 mm/month, 1971–2000 average).[5] Anecdotal evidence – supported by some (scientific) data from the Vancouver Island Health Authority – shows that the chemical and biological water contaminant levels are low during the rainy season, but are high to the point of exceeding

[4] I am not an advocate of the notion that every student has to learn the same curriculum content. Instead, a democratic and fair education, in my view, is one that assists each individual to develop its fullest access and participation in the context of societal relations with others. For some, this might mean developing the competence to solve differential equations already in high school, whereas for others, balancing income and expense ledgers may be what they will do with great proficiency.

[5] In July 2013, while working on this chapter, there was absolutely no rain.

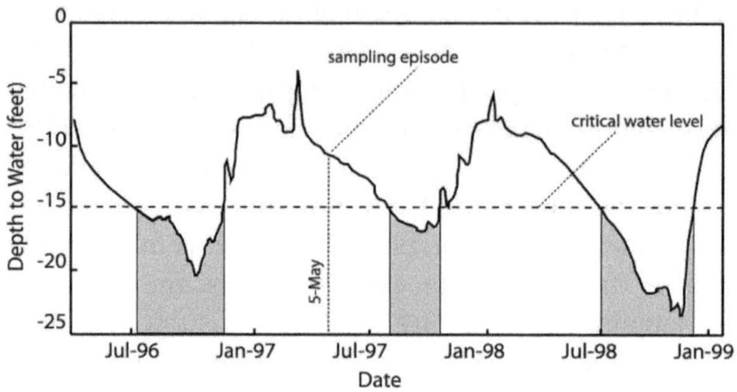

Fig. 12.2 The groundwater level in Mount Newton Valley and Senanus Drive, where the properties draw their water from groundwater wells. When the water falls below a critical level, the chemical and biological contaminants begin to exceed the acceptable norms stated in the Canadian drinking water guidelines (© Wolff-Michael Roth, used with permission)

the recommended levels of the Canadian drinking water guidelines.[6] The town council had hired an engineer to conduct its own water analyses. The engineering firm Lowen Consulting conducted one sampling episode on May 5, 1997. Arguing that this is a time of the year where "water levels are about average," its report concluded that "on average there are no contaminant problems." As an affected property owner pointed out during a public meeting, this conclusion does not address the fact that in the summer time the water levels are low and the contaminant levels do exceed the norms, which is equivalent to saying that the water is unusable – which, repeatedly, has led to water advisories issued by the health authority. Reported in the local newspaper during the summer months of several years running, residents were required to hire trucks for supplying them with drinking water. Not only were the contaminant levels high but also the water supplies had sunk to a trickle. Fig. 12.2 clearly shows that even though on average, there are more months during which the water levels are above the critical levels required for the contaminant levels to fall below the recommended ones, there are indeed several months per year – depending on the year – when the water level falls below those necessary to maintain the water at healthy levels.

Developing critical graphicacy may well mean developing competencies to participate in and contributing to the struggles in those public arenas where such issues are discussed (Roth et al. 2005). Not the mathematical properties of a graph import but the way in which certain graphs – such as the one presented in Fig. 12.2 – can be mobilized, qua inscription, to deconstruct town council decisions and engineering reports. For example, students could discuss the discursive construction of "on average," and the implications of high contaminant levels during a few summer months. Such graphs can be used to show that the problem of unusable water does not go away with the argument that on average, there is no problem.

---

[6] I have written an extensive analysis from a municipal engineering perspective for the engineering community (Roth 2008). After decades of wrangling, the water line was completed in 2013.

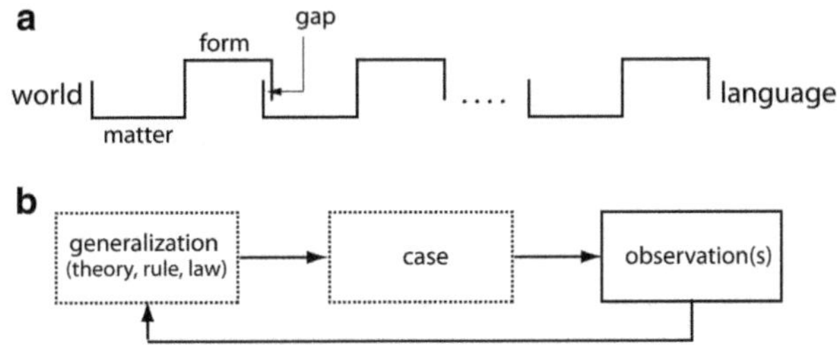

Fig. 12.3 (**a**) The chain of translations that leads scientists from the living world to statements of truth. (**b**) Abduction is a process of mutual implication and mutual constitution of concrete observation and a generalization

The figure shows that during the February 1996–January 1999 period, the water was unusable for up to 5 months.

My ethnographic data provided in Part II of this book show that scientists do not just interpret decontextualized data. They require familiarity with the natural setting and with the measurement process and the criteria that include or exclude some of these. This is similar to the residents of the Mount Newton Valley who knew from their decade-long personal experience that there were water problems, and whose familiarity with these problems jibes with the water level data, over which I laid the sampling episode and the critical water levels (Fig. 12.2). The resulting graphical representations are integral part of the entire research process, and familiarity with it is a requirement for interpreting them. Thus, the graphs have a part–whole function to the research in its entirety – or, more technically speaking, these are synecdoches of the research processes, that is, parts of the research process that point to the entirety of the process. We may look at the role of background familiarity in terms of two ideas developed in the preceding parts of the book. The first idea is that of the chain of translation, which scientists traverse to relate natural phenomena, on one end, with knowledge-claiming statements, on the other end (Fig. 12.3a). In Chap. 8 I suggest that data interpretation is abductive rather than inductive, which is equivalent to saying that scientists have at least some hypothesis of the thing (observation) that their graph (generalization) refers to (Fig. 12.3b). This type of relation, in fact, is true between any two neighboring matter–form configurations along the chain (Fig. 12.3a) – a fact that was observed in our research team's early struggles to relate the image of the objects on the microscopic slide to the graphs that resulted from the absorption measurements (Chap. 3). That is, to recover the relation between any two of the matter–form configurations, the reader has to be at least somewhat familiar with both configurations and with the practices that link any two configurations across the ontological gap that separates them (Fig. 12.3a).

There is convergent evidence from at least two studies at the middle school level and one at the elementary level, where students, confronted with some data, persisted in asking questions about the context within which the data were collected that students had been asked to analyze. In the absence of background information requested by students, the teacher "eventually abandoned the data analysis that was planned for this class session because the students continued to ask questions about the situation from which the data were generated" (Cobb and Tzou 2009, p. 162). The students did not want to engage in the data analysis until after being familiar with the context of the data collection. This study thereby reproduced findings of an earlier study, during which I noted that pairs of eight-grade students, asked to answer a peer's questions given the data that this peer had collected, requested more background information than was provided with the task (Roth 1996). Astonishingly perhaps, the students all had worked somewhere in the vicinity of this peer – though, admittedly, they did not know that the data provided had been collected by her peer.

Cobb and Tzou's episode points us to an important aspect of problem-solving practices in the everyday world where, as I suggest in Chap. 8, "persons-acting are free to transform, solve or resolve a problem, or abandon it in favor of other options. In the parlance of the [Adult Math Project], they 'own' their own problems" (Lave 1988, p. 156). Without articulation of the relation between graphing (a social practice) and the setting (of research) we have little to go by to explain how "cognition is constituted in dialectical relations among people acting, the contexts of their activity, and the activity itself" (p. 148). Dialectical relation here is equivalent to saying that there is a unity to the activity as a whole, which is the minimal unit to describe and explain the sense of any of its parts, including data and their graphical representation.

The results reported in Part II should encourage STEM educators to begin a debate concerning the experimental and ethical dimensions in data generation. Students need to learn to deal with a range of questions: Which measurements may be legitimately excluded from entering the data sources? On what basis are the decisions that distinguish between sources included and sources excluded? How does exclusion influence our theories of nature? What ethical implications are related to the question of excluding measurements from the data sources? From a nature of STEM perspective, students *ought to know* what scientists do and how their actions affect *what* we know about nature. Knowing what might affect the selection of data sources is as important for describing and explaining nature as describing and explaining the nature of the phenomenon (figure) against the overall variation within the data (ground). Future STEM education research, therefore, might be designed to investigate and (experimentally) tease out the role that familiarity with the entire inquiry process in general and the data generation process in particular plays in students' changing participation in changing activities, where science and discourses about the nature of science might be important resources in the emerging struggles over contentious issues.

Two Studies on Data Generation and Interpretation in the Face of Uncertainty

In this section, I review two studies that exhibit how mathematics educators in particular and STEM educators more generally may draw on the possibilities arising from the involvement of students in data collection or discussion of the source of data that they subsequently model mathematically. Such studies tend to exhibit the tremendous competencies and forms of knowledgeability students develop as they become familiar with the natural or social phenomena modeled and the mathematical forms used to articulate the models. That is, we might characterize these as studies that illustrate possible implications of the research among scientists (Part II).

A Focus on Context

Together with my colleague Michael Bowen, I had designed an eighth-grade science course that allowed students to become fluent in the deployment of mathematical inscriptions – based on what we had read in the social studies of science literature concerning the mathematization of experience (e.g. Roth et al. 1996; Roth and McGinn 1997). Our core study occurred during the year-terminating ecology unit. But already during the course of the year and to prepare the students for their ecology study, we progressively shifted the responsibility for learning to students. During the first unit on physics, we asked students to conduct specific experiments to answer equally specific research questions. During the subsequent chemistry unit, students started designing their own experiments to teacher-determined research questions. The third unit, on ecology, was characterized by student-designed research questions and investigations. Whereas all students evaluated the freedom during the ecology unit very positively in the end, a few did struggle in and with getting started and designing their first research question. Some were overwhelmed by the extent of their observations; others found too little to see in their "ecozone."[7] There was much independence, and the performance criteria were not as clear as in their other, lecture-style classes that they attended in this school (e.g. "Support your claims in a convincing manner"). Thus, some students did not know what to do and expect or where to start, as they felt thrown into a situation of uncertainty. However, the supporting structure described next, allowed students to gain confidence and to adjust quickly to the new learning environment.

At the time of the study, the students were engaged in a 10-week ecological study of the school's campus, a 50-acre lot with a variety of ecological zones. The teacher's goal of the ecology unit was for students to develop skills in designing independent research studies, making sense, and supporting findings in a convincing manner. This provided for many opportunities to observe mathematical practices that organizations such as the National Council of Teachers of Mathematics support: applying statistics in the phenomenal world; translating between inscriptions; making connections between mathematics, the phenomenal

[7] Ecozones are the broadest biogeographic divisions that are constructed of the planet Earth's land surfaces, accounting for the distributional patterns of organisms. In this unit, the term was part of a discourse characterizing the different ecological conditions on the school campus.

world, and science; and solving problems and modeling natural phenomena. During the ecology unit, each student group was asked to find out as much as they could about their own small plot ecozone about 35–40 m^2 in size. From these tasks, and to evolve convincing representations, there arose many opportunities for mathematizing their observations.

To create what we thought at the time to be the best conditions for learning during the open-inquiry field study, we imposed the following conditions: students were (a) to produce defensible arguments for the necessity of a research question and the soundness of the experimental design and (b) to present their results by means of inscriptions that would convince peers and teachers. In this, the learning environment incorporated one of the fundamental principles of scientific progress: Scientific practice has rhetorical aspect recognizable in scientists' efforts to support their positions. There are indications in students' conversations in the field and during the interviews that they were attuned to this attitude (How do you know that that's true?, How do you support your claims?, or Put lots of measurements, because he likes people to prove stuff). For example, after returning from the field back into the classroom, Jamie studied his notebook with the sequentially recorded data from his second field study. He suddenly exclaimed, "Look at this one: the higher the pH level, the higher the plant, but the lower the amount [of increase]; the taller the plant, the lesser the amount; and the lower the pH, the lower the height, but the more the amount." He proceeded to note this statement as a claim in his report. But about five minutes later, he urged his partner Mike that they had "to make a table which has to make a comparison which says, which compares the population density with the pH level and the growth" to be able to support this claim.

Unlike the students in the Cobb and Tzou (2009) study, who were given data (see below), the students in our courses designed their investigations, decided what would constitute good data for responding to their questions, collected the data, and then transformed them to serve as evidence for convincing peers – during the periodic peer group discussions we implemented – and teacher that whatever interesting phenomenon and pattern observed, the study provided good evidence for. Already during the data collection phase, the students engaged in attempts to make sense of measurements, patterns, and variations. When something did not fit, they immediately checked on measurements or collected additional data.

While working in the field on collecting data, students tended to take the opportunity to structure their ecozone in new ways (Fig. 12.4). They defined and identified aspects of the environment that they thought might correlate; and they refined their frames in the course of elaborating research problems. Students decided which aspects were important to them, researched these, and reported their findings. In this way, over the 10-week period, they evolved complex descriptions and explanations of ecozones and the relationships of biotic and abiotic variables. Because they spent a considerable amount of time in their ecozone (Fig. 12.4), they became deeply familiar with the animals, plants, microorganisms, and many physical and biological phenomena therein. They also developed tremendous familiarity with mathematizing natural phenomena. Table 12.1 shows an

Fig. 12.4 Two students working in the school yard collecting data to answer the research question whether there is a relationship between depth and the frequency of certain organisms (© Wolff-Michael Roth, used with permission)

Table 12.1 Frequency of six forms of inscription students used to mathematize phenomena

Type of inscription	Lab 1 ($n=16$)	Lab 2 ($n=17$)	Lab 3 ($n=20$)	Lab 4 ($n=10$)	Exam ($n=40$)
Descriptions using numbers	44	35	60	80	–
Map including measurements	38	41	70	70	65
Tables and lists	75	76	70	80	28
Averages	25	47	45	70	70
Graphs (bar, line, scatterplots)	56	59	60	90	45
Equations	26	18	20	30	–

Note: Most laboratory reports used more than one form of inscription. The counts represent the types of representation rather than the total number in each report. Thus, if a report contained to line graphs, we counted "1."

increasing deployment and use of mathematical inscriptions from their first to the final report of their investigation. It also shows that eighth-grade students took a while to reach levels of use of averages and equations (including %) that were comparable to most other representations. Whereas students increasingly used averages, however, the use of equations and percent calculations remained more limited.

On first sight, calculating averages might seem to be an easier operation than drawing graphs. Averages are pervasive in everyday life, including such areas as sports. The disposition to use averages could have been part of the cultural background that students bring to the classroom. Yet students used graphs more often than averages. However, the frequencies hide the fact that some of these representations were used inappropriately from the perspective of accepted conventions in the natural sciences. Our students did not initially distinguish between interval scales and categorical scales and used them interchangeably. They treated categorical data as interval data and vice versa (equidistant data points irrespective of actual x-values); they devised trend lines in graphs with categories on the abscissa values; they used point-to-point connections rather then trend lines; they used line graphs with test number as abscissa value rather than the appropriate variable – e.g. both the independent and dependent variables were plotted versus test numbers. It was evident that the students drew on their existing familiarity with inscriptions from subjects such as geography where they used bar graphs and point-to-point line graphs to represent monthly rainfall and temperature, respectively. This prior familiarity with the world was the ground upon which they subsequently established sophisticated data analysis practices that outstripped those that we observed among university graduates (BSc, MSc) enrolled in a post-graduate degree in science teacher education (Roth et al. 1998).

Students also evolved new representational forms, which, when these were shared with other students in the class became important new ways for students to mathematize their experience. For example, two girls measured both soil temperature and soil moisture on two occasions. They organized their first and second soil readings in a data table placing next to each pair of numbers an arrow up for an increase and an arrow down for a decrease in the measure. Then they constructed a pattern relating the arrows for the temperature difference with those of the soil moisture difference: "The arrows indicate whether the measurement went up the second time or down. Notice that every time the soil temperature went up, the moisture went down, and vice versa." Two boys produced a graph that related plant height to the amount of light these plants received. The students also indicated with each plotted point some special conditions of the site of data collection. In this way, they ended up with a graph that resembled a multidimensional scaling output in which objects are classified according to their weights on two or more characteristics. Two young gentlemen used this plot to classify their data points into sets with similar characteristics. A number of student pairs used various tiling methods to mathematize their ecozones, by classifying each tile according to the amount of light, the kind of plants, the variety of plants they contained. Or they used the tiles to represent distance from the nearby lake that aided them in constructing functional relationships between selected variables and distance from the lake.

We used students' answers to one examination question as an indication of how much they had appropriated about the mathematization of experience from their fieldwork. We provided the following description:

> You are in a forest ecozone with square boundaries that has a cut grass field along one edge. Discuss the methodology you would follow to determine the relationship between soil moisture, number of species of insects, insect population density, and average dandelion height as they might possibly be affected by nearness to the field. Use diagrams if that will help. Make sure that all sampling is done correctly and that all equipment and techniques are properly named and explained. Demonstrate *how* you would illustrate the numeric relationships you find (e.g. insect density & nearness to field).

As an indication for the degree of mathematization, we counted the types of representations students used or proposed in answering the problems (Table 12.1). We first investigated if and how students mathematized "nearness to the field." Of the 43 students who took the final examination, 33 % proposed the use of interval scales to measure distance ("staking out areas of 1 m apart," "measure [with a yardstick] how far that place is from the field," or "taking samples every meter"); 7 % suggested to determined the distance qualitatively by using three or more levels of distance (drawing showing three sections, "pick out three areas"); another 20 % suggested the qualitative distinction of "near" and "far" ("close and far away" "near the field and away from the field" "on one end of the field and also closer to the field"); 10 % suggested measuring distances, but in a context not related to the above problem ("if you got closer, it would be sunnier"), whereas the remaining 30 % did not at all address the question of "nearness to the field." When we compared the students' use of inscriptions on the last report with that on the exam, we noticed sharp drops in the uses of tables and graphs, but only marginal differences in the use of averages and maps.

In our effort to explain the students' performances we generated four plausible explanations that we do not consider to be mutually exclusive. First, while the students worked in a familiar context during the regular term, the problem we posed on the examination was decontextualized. From the students' perspective, the problem might have had little resemblance with their actual experience of framing a problem, collecting the data, and presenting their claims. To discuss these differences in the use of inscriptions between field experience and examination further, we proposed the notions of experience-near and experience-distant representations. Some inscriptions are closer to the natural phenomenon (e.g. maps for recording all the measurements, over lists, tables, totals) whereas others (e.g. means, scatter plots, line graphs, and equations) are more removed from the experience with natural phenomena. We proposed the adjectives *experience-near* and *experience-distant* to situate an inscription with respect to an individual's possibilities for engaging with a "raw" phenomenon. One would expect that students are more likely to use those inscriptions that are experience-near because they can be more easily related to the lived experience – similar to the iconic relations that they first produce between different inscriptions such a speed graph and the drawing of a race track. The adjective experience-distant then refers to inscriptions that stand for what scientists call "deep structure" because it is not immediately apparent upon seeing the phenomenon and is related to scientific explanations. Though this approach was plausible for the use of maps and averages on the examination, still left unexplained the low use of lists and ordered lists. Whereas we expected that the usage of graphs might be lowered because of their more experience-distant nature, we did not expect the same for the table/list category. However, the lack of actual measurements provided a context that did not encourage students to think of the organization of their data. We found further support for our categories of experience-near and experience-distant on the students' examination and in their work on word problems that we asked them to complete during the term. On these occasions, students attempted to re-contextualize word problems in terms of their field and other prior experience to have a meaningful framework within which the problem posed externally could be explained. That is, we found the same tendencies that I report in the ethnographic study of the scientists at work (Chap. 5).

Many students developed sophisticated means for dealing with variability in the data, which afforded unpacking the black boxes presented to them during experimental design tasks and examinations. This was shown in one group's discussion of

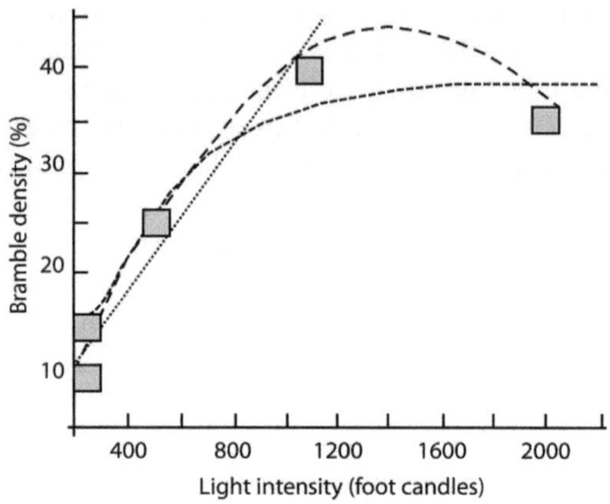

Fig. 12.5 One student plotted the data and discussed three possible trends through the five points (© Wolff-Michael Roth, used with permission)

a scatter plot that contained five data points relating the density of brambles, in percent coverage in each section of a plot of land, and the amount of light that was falling on the section (Fig. 12.5). Rod insisted that the data points approximated the linear trend, but Erica favored a parabolic relation. Ted suggested that the curve approximating a square root function would be the best choice. In the course of their discussion, they deliberated whether one or the other point might have been an outlier and how this would affect their conclusion.

The problematic was real to the students, because of the different positions held by individual group members. Rod and Erica favored different solutions. In and through their discussion they struggled towards a solution in which all three trends were viable alternatives to describe the data. The final statements by Erica and Ted illustrate this openness, the acceptance of three solutions, and the postponement of a decision to a point when further data would allow distinguishing between the three – which makes their discussion very much like the ones that I observed among the scientists in Part II. Thus, the scientific phenomenon was constituted through the interactive and descriptive work. It included the establishment of alternate solutions, the bearing of natural phenomena on particulars of the graph, and the invocation of mediating circumstances. In the present context, the trend lines took on tangible forms, which stood for a series of complex descriptions and relationships. Such descriptions and relationships could then be tested in the field through further data collection, data analysis, and data interpretation. The particular arena we had prepared allowed these students to move back and forth from the context of the ecozone to the graphs produced in the classroom or at home; and from the potential relationships (additional variables posited by the students) to the actual field tests. We had evidence that this movement between physical phenomena and mathematical representations encouraged students to develop familiarity with and well-integrated explanations of the phenomena and the inscriptions. The inscriptions were no more abstract than the soil, leaves, thermometers, soil corers, shovels, and notebooks they were dealing with and using while gathering data in the field.

Supporting Students' Learning About Data Creation

Another interesting study with respect to the relationship between data creation and graphical representations involved a group of students, which the researchers followed during their seventh and eighth grade (Cobb and Tzou 2009). Over the course of the year, students developed a good sense of how the "data creation process might affect their analysis" (p. 135). That is, these researchers explicitly addressed the issue that the ethnographic study in Part II of this book shows as endemic to scientific discovery work: data creation is not independent of the interpretation of the data or some other inscription that has been developed from it. The objective of the teaching experiment was to have students learn about distributions, with the goal for the seventh-graders to describe the distribution of univariate data. They were to use key words such as "center," "skewness," "spreadoutness," and "relative frequency." In addition, once in they had progressed to eighth grade, these students were to develop familiarity with bivariate distributions and the data as being distributed in a two-dimensional space of values. The study in the preceding subsection was one of the referents that the researchers used, which allowed them to be aware of the fact that "if the students did not collect data, the data values might be mere numbers for them rather than measures of attributes of a phenomenon" (p. 140). In fact, during a pilot test of their unit, the researchers had found out that precisely this happened leading to a situation where students "seldom reflected on the meaning of the results of [their] calculational procedure" (p. 140). They therefore held it as essential for the teacher to talk with students through the data generation process.[8] Following suggestions in the literature – which essentially realize pragmatic descriptions of activity – the researchers engaged students in discussion with the intent to make them familiar with the *with-what* (tool), *what-for*, and *for-whom* of the data, interpretation, and conclusions. Moreover, the teacher-researchers emphasized the translation processes that lead from the data to the representations ultimately used to interpret some research and to support the conclusions. In contrast to our own study described in the preceding subsection, Cobb and Tzou conjectured that it would not be necessary to have students collect the data themselves. They felt it sufficient for the teacher to talk students through the data generation process. The study among scientists, however, would suggest that Cobb and Tzou are not quite right in this respect.

In one task near the end of the first year of instruction, the teacher in the Cobb and Tzou study asked students to analyze data on the T-cell counts of AIDS patients enrolled in one of two treatment scenarios. The aim was to assist some medical officer in making a decision about the relative effectiveness of the treatments. The task asked students to analyze the data provided, to identify patterns that were relevant to the phenomenon at hand, and to produce a report that would provide the consulting advice. The students were to assume that the person reading the report

[8] Such talk, however, already delimits the messiness of the situation because language itself constitutes a frame that includes and excludes potentially useful aspects.

did not see or have access to the data, requiring the report to be written so that the reader nevertheless could intelligibly engage with it. Cobb and Tzou report that the students refused initiating analyses unless they were familiar with the data, the situation from which these were culled, and the needs of the ultimate user of the report they were to write. As a result of the unit, students became both savvy with respect to the clarification of the data and also keenly aware of the need to be familiar with the context of the data generation to insure the legitimacy of their conclusions. Thus,

> students became increasingly concerned with controlling extraneous variables when they contributed to data creation discussions … raised questions that related to issues of sampling and the representativeness of the data.… This contrasts sharply with the data creation discussions at the beginning of the seventh-grade design experiment in which the students neither asked clarifying questions nor attempted to ascertain the purpose of the analysis. (Cobb and Tzou 2009, p. 163)

As noted, the students in the Cobb and Tzou study did not generate the data themselves. They had access to the data description in a way similar to which doctors in training have access to cases, which are presented to them in discursive terms (i.e., files with descriptions of what someone else has noted). The entire situation is dependent on *what* the teacher present and how they present them. That is, students interact with a set of descriptions rather than being confronted, as the students in the our study, with natural phenomena, instrumentation, recording, and details of the initial transformation from the natural world to the first numbers. The ill-structured nature of data definition in one situation is not replicated when only descriptions are provided of the data collection process. These first steps appear to be integral in descriptions of the transformational change in ethnographies of field biologists at work (Latour 1993; Roth and Bowen 1999), and integral to what the graphs qua metonymies denote and consist in. These first steps of structuring the setting were also essential in the work of the experimental biologists that were part of the study in Part II. Thus, at the end of one particular long day of collecting data, Craig said that all the work had been in vain, for he apparently had not done during the dissection what he thought to have done. That is, already at this stage a selection was occurring in the data that the students in the Cobb and Tzou study would not have experienced and, therefore, an aspect that would not be part of their preparation of and personal relation with the data. We might anticipate, therefore, that students develop a symbolic mastery rather than mastery of the real thing – just as journalists tend to have a symbolic mastery of the (basketball, football, soccer, cricket, etc.) game they comment (upon). There are dangers in confusing the two forms of mastery, for it leads to situations where students are familiar with applying some routine procedure once the initial steps of selecting one over another have already been accomplished in the task setting. This first step appears to me as essential to the participation of mathematics students as the ironing of completed garments is to the participation of Vai tailor apprentices when they first enter the shop (Lave 1977). It is in this first step that the tailor apprentices are set up for making parts of the garments that are increasingly difficult. Further evidence for the need to collect and select data comes from studies that suggest the importance of learning to lose one's phenomenon as an important aspect of being familiar with the phenomenon (Garfinkel 2002).

Symbolic mastery refers to the competency to talk about some practice in terms of principles from the outside independent of what drives the practice from the inside. There is a transmutation of the actual mastery, displayed in real time and under the constraints of the situation, into theoretical mastery. But in this transmutation the practice ceases to exist "once people start asking whether it can be taught, as soon as they seek to base 'correct' practice on rules extracted, for the purposes of transmission, as in all academicisms, from the practices of earlier periods of their products" (Bourdieu 1980, p. 175). The (theoretical) representations of practice are at best partial, but generally are inadequate to practice. Thus,

> just as the teaching of tennis, the violin, chess, dancing or boxing decompose into positions, steps, or moves practices that integrate all these artificially isolated behavioral units into the unity of an organized and oriented practice, informants tend to give either general norms (always accompanied by exceptions) or remarkable "moves," because they cannot appropriate theoretically the practical matrix on the basis of which the moves are generated and which they do not possess other than in practice. (p. 174)

It is evident in this description that even practitioners themselves will fail to provide theoretical descriptions that do justice to their own practical mastery.

Integrating Across STEM Domains

In the preceding section, I provide brief descriptions of science and mathematics classrooms in which students developed high levels of data and graphing related competencies as they (a) either collected, processed, translated, reported, and used data for reporting purposes (b) or were presented with data the origin of which was described to them and that they could find out more about in discussions with their teacher. The curriculum intentions in each case were related to particular subjects, biology and mathematics, respectively. But mathematics and science can be taught together, and sometimes may have to be if students are to become competent, for example, when the description and explanation of a scientific phenomenon requires familiarity with the mathematical properties of graphical relations. For example, in one of my studies students developed mathematical and scientific practices related to the explanation of simple machines while designing and building prototypes of more complex machines (Roth et al. 1999). Their mathematical, conceptual-scientific, engineering design, and technological practices were intertwined and codependent because, for example, mechanical advantage, the related measurements and calculations, and the engineering design principles developed together in the production of machines that really provided an advantage over not using these. Thus, already while teaching high school physics I had the sense that students could not really explain motion phenomena unless they were familiar with the *mathematical* relations between position-time graphs, velocity-time graphs, and acceleration-time graphs. In their mathematics classes, the physics students covered integration and differentiation much later than they covered the motion curriculum in physics. Thus, I was setting up and encouraging students to explore these relations in close connection with studying motion phenomena. Students did explore the graphs they generated using the very first versions of graphing calculators (TI 81), early versions of the mathematical modeling program MathCAD running on the Macintosh SE computers that I had installed in the laboratory, and a statistics package (StatVIEW II). In the following, I provide a case study of one student, who initially failed to put it all together, as he told me one afternoon, but who, through some additional after-school tasks that he desired doing on his own, established deep explanations of the physical phenomenon and the underlying mathematics. The case is exemplary in the sense that the teacher of the mathematics course, who also was the head of department in our school, approached me a few months later asking what I had done with and to the physics students, who quite apparently had exhibited tremendous familiarity with the mathematical principles during their calculus classes.

Niko and his partner had completed an investigation in which they studied the motion of a cart that was pinned between two springs. As a result, they observed slightly damped oscillations. The two had collected motion data using a photo gate and a data collection program that permitted them to print out distance-time, velocity-time, and acceleration-time graphs and the corresponding data tables. Niko told me that he had difficulties interpreting the data and intelligibly hearing what his partner was saying. He wanted to return to the physics laboratory one afternoon and go through the experiment on his own. When he came to the laboratory, he brought with him his data and graphs. He set up the apparatus, and repeatedly shifted between moving the cart slowly from one side to the other, running the experiment in real time, gazing at his data tables and graphs, and using the blackboard as a scratch pad. Later, he ran a mathematical modeling program, MathCAD, to fit a curve through his velocity-time data to find the functional relationship of velocity and time. He generated graphs of the *derivative* and the *integral* of the velocity-time function by applying the respective operators available in MathCAD.

A few days later, Niko submitted a laboratory report which contained distance-time, velocity-time, and acceleration-time graphs to which he had added (a) color to highlight specific data points, (b) labels to identify specific data points by their coordinates, (c) hand drawn lines, and (d) explanatory text. For example, the following text accompanied the acceleration-time graph:

> All important points that I will be referring to have been labeled on the graph and have received asterisks beside them in the "table of values." Please understand that I have treated the air track like a number line. Motion to the right I will refer to as positive and motion to the left I will refer to as negative.
>
> *Data and transformations for the acceleration-time graph*
> The data plotted to make this graph was produced from the experiment. Since the derivative of a sine graph is a cosine graph, the acceleration-time graph should be a cosine graph, the acceleration-time graph should be a cosine graph. But it doesn't look like that because this acceleration-time graph is a derivative of a non-uniform velocity graph that has only

positive values. Again, this unexpected graph is due to the fact that the photo-gate system cannot measure direction.

This graph describes the cart's acceleration for one full cycle. Again, it would be pointless to find the regression line and its statistics, because the mean of a cosine graph is also the base line or 0.

INTERPRETATIONS, EXPLANATIONS, AND GENERALIZATIONS OF THE ACCELERATION-TIME GRAPH

The slope of the v-t graph at every maximum point is 0. Since the maximum point means maximum velocity, the acceleration at this point is 0 m/s/s as suggested by the slope at these points. The data between (.073, 1.64) and (.92, .11) represents the cart's positive, forward acceleration as it is approaching maximum velocity. The "y" value where the graph intersects the x-axis is the instant in time when the cart is no longer accelerating and has reached its maximum velocity.

All the negative data points between (.95, −.06) and (1.15, −1.85) represent the period after the maximum velocity. This is where the cart's inertia was propelling it forward in a positive direction and the spring on the left was accelerating the cart in the opposite, negative direction.

After the point (1.15, −1.85) the *a-t* graph abruptly has an extreme positive slope. This is due to the fact that the computer has now had to calculate and plot the derivative of a positive slope of the v-t graph. This is the disadvantage of the photo-gate not being able to measure direction.

The data between (1.42, 1.37) and (1.61, .03) represents the cart's negative acceleration as it approaches maximum velocity. The spring on the left is contracting and the spring on the right is expanding as time continues. At o acceleration where the graph intersects the x-axis, the force supplied by either spring is balanced.

From (1.64, −.24) to (1.88, −1.66) are the data points that indicate the cart's acceleration to its initial position in the negative direction as a result of its own inertia. At the same time, the spring on the right is accelerating the cart in the opposite positive direction. The cart actually comes to a complete stop for a brief instant at the end of this cycle, before it continues again.

This text is typical for those that the students in my classes submitted at the beginning of the second year of physics. It is apparent that Niko had sought to be familiar with the experiment, the data, and its implications. He made it clear that he wanted to communicate this familiarity to the reader of the report by making special provisions such as coloring important points of the graph which he discussed in the body of the text; by labeling these points with asterisks in the data table; by indicating his familiarity with positive and negative motion; by indicating his choice of treating the air track as number line; and by using number pairs to refer to data in the graphs. These preparations show how Niko drew on his familiarity with the everyday world, sometimes reaching as far back as to what he had learned in the elementary school where he used number lines to explain the direction of the basic arithmetic operations of addition and subtraction, or to middle school to describe the operations on negative numbers. Niko's new mathematical approaches further developed through the use of a symbol system all the while focusing on a natural phenomenon – the moving cart on the air track. As Niko described the motion of the cart, he made connections between several levels of conceptual abstraction. These levels included (a) the concrete objects and events with which he was dealing; (b) the description of motion in the form of data tables and graphs;

(c) the mathematical symbolic discourse; and (d) the explanatory, conceptual discourse of the physical events.

In his introduction, Niko used an inference based on his knowledge of derivatives ("Since the derivative of a sine graph is a cosine graph") that the acceleration-time graph should look like a cosine graph. He linked the observed discrepancy to the capacity of the photo-gate to measure speeds only, not direction, which precludes a measurement of velocity.

Niko observed the cart in real time and in slow motion. As he followed the cart's motion, he matched its velocity and acceleration with the corresponding graphs while indicating: "The '*y*' value where the graph intersects the *x*-axis is the instant in time when the cart is no longer accelerating and has reached its maximum velocity." In the next paragraph, he linked the graph and data with the explanation of the phenomena by invoking Newton's First Law ("(.95, −.06) and (1.15, −1.85) represent the period after the maximum velocity. This is where the cart's inertia was propelling it forward in a positive direction"), and connected both descriptions to the physical state of the spring ("and the spring on the left was accelerating the cart"). He continued by using a mathematical symbolic framework to write about the slopes of both acceleration- and velocity-time graphs, explaining the discrepancies between what one would expect and what he observed ("a-t graph abruptly has an extreme positive slope. This is due to the fact that the computer has now had to calculate and plot the derivative of a positive slope of the v-t graph"). Niko linked this discrepancy between his prior expectation and the graphs to the contingencies of the apparatus ("This is the disadvantage of the photo-gate not being able to measure direction.").

In this example we see a student coordinate and talk about multiple inscriptions pertaining to the same physical phenomenon. In fact, these inscriptions are related by transformations, which, mathematically, are expressed as operations d/dt and $\int dt$. The physical system (a frictionless cart suspended between two springs), the graphs and tables, and the mathematical equations describing the motion served as objects for the active coordinating work that links across the different ontological domains. Talking to himself silently (murmuring) and aloud while engaging in this work appeared to assist him making the connection. It was as if he were trying to explain to someone else what was going on, and, in so doing, he also came to explain it to himself. Niko's use of language for talking himself through a problem can be seen as a surrogate for communicating with a partner. From Part II we know that the doctoral student Shelby was often talking to the research associate Theo to familiarize himself with what was happening in the processing of the data, which the latter was familiar with because he was writing the software for the laboratory. Language provided a bridge, for it is at that level that familiarity with the natural world and familiarity with the world of inscriptions come to connect. It is in this way that the scientists recontextualized the data that their work had decontextualized before.

In the course of their work, Niko and his peers became familiar with the relations of associated mathematical inscriptions. This was also shown in the work that Niko

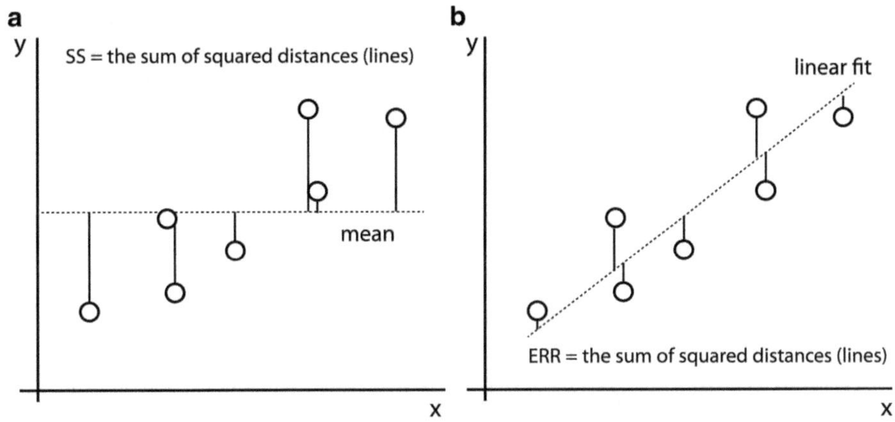

c $$\text{RSQ} = \frac{\text{explained}}{\text{total}} = \frac{\text{total - error}}{\text{total}} = \frac{\text{SS - ERR}}{\text{SS}}$$

Fig. 12.6 High school students learned about fitting functions using the least squares approach. Once they were familiar with using it, many were asking about how it actually worked. Typical inscriptions used were designed to render visible and intuitive some of the underlying concepts, including concepts such as the sum squares (© Wolff-Michael Roth, used with permission)

was doing. After describing and analyzing the graphs the physics students fitted a curve to their velocity-time graph. Under my occasional guidance, and with a graphics calculator, they first searched for a function that looked like the graph from their experiment. Students then adjusted the free parameters using a least squares method for finding the best fit. I initially had shown the students how to apply this method with a minimum of explanation. Niko, however, insisted on further explanations so that he could describe and explain the procedure of least square fitting prior to applying it. Again, I used drawings (i.e. further inscriptions) to talk in conceptual terms about what a least squares approach to fitting does and what the SS, ERR, and RSQ refer to in the visual (Fig. 12.6). Niko expressed satisfaction and proceeded to fitting a function to his data using MathCAD. Much like Shelby in Part II, he then used least square fit to seek a function that would best describe his data. He spent about an hour, deliberately changing one parameter in the velocity function at a time to find its effect on the fit. He continued investigating one parameter at a time. In this way he found out that the parameters in the function that his partner had identified describe the amplitude, decay constant, period, and phase shift. To complete the final step of the task, the Niko investigated both the integral and the derivative of the fitting function for the velocity-time data.

> When graphed, the dampening constant touches the crest of each half cycle indicating the rate at which friction is causing the amplitude of the v-t graph to constantly decrease as time continues.

By calculating ERR, the sum of the squared of the distance between the function and all of the data points, and SS, the sum of the squares of all of the distances between the data points and the mean, it is possible to determine a correlation coefficient according to

$$\mathrm{RSQ} = 1 - \mathrm{f(ERR, SS)}$$

If ERR is SS, you do not have a relationship. In this case, ERR $\neq$ SS and RSQ = .989, thus proving that there is a good relationship between the data of the velocity and the function v(t). The function describing the relationship between velocity and time is

$$v(t) = 0.51e^{-0.05t} \sin\left(\frac{2\pi t}{1.4} + 0.6\right)$$

INTERPRETATIONS, EXPLANATIONS, AND GENERALIZATIONS FOR THE V-T, A-T, AND D-T GRAPHS PRODUCED BY MATHCAD
Once the *v-t* function had been found, the *a-t* and the *d-t* graphs could also be found mathematically. The *a-t* graph is the derivative of the *v-t* graph. This means the numerical value for the slope of many tangents on the *v-t* graph were plotted as the *y*-values on the *a-t* graph. The distance time graph was found by using integration. Integration means that the area of rectangles between the *v-t* graph and the *x*-axis, with infinitesimally small amounts of time as their width were found and plotted against time to produce a *d-t* graph. These three graphs are the exact same as the three found by plotting the data from the experiment, except they have been found by using calculus.

CLAIMS
Non-uniform acceleration is a function of time, because its graph is not a horizontal line parallel to the x-axis.

By calculating the area beneath the velocity-time graph (integration), the distance-time graph can be found and by taking the derivative of the v-t graph, the acceleration time graph can be found.

The derivative of a sine graph is a cosine graph.

These excerpts bear witness for the concerns Niko had for finding words that describe each part of his investigation. The wordings he arrived at, which were his own as he had no written records of the conversations with peers or with his teacher (me) on these topics, articulated what he had done in the course of doing a fitting procedure developed out of the conversation which he had with me standing next to the chalkboard ("ERR, the sum of the squared distance between the function and all of the data points, and SS, the sum of the squares of all the distances between the data points and the mean"). He also indicated how the damping factor – the 0.05 in the exponent – relates to the experiment ("the rate at which friction") and how it affected the period function ("causing the amplitude of the *v-t* graph to constantly decrease").

Finally, even in his descriptions of the transformations from velocity function to distance and acceleration functions, Niko was not satisfied with simply stating derivative and integral. He developed full descriptions in which these concepts came to be related (e.g., "[The derivative is] the numerical value for the slope of many tangents on the *v-t* graph plotted as the *y*-values on the *a-t* graph" and "Integration means that the area of the rectangles between the *v-t* graph and the *x*-axis, with infinitesimally small amounts of time as their width were found and plotted against time to produce the *d-t* graph"). He concluded that the graphs arrived at through differentiation and integration were "the exact same as the three found by plotting the data from the experiment, except they have been found by calculus." In this way, like all other students who took physics, Niko became very familiar with the mathematical operations of differentiation (d/dx) and integration ($\int dx$) through the applications in physics before they treated these topics in their mathematics courses. Not only did the students return to me later to say how easy they found calculus, and about the comments the mathematics teacher apparently had made in class, but also the mathematics teacher came to me to find out how physics may be used to support the development of mathematical practices.

The complex of highly interrelated processes in which Niko was included (a) going from the natural phenomenon, the interfacing software generated data tables, and the graphs relating distance, velocity, and acceleration to time; (b) using the tables, to generate a new graph in MathCAD, which was used to fit a function to the data points; (c) producing the graph of the integral and derivative of the function found; (d) matching these graphs with the distance- and acceleration-time graphs; (e) placing the concrete objects and events of the interaction with the natural phenomena into an explanatory conceptual framework, which itself was linked to the distance- and acceleration-time graphs produced during the experiment; (f) using microcomputers to compress a lot of mathematical and physics experience into a short time; (g) finding a rule describing the natural phenomenon; and (h) using the rule to generate graphs that were themselves generated from investigating the natural phenomenon.

As he was linking inscriptions along the chain of translations (Fig. 12.3a), Niko was involved in many of the processes which curriculum standards tend to recommend for learning about functions and statistics: (a) he constructed a function to model real-world data; (b) he represented and analyzed relationships using tables, verbal rules, equations, and graphs; (c) he translated between tabular, symbolic, and graphical representations; (d) he analyzed the effects of parameter changes on the graph of a function; (e) he explored periodic real-world phenomena using sine and cosine functions; (f) he constructed and drew inferences from charts, tables, and graphs that summarized data from real-world situations; (g) he used curve fitting; and (h) he applied measures of variability and correlation. Niko did all this in the context of a physics course, so that the focus of his work was to evolve a better explanation of the phenomenon of damped harmonic oscillations. He did so without that these uses were explicit instructional purposes, but these uses evolved as needed in the given task. Niko simultaneously developed familiarity with mathematical processes and operations. In this example, it is impossible to separate becoming familiar with the scientific phenomenon and becoming familiar with using mathematical principles and discourse – just as I suggest above in the context of discussing Fig. 12.3. Thus, the natural phenomenon, the experimental condition, and the mathematical symbolic and experiential aspects became part of a whole.

The graphs, then, in fact indexed this whole and, therefore, bore metonymical relations much as this had been the case for the scientists in Part II of this book.

Niko had a great deal of control over what he was doing and why. He had chosen the particular natural phenomenon because he was interested in it. He selected the specific focus question, the phenomenon to be researched, the timelines to be set, and the analyses to be done. There was a great similarity between the experiences Niko has had in his physics course and the ones I describe in Part II. Much like the scientists described in the situation where they did not know what they were doing (Chap. 2), Niko and his fellow students dealt with problems where they had to decide what to do “when it is not clear what needs to be done to arrive at a solution” (Tobin 1990, p. 405). In fact, as seen in Chap. 2, it is unclear whether there is some solution and what the nature of the problem is. Much like the scientists, Niko was in a situation where he made the problems his own. In the process the students became increasingly savvy about how to deal with variability at different levels: both within the investigation itself, by becoming increasingly knowledgeable with respect to instrumental practices, and concerning the processing of measurements and using procedures for identifying trends or establishing best-fit curves. Thus, there is great similarity between the ultimate findings that appeared in Shelby’s dissertation and in the jointly authored article we published and in the report Niko and his peers wrote.

With respect to mathematical problems, the term *context* has been used to refer to three different issues (Roth 1996). First, from perspectives that consider all situations as text or recognize discourse as a fundamental aspect of knowing, *con*text is all additional text necessary to the familiarity with a mathematical problem, text, issue, discussion, and so on. Hearing or interpreting a text competently always already implies competently mastering a world of text such that our practical familiarity always envelops, accompanies, and concludes any text- or inscription-related interpretive effort, but without the interpretive effort, practical familiarity would not elaborated and explicated. This *con*text, however, is never completely explicit, for even intelligibly hearing or reading the story that embeds some mathematical issue requires competent literary practices. It often comes as implicit, practical familiarity with the working of the world and, as such, it goes without saying and therefore does not have to be spelled out. What goes without saying, however, is very much a function of an individual’s past experience; thus, the appropriate use of any text (e.g., instruction, information, plan) has to be elaborated and emerges as the result of contingent and situated inquiry in the textual world more generally.

Second, context may relate to some real-world phenomenon that can be modeled by a particular mathematical form (e.g., the notion of function). When students appropriate for their own use this mathematical form by engaging with the phenomenon, the latter can be considered that context which elaborates familiar use of the mathematical form. The traditional way of getting water from a well by means of a wheel and axle system with an attached rope and water bucket may serve as a physical context of a linear function in which the height of the bucket is coordinated with the number of cranks turned; and the growth of bacteria cultures may serve as a physical context for an exponential function. In both instances, changing participation (learning) arises from increasing familiarity with the phenomenon itself. No verbal description can purchase and serve as a shortcut.

A third way of using context is linked to the notions of setting and situation. Situations, characterized by social, physical, historical, spatial, and temporal aspects, constitute the contextualizing frames for ongoing work. The term *setting* is used to refer to the various physical sites of human activities, such as supermarket, psychological laboratory, kitchen, school, dairy factory, or scientific laboratory. In each of these arenas, different situations are possible that are distinguished by their associated practices (including mathematical practices). Thus, the work in a dairy factory includes preloading, loading, delivering, and keeping inventory. Arithmetic practices in the different situations are qualitatively different and yet peculiarly accurate. In their everyday settings, shoppers in supermarkets, dairy workers, child street vendors, street bookies, fish culturists, or electricians develop mathematical practices that are more successful than those that traditional school instruction can elicit or develop. These mathematical practices are aligned with many other practices that make for life in each setting, and this alignment is an important resource that leads to the robustness of everyday mathematics-related activity. When this alignment is not possible, mathematical performance drastically decreases because individuals can no longer draw on their familiar context as resource. Lack of familiarity, that is, referential isolation, can be used to explain why school-related mathematical practices are not transferred across the different arenas and settings wherein we find ourselves every day. Thus, students may become good at manipulate relatively unfamiliar symbols enough to do well on school tests. But these practices are relatively brittle, piecemeal, and useless in settings other than school.

References

Bourdieu, P. (1980). *Le sens pratique* [The logic of practice]. Paris: Les Éditions de Minuit.

Chin, C., & Chia, L.-G. (2004). Problem-based learning: Using students' questions to drive knowledge construction. *Science Education, 88*, 707–727.

Cobb, P., & Tzou, C. (2009). Supporting students' learning about data generation. In W.-M. Roth (Ed.), *Mathematical representation at the interface of body and culture* (pp. 135–170). Charlotte: Information Age Publishing.

Garfinkel, H. (2002). *Ethnomethodology's program: Working out Durkheim's aphorism*. Lanham, MD: Rowman & Littlefield.

Latour, B. (1993). *La clef de Berlin et d'autres leçons d'un amateur de sciences* [The key to Berlin and other lessons from a science lover]. Paris: Éditions de la Découverte.

Lave, J. (1977). Tailor-made experiments and evaluating the intellectual consequences of apprenticeship training. *Quarterly Newsletter of the Laboratory for Comparative Human Development, 1*, 1–3.

Lave, J. (1988). *Cognition in practice: Mind, mathematics and culture in everyday life*. Cambridge: Cambridge University Press.

Levinson, R. (2011). Science education from people for people: Taking a standpoint. *Studies in Science Education, 47*, 109–117.
Roth, W.-M. (1996). Where is the context in contextual word problems?: Mathematical practices and products in Grade 8 students' answers to story problems. *Cognition and Instruction, 14*, 487–527.
Roth, W.-M. (2008). Constructing community health and safety. *Municipal Engineer, 161*, 83–92.
Roth, W.-M. (2009). Limits to general expertise: A study of in- and out-of-field graph interpretation. In S. P. Weingarten & H. O. Penat (Eds.), *Cognitive psychology research developments* (pp. 1–38). Hauppauge: Nova Science.
Roth, W.-M., & Bowen, G. M. (1999). Digitizing lizards or the topology of vision in ecological fieldwork. *Social Studies of Science, 29*, 719–764.
Roth, W.-M., & Bowen, G. M. (2001). Professionals read graphs: A semiotic analysis. *Journal for Research in Mathematics Education, 32*, 159–194.
Roth, W.-M., & Bowen, G. M. (2003). When are graphs ten thousand words worth? An expert/expert study. *Cognition and Instruction, 21*, 429–473.
Roth, W.-M., & McGinn, M. K. (1997). Science in schools and everywhere else: What science educators should know about science and technology studies. *Studies in Science Education, 29*, 1–44.
Roth, W.-M., McGinn, M. K., & Bowen, G. M. (1996). Applications of science and technology studies: Effecting change in science education. *Science, Technology & Human Values, 21*, 454–484.
Roth, W.-M., McGinn, M. K., & Bowen, G. M. (1998). How prepared are preservice teachers to teach scientific inquiry? Levels of performance in scientific representation practices. *Journal of Science Teacher Education, 9*, 25–48.
Roth, W.-M., McGinn, M. K., Woszczyna, C., & Boutonné, S. (1999). Differential participation during science conversations: The interaction of focal artifacts, social configuration, and physical arrangements. *Journal of the Learning Sciences, 8*, 293–347.
Roth, W.-M., Pozzer-Ardenghi, L., & Han, J. (2005). *Critical graphicacy: Understanding visual representation practices in school science*. Dordrecht: Springer-Kluwer.
Tobin, K. (1990). Research on science laboratory activities: In pursuit of better questions and answers to improve learning. *School Science and Mathematics, 90*, 403–418.
Vygotskij, L. S. (2005). *Psychologija razvitija cheloveka* [Psychology of human development]. Moscow: Eksmo.
Wagenschein, M. (1999). *Verstehen lehren: Genetisch – Sokratisch – Exemplarisch* (5. Auflage) [Teaching understanding: Genetically, Socraticly, exemplarily, 5th ed.]. Weinheim: Beltz.

Part V
Epilogue

> Where it is impossible to poke around and dream, where the really interesting things remain in a box (in the watch, radio, television, receptacle) and simultaneously do not permit thinking of them mythically ("dreaming"), understanding is discouraged from the beginning. One is satisfied with words. (H. von Hentig, in Wagenschein 1999, p. 18)

In this book, uncertainty, graphs, and graphing in the discovery sciences are presented as an integral and irreducible part thereof. As shown in Chap. 8, a perspective of science as event*-in-the-making reframes the very ontology of subject, object, and tools, which themselves are phenomena in-the-making, the specific nature of which comes to be fixable and fixed only after the fact. In this epilogue, I extend my reflections on what we can learn in and from the discovery sciences for STEM education. A key point for me is what might be considered *authentic learning*, which fundamentally means immersion in situations of interest to the point that all detached consideration ceases to exist. What is characteristic of the sciences – as I experienced during my own training while completing a Masters degree in physics and as I am experiencing it as a learning scientist – is the absorption in the doing of science. This absorption is so deep that all separation of subject, object, and the deliberate actions and control of the former on and over the latter ceases to exist. In the course of such immersion, we may learn more STEM and more about STEM by engaging in one or a small number of exemplary investigations than by cursorily being confronted with a survey of topics, as this is characteristic of school science and mathematics in the way these are currently practiced. In the single chapter that constitutes this last part of the book, I reflect on Wagenschein's (1999) notion of the exemplary [exemplifying] method to the science and mathematics curriculum and on the notion of authentic learning. Here, I do not suggest that authentic learning in school science and classrooms attempts to copy what is happening in scientific laboratories but rather that in both we find forms of engagement that capture the agential subjects to the extent that they lose themselves in it the activity (doing science, doing schooling).

Reference

Wagenschein, M. (1999). *Verstehen lehren: Genetisch – Sokratisch – Exemplarisch* (5. Auflage) (Teaching understanding: Genetically, Socraticly, exemplarily) (5th ed.). Weinheim: Beltz.

Chapter 13
Discovery Science and Authentic Learning

> When trying to educate, in which sense can we take an object, a theme, or a problem within a curricular field as "exemplary and exemplifying" and what for? . . . The particular, into which one delves here, is not step or stage, it is mirror of the whole. (Wagenschein 1968, p. 12)

In Part IV of this book, I argue for a form of STEM education that allows students to engage in a small number of significant investigations that are really of interest to them and over which they have considerable control. In the introductory quotation of this chapter, the author – a physics and mathematics educator well-known within the German-speaking educational context – asks in which way one single object, phenomenon, or concept can be an example for becoming familiar with a whole that is exemplified in the particularity of the one. He emphasizes that the particular, such as one scientific experiment or one mathematical proof, is not important as something in itself or as a step or stage to something more advanced but because it constitutes a mirror of a whole, which is encountered in and through its particularization. Some readers may want to argue that if we engaged in investigating one phenomenon only, we would become hyper-specialists. But viewing the world through examples is the opposite of specialization: it is not intended to particularize but to see the whole of science and mathematics operating in and through the particular case, as if seeing the entire world reflected in a drop of water. Employing the rhetorical figure of hyperbole – again a mathematical expression – Wagenschein points out that a radically exemplary and exemplifying mathematics education could limit itself to the consideration of a single mathematical proof – such as that of the infinite continuation of the series of prime numbers – to learn a substantial amount, but not everything, that is characteristic of mathematics. All the while suggesting the hyperbolic nature of this example, and all the while rejecting it as a proposal for the organization of mathematics education, the author nevertheless states his conviction that in this one example alone, many students would become familiar with more mathematics than what they had tend to be at the

W.-M. Roth, *Uncertainty and Graphing in Discovery Work*,
DOI 10.1007/978-94-007-7009-6_13, © Springer Science+Business Media Dordrecht 2014

end of their academic high school training.[1] Wagenschein refers to other scholars, who, predating his own writings by 30 or 40 years, argued that (a) in the sole example of a particular nematode one could explain the essential aspects of biology or (b) with 5–10 selected animals, all essential phenomena, concepts, and laws of zoology could be learned and brought into relation. My example from the physics courses I had taught shows how in the course of the same investigation the student Niko became not only deeply familiar with the physical phenomenon of a damped oscillation of a cart suspended between two springs but also with the mathematical principles of integration and differentiation and with statistical procedures underlying the selection of a best-fitting curve (Chap. 12). Although I have never checked, I am convinced that there was probably sufficient evidence to show that his communication practices changed as his language competencies evolved in the course of talking himself through the experiment and then writing the report both for others as much as for himself.

Learning as Participation in a Discovery Science

In Part II of this book, I report one advanced research group in the discovery sciences and the experiment it conducted to find out about the changes in the visual pigment composition of coho salmon. Although the two professors involved – Craig and I – had designed the experiment and set up the relations with the participating fish hatcheries, it was a doctoral student (Shelby) who took charge of much of the data collection. This study turned out to be one of four studies that made up the body of his doctoral dissertation. Two more chapters of this body reported experiments on the visual composition of the same fish species, one concerning the dependency of the composition on hormone treatment and the other concerning changes in spectral absorption as a function of ontogenetic changes. The final experiment took a look at the visual pigment composition of another species, zebrafish, reporting evidence for a rhodopsin–porphyropsin exchange system similar to the one he had investigated in coho salmon. In all these experiments, the same equipment was used, which Craig had developed for the special purpose of investigating the photoreceptors in the retinal tissue of fish. As a result, much of what Shelby learned in and while developing as a research scientist – he went on to do postdoctoral work in several other countries – was in and through his participation in a research project that had slightly varying aspects. These research projects had as a constant the equipment that was used to identify

---

[1] Wagenschein refers to the *Abitur*, the final examination at the end of the academically streamed high school (i.e. *Gymnasium*, like the British grammar school) of the three-tiered German (Austrian, Swiss) educational system, where, at the time of his writing in the 1960s and 1970s, less than 15 % of the student population entered. (The others remained in the trades-oriented *Hauptschule,* which finished after eighth grade, or in the technically oriented *Mittelschule,* which finished after tenth grade.).

absorption of light in photoreceptors. In the course of Shelby's investigation, he also became deeply familiar with many of the mathematical aspects embodied in the software that Theo had written for the analysis.

The interview excerpts in Chap. 10 specifically show how Shelby learned while participating in *one* laboratory generally and in *one* project more specifically. His changing participation in a changing practice *was his learning*. His, as any graduate training, was premised on the idea that persons become scientists after having absolved the drudge and grunt work of prior schooling, including undergraduate studies at the college and university level (Traweek 1988). I knew Traweek had it about right because of two experiences. First, in my own training as a research physicist, the exciting part did not start until the moment when I began my thesis research; everything else before that I experienced as drudgery. Second, my high school physics students told me, when returning to the school after the first or second semester at the university, how backward and boring the undergraduate science laboratory sessions were when compared to what they had done in their physics courses. The real science begins with graduate school, and the underlying assumption is that by engaging in one real project during their doctoral work, individuals would come to sufficiently familiar with science for subsequently taking up postdoctoral research or faculty positions. In and through this one project, they learn (acquire) to write research articles and grant applications. The idea is not that the graduate students pursue studying the same organisms or using the same methods once they leave the laboratory in which they have trained – though many may do so. Rather, this one experience can be thought to be sufficiently exemplary and exemplifying that the (successful) individuals can be successful scientists in any nearby topic, using whatever method is required to achieve the goal.

In Shelby's case, during his postdoctoral work he continued to work in the general area of fish vision but changed the species he was researching and focusing on new dimensions such as the role of polarization, the differential sensitivity in the different parts of fish retina, and aspects of fish vision at the water–air interface. We may take Shelby's experience, as that of many others walking in the same shoes, as a start-off point to think about learning through engagement in a small number of projects that allow us to learn not just a narrow content domain but in fact ways of working, thinking, and coping with uncertainty in STEM-related fields. Specific facts are easily learned and read up on. But what is more difficult to develop and what elementary and high school students have the least opportunity in is the development of an orientation in and disposition to the different STEM fields, a scientific attitude, including an orientation to the stick-to-it-iveness that prolonged engagement requires. My point is not that all students should engage in STEM careers. Quite the opposite is the case. The project-based approach provides opportunities for those students not very interested in STEM to pursue something of their interest where STEM related practices *also* come in. As I know from my research experience, some students who initially may not begin with STEM concerns become interested through observing what their peers are doing and then orient themselves toward engaging in STEM oriented investigations or in using STEM principles to conduct their investigations.

In all the conversations we had with Shelby, getting a PhD diploma never was the topic of talk. At best, the video recordings and my personal experience while traveling with him suggest that he was rapt by the doing of science. Being rapt is another way of saying that we are caught up in the optimal experience of flow, from which we *emerge* as subject (subjectivities with identities) when we become conscious again of the object *as* object. Optimal experience and flow are probably the least attended to dimensions of human experience in the STEM-related learning sciences. In the following, I first articulate why, what, and how exemplary learning exemplifies and then elaborate on the issue of authentic science and flow.

Exemplary Learning Exemplifies

> On particular linguistic articulations one must not express oneself, as on other objects of learning, but from frequent, long-term, on-the-object-focused scientific exchange, it all of a sudden arises from the soul/mind like the light kindled by the jumping spark that sustains itself thereafter. (Plato, *The Seventh Letter*)

We have to become familiar with something before it reveals itself as such. Once we truly are familiar with the methods of the sciences or mathematics in one example and know how to get around in relevant contexts because of the relationship that binds the knowing subject, the object known, and the method of inquiry, then we can acquire, in principle, anything and everything else on our own (Wagenschein 1968).[2] This is so because nature always is there for every one of us – and science is not about what nature is but about how nature responds. STEM educators, and perhaps even more so the relevant policymakers and their advisers – who indeed have come through the courses taught in university education courses – frequently are concerned with the *coverage* of curriculum content. In the pertinent discussions, there appears to be little concern with the fact that most students will have forgotten what they nevertheless spilled onto paper during examination periods in their mathematics and science courses. Moreover, most students never develop a deeper familiarity with the nature of science, technology, engineering, or mathematics; and at the present time most scientists learn what real science is only during their graduate work while spending a lot of time in laboratories and interacting with other researchers (Traweek 1988).

A very different approach to science and mathematics education already exists in courses described in Chaps. 11 and 12, where eighth- and eleventh-/twelfth-grade students come to be familiar with mathematical principles while investigating natural phenomena. I had engaged in teaching in this way and supported other

[2] As a high school teacher and long before knowing about Wagenschein's writings, I used to say to my students: "Let's have some fun. And anything you need for the final, provincially determined exams we get you ready for during the last few weeks preceding those exams. At that time your understanding will be so deep that anything new can be hung onto it."

teachers in my department teaching in this manner long before I knew about M. Wagenschein science educator with a Ph.D in experimental physics who worked on problems in the didactic tradition of science and mathematics. To ask questions, we have to be familiar with a particular phenomenon and the experimental contexts. This became quite apparent to us in the context of the study described in Chap. 12 involving eighth-graders investigating ecological phenomena of their own interest. Although they began having difficulties articulating productive questions, they developed increasing numbers of and increasingly complex questions involving up to three dependent and three independent variables (Roth and Bowen 1993). Today I understand that the initial difficulties have their likely origin in their unfamiliarity with the plots that they investigated; and the more they were familiar, the more questions they had and investigated. Indeed, already during my first year of teaching and prior to having taken any courses in education or psychology, I asked my eighth- and ninth-grade science students to expand a particular investigation that their textbook described. Thus, rather than just doing one little investigation in how materials expand when heated, we expanded the investigation in many ways by asking questions such as "What if . . .?," "How can we test . . .?," "What do we have to do to . . .?" and then investigating whatever could be changed in the original investigation that the already inquiry-oriented textbook proposed.[3]

Wagenschein introduced the concept of *exemplary teaching* [exemplarisches Lehren]. The German adjective "exemplarisch" can be translated as "by example" and "being an example of. . . ." Thus, when students expand their participation in science by investigating a phenomenon to increasing depth, whether this phenomenon is from the natural world or a mathematical one, they become familiar not just with one topic but in fact they come to experience a way of participating in science, mathematics, engineering (design), and technology more generally. In and through investigating a specific phenomenon, students develop dispositions that go beyond the particular phenomenon because their engagement realizes the nature of the STEM domains more generally. This idea of the particular as a particular instance of the possible, and, thereby, as a manifestation of the whole can be found in a variety of fields as a form of analysis, including psychology, sociology, and epistemology (e.g. Bachelard 1950). Thus, for example, in the documentary method to the human sciences, each painting, each piece of art, or each novel is taken as a sign of the times, a concrete manifestation of the period, epoch, or Weltanschauung [worldview]. Although any one painting of a particular period – e.g. impressionism, expressionism, or cubism – differs from every other one, it also is a concrete manifestation of that period (Mannheim 2004). Each piece of art is a one-sided reflection of the same whole, the spirit of the times, the era. To take another example, each piece of art – e.g. a novel, a poem, a fable – is a concrete

[3] I was teaching in a small, completely isolated fishing village on the southern shores of the Labrador Peninsula. The school board had purchased kits of *Introductory Physical Sciences*, normally a one-year course, to be taught over the course of two years. The need to have enough material for a two-year course partially may have been the mother of invention. My inclination as an experimental physicist contributed to the ease with which I made adaptations to the curriculum.

manifestation of art in general; this, therefore, allows the analyst to extract a "psychology of art" from the analysis of a single piece of artwork (Vygotskij 2005). For the analyst the crucial issue is to treat the particular case as a (concrete) particular (instance) of the possible. This allows an identification and extraction of invariant properties, which would also be found by investigating any other case that therefore would constitute a possible testing ground for a generalization (Bourdieu 1992). A third example is from everyday life: Although we have never been taught the rules of making a queue, we do know how to participate in making queues – namely from the many concrete examples in our lived experience. Each queue, here, is a manifestation of the domain of possibilities for making a queue. There are invariants that can be found in every single queue even though we may not know these invariants as such. We learn to queue in the same way that we learn to speak in grammatically correct sentences without first learning formal grammar. Each (correct) sentence is a concrete example of grammar more generally. In the same way, I suggest we learn science by doing science. We do not have to prepare for it, just as we do not have to prepare – e.g. by studying books – to participate in making queues, and I am convinced and have experienced that we do not have to prepare students to engage in what are becoming fundamentally scientific ways of studying phenomena of their interest. In those few exemplary inquiries that we live through, we learn to do science even when we did not first study its grammar(s). I suspect that we can talk about (the grammars of) science once we are sufficiently familiar with it, just as we can talk about the grammar of our mother tongue once we already are quite familiar with speaking and writing it.

For Wagenschein, the problem is similar in the sense that in any investigation, whatever its particular, students may encounter aspects of a field – science, mathematics, engineering, or technology – that are invariant and characterize the field *as such*. The point of teaching, however, is not merely to expose students to examples but to draw the invariants from them. Just as we cannot learn grammar without already speaking our first or mother tongue, so we cannot extract general principles of mathematics and the mathematical field unless we have become sufficiently familiar to considerable depths so that the invariants typical of the mathematical can be made salient.

Wagenschein (1968) extensively discusses the example of the question, "Don't you find it curious [odd] that the radius can be marked off on the periphery exactly six times?" (p. 110). He explicitly questions answers that come forth too rapidly on the part of schooled individuals and points out that most university students in STEM fields do not have a good familiarity with what they are talking about. Showing how the problem reduces to an investigation in which three equilateral triangles are put together (Fig. 13.1a), and how two of these configurations yield the solution, Wagenschein then progresses to exemplify how this same approach provides the solution to the Thales problem, the fact that any triangle shaped by the diameter of a circle and a point on the semi-circle above it is a rectangle (Fig. 13.1b). In this example, children learn not only about generating solutions by means of translation of a geometrical object, but also aspects of mathematical doing, arguing, proving, and talking in general. Thus, a proof of the Pythagorean

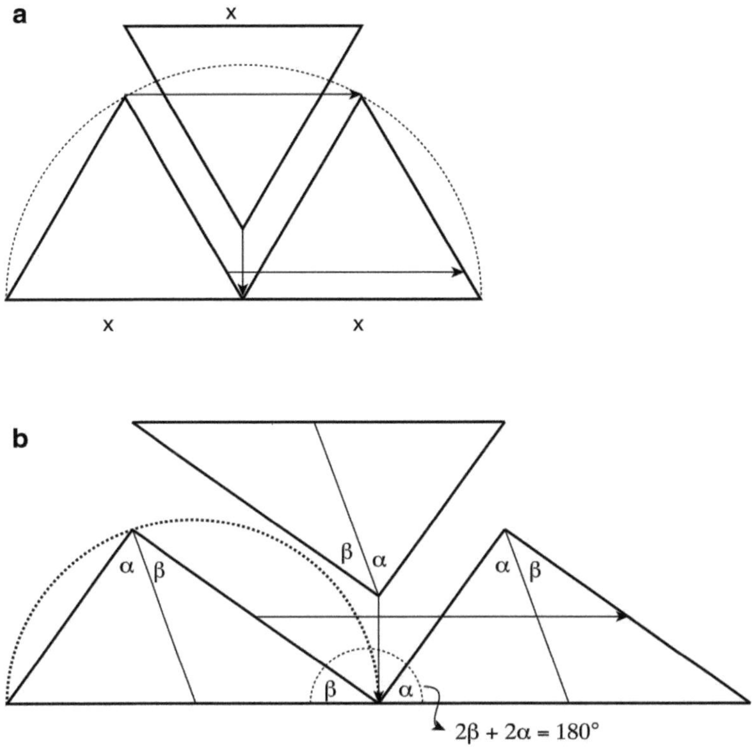

Fig. 13.1 (**a**) If an *equilateral triangle* is moved horizontally by exactly the length of its base x, then a copy of the same *triangle* can be pushed from above into the opening. (**b**) The same idea allows proving that the *triangle* defined by the diameter of a *circle* and a point on the periphery must be *rectangular* (i.e., $\alpha + \beta = 90°$) (© Wolff-Michael Roth, used with permission)

theorem – the square over the hypotenuse of a rectangular triangle is equal in size to the sum of the squares over the two shorter sides – can be derived from the same fundamental thought (Fig. 13.2). To get this doing, arguing, proving, and talking to have any substance, however, the children have to become deeply familiar with the phenomenon, explore it from many angles until they are sufficiently familiar with the domain to extract from the lived experience whatever is properly formal mathematical – just as they articulate grammatical rules only when they already are familiar with a language that is to be grammaticized and a language that they can use to do the grammaticalization.

In this situation, therefore, the particular is a concrete example of something more general, something that lies in the realm of the possible. It is therefore an exemplar that exemplifies something in and about mathematics. But unlike in everyday situations (e.g. queuing), where explicit rules tend to be unnecessary, the purpose of school mathematics (and science) would be to make both the rules explicit and relate these rules to the nature of mathematics (science).

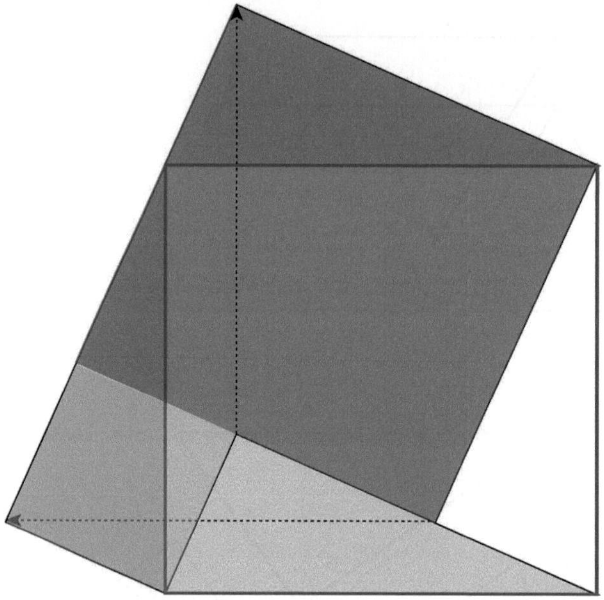

Fig. 13.2 A proof of the *Pythagorean triangle* based on the idea of translation. The *grey* (*lower*) *rectangular triangle* is moved upward by the length of its base yielding the big (*red*) *square*. Its copy, the *white triangle* to the right is translated to the left until it "sits" on the opposite side of the big *square*. (The *white* is identical to the *grey square* because of the same base length and the same two angles). Two new squares are created, one that has a length of one of the longer of the two remaining sides (*the top, reddish one*) and the other the length of the shortest side (*left, blue*). The viewer understands the proof precisely at the point that s/he sees it in this diagram (© Wolff-Michael Roth, used with permission)

The exemplary | exemplifying method is not built on the idea of a step-wise curriculum (many readers will recognize my squint towards the effort to establish *scope and sequence*), where some so-called more elementary matters and skills are taught prior to more advanced matters and skills. "Back to the basics," so resounds the rallying cry of those who think of knowing in bit-sized pieces of stuff kept in dark places of our long-term memory. The exemplary | exemplifying method is completely different, as it begins with some topic, which becomes the starting point for further inquiry into more basic, fundamental, or elementary phenomena and forms of thought and, simultaneously, to more complex and encompassing phenomena and forms of thought. As learners, we do not just accumulate stuff, but we look for what we need, in *this* situation and for *this* purpose to do *that* thing in particular. It is in this way that we are challenged by the singular and peculiar; and, in turn, we demand from it to reveal to us the more elementary and what only ultimately turns out to be the more simple.

Central to the pursuit of the exemplary | exemplifying method is this: the curriculum is not just something cognitive but something we live [*Erlebnis*] and live through [*Erfahrung*] mind, body, and all. A truly educating curriculum, in any individual STEM field, also is characterized by the fact that something is happening to us that we have not or could have anticipated. There is therefore also uncertainty in our changing participation, which is not entirely of our own making, and where we are confronted with the unknown. This is true for both students – who, with hindsight, will come to recognize that they have increased their degree of participation in some aspects of mathematics or science by immersing themselves – and teachers – who, with hindsight, come to describe that it is in and through teaching that they have become better teachers. That is, the exemplary | exemplifying curriculum reflects the totality of learners, the world and its possibilities seen through their eyes of the participants rather than through the eyes of the ideal knower, who frequently only sees what the child cannot do within his own world.[4] In such a more symmetric approach to the STEM curriculum, institutionally designated students and teachers alike are recognized as participants in societal relations that afford the learning of the respective Other (Roth and Radford 2010). This allows us to recognize that designated teachers learn and designated students teach.

Authentic Inquiry and Flow

Inquiry is authentic when it captures the person, who becomes subject at the same time as its object becomes object (Hegel 1807/1979). Inquiry is authentic when it transports us, when *experience* is equivalent to saying that something is happening to us, when we are not merely agents but patients in, to, and of *experience* (Roth and Jornet 2014). Authentic inquiry is another way of saying that STEM engages students in optimal experience. Investigating STEM education seldom provides us (learners) with an opportunity to study sustained engagement with a topic, in which we (learners) learn about ourselves as much as about some topic specifically and about STEM more generally. There also tends to be a strange relation between the theories that STEM education researchers adopt and employ and their own experiences, which do not quite fit into the theoretical models that are espoused and taught. We also tend to jump quickly to explicating gestures when asked to think about our own learning. These explications tend to obscure and hide what our lived experience has been at the instant when we did not (yet) quite know what we ultimately would know, immersed as we were in uncertainty. But there are occasions when a trace of an inquiry experience is left while engaging in some endeavor; and this trace reveals a lot about what engagement means, how we learn, how we are carried away when we are carried away, and so forth. From the continuous

[4] This is the case in Piaget's writings generally, which focuses on what children *cannot* do during the different stages in their lives, as measured against the telos, the scientific rationality that Piaget posits as the highest form of thought.

stream of experience has emerged *an* experience (Dewey 1938/2008). I have had many opportunities to observe such phenomena in my own life (e.g. Roth 2012d) and offer the following to exemplify the point that engagement in a STEM related inquiry may be transformative. The events recounted below not only changed my understanding of a physical phenomenon but also the ways in which I theorized knowing and learning in the sciences (e.g. Roth 2006b).

In 1999, during a stay in the neurosciences and cognitive sciences group of the Hanse-Wissenschaftskolleg (Institute of Advanced Studies), I analyzed a set of videotapes recorded during a 20-h curriculum on electricity in a tenth-grade physics course. Repeatedly, I found myself caught up doing the same investigations as the students whom I was observing, finding out that I was doing or seeing what the students were, even though it (sometimes) was not consistent with the physics that the instructors were teaching. I found myself in this puzzled situation even though I had successfully completed a Masters of Science degree in physics.[5] The experience of investigating the phenomena at the heart of the students' investigations, and my reflections, fundamentally changed my career as a learning scientist – both with respect to the topics I researched and to the method of investigation. From the experience had arisen *an* experience. In the following I reproduce the notes that I wrote in the course of one such investigation, which had captured me to such an extent that it took an entire night. The notes concern the investigation and reflections that I wrote. The notes speak to the experience of being captured, the almost frantic search for getting at the bottom of these issues, and the exemplary and exemplifying nature of an inquiry that has truly become my own. I have observed students of all ages so engrossed in their science inquiries that they forgot about time in the way I was in this situation.

In the following excerpts, taken in an unabridged way from the notes I kept during that night, show many features that one might find described in the literature on scientific inquiry, such as the description of one night's work in an observatory, where the astronomers "discover" the first optical pulsar (Garfinkel et al. 1981). We see the same concerns for repeatability at work, an orientation towards other science educators (Stefan von Aufschnaiter, Manuela Welzel, and their graduate students) who are to be confronted with the findings of my inquiry for the purpose of rethinking the teaching of physics and current theories of learning. We see in these notes evidence of how an object and topic grabs the inquiring subject (me), who, only in and through this exchange becomes a subject, sees himself through the eyes of others. The notes also exhibit evidence (a) for my emerging suspicion that (science and mathematics) educators do not give enough credit to their students, who act rationally in the worlds accessible to and accessed by them and (b) for my suspicion that we need to be thoroughly familiar with these worlds to be able to explain why, from the students' perspective, their doings and experiencing are rational rather than "wrong," "misconceived," and "contradictory." I frequently found myself doing what the students were doing in the videotapes I was analyzing,

[5] It had resulted in my first ever publication.

but which the science education colleagues who I was working with at the time where construing in terms of deficiency theories (e.g., students are wrong, incapable to see, poor students, or not so interested). In these notes, we see evidence of something that kept the writer (me) sufficiently interested to stay up an entire night, finding out about the passing of time only afterwards, after re-emerging from what has been "just process of inquiry." It is evident that this example, as the exemplary I exemplifying method more generally, is incompatible with the normal school organization that chops the school day and week into bit(e)-sized, temporally limited portions of 45, 50, or 60 min each.

The events concern a relatively simple investigation and its unavailable explanation. Even young students may sometimes encounter it. It involves an electroscope, the core of which is a jointed metal arrangement including a freely moving pointer. When electrical charges are placed on or brought near the arrangement, the pointer is deflected. Charges can be created by rubbing an overhead transparency on a piece of clothing made from cotton, wool, or silk. The presence of charges can be verified by means of a lamp, which, when brought into contact with the charged material, will glow. The notes also provide evidence for the role of inscriptions, some of which are depictions of an actual device and configurations of materials during the inquiry, others constitute moves toward abstraction and explications in terms of particle models. In addition, there are graphs next to physical situation that still permit to see where the chain of reference (Latour 1993) is built in the course of inquiry, before the apparent breaks are instituted that make the graphs appear to stand on their own. Finally, we observe equations in the final push toward abstract representations, or, conversely, a move to connect abstract notations to the concrete inquiry at hand.

The following notes are quite special in the sense that they are evidence of research in progress, research of a type that school students can do. These notes depict what was happening while it was happening, without any benefit of hindsight. This happening is authentic inquiry, authentic because it captured its subject.

Notes from One Night's Inquiry

June 11, 1999

A sleepless night. I am conducting electroscope experiments, pursuing the idea that the particle model might be inappropriate. The transparency is negatively charged, bulb glows away from hand. (a) The experiment also works when the transparency film is brought to the side of the electroscope. At that place, the charges should be sideward shifted (if the particle model is any good). (b) I can discharged it on the side of the transparency and still get the negative discharge, wherever I do it on the scope, near end to the film or far end. (c) I can go from side to needle and from the narrow end and still get the effec't.

It appears thus that unless the positive charges posited to be outside of the body itself, one cannot explain this. (a) If they were outside, then the electroscope was negative all over and I would have access only to the negative charges wherever I measure. (b) If they are not outside, the symmetry of the experiment would have positive charges on the needle.

The effect even occurs when the charges are seemingly pulled into the needle. When, for example, the film is brought near the upper part of the needle, but from the side so that one would expect the positive charges to be on the outside of this part of the needle. If it was charged positive and the negative charges were equally distributed throughout the electroscope as they should be (one can ground the system anywhere connected to the central suspension, and still get charging by means of induction).

Here, there should be positive charges on the upper left and bottom right of the needle, negative on the remainder of the electroscope, and we should get an attraction. (Perhaps the attraction from the film is stronger than that between the needle and the remainder of the scope, one could argue.) I can even ground between the needle and the film to discharge, always get negative, and get the induction.

So until now I have tested:

- Induction from top, discharging everywhere, even between film and plate. (metal plate, film through pants [seems to give me more consistently charges than metal on film])
- Induction from side (plate to get symmetry, film) charging from bottom left, bottom needle; top right, top left. Discharge between needle and film.

5–10 repetition for each. Making sure nothing has been touched, that the effect is repeatable. Testing charges on film, on scope.

I am aware that I want to have it consistent so that I can show it to Stefan and other physicists. It has to be a consistent phenomenon and hold up to potential criticism.

Actually it is good that it is night. I can see the slightest light flickering in the bulb. Sometimes even when the needle does not show anything, I can see light glowing weakly. I can see sparks from my pants to the foil or vice versa when I pull the transparency between my knees. Chills run down my back. It is like real discovery work. I am faintly aware of self, there seems to be a dynamic that has nothing to do with person, it is all process. There is no "I" no person who proudly would tell an achievement, not author, not Lansdowne Scholar. There is just process of inquiry.

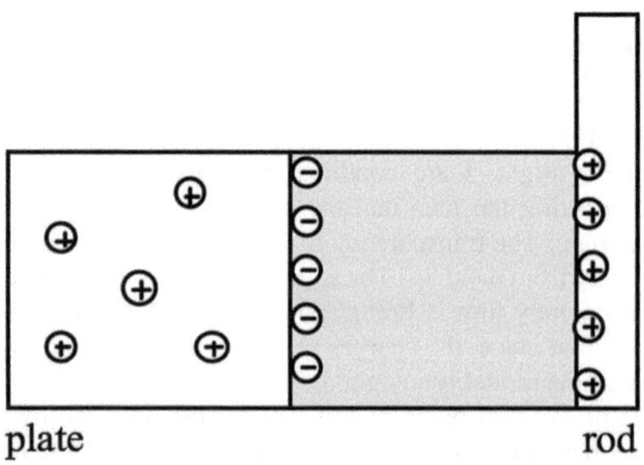

To me, it looks more and more as if the particle model (the part that makes the charges move around) actually does not explain the phenomenon. (a) One possibility is that the positive charges are thought to be somewhere outside of the scope, on the boundary but no longer effecting the scope with its charges, so that the metal is negatively charged throughout.

[[[[I am not sure such an explanation would fly. One would have to draw recourse to mereology and explain how it came that the positive charges moved outside, became part of another whole, now inaccessible to being shown. I tried something else earlier in the night, with another area, the air column which could be identified in terms of one of the surfaces as in my diagrams of the charging by induction and separation drawn yesterday. See diagram. But the direction of the charges does not make sense.

When I think about it now, why should the electron holes have a homogeneous density in the material, whereas there is such a high concentration on the boundary? I guess one could say that in this way, the material is "shielded" and there can be equi-distribution of charges. There can be no field inside the material either. So the electrons to be shielding would have to sit outside, really on a boundary inaccessible to the material on the inside. This would then explain the situation. But in a physical metaphor, it is difficult to accept that one could not get at the electrons on the surface. They would have to be above the surface. Thus, even if I wanted to do a test, the electrons would be said to float on the surface. Good. But the argument would not hold for the lack of electrons, the holes to be outside the material. Or on its surface?

I remember something. There cannot be electrical fields inside a material. Thus, on the surface we have to have an arrangement such that the field inside can be zero. Mereologically,[6] it is as if approaching the rod now brings in an area to be considered in the equation. We cannot say there is nothing, but there is a field and air column. The system encompasses all three, the plate, the rod, and the column. Now, the charges on the surface become part of the middle part, the column though they are on the boundary. But as Smith (1997) writes, boundaries are funny beasts, there can be two shortest (coincident) lines between two points, they are at the boundary of categories. So now the electrons in my figure belong to the middle part, the field, as do the protons (electron holes) on the right. So this means, the holes have to be actually on the surface of the material and would rearrange if I brought a testing medium in between the film and the surface of the plate.

In effect, this then is consistent with the view that the potential has been moved up in the metal as described [earlier] below. The two models are consistent again.]]]]]

(b) Another alternative I thought up is if one used a potential energy model. Charging film changes the potential of the scope. In the present case, the potential goes down such that electrons are flowing off the electroscope when grounded.

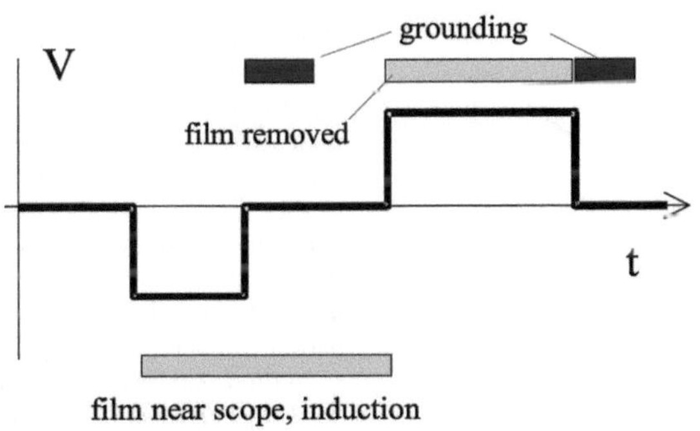

[6] Note added at the time of this writing: Mereology denotes the formal study of the relation between wholes and their parts. Mereology belongs to philosophy and mathematical logic. It finds its study and applications in logic, ontology, linguistics, engineering, computer science, and artificial intelligence. Part–whole relations are important in cultural-historical activity theory as well, where the minimum unit of analysis *manifests* itself in different ways, each of which is taken, by non-dialectically thinking, as "part" or "element."

When the film is taken away, the potential moves back up and now will be above zero. In this case, we would observe "positive" charge on the scope after the film has been removed. The diagram shows this. During the first grounding, the charges observed would be the same as on the film, negative (away from body), and the opposite during the second grounding.

It is 5 am. I find that I have spent the last three hours thinking through this problem rather than sleeping. I have not seen the time gone by and, if I had to say how much time I spent, would have given it at best 30 minutes. I have a sense of exhilaration. I want to talk to a physicist or several to talk more about the particle model and why it seems inappropriate. And ask them what they think about the alternatives. I want to know more about whether someone has already come up with an explanation, I thought that the old timers like Faraday who have been so consistent in their attention to the detail would certainly have attended to something like this.

After writing this, I had gone back to write the stuff in the square parenthesis. 45 minutes have passed, and again I have not seen them passing. I am down a bit because I seem to have rediscovered what someone already explained. But why would Manuela, Stefan, and André not have known about this? Why did she tell me that the particle model broke down at this point? And why would they have tried and not come to a similar conclusion as I did?

I remember now that there was something in my second year intro physics course about electrons arranging themselves such that, for example, around openings in a conductor there would be no fields inside. This is how the professor explained shielding. (Perhaps it was electrodynamics course – likely not.)

Anatomy of discovery work: A phenomenological inquiry

My own "discoveries" (compare to students' discoveries)

When I look at what students have done in their discovery work and what I have done one might say that I was more methodical. But then I also had more experience, and at least as many attempts at doing it. I had the night in my favor, a longer period of thinking about it, had seen them do experiments with the material. I was also had a different motivation, seeking an explanation for this phenomenon. This is already the second long day/night session spent on this very problem.

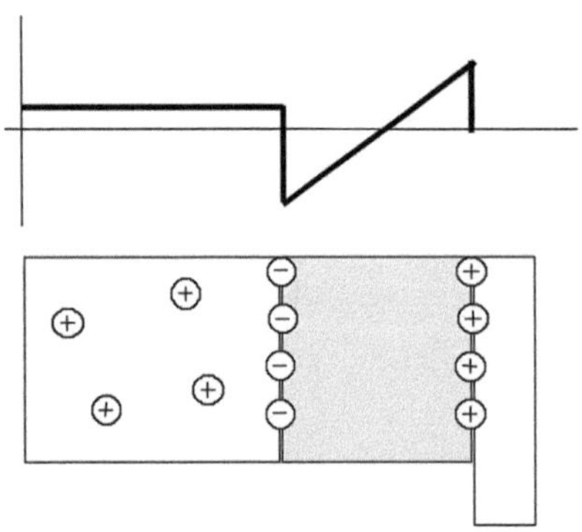

I was more methodical now because I had a specific thing I wanted to check. I already knew about a lot, and had a lot of experience with the charging, electroscope, etc. I now pursued more systematically the issue with the non-accessibility of the charges on the surface and the explanation of the effect. I was also tinkering with two alternative models, mereology and the electricians model that made immediate sense to me, and where I did not have to invoke the particle model.

6:06 I am making some coffee. I am thinking that even if there is no field, there is still a high concentration of the charges on the surface. But when the surface belongs to the metal then we should have some equilibration. But it could be such that, in terms of the potential, the charges "on the surface" already belong to the field on the outside. That is, we would get the following distribution of the potential: Diagram 1.

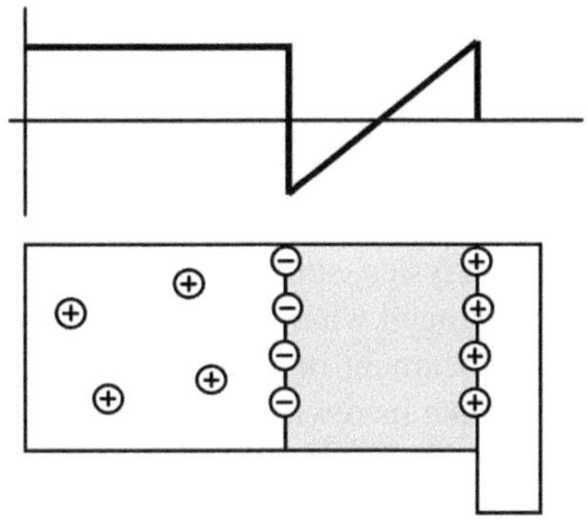

As I am making the drawing, I want to have the two triangles symmetric and measure them out, that is, I make the same height. I wonder how the height or area on the left hand side has to compare to the other areas and/or heights. Because of the number of charges according to the atomic model on the surface of the rod and the plate, I expect the potential to be the same. It's probably the same but opposite, for otherwise they do not add to zero on the surface and there would therefore be a field. Also, the number of charges is equal on the surface to the opposite in the material. I wonder how the charges are related to voltage and potential. I try dimensional analysis but don't get far because I don't remember them Q [As] = I/t. VI = J/s [W]. Actually, V = [J]/[As]Q; V = E/Q. This is the energy of accelerated charge relation.

I am thinking that perhaps because the charges are distributed in the plate, we can't have a charge equilibrium on the boundary. So it could well be that the potential has to be the same leading to Diagram 2.

Mereologically it would make sense then that the charges on the surface are no longer part, because they correspond to the second potential and to the yellow area.

It is 6:50. I am just not noticing the time passing.

Reflections on Authentic Learning

In Part II of this book, and especially in Chaps. 2 and 3, we see scientists fully immersed in what they do. There are no signs of metacognitive activity, an often-postulated characteristic of good learning. There is an unfolding in the happening that transcends the individual researcher, giving the observer the impression of a frantic happening oriented towards getting the activity back on track by restoring the equipment that currently is not working. We also notice the openness toward the future, which forces scientists to act prior to knowing the effects of their actions and prior to being able to explain what they will eventually come to know. Scientists do not know where they end up. This is in contrast to much of school science and mathematics where the end points of students' involvement is predetermined by the curriculum in terms of the objectives that they are to obtain – or fail to obtain with repercussions on their grades. The preceding notes, which capture one night's worth of work on a particular problem that I had chosen for myself: describing and explaining what students are doing and articulating possible reasons for the trajectories of their changing participation. In fact, the "problem" never was *chosen* as such, but I found myself inquiring something that held my attention for an entire night. Trying to describe and explain what students were doing dispositioned me to do what they had been asked to do by their teacher. This put me into a situation where my very training as a physicist came to be challenged. More than anything, we might read these notes as documenting a struggle towards getting a greater foothold or grip on something in a particular area of inquiry. Much as Wagenschein (1968) noted, most of us who do Masters or PhD degrees in a science, even the most successful ones among us, tend to acquire routine ways of responding in many areas of their discipline without having the time or interest to develop deep familiarity in and with every single one.[7] During that night, I became aware that even though I had a Masters degree in physics and even though I had taught physics and had spent years studying how students learn physics, some of the most basic and first experiments of any physics students challenged what I knew.

The preceding notes constitute an interesting document of science-in-the-making, unfiltered by after-the-fact explications that make us rewrite the history of our inquiries subsequent to or at the end of an inquiry. It is not "big" science in the context of the inquiries that make the headlines today (e.g. the work at CERN in search of the Higgs boson); but it is "big" science in the sense of its capacity to capture the attention even of someone who already has an advance degree in science. As I show in Part II, Shelby's account change in the way his own research appeared to him. In the context of science and mathematics education, this may be considered authentic in the sense that the investigation truly captures and engages the whole person rather than that it has anything to do with the content and procedures of the science conducted in a scientific research laboratory. But there are some essential similarities concerning the involvement that have little to do with detached and disinterested pursuit of knowledge. I remember the year during which I did the research for my Masters thesis, which was essentially characterized by such forms of involvement that also feature in the preceding account of one night's work. Being captured, delving in the midst of a moving torrent rather than doing the routine work on an assembly line are much better ways of describing what was happening to me – on both occasions, the thesis and the preceding investigation of static electricity.[8]

I saw the same kind of involvement again while studying fourth- and fifth-grade students in a design engineering curriculum, where students, after having worked on their projects for an entire morning (including their recess) and after having been called to order, expressed astonishment about the fact that the science lesson already had ended: "Oh, we are already done with science for today?" It is that same kind of passionate involvement in the identification of problems and search for answers that characterizes what Shelby was telling those who interviewed him about our scientific research project. We can observe traces of such involvement in his after-the-fact account of the work surrounding what turned out to be a methodological change that had significant impact on the data collected. Notably, Shelby noted that he was not thinking about the impact; from his a posteriori perspective,

---

[7] At the end of my Masters degree work, I "complained" to a physics professor that the studies had provided us with nothing more than a set of tools that can be blindly applied without worrying about deeper connections, understandings, and philosophical implications. He responded that this is all that a university program in physics is to do: transmit a set of knowledge and skills.

[8] With the right disposition, even assembly line work can lead to the flow type experiences discussed here (e.g. Csikszentmihalyi 1990).

he already assumed that he could have deliberated the impact of the work when in fact there was nothing he had to go by to assume that there would be a difference when sampling photoreceptor cells a little out of the dorsal area of the retina where Sam and he had come to sample. This method of sampling specific areas of the retina had emerged only in the course of this project, as our team in its previous studies and publications had not differentiated the locations from where it taken sample cells.[9]

> There was a problem with our um methodology over the summer-time when Sam started with me we started dissecting differently and that was, it had some serious implications for the results unfortunately, ended up . . . well I can show you a graph of what happened but where we took the retina from yeah and so we were starting to take a strip um because at the same time we were doing um where is it here (*looking for graph*). Yea, so instead of just getting the dorsal [part of the retina] and then taking a small corner of the dorsal we started to take a strip because we were hoping to get more cones. And that was particularly applicable to the stuff that we were doing with the T3 T4 [hormone] treatment because we wanted to record uh the change in porphyropsin rhodopsin in rods and in cones. And so by changing the methodology we didn't really, we didn't think that it would, well *we didn't really think about it I guess enough to think that it would have much of an impact* and because we don't do our analysis until a lot later we didn't find out until too late. And it's really a shame because it meant that I lost a whole segment of my analysis of my data set which is really crappy. So basically what happened was, so we were originally taking the eye and cutting it in half and then cutting this in half, this dorsal section, and taking this little corner and that's all we'd take. But in order to get more cones we started taking a strip all the way across here, which took us out of just the dorsal, it took us into the central area of the eye. And so what that did, it caused a change cause there's variation across the retina in the percent porphyropsin. And yea, I didn't really suspect that there would be that much of a change across this diff, across that difference, but what I ended up doing is there's the data set there (*shows data*) so what we have is this summer's data, these four data points are sitting much higher than they should've been. So the regular trajectory would've been to come down here increase in the wintertime and then decrease again towards the summer, but because of this increase porphyropsin um from the sampling technique it's caused an increase. So if we look at, there it's spread out more you can see more clearly there, that's much higher than we've had any time before in Robertson Creek and I don't think that's environmentally significant I think it's methodo- methodologically.

If there is anything I would want students to experience in STEM education classes, then it is the exhilaration that comes from completely immersing oneself into an inquiry to the point that only flow is left, where the students realize only later that time has passed without their being aware of it. These are experiences where issues of identity completely disappear, because the subject | object distinction itself disappears (e.g. Roth 2012d). The is, experiences such as those that leave traces such as the preceding notes challenge any notion of identity. In fact, there are explicit signs that there is no awareness of situation or individuality. For instance, in my description we find this statement: "I am faintly aware of self, there seems to be

[9] Many years later, Shelby would publish an article in which he reviewed the literature and discussed why different regions of the retina have different spectral sensitivities, that is, were characterized by different porphyropsin (A_2) to rhodopsin (A_1) ratios. In the quotation Shelby tells the first emergence of a problematic that would subsequently turn into another investigation.

a dynamic that has nothing to do with person, it is all process. There is no 'I,' no person who proudly would tell an achievement, not author, not Lansdowne Scholar. There is just process of inquiry." It is precisely such absorption that questions constructivist conceptions of engagement, the separation of the investigating subject and investigated object, and the relation of the two by means of (detached) (mental, practical) actions that operate upon the object or phenomenon. This is so because re-presentations or constructions stand between us and our experience, mediating it, whereas the evidence I provide here point to the unmediated nature of experience of engagement in science. It is this form of engagement that deserves the adjective *authentic*.

In an experience such as that implicit in the preceding subsection, as in the inquiries of the scientists described in Part II of this book, what the phenomenon is often remains uncertain until it reveals itself, after the fact so to speak and with hindsight. It has been there all along, we might subsequently say, we had just failed to see it. A phenomenon is a phenomenon only when it shows up consistently, when we can make it appear doing something rather than something else. Whereas science and mathematics students tend to be given tasks so that they discover ("construct") some phenomenon in the way it is handled in the sciences and mathematics, describing and perhaps explaining the conditions under which the phenomenon is not seen is even more important to practitioners. Thus, for example, Craig and Theo talked about the need to work with fresh minimum essential medium (MEM) of the correct concentration. If one or the other of these two parameters is not given, then the phenomenon – light absorption in photoreceptors – might be lost. This aspect of science – how to lose a phenomenon – is even more important to the familiarity with the nature of science than actually getting it on the first try (Garfinkel 2002). It is not surprising to find, therefore, that both my ethnographic account of a discovery science and my research notebook entry quoted above contain evidence of efforts to assure the phenomenon can be obtained in a consistent and repeatable way. Sometimes variation of conditions is intentionally produced in the attempt to see whether the phenomenon disappears; sometimes scientists stumble across a situation where their phenomenon disappears, requiring them to engage in a search to make it return. When this disappearance happens only once, scientists may not pursue searching for the reasons of this non-appearance of the phenomenon – as the missing images in Chap. 2. But when something consistent interferes with the phenomenon, such as the troublesome "ringing" that oftentimes interfered with the absorption curves, then scientists strive to find the reasons for the trouble for the purpose of eliminating the source of the disturbance – e.g. by purchasing a different light source in the hope that the problem had been arising from the original one.

Authentic inquiry "grabs" the individual to the point that there is no more separation of activity into subject and object. In fact, in the perspective of the event*-in-the-making (Chap. 9), subject and object are subject*-in-the-making and object*-in-the-making as much as the event*. The subject then is dealt with as integral and irreducible part of the happening; the subject is captured and carried away by a happening* that is not yet defined or definable as a specific event. Being captured and the subsequent separation of subject and object also is a topic for Wagenschein (1968) in the pursuit of mathematics and science as part of an education [*Bildung*] rather than schooling.

> If we understand the educational process as implying a comprehending/captured comprehension [*ergriffenes Ergreifen*], leading to a separation/exchange [*Auseinander-Setzung*] of/between subject and object . . . then we no longer [in physics] have a purely physical curriculum, and, in fact, it cannot only be in this way, if it is to educate. (p. 20)

A truly educational STEM curriculum – educational because it engages students in a way so that they participate in a participating way – also leads to the constitutive formation of subject and object. The German *ergriffenes* not only has the sense of comprehended but also the sense of being captured and being moved. There is only flow and the subject emerges together with its object. Wagenschein emphasizes that a subject-matter-focused curriculum only schools but does not educate students. It is in the process of exchange that subject and object come to be ex-posed, posited one over and against the other. This exposition and exchange involves the whole person, and the whole person is not moved by a decontextualized and decontextualizing science, mathematics, engineering, or technology curriculum.

References

Bachelard, G. (1950). *La dialectique de la durée* [The dialectic of duration]. Paris: Presses Universitaires de France.

Bourdieu, P. (1992). The practice of reflexive sociology (The Paris workshop). In P. Bourdieu & L. J. D. Wacquant (Eds.), *An invitation to reflexive sociology* (pp. 216–260). Chicago: University of Chicago Press.

Csikszentmihalyi, M. (1990). *Flow: The psychology of optimal experience*. New York: Harper & Row.

Dewey, J. (2008). *The later works vol. 13: Experience and education*. In J.-A. Boydston (Ed.), Carbondale: Southern Illinois University Press. (First published in 1938)

Garfinkel, H. (2002). *Ethnomethodology's program: Working out Durkheim's aphorism*. Lanham: Rowman & Littlefield.

Garfinkel, H., Lynch, M., & Livingston, E. (1981). The work of a discovering science construed with materials from the optically discovered pulsar. *Philosophy of the Social Sciences, 11*, 131–158.

Hegel, G. W. F. (1979). *Phänomenologie des Geistes* (Phenomenology of spirit). Frankfurt/M: Suhrkamp. (First published in 1807).

Latour, B. (1993). *La clef de Berlin et d'autres leçons d'un amateur de sciences* (The key to Berlin and other lessons from a science lover). Paris: Éditions de la Découverte.

Mannheim, K. (2004). Beiträge zur Theorie der Weltanschauungs-Interpretation (Contributions to the theory of worldview interpretation). In J. Strübing & B. Schnettler (Eds.), *Methodologie interpretativer Sozialforschung: Klassische Grundlagentexte* (pp. 103–153). Konstanz: UVK.

Roth, W.-M. (2006). *Learning science: A singular plural perspective*. Rotterdam: Sense Publishers.

Roth, W.-M. (2012). *First person methods: Towards an empirical phenomenology of experience*. Rotterdam: Sense Publishers.

Roth, W.-M., & Bowen, G. M. (1993). An investigation of problem solving in the context of a grade 8 open-inquiry science program. *Journal of the Learning Sciences, 3*, 165–204.

Roth, W.-M., & Jornet, A. G. (2013). Situated cognition. *WIREs Cognitive Science, 4*, 463–478.

Roth, W.-M., & Jornet, A. (2014). Towards a theory of experience. *Science Education, 98*, 106–126.

Roth, W.-M., & Radford, L. (2010). Re/thinking the zone of proximal development (symmetrically). *Mind, Culture, and Activity, 17*, 299–307.

Smith, B. (1997). Boundaries: An essay in mereotopology. In L. Hahn (Ed.), *The philosophy of Roderick Chisholm* (pp. 534–561). La Salle: Open Court.

Traweek, S. (1988). *Beamtimes and lifetimes: The world of high energy physicists*. Cambridge, MA: MIT Press.

Vygotskij, L. S. (2005). *Psychologija razvitija cheloveka* (Pyschology of human development). Moscow: Eksmo.

Wagenschein, M. (1968). *Verstehen lehren: Genetisch – Sokratisch – Exemplarisch* (Teaching understanding: Genetically, Socraticly, exemplarily). Weinheim: Beltz.

Appendix

Transcription Conventions

For the transcriptions, I follow a commonly used system based on conversation analysis adapted for the inclusion of prosodic features. In the rules implemented here, everything is written in small letters and sound words that run into each other are transcribed that way unless the run-in sign "=" is used when it would be difficult to distinguish pronunciation (e.g., "a = one"). The transcription is phonetic such that if a participant pronounces the words "this" or "that" in the way a French or German speaker often does, that is, with a soft "d" or "s," the transcription will read something like "ze other one is to read dis ze whole branch."

Notation	*Description*	*Example*
`(0.14)`	Time without talk, in seconds	`more ideas. (1.03) just`
`(.)`	Pause in speech less than 0.10 seconds	`011 C: o:kAY (.) could be a double cone`
`((`*`turns`*`))`	Verbs and descriptions in double parentheses and italics are transcriber's comments	`((`*`modifies graph`*`))`
*	Asterisks marks the instant in speech that corresponds to the video image on the right	
`(??)`	Marks inaudible words, about one word per question mark	`042 T: <<p>but dat (??) in the positions here`
`::`	Colons indicate lengthening of phoneme, about 1/10 of a second per colon	`si::ze`

W.-M. Roth, *Uncertainty and Graphing in Discovery Work*,
DOI 10.1007/978-94-007-7009-6, © Springer Science+Business Media Dordrecht 2014

Notation	*Description*	*Example*
[ ]	Square brackets in consecutive lines indicate overlap	011 C: o:kAY (.) could be a double cone sidewa[ys]. 012 T: [yea]
<<f> >	Forte, words are uttered with louder than normal speech volume	<<f>um>
<<p> >	Piano, lower than normal speech volume	042 T: <<p>but dat (??)
<<pp> >	Pianissimo, much lower speech volume	009 T: <<pp>under way>
<<dim> >	Diminuendo, becoming weaker	<<dim>i donno>
prETty	Capital letters indicate louder than normal talk indicated in small letters.	looks prETty grEEN to mE
hh	Noticeable out-breath	
.h	Noticeable in-breath	021 T: <<dim>hu hu hu hu hu> .hhfs
–,?;.	Punctuation is used to mark movement of pitch (intonation) toward end of utterance, flat, slightly and strongly upward, and slightly and strongly downward, respectively	C: okay; save that. (0.27) do you want me to blEACH it?.
=	Equal sign indicates that the phonemes of different words are not clearly separated	i=ll
`,´,ˇ	Diacritic indicates movement of pitch within the word that follows – down, up, down up	ˇsimilar.

Index

DOI 10.1007/978-94-007-7009-6, © Springer Science+Business Media Dordrecht 2014

MIX
Papier aus verantwortungsvollen Quellen
Paper from responsible sources
FSC® C105338

If you have any concerns about our products,
you can contact us on
ProductSafety@springernature.com

In case Publisher is established outside the EU,
the EU authorized representative is:
Springer Nature Customer Service Center GmbH
Europaplatz 3, 69115 Heidelberg, Germany

Printed by Libri Plureos GmbH
in Hamburg, Germany